Advances in Experimental Medicine and Biology

Volume 1354

Advances in Experimental Medicine and Biology provides a platform for scientific contributions in the main disciplines of the biomedicine and the life sciences. This series publishes thematic volumes on contemporary research in the areas of microbiology, immunology, neurosciences, biochemistry, biomedical engineering, genetics, physiology, and cancer research. Covering emerging topics and techniques in basic and clinical science, it brings together clinicians and researchers from various fields.

Advances in Experimental Medicine and Biology has been publishing exceptional works in the field for over 40 years, and is indexed in SCOPUS, Medline (PubMed), Journal Citation Reports/Science Edition, Science Citation Index Expanded (SciSearch, Web of Science), EMBASE, BIOSIS, Reaxys, EMBiology, the Chemical Abstracts Service (CAS), and Pathway Studio.

2020 Impact Factor: 2.622

More information about this series at https://link.springer.com/bookseries/5584

Guoyao Wu
Editor

Recent Advances in Animal Nutrition and Metabolism

Springer

Editor
Guoyao Wu
Department of Animal Science
Texas A&M University
College Station, TX, USA

ISSN 0065-2598 ISSN 2214-8019 (electronic)
Advances in Experimental Medicine and Biology
ISBN 978-3-030-85685-4 ISBN 978-3-030-85686-1 (eBook)
https://doi.org/10.1007/978-3-030-85686-1

This Springer imprint is published by the registered company Springer Nature Switzerland AG
The registered company address is: Gewerbestrasse 11, 6330 Cham, Switzerland

Contents

1 Nutrition and Metabolism: Foundations for Animal Growth, Development, Reproduction, and Health

Guoyao Wu

Abstract

Consumption of high-quality animal protein plays an important role in improving human nutrition, growth, development, and health. With an exponential growth of the global population, demands for animal-sourced protein are expected to increase by 60% between 2021 and 2050. In addition to the production of food protein and fiber (wool), animals are useful models for biomedical research to prevent and treat human diseases and serve as bioreactors to produce therapeutic proteins. For a high efficiency to transform low-quality feedstuffs and forages into high-quality protein and highly bioavailable essential minerals in diets of humans, farm animals have dietary requirements for energy, amino acids, lipids, carbohydrates, minerals, vitamins, and water in their life cycles. All nutrients interact with each other to influence the growth, development, and health of mammals, birds, fish, and crustaceans, and adequate nutrition is crucial for preventing and treating their metabolic disorders (including metabolic diseases) and infectious diseases. At the organ level, the small intestine is not only the terminal site for nutrient digestion and absorption, but also intimately interacts with a diverse community of intestinal antigens and bacteria to influence gut and whole-body health. Understanding the species and metabolism of intestinal microbes, as well as their interactions with the intestinal immune systems and the host intestinal epithelium can help to mitigate antimicrobial resistance and develop prebiotic and probiotic alternatives to in-feed antibiotics in animal production. As abundant sources of amino acids, bioactive peptides, energy, and highly bioavailable minerals and vitamins, animal by-product feedstuffs are effective for improving the growth, development, health, feed efficiency, and survival of livestock and poultry, as well as companion and aquatic animals. The new knowledge covered in this and related volumes of Adv Exp Med Biol is essential to ensure sufficient provision of animal protein for humans, while helping reduce greenhouse gas emissions, minimize the urinary and fecal excretion of nitrogenous and other wastes to the environment, and sustain animal agriculture (including aquaculture).

Keywords

Animal protein · Biomedicine · Health · Disease · Intestine · Diet

Abbreviations

AAs	Amino acids
AEMB	Adv Exp Med Biol
IUGR	Intrauterine growth restriction

G. Wu (✉)
Department of Animal Science, Texas A&M University, College Station, TX, USA
e-mail: g-wu@tamu.edu

G. Wu (ed.), *Recent Advances in Animal Nutrition and Metabolism*, Advances in Experimental Medicine and Biology 1354,
https://doi.org/10.1007/978-3-030-85686-1_1

NRC National Research Council
SDEP Spray-dried egg product

1.1 Introduction

There are approximately 1.032 million animal species in nature, which include mammals, 4,000; birds, 9,000; fish and lower chordates, 18,800; crustaceans, 45,000; reptiles, 6,300; amphibians, 4,200; and insects, 0.9 (Wilson 1992). As important parts of the ecosystem, animals ingest nutrients for survival, growth, development, reproduction, and health. Despite their vast diversities as two broad groups (vertebrates and invertebrates), animals exhibit greater similarities in physiology, metabolism, and nutrition than differences (Wu 2018). In human civilization, there has been a rich history of studies investigating the dietary requirements of farm (e.g., livestock, poultry, fish, shrimp, and crabs), companion (e.g., cats, dogs, and horses), and laboratory (e.g., rats and mice) animals for energy, amino acids (AAs), lipids, vitamins, minerals, and water during their life cycles under various physiological and pathological conditions (Baker 2008; Baldwin 1995; Bauman et al. 2011; Bazer et al. 2011, 2015, 2018, 2021; Beitz 1985; Bergen 2007, 2021; Burrin and Mersmann 2005; Halloran et al. 2021; Matthews et al. 2016; NRC 2002; Webb et al. 1992; Wu et al. 2022; Zhang et al. 2015; Zhu et al. 2022). This is because animals contribute to (a) the production of high-quality foods (e.g., meats, eggs, and milk) for human consumption, as well as raw materials such as wool and leather for clothing and accessories for humans; (b) the companionship and well-being of humans; (c) the development of new biotechniques to efficiently produce proteins and other biomolecules; and (d) advancing biomedical research to prevent and treat inborn, metabolic, and infectious diseases of humans and other animals. Review articles of select topics on nutrition and metabolism in animals of both agricultural and biomedical importance are highlighted in this volume of *Advances in Experimental Medicine and Biology* (AEMB) to benefit our readers (Bergen 2022, Dai et al. 2022; Hay et al. 2022; Jia et al. 2022; Li et al. 2022; Li and Wu 2022; Monzani et al. 2022; Moses et al. 2022; Mu et al. 2022; Reynolds et al. 2022; Smith and Govoni 2022; Stenhouse et al. 2022a, b; Swaggerty et al. 2022; Wu et al. 2022; Zhu et al. 2022).

1.2 Production of High-Quality Food Protein and Fiber (Wool) by Animal Agriculture for Human Consumption, Health and Well-Being

Animal agriculture (including aquaculture) plays an important role in providing high-quality food protein (e.g., milk, eggs, and meat) for human consumption to optimize human growth, development, health, and well-being, although a mix of complementary plant proteins through an adequate understanding of protein nutrition may also result in a healthy nutritional state (Grillenberger et al. 2003; Murphy and Allen 2003; Wu et al. 2022). Animal proteins generally contain adequate and balanced amounts of all proteinogenic (protein-creating) AAs for human consumption. In addition, animal-sourced foods provide taurine, carnosine (β-alanyl-L-histidine), anserine (β-alanyl-L-1-methylhistidine), and creatine; they are nonproteinogenic nutrients that possess antioxidative properties and are crucial for energy metabolism in tissues, but are absent from plant-sourced foods (Wu 2020b). As a part of healthy diets for humans, animal-sourced foods also contain highly bioavailable essential minerals (e.g., iron, zinc, copper, manganese, and selenium) and vitamins (Murphy and Allen 2003). This is in contrast to a myth that there are virtually no nutrients in animal-based foods that are not better provided by plants. Furthermore, animal by-products from the rendering industry are major sources of protein, AAs, bioactive peptides, lipids, minerals, and vitamins in the diets of livestock, poultry, fish, and crustacean, and in pet foods (Li and Wu 2022; Li et al. 2021e; Wilkinson and Meeker 2021). Improved animal nutrition can enhance the quality of foods for human consumption, whereas healthy companion animals contribute to the well-being of their owners. Finally, wool (textile fiber; cysteine-rich α-keratin proteins) produced by sheep and other animals (including cashmere and mohair from goats) is used to manufacture cloths and related daily life products (Cao et al. 2021). Production of proteins by animals requires sufficient provision and optimum utilization of dietary AAs and other nutrients (Bergen 2021; Gilbreath et al. 2021; He et al. 2021a; Li et al. 2021a, b; Zhang et al. 2021). This indicates a close link between animal and human nutrition.

According to the Food and Agriculture Organization of the United Nations (FAO 2021), the global numbers of livestock and poultry have increased by 3.6-fold over the past 60 years to about 33 billion head in 2019 (Table 1.1). Furthermore, the global aquaculture production has increased over the past decade at an average rate of about 3.5% per year to be approximately 120 billion kg in 2019 and provides more than 50% of fish filets for human consumption (FAO 2020). Both extensive (e.g., grazing pasture) and intensive (e.g., indoor housing) systems are currently used to raise ruminants (e.g., cattle, sheep, and goats) and nonruminants (e.g., swine and poultry) worldwide (Bazer et al. 2020). Aquatic animals are raised through both the outdoor ponds and the indoor recirculating aquaculture systems (Ebeling and Timmons 2012). Farm animals are biological transformers that convert

Table 1.1 Global stocks of livestock species and poultry in 1961 and 2019[a]

Species	Year	World	Africa	Australia	Brazil	Canada	China	Europe	India	Mexico	USA
Cattle	1961	942	123	17.3	56.0	10.7	49.5	192	176	16.5	97.7
	2019	1510	361	24.7	215	11.5	63.5	117	193	35.2	94.8
Chickens	1961	3906	274	19.9	132	70.0	541	1329	108	62.6	751
	2019	25,915	2043	112	1467	171	5247	2020	808	581	1972
Ducks	1961	193	6.23	0.178	2.68	0.398	100	25.8	6.70	1.00	3.50
	2019	1,177	16.3	1.31	3.42	1.53	720	77.1	33.5	8.53	7.37
Geese + GF	1961	36.6	3.88	–	–	0.314	16.5	13.5	–	–	–
	2019	362	26.0	–	–	0.326	312	15.0	–	–	–
Goats	1961	349	94.2	0.04	4.90	0.012	51.3	22.5	60.9	8.93	3.47
	2019	1090	459	3.90	11.3	0.301	137	16.1	149	8.79	2.62
Horses	1961	62.2	3.49	0.598	4.41	0.555	6.59	22.0	1.33	4.05	2.37
	2019	59.0	7.40	0.222	5.85	0.399	3.67	4.70	0.342	6.38	10.7
Pigs	1961	406	5.67	1.61	25.6	5.00	85.6	168	5.18	5.99	55.6
	2019	850	42.7	2.32	40.6	14.4	316	187	9.06	18.4	78.7
Rabbits + hares	1961	98.0	2.78	–	0.550	–	16.1	76.6	–	0.095	–
	2019	300	16.0	–	0.161	–	233	9.63	–	1.40	–
Sheep	1961	994	135	153	14.0	0.757	61.6	267	40.2	5.85	32.7
	2019	1240	408	65.8	19.7	0.828	163	128	74.3	8.71	5.23
Turkeys	1961	204	1.21	0.17	0.698	3.50	0.313	80.5	–	6.00	108
	2019	428	33.3	1.11	34.6	5.70	0.090	68.0	–	3.79	229

[a]Adapted from FAO (2021). Values are $\times 10^6$ head

GF = guinea fowl; USA = United States of America; "–" = Data are not available materials not consumed by humans (e.g., forages; by-products of plants such as pasture grasses, alfalfa, clovers, hays, straw, and silages; and rendered animal by-products) into high-quality foods (Wilkinson and Meeker 2021; Wu 2018). As documented in this volume of AEMB, recent research on dietary requirements for nutrients (particularly AAs) is expected to enhance the efficiency of animal agriculture globally and alleviate its potential adverse effects on the environment. The significance of animal agriculture is indicated by the fact that this enterprise accounts for 50–75% and 25–40% of the total amount of agricultural output in industrialized and developing countries, respectively (Wu et al. 2014c). Animal-sourced foods can prevent protein deficiency in children and adults (including the elderly and hospitalized patients) worldwide, particularly those living in underdeveloped nations (Wu 2021). In 2016, about 815 million people (10.7% of the world population) had chronic deficiencies of nutrients, particularly protein, vitamins, and microminerals (FAO 2018), and globally 150 million children under five years of age were stunted in their growth (UNICEF 2018). Maternal malnutrition during gestational and neonatal periods affects not only the first generation of offspring, but also subsequent generations through epigenetic-mediated mechanisms (Del Curto et al. 2013; Wang et al. 2012). Preventing hunger and malnutrition will be an even greater challenge as the global population is expected to grow exponentially from 7.9 billion people in 2021 to 9.6 billion by 2050 (United Nations 2021). The demands for animal-sourced protein are expected to increase by 60% between 2021 and 2050 for supporting optimal human growth and mitigating sarcopenia in the elderly. Thus, animal agriculture plays an important role in providing animal-sourced food as part of a healthy diet for humans, while contributing to scientific, economic, and social developments worldwide.

1.3 Animals as Models for Biomedical Research and as Bioreactors for Producing Therapeutic Proteins

Biologically, humans are members of the animal kingdom. As noted previously, pigs, sheep, cattle, chickens, and fish are agriculturally important domestic animal species. There is growing interest in their use as animal models for nutrition research worldwide. This is because the nature of medical research often involves invasive tissue collections and surgical procedures and may result in potential harmful effects (Bergen 2021; Govoni et al. 2019; Ireland et al. 2008; Jia et al. 2021; Odle et al. 2017; Reynold et al. 2019, 2022; Smith and Govoni 2022; Webb et al. 1992; Wu and Knabe 1994). Thus, it is neither ethical nor practical to conduct such studies with humans in the fetal, infant, or adult stages of life. Similarities, major differences, as well as advantages and disadvantages of porcine, ovine, bovine, avian, and fish models are summarized in Table 1.2. Examples of using animal models to pursue nutrition research are highlighted in Table 1.3. Of particular note, results of those studies have aided in delineating the mechanisms for the following physiological or pathological features in humans: (1) intrauterine growth restriction (IUGR); (2) arginine deficiency and hyperammonemia in preterm infants; (3) abundance of polyamines as well as glutamine and proline in milk; (4) the obligatory role of dietary AAs in intestinal integrity; (5) hyperglycemia-induced endothelial dysfunction; (6) ammonia toxicity in patients with N-acetylglutamate deficiency; (7) extrahepatic urea synthesis in the small intestine of post-weaning mammals; (8) the susceptibility of neonates to intestinal disease; and (9) biomarkers for intestinal adaptation in preterm neonates (Rhoads et al. 2005; Rhoads and Wu 2009; Wu and Morris 1998; Wu et al.

Table 1.2 Similarities, major differences, as well as advantages and disadvantages of pig, sheep, cattle, chicken, and fish models for biomedical research on human nutrition, metabolism, and health

Animal model	Similarities to humans	Major differences than humans	Advantages	Disadvantages
Pig	Anatomy, physiology, digestion, food intake, and the metabolism of nutrients (including AAs, glucose, and minerals); monogastric omnivores; mechanisms for nutrient transport; the regulation of blood flow	*Pregnancy:* Time of implantation, type of placentation, fetal fluid compartments, gestational length, and number of offspring; *Nutrition:* Amounts of nutrients (fat and iron) stored in the body at birth; *Metabolism:* Synthesis of ascorbic acid and the site of FA synthesis	Provides adequate tissue samples for various biochemical assays; convenient to work with both fetal and postnatal pigs; sensitive to dietary intakes of AAs and other nutrients; a large database in the literature on pigs	The growth of fetal pigs is not highly sensitive to maternal deficiency of protein; high risk for abortion in pregnant pigs after surgical catheterization of blood vessels; no BAT; litter-bearing
Sheep	Metabolic pathways for nutrient utilization, the regulation of blood flow and thermogenesis, mechanisms for nutrient transport, the number of offspring, maternal size, embryogenesis, BAT and thermogenesis, and birth weight	*Pregnancy:* Time of implantation, type of placentation, fetal fluid compartments, pregnancy recognition signal, and gestational length; *Digestion:* Fermentation of nutrients in the rumen; *Metabolism:* Intestinal catabolism of BCAAs; the metabolism of glucose, ammonia, and SCFAs; gestational length; the site and substrates for FA synthesis	Well-established animal model for studying human pregnancy, Convenient to work with sheep, pregnant ewes are well adaptable to surgical procedures of placing catheters into maternal and fetal blood vessels, and a large database in the literature on sheep	Extensive fermentation of dietary nutrients in the rumen, most dietary nutrients must be protected from rumen degradation, sensitive to ketosis and copper toxicity, a high rate of mortality during late pregnancy for ewes with ≥ 3 fetuses
Cattle	Metabolic pathways for nutrient utilization, the regulation of blood flow and thermogenesis, mechanisms for nutrient transport, number of offspring, embryogenesis, gestational length, as well as BAT and thermogenesis	*Pregnancy*: Time of implantation, type of placentation, fetal fluid compartments, and pregnancy recognition signal; *Digestion*: Fermentation of nutrients in the rumen; *Metabolism*: Intestinal catabolism of BCAAs, the metabolism of glucose, ammonia, and SCFAs; the site and substrates for FA synthesis	Well-established animal model for studying human citrullinemia[a], easiness for in vitro and in vivo gene transfer studies, historically used for research on infectious infectious diseases (e.g., tuberculosis, cowpox[b]), and production of interferon γ	Extensive fermentation of dietary nutrients in the rumen, most dietary nutrients must be protected from rumen degradation, sensitive to ketosis and rumen bloat (excessive accumulation of gases in the rumen), inconvenient to work with adult cattle, and high costs of doing experiments
Chicken	Digestion and absorption of nutrients in the small intestine, the metabolism of most nutrients, monogastric omnivores, the regulation of blood flow, angiogenesis, mechanisms for nutrient transport, the site and	*Embryo:* Hatching of eggs in birds versus mammalian embryos in the uterus, different length of embryonic development; *Digestion:* Proventriculus and gizzard, the expression of digestive enzymes in	Provides adequate tissue samples for various biochemical assays; convenient to work with both pre- and post-hatching birds, sensitive to dietary intakes of AAs and other nutrients; unique to study gout, retinal degeneration[d],	Different mechanisms for ammonia detoxification, no placenta for embryonic growth and development, different paths for the absorption of lipids and lipid-soluble vitamins, the lack of muscle GLUT4, and mammalian

(continued)

Table 1.2 (continued)

Animal model	Similarities to humans	Major differences than humans	Advantages	Disadvantages
	substrates for FA synthesis, and BAT[c]	the small intestine, and lipid absorption via the portal vein *Metabolism:* Uric acid synthesis, the lack of hepatic gluconeogenesis from AAs, and high metabolic rate	and angiogenesis, naturally "diabetic", and spontaneously develops ovarian cancer	antibodies are generally not applicable to birds
Rodent[e]	Anatomy, physiology, digestion, and nutrition; metabolic pathways for nutrient utilization, monogastric omnivores; mechanisms for nutrient transport; the regulation of blood flow; as well as BAT and thermogenesis	*Pregnancy:* Time of implantation, type of placentation[f], fetal fluid compartments, gestational length, and number of offspring; *Digestion:* Fast gastric emptying and short GI transit time (reaching post-absorptive state at 6 h after feeding versus 12 h in humans) *Metabolism:* High metabolic rate (6–10 times that in humans)	Provides adequate tissue samples for various biochemical assays; convenient to work with both fetal and postnatal rodents; sensitive to dietary intakes of AAs and other nutrients; a large database in the literature on rodents; widely available; low cost	A short period of gestation (21 days), ability to synthesize vitamin C, high food intake per kg body weight, rapid aging processing, difficulties in performing surgeries on rodents, litter-bearing, small size, and very different metabolic rates between rodents and humans
Fish	Digestion and absorption of nutrients in the small intestine, metabolic pathways for nutrient utilization, mechanisms for nutrient transport, intestinal AA metabolism, and the secretion of digestive enzymes	*Embryo:* Hatching of eggs in fish vs. mammalian embryos in the uterus, different length of embryonic development; *Digestion:* Some fish lack a stomach; a short gut; requires a long time for nutrient digestion and absorption; *Metabolism:* Most fish lack the urea cycle and have high dietary requirements for AAs; intolerable to dietary starch; limited oxidation of glucose and lipids in muscle; unique roles of the head kidneys[g]	Big fish provide adequate tissue samples for various biochemical assays; convenient to work with both pre- and post-hatching fish, sensitive to dietary intakes of starch and other nutrients; unique to study starch-intolerable disease in the liver; useful for studying the usual absence of tumors in the small intestine; zebrafish is widely used in biomedical research	Different mechanisms for ammonia removal, no placenta for embryonic growth and development, different metabolic patterns for glucose and fatty acids in fish than in mammals, requires intensive labor, mammalian antibodies are generally not applicable to fish, difficult to measure food intake, and small size for juvenile fish

[a]A disease due to the deficiency of argininosuccinate synthase (a urea cycle enzyme)
[b]The first vaccine against the cowpox virus
[c]Deveopment of brown adipose tissue in response to cold challenge
[d]Mutation in the photoreceptor guanylate cyclase
[e]Rats and mice
[f]Labyrinthine-type in rats (an intricate structure of interconnecting passages) versus the villous-type in humans
[g]A tissue that contains cytokine-producing lymphoid cells of the immune system; endocrine cells that secret cortisol, catecholamines, and thyroid hormones; and hematopoietic stem cells that are capable of hematopoiesis (the production of new blood cells)
AAs = amino acids; BAT = brown adipose tissue, BCAAs = branched-chain amino acids; FA = fatty acid; GI = gastrointestinal; GLUT4 = glucose transporter 4; SCFAs = short-chain fatty acids

Table 1.3 Use of animal models for nutrition and biomedical research

Animal	Nutrition and biomedical research	References
Pig	Fetal and postnatal development of intestine and other tissues	Buddington et al. (2012), Dekaney et al. (2001)
	Effects of maternal nutrition on placental and fetal growth	Ji et al. (2017), NRC et al. (2012), Wu et al. (1996, 2006)
	Fetal composition of AAs and other nutrients in the body	Kim et al. (2009), Pond et al. (1969), Wu et al. (1999)
	Tissue-specific and whole-body metabolism of amino acids	Blachier et al. (2013), Reeds et al. (1996, 1997)
	Mammary gland synthesis of bioactive products	Hurley (2019), O'Quinn et al. (2002)
	Nutrient requirements of neonates and adults	Davis et al. (2002, 2008, 2010), Wang et al. (2014)
	Development of infant formulas and TPN solutions	Brunton et al. (1999), Jamin et al. (2010), Odle et al. (2017)
	Regulation of vasodilator production by endothelial cells	Wang et al. (2011), Wu et al. (2001)
	Cardiovascular diseases and responses to exercise and diets	Tsang et al. (2016); Walters et al. (2017)
	Novel biochemical pathways and metabolic defects	Wu (1997), Wu et al. (1994, 2000, 2004, 2005)
	Development of immune systems[a]	Furukawa et al. (2020), Johnson et al. (2006), Wu (1996)
	Prevention and treatment of IUGR in mammals	Ashworth (1991), Rehfeldt et al. (2004), Wu et al. (2006)
	Production of recombinant proteins and antimicrobials	Hay et al. (2022), Monzani et al. (2022)
	Microbial development in the small and large intestines	Dai et al. (2022); Mu et al. (2022); Ren et al. (2020)
Sheep	Composition of AAs and PAs in fetal fluids and placentae	Kwon et al. (2003a, b; 2004a, b)
	Expression of AA and sugar transporters in the conceptus	Gao (2020), Moses et al. (2022)
	Syntheses of NO and PAs in placentae	Kwon et al. (2003a, b), Wang et al. (2015a, b, 2016)
	Prevention of IUGR in underfed dams through Arg supplementation	Gilbreath et al. (2021), Lassala et al. (2009, 2010, 2011), McCoard et al. (2013, 2014, 2016), Sales et al. (2016)
	Growth and lactation performance	Reynolds et al. (2019), Wu et al. (2022)
	Skeletal muscle growth, development, and adaptation	Gonzalez et al. (2020), Govoni et al. (2019)
Cow	Transgenic cattle with desirable production traits[b]	Hay et al. (2022), Monzani et al. (2016, 2022)
	As bioreactors to produce recombinant proteins[c]	Hay et al. (2022), Monzani et al. (2016; 2021)
	Develop therapeutic treatment of citrullinemia[d]	Harper et al. (1986), Lee et al. (1999)
Chicken	Elucidate mechanisms responsible for gout, retinal degeneration and detachment, diabetes, and ovarian cancer, as well as their prevention and treatment	Cebulla et al. (2012), Lim et al. (2012), Larger et al. (2004), Ulshafer and Allen (1985)
	Study cardiovascular development and angiogenesis	Vilches-Moure (2019) and Ziche et al. (1994)

(continued)

Table 1.3 (continued)

Animal	Nutrition and biomedical research	References
Fish	Study mechanisms for the utilization of high dietary protein	Ballantyne (2001), Li et al. (2020a, d, e, f)
	Study mechanisms for glycogenic hepatopathy	Li et al. (2020a, b; 2022)
	Study mechanisms for hepatic steatosis	Li et al. (2022)
	Study mechanisms for black skin syndrome	Li et al. (2021a, b, c, d)
	Study mechanisms for the absence of cancer in the intestine	Jia et al. (2021)
	Study mechanisms for Arg deficiency in growth and survival	Li et al. (2021a, b, 2022)

[a]Including intestinal intraepithelial lymphocytes and Peyer's patches
[b]Including resistance to diseases and improved growth and lactation performance
[c]Including tissue-type plasminogen activator, recombinant human growth hormone, recombinant human albumin, recombinant anti-CD20 monoclonal antibody, human lactoferrin, α-lactalbumin, myelin basic protein, and human bile salt-stimulating
[d]A rare Holstein and Holstein–Friesian-specific metabolic genetic disorder of cattle
AAs = amino acids; Arg = arginine; IUGR = intrauterine growth restriction; NO = nitric oxide; PAs = polyamines (putrescine, spermidine, and spermine); TPN = total parenteral nutrition

2004, 2014a). Examples of discovery research using ovine models has greatly advanced the following aspects of AA nutrition research: ((1) the unusual high abundance of the arginine-family AAs and serine in fetal fluids (Kwon et al. 2003a, b), (2) the expression of AA and sugar transporters in the conceptus (Gao 2020; Huang et al. 2018; Moses et al. 2022; Reynolds et al. 2022; Satterfield et al. 2010); (3) the syntheses of nitric oxide and polyamines (key regulators of angiogenesis) in placentae (Kwon et al. 2003a, b); and (4) the prevention of IUGR in underfed dams by increasing the availability of AAs to the fetus through either the dietary realimentation or modulation of the uterine arginine-nitric oxide pathway (Gilbreath et al. 2021; Lassala et al. 2009, 2010, 2011; Satterfield et al. 2012, 2013). In additions, chickens are useful models to study mechanisms responsible for gout, retinal degeneration and detachment, diabetes, and ovarian cancer (Cebulla et al. 2012; Lim et al. 2012; Larger et al. 2004; Ulshafer and Allen 1985, as well as cardiovascular development (Vilches-Moure 2019) and angiogenesis (Ziche et al. 1994). Finally, fish can be used to investigate mechanisms for the utilization of high amounts of dietary protein, glycogenic hepatopathy, hepatic steatosis, black skin syndrome, and intestinal carcinogenesis (Jia et al. 2021; Li et al. 2021a, b, 2022).

Over the past two decades, genetic engineering techniques [e.g., recombinant DNA technology and genome editing (Fig. 1.1)] have been used to generate transgenic cattle that possess desirable production traits (including resistance to diseases and improved growth and lactational performance) and produce recombinant proteins with nutritional and therapeutic values (Hay et al. 2022; Monzani et al. 2016, 2022). Those proteins include tissue-type plasminogen activator, recombinant human growth hormone (nutrient metabolism), recombinant human albumin, recombinant anti-CD20 monoclonal antibody, human lactoferrin (antimicrobial in the small intestine), α-lactalbumin (milk protein), myelin basic protein (neurological development), and human bile salt-stimulating lipase (lipid digestion). In addition, transgenic pigs have been generated to produce porcine growth hormone, carbohydratases, phytase, antimicrobials, and anti-viral antibodies (Wu and Bazer 2019). Advanced biotechnology holds great promise for conserving the diverse breeds of animals, enhancing their food efficiency and productivity, and developing new alternatives to in-feed antibiotics in the future.

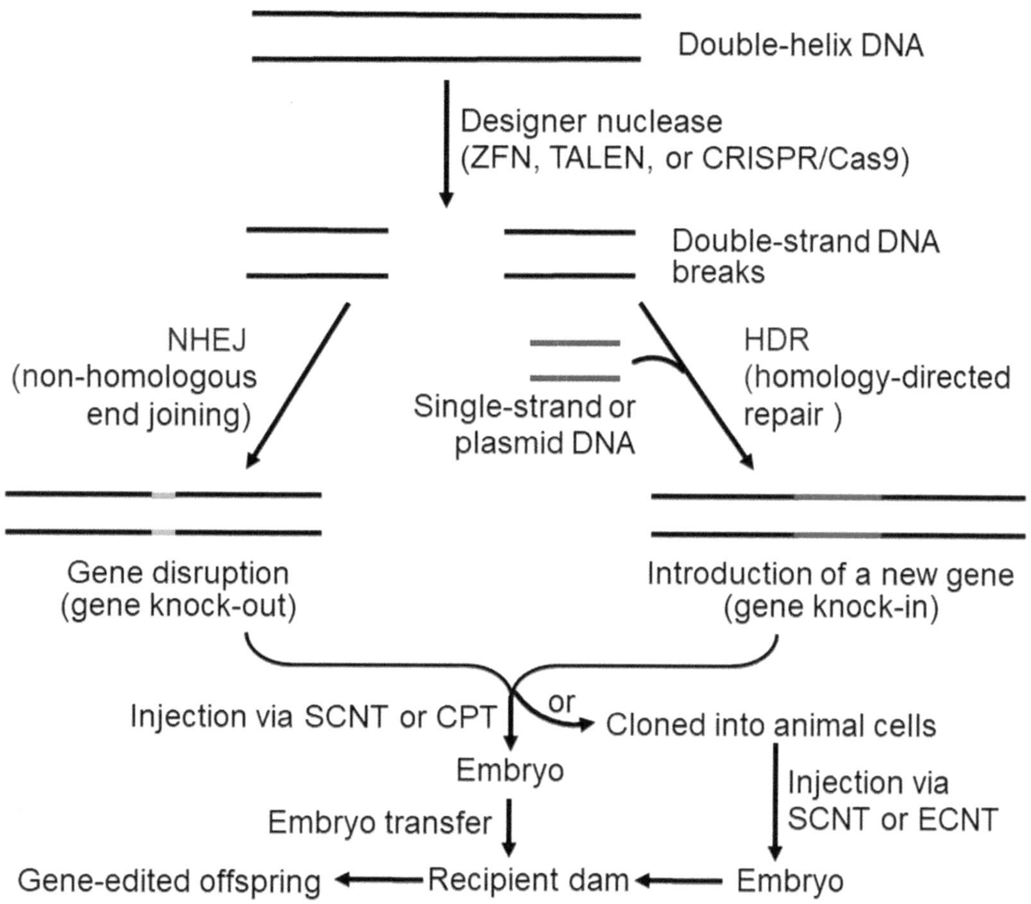

Fig. 1.1 Gene (genome) editing of animals using the zinc finger nuclease (ZFN), transcription activator-like effector nuclease (TALEN), or clustered regularly interspaced short palindromic repeats-associated nuclease-9 (CRISPR/Cas9) technique. A designer nuclease (ZFN, TALEN or CRISPR/Cas9) cleaves a DNA molecule to generate a double-strand break (DSB) at a desired genomic locus. Thereafter, one of two endogenous repair mechanisms may repair the DSB DNA: non-homologous end joining (NHEJ) and the homology-directed repair (HDR). In the NHEJ pathway, the two ends of the DSB DNA are brought together and ligated without a homologous template for repair, which often inserts or deletes nucleotides (indels) to cause gene disruption (knockout). The HDR pathway requires the provision of an exogenous DNA template along with a site-specific genome editing nuclease to repair the DSB DNA, thereby causing the knock-in of a desired sequence of DNA into the genome of an embryo or animal cells. Because of its more precise targeting of genes, CRISPR/Cas9 is gaining momentum in life sciences as the preferred editor of gene editing of livestock species. Reproduced from Wu and Bazer (2019), with permission

1.4 Nutritional Requirements, Deficiencies, and Health of Animals

All animals have dietary requirements for energy, AAs, lipids, carbohydrates, minerals, vitamins, and water in their life cycles (Greene 2016; NRC 2002, 2012; Wu 2018). All nutrients interact with each other, intestinal microbes, and the environment to influence growth, development, and health of animals (Fig. 1.2). Grazing ruminants without protein or mineral supplements exhibit suboptimal growth and lactational performance (Bergen et al. 2021; Cao et al. 2021; Gilbreath et al. 2021; Govoni et al. 2019).

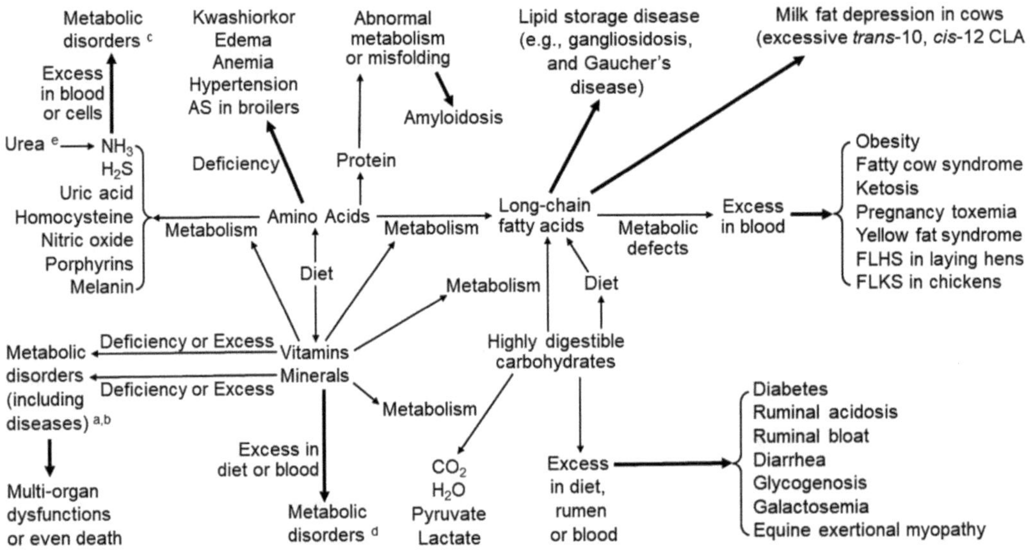

Fig. 1.2 Metabolic disorders (including metabolic diseases) in ruminant and nonruminant animals. Defective biochemical pathways due to deficiencies in their enzymes, coenzymes, or cofactors can cause abnormal nutrient metabolism (either inherited or acquired), resulting in multi-organ dysfunctions, diseases and even death in livestock and poultry. Reproduced from Wu (2020c), with permission. [a]Disorders (including diseases) caused by the deficiency of a vitamin include polioencephalomalacia in ruminants and beriberi (thiamin), pellagra (niacin), burning-foot syndrome (pantothenic acid), neural tube defects (folate), scurvy (vitamin C), photophobia (riboflavin), xerophthalmia and keratomalacia (vitamin A), rickets and osteomalacia (vitamin D), liver steatosis (choline), myopathy and liver necrosis (vitamin E), hemorrhage (vitamin K), and infertility (vitamins A and E). [b]Disorders (including diseases) caused by the deficiency of a mineral include milk fever in cows, rickets, and osteomalacia (calcium), grass tetany (magnesium), anemia and hemorrhage (iron), ammonia toxicity (manganese), Keshan's disease (selenium), goiter (iodine), dental caries (fluorine), Menke's and Wilson's diseases (copper), and infertility (phosphorus). [c]Disorders (including diseases) caused by an excessive production of amino acid metabolites include hyperhomocysteinemia, hyperammonemia, gout, melanosis, and porphyria. [d]Disorders (including diseases) caused by an excessive lipid-soluble vitamin include hypervitaminosis A, D, E, and K, whereas diseases caused by an excessive mineral include polioencephalomalacia in ruminants (sulfate), hypertension (sodium), hyperkalemic periodic paralysis in horses (potassium), copper toxicity, and selenosis (selenium). [e]This reaction is catalyzed by urease in the rumen fluid of ruminants and the intestine of all animals. As, ascites syndrome; CLA, conjugated linoleic acid; FLHS, fatty liver hemorrhagic syndrome; FLKS, fatty liver and kidney syndrome

Likewise, nonruminants (e.g., swine and poultry) fed typical corn- and soybean meal-based diets without dietary supplementation with deficient AAs cannot achieve their maximum genetic potential for growth and egg production (He et al. 2021a; Zhang et al. 2021). Thus, because protein is the most abundant dry matter component in growing animals (e.g., about 65–67% in skeletal muscle on the dry matter basis) and a major nutrient in human foods (e.g., meat, eggs, and milk) and feedstuffs for animal diets, protein nutrition and metabolism has been an active research area in animal nutrition over the past century (Baker 2008; Bergen 2007, 2008; Firkins et al. 2007). The pages of AEMB have highlighted recent advances in the functions of AAs in the different organ systems of animals (Beaumont and Blachier 2020; Chen et al. 2020; Durante 2020; Flynn et al. 2020; Gao 2020; He and Wu 2020; Hou et al. 2020; Li et al. 2020a; Sandoval et al. 2020; Solano 2020; Wu 2021), the inter-organ metabolism of AAs (He et al. 2021b; Posey et al. 2021; Ryan et al. 2021), and the roles of AAs in gene expression and cell signaling (Halloran et al. 2021; Paudel et al. 2021; Sah et al. 2021; Shen et al. 2021; Yang et al. 2021). Additional AEMB papers focus on the nutrition and metabolism of humans and other animals, including aquatic, companion, zoo animals (Che et al. 2021; Herring et al. 2021; Jia et al. 2021; Li et al. 2022; Oberbauer and Larsen 2021; Sarkar et al. 2021; Wu et al. 2021). Compelling evidence indicates that animals (including humans) have dietary requirements for proteinogenic AAs that are not synthesized de novo (e.g., leucine and lysine) and proteinogenic AAs that are synthesized de novo in animal cells (e.g., glutamate, glutamine, glycine, and proline; Wu 2021). Some of the AAs, known as functional AAs in nutrition (Wu 2010), participate in and regulate key metabolic pathways to improve the health, survival, growth, development, lactation, and reproduction of animals. Notably, these two new nutritional concepts are now transforming the feeding practices for animals (including livestock, poultry, and fish) worldwide (Chalvon-Demersay et al. 2021; Li et al. 2021a, b; Rodrigues et al. 2021; Rossi et al. 2021). In addition, this volume of AEMB includes articles on mechanisms for the transport of water (Zhu et al. 2021), fructose (a major sugar in the conceptuses of ungulates; Moses et al. 2022), calcium, and phosphorus (Stenhouse et al. 2022b) in adult animals, particularly in pregnant dams with developing conceptuses. Adequate nutrition is crucial for preventing and treating metabolic disorders (including metabolic diseases) and infectious diseases in all species of animals (Li et al. 2007; Wu 2020c; Table 1.4).

The small intestine is not only the terminal site for nutrient digestion and absorption, but also plays an important role in AA metabolism (Fig. 1.3). Notably, the synthesis of citrulline from glutamine, glutamate, and proline occurs in the enterocytes (the columnar absorptive epithelial cells of the small intestine) of most mammals, including humans, pigs, rats, mice, cattle, and sheep (Blachier et al. 2013; Wu et al. 2021, 2022; Zhang et al. 2021), but not in any cells of chickens (He et al. 2021a, b). As for pigs (Zhang et al. 2021), the proximal intestine of fish (e.g., hybrid striped bass and largemouth bass; Li et al. 2020d, e, f; Jia et al. 2021) and crustaceans (Li et al. 2021b) extensively oxidizes glutamate, glutamine, and aspartate to provide most of the energy needed by the tissue. In addition, the gut intimately interacts with a diverse community of intestinal antigens and bacteria to influence gut and whole-body health (Ren et al. 2020; Wang et al. 2020). Compared with humans, farm animals are at greater risks for infections by pathogens such as bacteria, fungi, parasites, and viruses. Thus, maintaining a healthy gut is essential to the survival, growth, and reproduction of the animals (Ren et al. 2020). Since the 1950s, sub-therapeutic levels of feed antibiotics have been included in conventional diets to improve the growth performance and feed efficiency of swine and poultry. However, due to the development and spread of bacteria resistant to antibiotics, feed antibiotics have been banned in many countries (e.g., the European Union, the U. S., and China) and are being phased out in many other nations. Some bacteria are resistant to one class of antibiotics, and others are resistant to multiple antibiotics, thereby posing a serious global health concern (Koch et al. 2017). Several comprehensive articles in this volume of AEBM highlight the species and metabolism of intestinal microbes, as well as their interactions with the intestinal immune systems and the host intestinal epithelium in swine and poultry (Dai et al. 2022; Mu et al. 2022; Swaggerty et al. 2022). This new knowledge can help to mitigate antimicrobial resistance through the development of prebiotic and probiotic alternatives to in-feed antibiotics in animal production systems.

1.5 Protein Foodstuff Sources for Animals

Protein is the most expensive nutrient in animal diets (Gatlin et al. 2007; Kim et al. 2009; Li and Wu 2020; Li et al. 2020b, c; Wu et al. 2014a). Plant- and animal-sourced foodstuffs are the sources of protein and AAs for omnivores, whereas carnivores consume animal carcasses or animal-derived products. As reviewed in this volume of AEBM (Jia et al. 2022; Li and Wu 2022; Li et al. 2021e), the composition of most AAs differ substantially between plant and animal proteins. Animal-sourced feedstuffs are generally superior to plant-sourced ones for the growth and health of

Table 1.4 Metabolic disorders (including metabolic diseases) in ruminant and nonruminant animals due to nutrient deficiencies

Nutrient	Metabolic disorder (or disease)	Cause	Major syndromes
Glucose	Type 1 diabetes	Lack of insulin secretion from pancreatic β-cells, impaired use of glucose	Hyperglycemia, retinal damage, blindness, ketosis, impaired blood flow (leading to amputation), muscle loss and weakness, and dyslipidemia
	Type 2 diabetes	Insulin resistance (impaired) insulin signaling, or obesity), impaired use of glucose	Hyperglycemia, retinal damage, blindness, impaired blood flow (leading to amputation), excessive glycogen in muscle, and dyslipidemia
	Hypoglycemia	Inadequate glucose provision or synthesis	Brain damage, coma, and death
	Ruminal acidosis in ruminants	High intake of starch and monosaccharides, a sudden decrease in ruminal fluid pH to <5.5 due to rapid lactate production	Inhibits the growth of cellulolytic bacteria and acetate-producing bacteria, but promotes the growth of propionate-producing bacteria in the rumen; reduces roughage digestion and acetate production
	Ruminal bloat in ruminants	Excessive production of gasses (primarily CO_2 and CH_4), formation of foams	Internal pressure on vital organs, multi-organ (e.g., circulatory and respiratory) dysfunction, and eventually death of the animals
	Equine exertional myopathy (Monday morning disease)	Excessive lactate production in skeletal muscles of an exercising horse	Muscle cramps and damage, recumbency, and an inability to stand
	Glycogenosis	Excessive accumulation of glycogen in the liver and muscle; high starch intake	Enlarged and watery liver, reduced food intake, reduced growth, and reduced feed efficiency
	Diarrhea in sucrose-fed neonatal pigs	Natural absence of sucrase from the small-intestinal mucosa at birth (detectable only after 7 days of age)	Diarrhea, dehydration, reduced food intake, reduced growth, reduced feed efficiency, and death
	Diarrhea in lactose-fed chicks	Natural absence of lactase from the GI tract of 1- to 7-day-old or older chicks	Diarrhea, dehydration, reduced food intake, reduced growth, reduced feed efficiency, and death
	Galactosemia	Deficiency of galactokinase or galactose-1-phosphate uridylyltransferase	Enlarged liver, cirrhosis of liver, renal failure, cataracts, vomiting, seizure, and brain damage; chickens are very susceptible to the disease
Lipids	Essential fatty acid deficiency syndrome	Deficiency of linoleic acid (ω6) or α-linolenic acid (ω3) in all animals; deficiency of arachidonic acid (ω6) in cats	Skin lesions, growth restriction, reproductive failure, increased susceptibility to infection, thrombocytopenia, impairment of neurological development, poor wound healing, and alopecia
	Obesity	Chronic excessive energy intake relative to energy expenditure	Excessive fat deposition in the body, leading to obesity, dyslipidemia, insulin resistance, and type 2 diabetes mellitus
	Bovine fatty liver syndrome (usually in dairy cows during early lactation)	Excessive mobilization of adipose tissue during early lactation, hepatic uptake of fatty acids from the blood	Peri-parturient metabolic, health and production problems (e.g., milk fever, ketosis, hepatic dysfunction, low feed intake, mastitis, metritis, and impairments of lactation and reproduction)
	Ketosis in lactating dairy cows (usually during early lactation)	Excessive mobilization of adipose tissue due to low feed intake, excessive KB production by the liver)	Reduced feed intake, reduced milk production, body weight loss, skeletal muscle loss and weakness, abnormal

(continued)

Table 1.4 (continued)

Nutrient	Metabolic disorder (or disease)	Cause	Major syndromes
			behavior (e.g., walking in circles), and reduced blood pH
	Low-fat milk syndrome (milk fat depression) in dairy cows	Low intake of fiber and high intakes of starch and unsaturated fatty acids; high amount of CLA in the rumen	A reduction in the concentrations of milk fats (up to 50% or more) with little or no change in concentrations of lactose or protein in milk, and reductions in ruminal pH and Ac/Prop ratio
	Pregnancy toxemia[a] (prevalent in ewes and does with multiple fetuses in late pregnancy)	Excessive mobilization of adipose tissue due to low feed intake, excessive KB production by the liver	Reduced feed intake, reduced blood pH, reduced fetal growth, skeletal muscle loss and weakness, abnormal behavior (e.g., walking in circles), and maternal and fetal death
	Fatty liver hemorrhagic syndrome in laying hens	Excessive energy intake, particularly in heat stress (often in prolific laying hens housed in cages)	High amount of abdominal fats, and possibly pale combs; the enlarged liver is prone to damage and bleeding. Hemorrhage often occurs when a hen is straining to lay her egg. A high rate of mortality
	Fatty liver and kidney in chickens	Biotin deficiency, and thus impaired carboxylation[b]	The liver and kidneys are pale and swollen, and contain high lipid deposits; lactic acidosis; death
	Yellow fat disease[c] (steatitis or pansteatitis)	Excessive intake of oxidized unsaturated fatty acids, and reduced intakes of vitamin E	Marked inflammation of white adipose tissue, lipid peroxidation, and the deposition of yellow-wish pigment in adipocytes; possibly myopathy
	Gangliosidoses	A lack of or low activity of enzymes to hydrolyze GM	Excessive lysosomal accumulation of lipids known as gangliosides in tissues
	Gaucher's disease	Deficiency of β-glucocere-brosidase to degrade GCB	Excessive lysosomal accumulation of lipids known as glucocerebroside in tissues
Amino acids (AAs)	Amyloidosis	Abnormal synthesis, degradation, or folding of extracellular amyloidd	Deposition of amyloid fibrils (firm and solid extracellular substances) in tissues; organ failure, and death
	Kwashiorkor	Dietary protein deficiency	Weakness, poor health, stunting, anemia, muscle wasting, calcium and bone losses, edema, reduced number of red blood cells, and impaired immunity
	Hyperammonemia	Excessive ammonia in blood, an Arg or UCE deficiency in mammals; excess AA intake	Impaired flow of blood to the brain, tissue damage, oxidative stress, impaired Krebs cycle, pregnancy loss, coma, and death
	Gout	Excessive urate production, an Arg or UCE deficiency in mammals; excess AA intake	High amount of uric acid crystallizes in joints, tendons, and surrounding tissues, resulting in red, tender, hot, and swollen tissues; pain
	Hyperhomocysteinemia	Excessive homocysteine in blood due to low intakes of vitamins[e] or high SAA intake	Deficiency of NO, impaired flow of blood to tissues, high risk for

(continued)

Table 1.4 (continued)

Nutrient	Metabolic disorder (or disease)	Cause	Major syndromes
			cardiovascular disease, oxidative stress, neural tube defect, and death
	Melanosis	Deposition of a brownish-black pigment in tissue	Negatively affect sensory characteristics of meat and its marketability, no adverse effect on health
	Porphyria	A defect in heme synthesis; accumulation of porphyrins	Abdominal pain, vomiting, seizures, constipation, high blood pressure, tachycardia, and paralysis
	Ascites syndrome	Pulmonary arterial hypertension due to NO deficiency	Accumulation of fluid in ventral hepatic, peritoneal, or pericardial spaces; hypertension
Vitamins	Beriberi	Deficiency of thiamin[f]	Muscular and nerve degeneration, edema, brain dysfunction, edema, leg disorders, and paralysis
	Photophobia	Deficiency of riboflavin (an essential precursor of FMN and FAD)	Photophobia, itching or teary eyes, loss of visual acuity, lesions in mouth corners, dermatitis, nerve degeneration, and burning mouth or burning tongue
	Pellagra	Deficiency of niacin (an essential precursor of NAD and NADP)	Dermatitis, diarrhea, dementia, and death
	Burning-foot syndrome	Deficiency of pantothenate	Skin lesions, loss of hair, depression, and fatigue
	Neural tube defects	Deficiency of folate	Anemia, homocysteinemia, birth defects, and stunting
	Scurvy (bleeding gum)	Deficiency of ascorbate	Bleeding gums, and abnormal connective tissue
	Liver steatosis	Deficiency of choline	Hepatic steatosis, fatty liver, and CNS dysfunction
	Xerophthalmia and keratomalacia	Deficiency of vitamin A (hypovitaminosis A)	Defective night vision, keratinization of epithelial tissues, decreased mucous secretion, and blindness
	Rickets and osteomalacia	Deficiency of vitamin D (hypovitaminosis D)	Abnormal bone structure, osteomalacia, and the impaired intestinal absorption of calcium and phosphate
	Nutritional myopathy and liver necrosis	Deficiency of vitamin E (hypovitaminosis E)	Oxidative stress, muscular degeneration, hepatic necrosis in pigs, exudative diathesis in chicks, white muscle disease in lambs and calves, anemia
	Hemorrhage	Deficiency of vitamin K (hypovitaminosis K)	Hemorrhage, loss of blood, anemia, abnormal bone growth, immune dysfunction, and death
Minerals	Electrolyte imbalance and osmotic disorders	Deficiencies and imbalance of Na^+, K^+ and Cl^-	Reductions in extracellular osmolarity and hydration, muscle weakness, spasms, tetany, paralysis, numbness, cardiac rhythm abnormalities, cardiac arrest, and death
	Anemia	Deficiency of iron	Fatigue, weakness, poor health stunting, pale skin, dizziness, reduced number of

(continued)

Table 1.4 (continued)

Nutrient	Metabolic disorder (or disease)	Cause	Major syndromes
			red blood cells, impaired immunity, cold feet, and the shortness of breath
	Milk fever (in lactating cows)	Deficiency of calcium (Ca^{2+} in blood <1.25 mM)	Muscular spasms, paralysis, unconsciousness, an inability to stand, and reduced milk production
	Phosphorus deficiency syndrome	Deficiency of phosphorus	Hemolysis, fatigue, infertility, reduced myocardial contractility, muscle weakness, respiratory failure, tremors, ataxia, anorexia, nausea, and vomiting
	Grass tetany (mainly in grazing ruminants)	Deficiency of magnesium (Mg^{2+} in blood <0.4 mM)	Skeletal muscle cramps, seizures, paralysis, death; and the abnormal metabolism of all nutrients
	Keshan's disease	Deficiency of selenium	Oxidative disorders (including myopathy), and death
	Goiter	Deficiency of iodine	Enlarged thyroid, dry and scaly skin, stunting, excessive fat deposition in the body, impaired reproduction, and neurological dysfunction
	Dental caries	Deficiency of fluorine	Tooth decay, dental caries, and osteoporosis
	Menke's disease	Impaired absorption of dietary copper due to a defect of P-type ATPase on the BM of enterocytes	Accumulation of copper in the intestine and the kidneys; decreased concentrations of copper in the liver, serum, and brain; abnormal (easily broken) and steely (kinky) hair
	Wilson's disease	Impaired efflux of copper from hepatocytes to blood due to a defect of P-type ATPase in these cells	Accumulation of copper in the liver and the other tissues; decreased concentrations of copper in serum; a green or golden pigmented ring around the corner of eyes (copper deposition)

[a]Pregnancy toxemia also occurs in other mammals, such as beef cows, dairy cows, dogs, and mares during pregnancy

[b]Biotin-, ATP- and HCO_3^--dependent carboxylation reactions catalyzed by pyruvate carboxylase, acetyl-CoA carboxylase, propionyl-CoA carboxylase, and ß-methylcrotonyl-CoA carboxylase

[c]The disease occurs most commonly in nonruminants (e.g., cats, dogs, ferrets, fish, foals, horses, mink, pigs, poultry, rabbits, rats, and reptiles) and rarely in ruminants

[d]Amyloid is a protein produced by reticuloendothelial cells, histiocytes, and plasma cells and contains less than 5% carbohydrates

[e]Vitamins (folate, vitamin B_6, and vitamin B_{12}) and betaine play an important role in one-carbon metabolism and in the methylation reaction that is required to convert homocysteine into methionine

[f]Thiamin diphosphate participates in the oxidative decarboxylation of α-ketoacids for ATP production and in transketolase reactions for generation of NADPH

AA, amino acid; Ac, acetate; Arg, L-arginine; BM, basolateral membrane; CLA, *trans*-10, *cis*-12 conjugated linoleic acid; CNS, central nervous system; GCB, glucocerebroside; GI, gastrointestinal; GM, ganglioside monosialic acid; KB, ketone bodies (acetoacetate, β-hydroxybutyrate, and acetone); Prop, propionate; SAA, sulfur-containing amino acids; UCE, urea cycle enzyme livestock, poultry, fish, crustaceans, fur-bearing animals, zoo animals, and companion animals. This is due, in part, to the higher protein quality (based on the quantity, ratios, and digestibilities of proteinogenic AAs) of animal-sourced as compared to plant-sourced feedstuffs (Moughan 2003; Wu 2018). Of note, the content of AAs is different among the animal-sourced feedstuffs manufactured by the rendering industry (Li et al. 2021e; Wilkinson and Meeker 2021). Some animal products [e.g., peptones (partial protein hydrolysates)] may be mixed with carrier plant proteins during the manufacturing process. For example, PEP2+, Peptone 50 and PEP-NS are all porcine intestinal mucosa products [containing large amounts of bioactive molecules (Li and Wu 2022)] that have been co-dried with different sources of plant proteins. PEP2+ is co-dried with enzymatically processed vegetable proteins, Peptone 50 with a vegetable protein, and PEP-NS with byproducts from corn wet milling (Myers et al. 2014). The high abundances of biosynthesizable AAs (including glycine, serine, and proline) as well as methionine and cysteine in animal proteins can reduce the energetic and precursor AA costs of AA syntheses in animals, while improving their antioxidative and immune defense systems. In addition, animal-sourced feedstuffs usually contain functional antioxidative compounds (e.g., taurine, creatine, carnosine, and anserine) that are all absent from plants, as noted previously (Hou et al. 2019; Wu 2020b; Wu et al. 2014c). Those unique nutrients are critical and irreplaceable in diets of carnivores and possibly many omnivorous species (Li et al. 2021e). Furthermore, spray-dried animal plasma (Boyer et al. 2015) and spray-dried egg products (Pereira et al. 2019) provide large amounts of immunoglobulins that can neutralize invading pathogens, thereby improving immune responses in animals.

Let's take the spray-dried egg product (SDEP) as an example of a high-quality protein source (Li and Wu 2020). This feedstuff provides high-quality protein that contains an abundant amount of all AAs in desirable ratios relative to the dietary requirements of all animal species. In addition, this feedstuff is particularly rich in cysteine, methionine, and serine. For example, SDEP contains about 100% more cysteine than black soldier fly larvae meal and fishmeal, about 70% more methionine than black soldier fly larvae meal and spray-dried enzymes-treated porcine mucosal tissues, as well as 46% more methionine than poultry by-product (Table 1.5). Furthermore, SDEP contains 64% to 130% more serine than black soldier fly larvae meal, chicken by-product meal, fishmeal, spray-dried enzymes-treated porcine mucosal tissues, and poultry by-product. Among all analyzed feedstuffs, SDEP contains the highest content of serine (about 9% of protein), compared with 4% to 5% in chicken by-product meal, fishmeal, and poultry by-product. Cysteine is essential for the synthesis of protein and glutathione (a major antioxidant), whereas both methionine and serine are methyl-group donors that actively participate in one-carbon metabolism and methylation to support the syntheses of essential molecules (e.g., DNA, creatine, and polyamines; Bazer et al. 2021; Seo et al. 2021), whole-body energy metabolism, and cell growth, while reducing the concentrations of homocysteine (a highly toxic substance that induces oxidative stress, impairs blood flow, and even causes death in animals) in plasma. Like all animal-sourced feedstuffs, SDEP is an excellent source of taurine (a major antioxidant and a major osmolyte in animal cells). Compared with SDEP, plant-sourced feedstuffs are severely deficient in cysteine, methionine, and serine, and do not contain taurine. For example, SDEP contains 95%, 234%, and 152% more cysteine, methionine, and serine, respectively than even soybean meal (a common source of protein in diets for swine and poultry). Inclusion of SDEP in diets alone or in combination with other animal-sourced feedstuffs can substantially reduce or even completely eliminate the need for dietary supplementation with synthetic cysteine and methionine products (Li et al. 2021e). The highly abundant functional AAs in SDEP play important roles in improving the growth, development, health, feed efficiency, and survival of all animals, while helping to reduce greenhouse gas emissions from livestock, minimize the urinary and fecal excretion of nitrogenous and other wastes to the environment, and sustain animal agriculture worldwide.

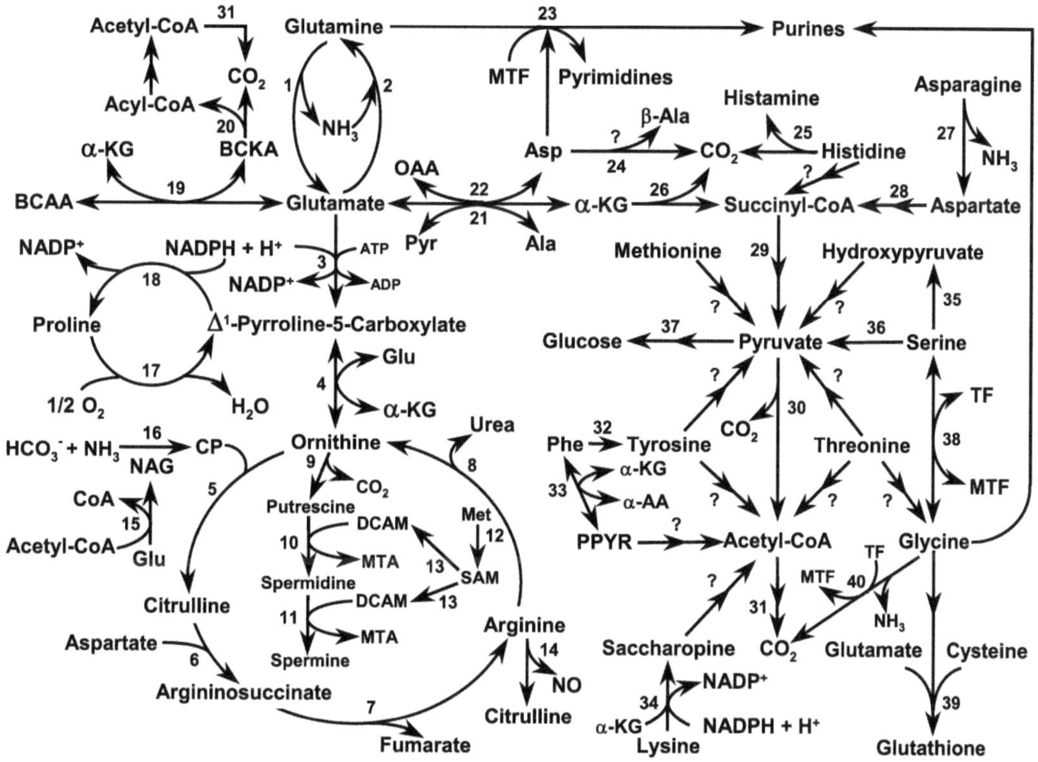

Fig. 1.3 Intestinal mucosal amino acid metabolism. Enzymes which catalyze the indicated reactions are: (1) phosphate-dependent glutaminase; (2) glutamine synthetase; (3) pyrroline-5-carboxylate synthase; (4) ornithine aminotransferase; (5) ornithine carbamoyltransferase; (6) argininosuccinate synthase; (7) argininosuccinate lyase; (8) arginase; (9) ornithine decarboxylase; (10) spermidine synthase; (11) spermine synthase; (12) *S*-adenosylmethionine synthase; (13) S-adenosylmethionine decarboxylase; (14) nitric oxide synthase; (15) N-acetylglutamate synthase; (16) carbamoylphosphate synthase-I; (17) proline oxidase; (18) pyrroline-5-carboxylate reductase; (19) branched-chain amino acid transaminase; (20) branched-chain α-ketoacid dehydrogenase; (21) alanine transaminase; (22) aspartate transaminase; (23) purine- and pyrimidine-synthesizing enzymes; (24) aspartate decarboxylase; (25) histidine decarboxylase; (26) α-ketoglutarate dehydrogenase; (27) asparaginase; (28) via aspartate transaminase and Krebs cycle enzymes; (29) possibly via NADP-linked malic enzyme, phosphoenolpyruvate carboxykinase/pyruvate kinase, and oxaloacetate decarboxylase; (30) pyruvate dehydrogenase; (31) via Krebs cycle enzymes; (32) phenylalanine hydroxylase; (33) phenylalanine transaminase; (34) lysine: α-ketoglutarate reductase; 35 serine transaminase; (36) serine dehydratase; (37) via enzymes of gluconeogenesis; (38) serine hydroxymethyltransferase; (39) glutathione-synthesizing enzymes; and (40) glycine cleavage system. The symbol "?' denotes unknown reactions in the intestinal mucosa; however, oxidation of methionine or cysteine to CO_2 is negligible in porcine, ovine, bovine, and chicken enterocytes, and there is no detectable oxidation of histidine, lysine, phenylalanine, threonine, tryptophan, and tyrosine to CO_2 in these cells. Ala, alanine; Asp, aspartate; BCAA, branched-chain amino acids; BCKA, branched-chain α-ketoacid; CoA, coenzyme A; CP, carbamoyl phosphate; DCAM, decarboxylated *S*-adenosylmethionine; α-KG, α-ketoglutarate; Met, methionine; MTA, methylthioadenosine; MTF, N^5,N^{10}-methylenetetrahydrofolate; NAG, N-acetylglutamate; OAA, oxaloacetate; PPYR, phenylpyruvate; Pyr, pyruvate; TF, tetrahydrofolate. Reproduced from Wu et al. (2005) Elsevier, with permission

Table 1.5 Content of total cysteine, methionine, and serine in animal-sourced feedstuffs[a]

Amino acid	Animal-sourced feedstuffs											
	BSFM	CBPM	CVD	Feather meal	FM-M	FM-P	FM-SE	SDPM	PBM (PFG)	SDPP	SDEP	SBM
g/kg feed (as-fed basis)												
Cys	6.89	10.68	12.17	41.70	6.74	7.15	6.85	9.85	10.41	26.50	13.69	7.01
Met	12.01	14.28	15.81	7.53	19.70	21.62	20.31	12.01	13.83	19.68	20.13	6.02
Ser	23.40	30.51	64.49	89.22	25.18	29.14	32.82	33.72	26.22	49.91	53.91	21.43
% more or less abundant in SDEP												
Cys	+99	+28	+12	−67	+103	+91	+100	+39	+32	−48	–	+95
Met	+68	+41	+27	+167	+2	−7	−1	+68	+46	+2	–	+234
Ser	+130	+77	−16	−40	+114	+85	+64	+60	+106	+8	–	+152

[a]Adapted from Li P and Wu G (2020) *Amino Acids* 52:523–542

BSFM black soldier fly larvae meal; *CBPM* chicken by-product meal; *CVD* chicken visceral digest; *FM-M* fishmeal (United States Menhaden); FM-P *fishmeal* (Peruvian anchovy); *FM-SE* fishmeal (Southeast Asian miscellaneous marine fishes); *PBM (PFG)* poultry by-product meal (pet-food grade); *SBM* soybean meal; *SDEP* spray-dried egg product; *SDPM* spray-dried peptone from enzymes-treated porcine mucosal tissues; *SDPP* spray-dried poultry plasma

" + " and " − " denote more and less abundant in SDEP, respectively, as compared with the analyzed feedstuff

1.6 Conclusion

Optimal growth, development, and health of animals depend on the adequate intakes of dietary nutrients (including AAs, carbohydrates, lipids, minerals, vitamins, and water). High-quality animal protein provides sufficient amounts and proper ratios of all AAs. Animal-sourced foodstuffs also contain (1) antioxidant nutrients (e.g., taurine, creatine, carnosine, and anserine) that are absent from plants, and (2) highly bioavailable minerals (e.g., calcium, phosphorus, iron, zinc, copper, manganese, and selenium). Through converting low-quality feedstuffs and forages into high-quality protein, farm animals do not compete with humans for food, and animal agriculture plays an important role in scientific, social and economic developments in both developed and developing nations. An adequate understanding of nutrition, metabolism, and biotechnologies is the necessary foundation for improving the growth performance, feed efficiencies, productivity, and health of livestock, poultry, fish, and crustaceans. The overall goal is to produce sufficient animal protein, sustain animal agriculture (including aquaculture), and mitigate its potential undesired effects on the environment. The new knowledge of animal nutrition and metabolism that is thoroughly covered in this and related volumes of AEMB is essential for achieving this noble goal for animal agriculture in the present and well into the future. These advances also have important implications for improving human nutrition and health particularly during the current global COVID-19 pandemic.

Acknowledgements Work in the author's laboratory was supported by Agriculture and Food Research Initiative Competitive Grants (2014-67015-21770, 2015-67015-23276, 2016-67015-24958, 2018-505706-95720, and 2021-67015-34534) from the USDA National Institute of Food and Agriculture, and Texas A&M AgriLife Research (H-8200). The author thanks Dr. Fuller W. Bazer for helpful comments on this manuscript.

References

Ashworth CJ (1991) Effect of pre-mating nutritional status and post-mating progesterone supplementation on embryo survival and conceptus growth in gilts. Anim Reprod Sci 26:311–321

Baker DH (2008) Animal models in nutrition research. J Nutr 138:391–396

Baldwin RL (1995) Modeling digestion and metabolism. Chapman & Hall, London, UK

Ballantyne JS (2001) Amino acid metabolism. Fish Physiol 20:77–107
Bauman DE, Harvatine KJ, Lock AL (2011) Nutrigenomics, rumen-derived bioactive fatty acids, and the regulation of milk fat synthesis. Annu Rev Nutr 31:299–319
Bazer FW, Wu G, Johnson GA, Kim JY, Song GW (2011) Uterine histotroph and conceptus development: select nutrients and secreted phosphoprotein 1 affect MTOR cell signaling in ewes. Biol Reprod 85:1094–1107
Bazer FW, Johnson GA, Wu G (2015) Amino acids and conceptus development during the peri-implantation period of pregnancy. Adv Exp Med Biol 843:23–52
Bazer FW, Burghardt RC, Johnson GA, Spencer TE, Wu G (2018) Mechanisms for the establishment and maintenance of pregnancy: synergies from scientific collaborations. Biol Reprod 99:225–241
Bazer FW, Lamb GC, Wu G (2020) Animal agriculture: challenges, innovations, and sustainability. Elsevier, New York, pp 1–541
Bazer FW, Seo H, Johnson GA, Wu G (2021) One-carbon metabolism and development of the conceptus during pregnancy: lessons from studies with sheep and pigs. Adv Exp Med Biol 1285:1–15
Beaumont M, Blachier F (2020) Amino acids in intestinal physiology and health. Adv Exp Med Biol 1265:1–20
Beitz DC (1985) Physiological and metabolic systems important to animal growth: an overview. J Anim Sci 61(Suppl 2):1–20
Bergen WG (2007) Contribution of research with farm animals to protein metabolism concepts: a historical perspective. J Nutr 137:706–710
Bergen WG (2008) Measuring in vivo intracellular protein degradation rates in animal systems. J Anim Sci 86(Suppl 14):E3–E12
Bergen WG (2021) Amino acids in beef cattle nutrition and production. Adv Exp Med Biol 1285:29–42
Bergen WG (2022) Pigs (*Sus Scrofa*) in biomedical research. Adv Exp Med Biol 1354:335–343
Blachier F, Yin YL, Wu G, Andriamihaja M (2013) Terminal digestion of polypeptides and amino acid absorption by the pig intestine epithelial cells during development. In: Wu G, Yin YL, Blachier F (eds) Nutritional and physiological functions of amino acids in pigs. Springer, New York, pp 51–57
Boyer PE, D'Costa S, Edwards LL, Milloway M, Susick E, Borst LB, Thakur S, Campbell JM, Crenshaw JD, Polo J, Moeser AJ (2015) Early-life dietary spray-dried plasma influences immunological and intestinal injury responses to later-life Salmonella typhimurium challenge. Br J Nutr 113:783–793
Brunton JA, Bertolo RF, Pencharz PB, Ball RO (1999) Proline ameliorates arginine deficiency during enteral but not parenteral feeding in neonatal piglets. Am J Physiol 277:E223–E231
Buddington RK, Sangild PT, Hance B, Huang EY, Black DD (2012) Prenatal gastrointestinal development in the pig and responses after preterm birth. J Anim Sci 90(Suppl 4):290–298
Burrin DG, Mersmann HJ (2005) Biology of metabolism of growing animals. Elsevier, New York
Cao Y, Yao J, Sun X, Liu S, Martin GB (2021) Amino acids in the nutrition and production of sheep and goats. Adv Exp Med Biol 1285:63–79
Cebulla CM, Zelinka CP, Scott MA, Lubow M, Bingham A, Rasiah S, Mahmoud AM, Fischer AJ (2012) A chick model of retinal detachment: cone rich and novel. PLoS One 7:e44257
Chalvon-Demersay T, Luise D, Le Floc'h N, Tesseraud S, Lambert W, Bosi P, Trevisi P, Beaumont M, Corrent E (2021) Functional amino acids in pigs and chickens: implication for gut health. Front Vet Sci 8:663727
Che DS, Nyingwa PS, Ralinala KM, Maswanganye GMT, Wu G (2021) Amino acids in the nutrition, metabolism, and health of domestic cats. Adv Exp Med Biol 1285:217–231
Chen JQ, Jin Y, Yang Y, Wu ZL, Wu G (2020) Epithelial dysfunction in lung diseases: effects of amino acids and potential mechanisms. Adv Exp Med Biol 1265:57–70
Dai ZL, Wu ZL, Zhu WY, Wu G (2022) Amino acids in microbial metabolism and function. Adv Exp Med Biol 1354:127–143
Davis TA, Fiorotto ML, Burrin DG, Reeds PJ, Nguyen HV, P.R. Beckett PR, Vann RC, O'Connor PMJ (2002) Stimulation of protein synthesis by both insulin and amino acids is unique to skeletal muscle in neonatal pigs. Am J Physiol 282:E880-890
Davis TA, Suryawan A, Orellana RA, Nguyen HV, Fiorotto ML (2008) Postnatal ontogeny of skeletal muscle protein synthesis in pigs. J Anim Sci 86(Suppl 14):E13-18
Davis TA, Suryawan A, Orellana RA, Fiorotto ML, Burrin DG (2010) Amino acids and insulin are regulators of muscle protein synthesis in neonatal pigs. Animal 4:1790–1796
Dekaney CM, Wu G, Jaeger LA (2001) Ornithine aminotransferase messenger RNA expression and enzymatic activity in fetal porcine intestine. Pediatr Res 50:104–109
Del Curto H, Wu G, Satterfield MC (2013) Nutrition and reproduction: links to epigenetics and metabolic syndrome in offspring. Curr Opin Clin Nutr Metab Care 16:385–391
Durante W (2020) Amino acids in circulatory function and health. Adv Exp Med Biol 1265:39–56
Ebeling JM, Timmons MB (2012) Recirculating aquaculture systems. In: Tidwell JH (ed) Aquaculture production systems. Wiley, Oxford, pp 245–277
FAO (2018) Food and Agriculture Organization of the United Nations. The State of World Fisheries and Aquaculture. www.fao.org/documents/card/en/c/19540EN. Accessed 01.11.2018
FAO (2020) The state of world fisheries and aquaculture 2020. Food and Agriculture Organization of the United Nations, Rome
FAO (2021) Food and agriculture organization of the united nations. Livestock primary. http://www.fao.org/faostat/en/#data/QL/visualize. Accessed 05.31.2021

Firkins JL, Yu Z, Morrison M (2007) Ruminal nitrogen metabolism: perspectives for integration of microbiology and nutrition for dairy. J Dairy Sci 90 (E Suppl): E1–E16

Flynn NE, Shaw MH, Becker JT (2020) Amino acids in health and endocrine function. Adv Exp Med Biol 1265:97–109

Furukawa M, Ito S, Suzuki S, Fuchimoto D, Onishi A, Niimi K, Usami K, Wu G, Bazer FW, Ogasawara K, Watanabe K, Aso H, Nochi T (2020) Organogenesis of ileal Peyer's patches is initiated prenatally and accelerated postnatally with comprehensive proliferation of B cells in pigs. Front Immunol 11:604674

Gao H (2020) Amino acids in reproductive nutrition and health. Adv Exp Med Biol 1265:111–131

Gatlin DM III, Barrows FT, Brown P, Dabrowski K, Gaylord TG, Hardy RW, Herman E, Hu G, Krogdahl Å, Nelson R, Overturf K, Rust M, Sealey W, Skonberg D, Souza EJ, Stone D, Wilson R, Wurtele E (2007) Expanding the utilization of sustainable plant products in aquafeeds—a review. Aquac Res 38:551–579

Gilbreath KR, Bazer FW, Satterfield MC, Wu G (2021) Amino acid nutrition and reproductive performance in ruminants. Adv Exp Med Biol 1285:43–61

Gonzalez ML, Busse NI, Waits CM, Sally E Johnson SE (2020) Satellite cells and their regulation in livestock. J Anim Sci 98:skaa081

Govoni KE, Reed SA, Zinn SA (2019) Poor maternal nutrition during gestation: effects on offspring whole-body and tissue-specific metabolism in livestock species. J Anim Sci 97:3142–3152

Greene LW (2016) Assessing the mineral supplementation needs in pasture-based beef operations in the Southeastern United States. J Anim Sci 94:5395–5400

Grillenberger M, Neumann CG, Murphy SO, Bwibo NO, van't Veer P, Hautvast JG, West CE (2003) Food supplements have a positive impact on weight gain and the addition of animal source foods increases lean body mass of Kenyan schoolchildren. J Nutr 133:3957S–3964S

Halloran KM, Stenhouse C, Wu G, Bazer FW (2021) Arginine, agmatine and polyamines: key regulators of conceptus development in mammals. Adv Exp Med Biol 1332:85–105

Harper PAW, Healy PJ, Dennis JA, O'Brien JJ, Rayward DH (1986) Citrullinaemia as a cause of neurological disease in neonatal Friesian calves. Aust Vet J 63:378–379

Hay AN, Farrell K, Leeth CM, Lee K (2022) Use of genome editing techniques to produce transgenic farm animals. Adv Exp Med Biol 1354:279–297

He W, Wu G (2020) Metabolism of amino acids in the brain and their roles in regulating food intake. Adv Exp Med Biol 1265:167–185

He WL, Li P, Wu G (2021a) Amino acid nutrition and metabolism in chickens. Adv Exp Med Biol 1285:109–131

He WL, Furukawa K, Toyomizu M, Nochi T, Bailey CA, Wu G (2021b) Interorgan metabolism, nutritional impacts, and safety of dietary L-glutamate and L-glutamine in poultry. Adv Exp Med Biol 1332:107–128

Herring CM, Bazer FW, Wu G (2021) Amino acid nutrition for optimum growth, development, reproduction, and health of zoo animals. Adv Exp Med Biol 1285:233–253

Hou YQ, He WL, Hu SD, Wu G (2019) Composition of polyamines and amino acids in plant-source foods for human consumption. Amino Acids 51:1153–1165

Hou YQ, Hu SD, Li XY, He WL, Wu G (2020) Amino acid metabolism in the liver: nutritional and physiological significance. Adv Exp Med Biol 1265:21–37

Huang J, Jia Y, Li Q, Burris WR, Bridges PJ, Matthews JC (2018) Hepatic glutamate transport and glutamine synthesis capacities are decreased in finished vs. growing beef steers, concomitant with increased GTRAP3-18 content. Amino Acids 50:513–525

Hurley WL (2019) Mammary gland development in swine: embryo to early lactation. Animal 13:S11-19

Ireland JJ, Roberts RM, Palmer GH, Bauman DE, Bazer FW (2008) A commentary on domestic animals as dual-purpose models that benefit agricultural and biomedical research. J Anim Sci 86:2797–2805

Jamin A, D'Inca R, Le Floc'h N, Kuster A, Orsonneau J-L, Darmaun D, Boudry G, Le Huërou-Luron I, Sève B, Gras-Le Guen C (2010) Fatal effects of a neonatal high-protein diet in low-birth-weight piglets used as a model of intrauterine growth restriction. Neonatology 97:321–328

Ji Y, Wu ZL, Dai ZL, Wang XL, Li J, Wang BG, Wu G (2017) Fetal and neonatal programming of postnatal growth and feed efficiency in swine. J Anim Sci Biotechnol 8:42

Jia S, Li XY, He WL, Wu G (2021) Oxidation of energy substrates in tissues of fish: metabolic significance and implications for gene expression and carcinogenesis. Adv Exp Med Biol 1332:67–83

Jia S, Li XY, He WL, Wu G (2022) Protein-sourced feedstuffs for aquatic animals in nutrition research and aquaculture. Adv Exp Med Biol 1354:237–261

Johnson IR, Ball RO, Baracos VE, Field CJ (2006) Glutamine supplementation influences immune development in the newly weaned piglet. Dev Comp Immunol 30:1191–1202

Kim SW, Hurley WL, Wu G, Ji F (2009) Ideal amino acid balance for sows during gestation and lactation. J Anim Sci 87:E123-132

Koch BJ, Hungate BA, Price LB (2017) Food-animal production and the spread of antibiotic resistance: the role of ecology. Front Ecol Environ 15:309–318

Kwon H, Spencer TE, Bazer FW, Wu G (2003a) Developmental changes of amino acids in ovine fetal fluids. Biol Reprod 68:1813–1820

Kwon H, Wu G, Bazer FW, Spencer TE (2003b) Developmental changes in polyamine levels and synthesis in the ovine conceptus. Biol Reprod 69:1626–1634

Kwon H, Wu G, Meininger CJ, Bazer FW, Spencer TE (2004a) Developmental changes in nitric oxide synthesis in the ovine conceptus. Biol Reprod 70:679–686

Kwon H, Ford SP, Bazer FW, Spencer TE, Nathanielsz PW, Nijland MJ, Hess BW, Wu G (2004b) Maternal undernutrition reduces concentrations of amino acids and polyamines in ovine maternal and fetal plasma and fetal fluids. Biol Reprod 71:901–908

Larger E, Marre M, Corvol P, Gasc JM (2004) Hyperglycemia-induced defects in angiogenesis in the chicken chorioallantoic membrane model. Diabetes 53:752–761

Lassala A, Bazer FW, Cudd TA, Li P, Li XL, Satterfield MC, Spencer TE, Wu G (2009) Intravenous administration of L-citrulline to pregnant ewes is more effective than L-arginine for increasing arginine availability in the fetus. J Nutr 139:660–665

Lassala A, Bazer FW, Cudd TA, Datta S, Keisler DH, Satterfield MC, Spencer TE, Wu G (2010) Parenteral administration of L-arginine prevents fetal growth restriction in undernourished ewes. J Nutr 140:1242–1248

Lassala A, Bazer FW, Cudd TA, Datta S, Keisler DH, Satterfield MC, Spencer TE, Wu G (2011) Parenteral administration of L-arginine enhances fetal survival and growth in sheep carrying multiple pregnancies. J Nutr 141:849–855

Lee B, Dennis JA, Healy PJ, Mull B, Pastore L, Yu H, Aguilar-Cordova E, O'Brien W, Reeds P, Beaudet AL (1999) Hepatocyte gene therapy in a large animal: a neonatal bovine model of citrullinemia. Proc Natl Acad Sci USA 96:3981–3986

Li P, Wu G (2020) Composition of amino acids and related nitrogenous nutrients in feedstuffs for animal diets. Amino Acids 52:523–542

Li P, Wu G (2022) Functional molecules of intestinal mucosal products in animal nutrition and health. Adv Exp Med Biol 1354:263–277

Li P, Yin YL, Li DF, Kim SW, Wu G (2007) Amino acids and immune function. Br J Nutr 98:237–252

Li XY, Zheng SX, Wu G (2020a) Amino acid metabolism in the kidneys: nutritional and physiological significance. Adv Exp Med Biol 1265:71–95

Li XY, Zheng SX, Ma XK, Cheng KM, Wu G (2020b) Effects of dietary starch and lipid levels on the protein retention and growth of largemouth bass (*Micropterus salmoides*). Amino Acids 52:999–1016

Li XY, Zheng SX, Ma XK, Cheng KM, Wu G (2020c) Effects of dietary protein and lipid levels on growth performance, feed utilization, and liver histology of largemouth bass (*Micropterus salmoides*). Amino Acids 52:1043–1061

Li XL, Zheng SX, Jia SC, Song F, Zhou CP, Wu G (2020d) Oxidation of energy substrates in tissues of largemouth bass (*Micropterus salmoides*). Amino Acids 52:1017–1032

Li XY, Zheng SX, Han T, Song F, Wu G (2020e) Effects of dietary protein intake on the oxidation of glutamate, glutamine, glucose and palmitate in tissues of largemouth bass (*Micropterus salmoides*) Amino Acids 52:1491–1503

Li XL, Zheng SX, Wu G (2020f) Nutrition and metabolism of glutamate and glutamine in fish. Amino Acids 52:671–691

Li XY, Zheng SX, Wu G (2021a) Nutrition and functions of amino acids in fish. Adv Exp Med Biol 1285:133–168

Li XY, Han T, Zheng SX, Wu G (2021b) Nutrition and functions of amino acids in aquatic crustaceans. Adv Exp Med Biol 1285:169–198

Li XY, Zheng SX, Ma XK, Cheng KM, Wu G (2021c) Use of alternative protein sources for fishmeal replacement in the diet of largemouth bass (*Micropterus salmoides*). Part I: effects of poultry by-product meal and soybean meal on growth, feed utilization, and health. Amino Acids 53:33–47

Li XY, Zheng SX, Cheng KM, Ma XK, Wu G (2021d) Use of alternative protein sources for fishmeal replacement in the diet of largemouth bass (*Micropterus salmoides*). Part II: effects of supplementation with methionine or taurine on growth, feed utilization, and health. Amino Acids 53:49–62

Li P, He WL, Wu G (2021e) Composition of amino acids in foodstuffs for humans and animals. Adv Exp Med Biol 1332:189–209

Li XY, Han T, Zheng SX, Wu G (2022) Hepatic glucose metabolism and its disorders in fish. Adv Exp Med Biol 1354:207–236

Lim W, Kim HS, Jeong W, Ahn SE, Kim J, Kim YB, Kim MA, Kim MK, Chung HH, Song YS, Bazer FW, Han JY, Song G (2012) SERPINB3 in the chicken model of ovarian cancer: a prognostic factor for platinum resistance and survival in patients with epithelial ovarian cancer. PLoS One 7:e49869

Matthews JC, Huang J, Rentfrow G (2016) High-affinity glutamate transporter and glutamine synthetase content in longissimus dorsi and adipose tissues of growing Angus steers differs among suckling, weanling, backgrounding, and finishing production stages. J Anim Sci 94:1267–1275

McCoard S, Sales F, Wards N, Sciascia Q, Oliver M, Koolaard J, van der Linden D (2013) Parenteral administration of twin-bearing ewes with L-arginine enhances the birth weight and brown fat stores in sheep. Springerplus 2:684

McCoard AS, Wards N, Koolaard J, Salerno MS (2014) The effect of maternal arginine supplementation on the development of the thermogenic program in the ovine fetus. Anim Prod Sci 54:1843–1847

McCoard SA, Sales FZ, Sciascia QL (2016) Amino acids in sheep production. Front Biosci E8:264–288

Monzani PS, Adona PR, Ohashi OM, Meirelles FV, Wheeler MB (2016) Transgenic bovine as bioreactors: challenges and perspectives. Bioengineered 7:123–131

Monzani PS, Adona PR, Long SA, Wheeler MB (2022) Cows as bioreactors for the production of nutritionally and biomedically significant proteins. Adv Exp Med Biol 1354:299–314

Moses RM, Kramer AC, Seo H, Wu G, Johnson GA, Bazer FW (2022) A role for fructose metabolism in development of sheep and pig conceptuses. Adv Exp Med Biol 1354:49–62

Moughan PJ (2003) Amino acid availability: aspects of chemical analysis and bioassay methodology. Nutr Res Rev 16:127–141

Mu C, Pi Y, Zhang C, Zhu WY (2022) Microbiomes in the intestine of developing pigs: implications for nutrition and health. Adv Exp Med Biol 1354:161–176

Murphy SP, Allen LH (2003) Nutritional importance of animal source foods. J Nutr 133:3932S-3935S

Myers AJ, Goodband RD, Tokach MD, Dritz SS, DeRouchey JM, Nelssen JL (2014) The effects of porcine intestinal mucosa protein sources on nursery pig growth performance. J Anim Sci 92:783–792

National Research Council (NRC, 2002) Scientific Advances in Animal Nutrition. National Academies Press, Washington, DC

National Research Council NRC (2012) Nutrient Requirements of Swine. National Academies Press, Washington, DC

Oberbauer AM, Larsen JA (2021) Amino acids in dog nutrition and health. Adv Exp Med Biol 1285: 199–216

Odle J, Jacobi SK, Boyd RD, Bauman DE, Anthony RV, Bazer FW, Lock AL, Serazin AC (2017) The potential impact of animal science research on global maternal and child nutrition and health: a landscape review. Adv Nutr 8:362–381

O'Quinn PR, Knabe DA, Wu G (2002) Arginine catabolism in lactating porcine mammary tissue. J Anim Sci 80:467–474

Paudel S, Wu G, Wang XQ (2021) Amino acids in cell signaling: regulation and function. Adv Exp Med Biol 1332:17–33

Pereira EPV, van Tilburg MF, Florean EOPT, Guedes MIF (2019) Egg yolk antibodies (IgY) and their applications in human and veterinary health: A review. Int Immunopharmacol 73:293–303

Pond WG, Strachan DN, Sinha YN, Walker EF Jr, Dunn JA, Barnes RH (1969) Effect of protein deprivation of swine during all or part of gestation on birth weight, postnatal growth rate and nucleic acid content of brain and muscle of progeny. J Nutr 99: 61–67

Posey EA, Bazer FW, Wu G (2021) Amino acids and their metabolites for improving human exercising performance. Adv Exp Med Biol 1332:151–166

Reeds PJ, Burrin DG, Jahoor F, Wykes L, Henry J, Frazer EM (1996) Enteral glutamate is almost completely metabolized in first pass by the gastrointestinal tract of infant pigs. Am J Physiol 270:E413–E418

Reeds PJ, Burrin DG, Stoll B, Jahoor F, Wykes L, Henry J, Frazer ME (1997) Enteral glutamate is the preferential source for mucosal glutathione synthesis in fed piglets. Am J Physiol 273:E408–E415

Rehfeldt C, Nissen PM, Kuhn G, Vestergaard M, Ender K, Oksbjerg N (2004) Effects of maternal nutrition and porcine growth hormone (pGH) treatment during gestation on endocrine and metabolic factors in sows, fetuses and pigs, skeletal muscle development, and postnatal growth. Domest Anim Endocrinol 27:267–285

Ren WK, Bin P, Yin YL, Wu G (2020) Impacts of amino acids on the intestinal defensive system. Adv Exp Med Biol 1265:133–151

Reynolds LP, Borowicz PP, Caton JS, Crouse MS, Dahlen CR, Ward AK (2019) Developmental programming of fetal growth and development. Vet Clin North Am Food Anim Pract 35:229–247

Reynolds LP, McLean KJ, Kacie L, McCarthy KL, Diniz WJS, Menezes ACB, Forcherio JC, Scott RR, Ward AK, Dahlen CR, Caton JS (2022) Nutritional regulation of embryonic survival, growth and development. Adv Exp Med Biol 1354:63–76

Rhoads JM, Wu G (2009) Glutamine, arginine, and leucine signaling in the intestine. Amino Acids 37:111–122

Rhoads JM, Plunkett E, Galanko J, Lichtman S, Taylor L, Maynor A, Weiner T, Freeman K, Guarisco JL, Wu G (2005) Serum citrulline correlates with enteral tolerance and bowel length in infants with short bowel syndrome. J Pediatr 146:542–547

Rodrigues LA, Wellington MO, González-Vega JC, Htoo JK, Van Kessel AG, Columbus DA (2021) Functional amino acid supplementation, regardless of dietary protein content, improves growth performance and immune status of weaned pigs challenged with Salmonella Typhimurium. J Anim Sci 99:skaa365

Rossi W Jr, Allen KM, Habte-Tsion H-M, Meesala K-M (2021) Supplementation of glycine, prebiotic, and nucleotides in soybean meal-based diets for largemouth bass (*Micropterus salmoides*): effects on production performance, whole-body nutrient composition and retention, and intestinal histopathology. Aquaculture 532:736031

Ryan P, Riechman SE, Fluckey JD, Wu G (2021) Interorgan metabolism of amino acids in health and disease. Adv Exp Med Biol 1332:129–149

Sah N, Wu G, Bazer FW (2021) Regulation of gene expression by amino acids in animal cells. Adv Exp Med Biol 1332:1–15

Sales F, Sciascia Q, van der Linden DS, Wards NJ, Oliver MH, McCoard SA (2016) Intravenous maternal arginine administration to twin-bearing ewes, during late pregnancy, is associated with increased fetal muscle mTOR abundance and postnatal growth in twin female lambs. J Anim Sci 94:2519–2531

Sandoval C, Wu G, Smith SB, Dunlap KA, Satterfield MC (2020) Maternal nutrient restriction and skeletal muscle development: consequences for postnatal health. Adv Exp Med Biol 1265:153–165

Sarkar TR, McNeal CJ, Meininger CJ, Niu YB, Mallick BK, Carroll RJ, Wu G (2021) Dietary intakes of amino acids and other nutrients by adult humans. Adv Exp Med Biol 1332:211–227

Satterfield MC, Gao HJ, Li XL, Wu G, Johnson GA, Spencer TE, Bazer FW (2010) Select nutrients and their associated transporters are increased in the ovine uterus following early progesterone administration. Biol Reprod 82:224–231

Satterfield MC, Dunlap KA, Keisler DH, Bazer FW, Wu G (2012) Arginine nutrition and fetal brown adipose tissue development in diet-induced obese sheep. Amino Acids 43:1593–1603

Satterfield MC, Dunlap KA, Keisler DH, Bazer FW, Wu G (2013) Arginine nutrition and fetal brown adipose tissue development in nutrient-restricted sheep. Amino Acids 45:489–499

Seo H, Johnson GA, Bazer FW, Wu G, McLendon BA, Kramer AC (2021) Cell-specific expression of enzymes for serine biosynthesis and glutaminolysis in farm animals. Adv Exp Med Biol 1285:17–28

Shen JZ, Wu G, Guo SD (2021) Amino acids in autophagy: regulation and function. Adv Exp Med Biol 1332:51–66

Smith BI, Govoni KE (2022) Use of agriculturally important animals as models in biomedical research. Adv Exp Med Biol 1354:315–333

Solano F (2020) Metabolism and functions of amino acids in the skin. Adv Exp Med Biol 1265:187–199

Stenhouse C, Seo H, Wu G, Johnson GA, Bazer FW (2022a) Insights into the regulation of implantation and placentation in humans, rodents, sheep, and pigs. Adv Exp Med Biol 1354:25–48

Stenhouse C, Suva LJ, Gaddy D, Wu G, Bazer FW (2022b) Phosphate, calcium, and vitamin D: key regulators of fetal and placental development in mammals. Adv Exp Med Biol 1354:77–107

Swaggerty CL, Bortoluzzi C, Lee A, Eyng C, Pont GD, Kogut MH (2022) Potential replacements for antibiotic growth promoters in poultry: interactions at the gut level and their impact on host immunity. Adv Exp Med Biol 1354:145–159

Tsang HG, Rashdan NA, Whitelaw CBA, Corcoran BM, Summers KM, MacRae VE (2016) Large animal models of cardiovascular disease. Cell Biochem Func 34:113–132

Ulshafer RJ, Allen CB (1985) Hereditary retinal degeneration in the Rhode Island Red chicken: ultrastructural analysis. Exp Eye Res 40:865–877

United Nations (2021) https://www.un.org/development. Accessed June 20, 2021

United Nations Food and Agriculture Organization (FAO 2017) The State of Food Security and Nutrition in the World. http://www.fao.org/3/a-I7695e.pdf. Accessed 01.11.2018

United Nations Food and Agriculture Organization (FAO 2020a) http://www.fao.org/faostat/en/#data/QA. Accessed May 31, 2021

UNICEF, World Health Organization (FAO), and The World Bank (2018) Levels and trends in child malnutrition. http://data.unicef.org/wp-content/uploads/2018/05/JME-2018-brochure.pdf. Accessed 01.11.2018

Vilches-Moure JG (2019) Embryonic chicken (*Gallus gallus domesticus*) as a model of cardiac biology and development. Comp Med 69:184–203

Walters EM, Wells KD, Bryda EC, Schommer S, Prather RS (2017) Swine models, genomic tools and services to enhance our understanding of human health and diseases. Lab Anim 46:167–172

Wang CY, Yeh HI, Chang TJ, Hsiao HJ, Tsai MS, Tsai SM, Liu PA (2011) Attenuation of nitric oxide bioavailability in porcine aortic endothelial cells by classical swine fever virus. Arch Virol 156:1151–1160

Wang JJ, Wu ZL, Li DF, Li N, Dindot SV, Satterfield MC, Bazer FW, Wu G (2012) Nutrition, epigenetics, and metabolic syndrome. Antioxid Redox Signal 17:282–301

Wang WW, Dai ZL, Wu ZL, Lin G, Jia SC, Hu SD, Dahanayaka S, Wu G (2014) Glycine is a nutritionally essential amino acid for maximal growth of milk-fed young pigs. Amino Acids 46:2037–2045

Wang XQ, Johnson GA, Burghardt RC, Wu G, Bazer FW (2015a) Uterine histotroph and conceptus development. I. Cooperative effects of arginine and secreted phosphoprotein 1 on proliferation of ovine trophectoderm cells via activation of the PDK1-Akt/PKB-TSC2-MTORC1 signaling cascade. Biol Reprod 92:51

Wang XQ, Burghardt RC, Romero JJ, Hansen TR, Wu G, Bazer FW (2015b) Functional roles of arginine during the peri-implantation period of pregnancy. III. Arginine stimulates proliferation and interferon tau production by ovine trophectoderm cells via nitric oxide and polyamine-TSC2-MTOR signaling pathways. Biol Reprod 92:75

Wang XQ, Johnson GA, Burghardt RC, Wu G, Bazer FW (2016) Uterine histotroph and conceptus development. II. Arginine and secreted phosphoprotein 1 cooperatively stimulate migration and adhesion of ovine trophectoderm cells via focal adhesion-MTORC2 mediated cytoskeleton reorganization. Biol Reprod 95:71

Wang B, Sun SQ, Liu MY, Chen H, Liu N, Wu ZL, Wu G, Dai ZL (2020) Dietary L-tryptophan supplementation regulates colonic serotonin homeostasis and inhibits gut inflammation in mice with dextran sodium sulfate-induced colitis. J Nutr 150:1966–1976

Webb KE Jr, Matthews JC, DiRienzo DB (1992) Peptide absorption: a review of current concepts and future perspectives. J Anim Sci 70:3248–3257

Wilkinson AD, Meeker DL (2021) How agricultural rendering supports sustainability and assists livestock's ability to contribute more than just food. Anim Front 11:24–34

Wilson EO (1992) The Diversity of Life. Harvard University Press, Cambridge, MA

Wu G (1996) Effects of concanavalin A and phorbol myristate acetate on glutamine metabolism and proliferation of porcine intraepithelial lymphocytes. Comp Biochem Physiol A 114:363–368

Wu G (1997) Synthesis of citrulline and arginine from proline in enterocytes of postnatal pigs. Am J Physiol 272:G1382–G1390

Wu G (2010) Functional amino acids in growth, reproduction and health. Adv Nutr 1:31–37

Wu G (2018) Principles of Animal Nutrition. CRC Press, Boca Raton, Florida

Wu G (2020a) Important roles of dietary taurine, creatine, carnosine, anserine and 4-hydroxyproline in human nutrition and health. Amino Acids 52:329–360

Wu G (2020b) Management of metabolic disorders (including metabolic diseases) in ruminant and non-ruminant animals. In: Lamb GC, Wu G, Bazer FW (eds) Animal agriculture: challenges, innovations, and sustainability. Elsevier, New York, pp 471–492

Wu G (2020c) Metabolism and functions of amino acids in sense organs. Adv Exp Med Biol 1265:201–217

Wu G (2021) Amino acids: biochemistry and nutrition, 2nd edn. CRC Press, Boca Raton, Florida

Wu G, Knabe DA (1994) Free and protein-bound amino acids in sow's colostrum and milk. J Nutr 124:415–424

Wu G, Morris SM Jr (1998) Arginine metabolism: nitric oxide and beyond. Biochem J 336:1–17

Wu G, Bazer FW (2019) Application of new biotechnologies for improvements in swine nutrition and pork production. J Anim Sci Biotechnol 10:28

Wu G, Knabe DA, Flynn NE (1994) Synthesis of citrulline from glutamine in pig enterocytes. Biochem J 299:115–121

Wu G, Bazer FW, Tuo W, Flynn SP (1996) Unusual abundance of arginine and ornithine in porcine allantoic fluid. Biol Reprod 54:1261–1265

Wu G, Ott TL, Knabe DA, Bazer FW (1999) Amino acid composition of the fetal pig. J Nutr 129:1031–1038

Wu G, Flynn NE, Knabe DA (2000) Enhanced intestinal synthesis of polyamines from proline in cortisol-treated piglets. Am J Physiol 279:E395–E402

Wu G, Haynes TE, Yan W, Meininger CJ (2001) Presence of glutamine:fructose-6-phosphate amidotransferase for glucosamine-6-phosphate synthesis in endothelial cells: effects of hyperglycaemia and glutamine. Diabetologia 44:196–202

Wu G, Jaeger LA, Bazer FW, Rhoads JM (2004) Arginine deficiency in premature infants: biochemical mechanisms and nutritional implications. J Nutr Biochem 15:442–451

Wu G, Bazer FW, Hu J, Johnson GA, Spencer TE (2005) Polyamine synthesis from proline in the developing porcine placenta. Biol Reprod 72:842–850

Wu G, Bazer FW, Wallace JM, Spencer TE (2006) Intrauterine growth retardation: implications for the animal sciences. J Anim Sci 84:2316–2337

Wu G, Bazer FW, Dai ZL, Li DF, Wang JJ, Wu ZL (2014a) Amino acid nutrition in animals: protein synthesis and beyond. Annu Rev Anim Biosci 2:387–417

Wu G, Fanzo J, Miller DD, Pingali P, Post M, Steiner JL, Thalacker-Mercer AE (2014b) Production and supply of high-quality food protein for human consumption: sustainability, challenges and innovations. Ann NY Acad Sci 1321:1–19

Wu G, Bazer FW, Cross HR (2014c) Land-based production of animal protein: impacts, efficiency, and sustainability. Ann NY Acad Sci 1328:18–28

Wu G, Meininger CJ, McNeal CJ, Bazer FW, Rhoads JM (2021) Role of L-arginine in nitric oxide synthesis and health in humans. Adv Exp Med Biol 1332:167–188

Wu G, Bazer FW, Satterfield MC, Gilbreath KR, Erin A. Posey, and Sun YX (2022) L-Arginine nutrition and metabolism in ruminants. Adv Exp Med Biol 1354:177–206

Yang Y, He Y, Jin Y, Wu G, Wu ZL (2021) Amino acids in endoplasmic reticulum stress and redox signaling. Adv Exp Med Biol 1332:35–49

Zhang S, Wong EA, Gilbert ER (2015) Bioavailability of different dietary supplemental methionine sources in animals. Front Biosci E7:478–490

Zhang Q, Hou YQ, Bazer FW, He WL, Posey EA, Wu G (2021) Amino acids in swine nutrition and production. Adv Exp Med Biol 1285:81–107

Zhu C, Jiang ZY, Johnson GA, Bazer FW, Wu G (2022) Nutritional and physiological regulation of water transport in the conceptus. Adv Exp Med Biol 1354:109–125

Ziche M, Morbidelli L, Masini E, Amerini S, Granger HJ, Maggi CA, Geppetti P, Ledda F (1994) Nitric oxide mediates angiogenesis in vivo and endothelial cell growth and migration in vitro promoted by substance P. J Clin Invest 94:2036–2044

2 Insights into the Regulation of Implantation and Placentation in Humans, Rodents, Sheep, and Pigs

Claire Stenhouse, Heewon Seo, Guoyao Wu, Gregory A. Johnson, and Fuller W. Bazer

Abstract

Precise cell-specific spatio-temporal molecular signaling cascades regulate the establishment and maintenance of pregnancy. Importantly, the mechanisms regulating uterine receptivity, conceptus apposition and adhesion to the uterine luminal epithelia/superficial glandular epithelia and, in some species, invasion into the endometrial stroma and decidualization of stromal cells, are critical prerequisite events for placentation which is essential for the appropriate regulation of feto-placental growth for the remainder of pregnancy. Dysregulation of these signaling cascades during this critical stage of pregnancy can lead to pregnancy loss, impaired growth and development of the conceptus, and alterations in the transplacental exchange of gasses and nutrients. While many of these processes are conserved across species, significant variations in the molecular mechanisms governing maternal recognition of pregnancy, conceptus implantation, and placentation exist. This review addresses the complexity of key mechanisms that are critical for the establishment and maintenance of a successful pregnancy in humans, rodents, sheep, and pigs. Improving understanding of the molecular mechanisms governing these processes is critical to enhancing the fertility and reproductive health of humans and livestock species.

Keywords

Conceptus · Endometrium · Implantation · Placentation · Pregnancy

List of Abbreviations

AKR1C1	Aldo-keto reductase family 1 member C1
BH4	Tetrahydrobiopterin
CE	Chorionic epithelium
CGB	Chorionic gonadotrophin beta
CL	Corpus luteum
CST3	Cystatin C
CTB	Cytotrophoblasts
CTSL	Cathepsin L
CYPs	Cytochrome P450 mixed-function oxidases
E2	Estradiol
ECM	Extracellular Matrix
ESR1	Estradiol receptor
eEVT	Endovascular extravillous trophoblast
EVT	Extravillous trophoblast
FGF	Fibroblast growth factor
FGFR	Fibroblast growth factor receptor
GCH1	GTP cyclohydrolase 1

C. Stenhouse · H. Seo · G. Wu · G. A. Johnson · F. W. Bazer (✉)
Department of Animal Science and Department of Veterinary Integrative Biosciences, Texas A&M University, College Station, TX 77843, USA
e-mail: fbazer@cvm.tamu.edu

G. Wu (ed.), *Recent Advances in Animal Nutrition and Metabolism*, Advances in Experimental Medicine and Biology 1354,
https://doi.org/10.1007/978-3-030-85686-1_2

GE	Glandular epithelium
GLYCAM1	Glycosylation dependent cell adhesion molecule 1
GNRH	Gonadotropin-releasing hormone
GNRHR	Gonadotropin-releasing hormone receptor
HBEGF	Heparin-binding epidermal growth factor
HGF	Hepatocyte growth factor
HIF2A	Hypoxia inducible factor 2 α
HLA	Human leukocyte antigen
ICM	Inner cell mass
iEVT	Interstitial extravillous trophoblast
IFND	Interferon delta
IFNG	Interferon gamma
IFNT	Interferon tau
IL1B	Interleukin-1 beta
IRF	Interferon regulatory factor
ISG	Interferon stimulated gene
LE	Luminal epithelium
LGALS15	Galectin 15
LH	Luteinizing hormone
LHCGR	Luteinizing hormone/choriogonadotropin receptor
MAPK	Mitogen activated protein kinases
MMP	Matrix metalloproteinases
MTORC	Mechanistic target of rapamycin
MUC1	Mucin 1
OXT	Oxytocin
OXTR	Oxytocin receptor
P4	Progesterone
PA1	Plasminogen activator inhibitor
PGF	Prostaglandin F2α
PGFM	Prostaglandin F2α metabolite
PGR	Progesterone receptor
PI3K	Phosphoinositide-3 kinase
PRL	Prolactin
PTGFR	Prostaglandin F2α receptor
PTGS2	Prostaglandin synthase 2
sGE	Superficial glandular epithelium
SLC	Solute carrier family
SPP1	Secreted phosphoprotein 1
STAT1	Signal transducer and activator of transcription 1
STB	Syncytiotrophoblast
TGC	Trophoblast giant cell
TGFB	Transforming growth factor beta
TIMPs	Tissue inhibitors of matrix metalloproteins
TPA	Tissue-type plasminogen activator
UF	Uteroferrin
UPA	Urokinase-type plasminogen activator
WNT	Wingless-related integration site
20α-OHP	20α-Hydroxyprogesterone
20αHSD	20α-Hydroxysteroid dehydrogenase

2.1 Introduction

The establishment and maintenance of pregnancy requires a series of complex and concerted events to maximize the likelihood of a successful outcome of pregnancy (Bazer et al. 2011). This intricate process is reliant upon precise cell-specific spatio-temporal molecular signaling between the developing conceptus (embryo and associated placental membranes) and the uterine endometrium during conceptus elongation, adhesion, apposition, attachment, and formation of a functional placenta. While many of these processes are conserved across species, significant variations in the molecular mechanisms governing maternal recognition of pregnancy, conceptus implantation, and placentation exist. Dysregulation of these signaling cascades during this critical stage of pregnancy can lead to pregnancy loss, impaired growth and development of the conceptus, and alterations in the transplacental exchange of gasses and nutrients (Bazer and Johnson 2014; Wu 2022). This review addresses the complexity of key mechanisms that are critical for the establishment and maintenance of a successful pregnancy in humans, rodents, sheep, and pigs. Improving understanding of the molecular mechanisms governing these processes is critical to enhancing the fertility and reproductive health of humans and livestock species.

2.2 Fertilization and Embryonic Development to the Blastocyst Stage in Mammals

In all mammals, fertilization occurs in the oviduct when oocytes from the mammalian ovary fuse with sperm from the male to produce a one-cell embryo (the zygote) with a diploid (maternal and paternal) set of chromosomes. The zygote and embryo in the early stages of development are within the zona pellucida; a protective membrane composed mostly of glycoproteins (Wassarman 1988). The zygote goes through rapid cell division to the 2-cell, 4-cell, 8-cell, and so on to form a 32- to 64-cell stage embryo known as a morula. Thereafter, in addition to continued proliferation of cells, the cells segregate and differentiate to form either the embryonic disc (also known as the inner cell mass) or trophectoderm, and those entities surround a blastocoel into which nutrients are transported. The free-floating blastocyst must emerge, a process known as "hatching" from the zona pellucida (Wassarman and Litscher 2008). This involves proteases and glycosidases that allow rupture of the zona pellucida and emergence of the blastocyst so that it can expand in preparation for implantation on the epithelial lining of the uterine lumen.

Here, we will discuss this process in more detail for humans and rodents in which the blastocyst, in a spherical form, invades into the uterine endometrium to gain access to maternal blood for the exchange of nutrients and gases, i.e., carbon dioxide and oxygen. In contrast, blastocysts of sheep and pigs undergo a rapid and unique transformation from a spherical form to a tubular form and then a greatly elongated filamentous form without invading into the endometrium. Blastocysts elongate from 10 mm spheres to 250 mm filamentous forms in sheep and up to 1,000 mm elongated forms in pigs (Bazer 2013). Elongation of the trophectoderm is essential for creating a large surface area for uptake of nutrients and exchange of gases. The following sections will provide details of species-specific mechanisms for pregnancy recognition signaling, conceptus development, and implantation.

2.3 Pre-Implantation Conceptus Development in Humans, Sheep, Pigs, and Rodents

2.3.1 Pregnancy Recognition Signaling

For pregnancy to be established and maintained, the conceptus (embryo and its extra-embryonic membranes) must provide a hormonal signal for maternal recognition of pregnancy (Spencer and Bazer 2004). The estrous cycle of subprimate species is uterine-dependent because the uterus is the source of prostaglandin $F_{2\alpha}$ (PGF), the luteolytic hormone responsible for functional and structural regression of the ovarian corpus luteum (CL) (Stouffer 1988). With regression of the CL and decreasing concentrations of progesterone (P4) in the circulation, the estrous or menstrual cycle begins anew. In primates, however, the menstrual cycle is uterine-independent as luteolytic PGF is from an intra-ovarian source (Stouffer et al. 2014). While differences in the maternal recognition of pregnancy signals exist among species, they are from conceptus trophectoderm to maternal uterus or ovarian corpora lutea (CL). These signals from the conceptus are either anti-luteolytic, i.e., they prevent the release of luteolytic PGF from the uterus, or they are luteotrophic and act directly on the CL to prevent luteolysis (Spencer and Bazer 2004).

2.3.2 Pregnancy Recognition Signaling in Humans

Pregnancy recognition signals are required to extend the lifespan of the CL that produces P4, the hormone required for the establishment and maintenance of pregnancy. For primates, the CL is the sole source of P4 until the time of the luteal-placental shift when production of P4 by the placenta is sufficient to support pregnancy (Stouffer and Hearn 1998; Fazleabas et al. 2004). Three to four days following ovulation in primates, morula stage embryos enter the uterus, hatch from the zona pellucida, and initiate implantation, with trophectoderm cells attaching to uterine luminal epithelium (LE) 7–9 Days post-ovulation in humans. In primates, the production of chorionic gonadotrophin beta (CGB) by trophectoderm of blastocyst signals maternal recognition of pregnancy and acts via the receptor for Luteinizing Hormone (LHCGR) (Srisuparp et al. 2001). In all primates, CGB is detectable in maternal blood around the time of implantation, with a peak in, concentrations during the first trimester, before decreasing during late gestation. The expression of *CGB* mRNA has been detected in both 8-cell embryos and hatched blastocysts from women (Bonduelle et al. 1988; Syrkasheva et al. 2017).

In the human trophoblast, gonadotropin-releasing hormone 1 (luteinizing-releasing hormone) (GNRH1) from the uterus is believed to regulate the production of CGB, as receptors for GNRH1 (GNRHR) are expressed by trophectoderm. GNRH agonists and antagonists enhance and suppress secretion of CGB, respectively, suggesting a pivotal regulatory role of GNRH in the production of CGB. Additionally, inhibin, activin, steroids, and P4, from the ovary and/or placenta, may regulate production of CGB. The secretion of CGB is critical for CL maintenance in early pregnancy; however, around the time of the luteal-placental shift in P4 production, the production of CGB decreases. Immunization of primates with modified forms of CGB results in infertility, but the immunized animals continue to exhibit normal menstrual cycles. Further, administration of exogenous CGB increases the production of P4 while extending the lifespan of the CL in women.

2.3.3 Pregnancy Recognition Signaling in Rodents

Laboratory rodents are spontaneously ovulating, non-seasonal, polyestrous mammals, with short-generation intervals, making them valuable and extensively utilized animal models for studies of conceptus growth and development. Rodents have estrous cycles of 4–5 days in length that include proestrus (12–14 h), estrus (25–27 h), metestrus (6–8 h), and diestrus (55–57 h) (Freeman et al. 1974; Soares et al. 2007). The CL of cyclic rats and mice, initially secrete P4 for two days. Pulses of prolactin from the anterior pituitary gland are produced following vaginal stimulation to allow the CL to become fully functional and produce P4. Newly formed CL are maintained in rats through metestrus of the following cycle, after which time the luteal cells undergo apoptosis, blood vessels degenerate, and leukocytes infiltrate the CL to remove cellular debris. In cyclic rodents, P4 secreted by the CL is metabolized by aldo–keto reductase family 1 member C1 (AKR1C1; also known as 20α-hydroxysteroid dehydrogenase [20α-HSD]) to 20α-hydroxyprogesterone (20α-OHP). The uterine decidual reaction required for the establishment of pregnancy is not induced by 20α-OHP. The secretion of 20α-OHP by the CL declines during diestrus, until the onset of proestrus when estrus and ovulation begin a new cycle.

The CL must continue to produce P4 until Day 17 of the 20–22 Days gestation period in rats, mice, and hamsters. The production of P4 is necessary for implantation, induction of the uterine decidual reaction, placentation, and a successful pregnancy (Soares et al. 2007). Two endocrine events are critical for the establishment and maintenance of pregnancy. Mating induces diurnal and nocturnal surges the secretion of prolactin (PRL) from lactotroph cells in the anterior pituitary. This increases the expression of LHCGR on luteal cells and suppresses AKR1C1 (20α-HSD) activity in the CL that prevents the conversion of P4 to 20α-OHP (Gunnet and Freeman 1983; Soares et al. 2007). Second, maintenance of pregnancy beyond Day 12 in rodents is reliant upon implantation, conceptus development, and the production of lactogenic hormones by both the uterine decidua and placenta (Soares 2004). Prolactin and placental lactogen are members of the lactogenic family of hormones, which act in a luteotrophic manner in mice and rats to ensure CL maintenance and its continued secretion of P4 required for the establishment and maintenance of pregnancy to full term (Soares et al. 2006, 2007).

2.3.4 Pregnancy Recognition Signaling in Sheep

In sheep, regulation of the estrous cycle is dependent upon the production of PGF by the uterine epithelia (Bazer et al. 1994; Thatcher et al. 1995). P4 acts upon the uterine epithelia during diestrus to increase phospholipid stores and expression of prostaglandin synthase 2 (PTGS2), both of which are necessary for mobilization of arachidonic acid by phospholipase A2 and conversion of arachidonic acid by PTGS2 to PGF. Importantly, P4 acts upon the uterus to down-regulate the expression of the progesterone receptors (PGR) which increases the expression of receptors for estradiol (ESR1) and oxytocin (OXTR) in uterine LE and superficial glandular epithelia (sGE) initially, and later in glandular epithelia (GE) and stromal cells (Spencer and Bazer 2004). These alterations in uterine epithelial gene expression are critical events in activation of the luteolytic mechanism for the production of luteolytic pulses of PGF by the uterus in sheep. Estradiol (E2), acting via ESR1, induces expression of phospholipase A2 that mobilizes arachidonic acid for conversion to PGF. The posterior pituitary and CL release pulses of OXT which binds to the OXTR to induce the pulsatile release of PGF. The pulsatile production of PGF induces regression of CL by Day 16 of the estrous cycle.

In ruminants, the signal for maternal recognition of pregnancy is interferon tau (IFNT) (Bazer 2013). IFNT, a Type I interferon, is produced by the mononuclear trophectoderm cells of the conceptus during the peri-implantation period of pregnancy as the conceptus undergoes morphological transition from spherical, to tubular, and filamentous forms (Bazer et al. 2018). Secretion of ovine IFNT begins on about Day 10 and increases to Day 16, then decreases to Day 21 after which production ceases and the IFNT gene is no longer expressed by the conceptus. In summation, IFNT silences transcription of ESR1 to preclude estrogen receptor α interactions with SP1 and/or AP-1 that otherwise stimulate oxytocin receptor expression in uterine LE/sGE to abrogate the oxytocin-dependent pulsatile release of luteolytic PGF (Fig. 2.1) [reviewed by (Bazer et al. 2015b)]. Thus, the CL is maintained to produce P4, the hormone of pregnancy (Fleming et al. 2006). Interferon tau also has potent antiviral, antiproliferative, and immunomodulatory activities characteristic of other Type I interferons (Bazer et al. 2015b).

The loss of expression of PGR in uterine epithelia cells on Days 12–13 of pregnancy is essential for activation of key events required for the establishment of pregnancy (Bazer et al. 2009). The loss of PGR by endometrial epithelia is required for implantation that is dependent on the loss of expression of some genes, such as mucin 1 (MUC1), on the surface of uterine LE, that would otherwise block implantation. In addition to down-regulation of expression of PGR and ESR1 by uterine epithelia being a prerequisite for uterine receptivity to conceptus implantation, this is also critical for up-regulation of the expression of many genes including those for secretory proteins and nutrient transporters for transport of glucose and amino acids into the uterine lumen to support growth and development of the conceptus. Down-regulation of PGR in uterine epithelia allows P4 to act on PGR-positive uterine stromal cells, upregulating the expression of progestamedins, i.e., fibroblast growth factor 7 (FGF7) and -10 (FGF10), and hepatocyte growth factor (HGF). The progestamedins exert paracrine effects on uterine epithelia and conceptus trophectoderm which express receptors for FGF7 and FGF10 (*FGFR2IIIb)* and HGF (MET; protooncogene *MET)* (Igarashi et al. 1998; Chen et al. 2000a, b). Whilst the expression of many IFNT-stimulated genes (ISGs) are known to be induced by P4 and stimulated by IFNs (Bazer et al. 2008), it is not known if progestamedins and IFNs act on uterine epithelial cells via non-classical cell signaling pathways, independent of PGR and signal transducer and activator of transcription 1 (STAT1) to alter gene expression and uterine receptivity to implantation [reviewed by (Bazer et al. 2015b)]. Type I IFNs may bind to the same receptor, but activate unique cell-specific

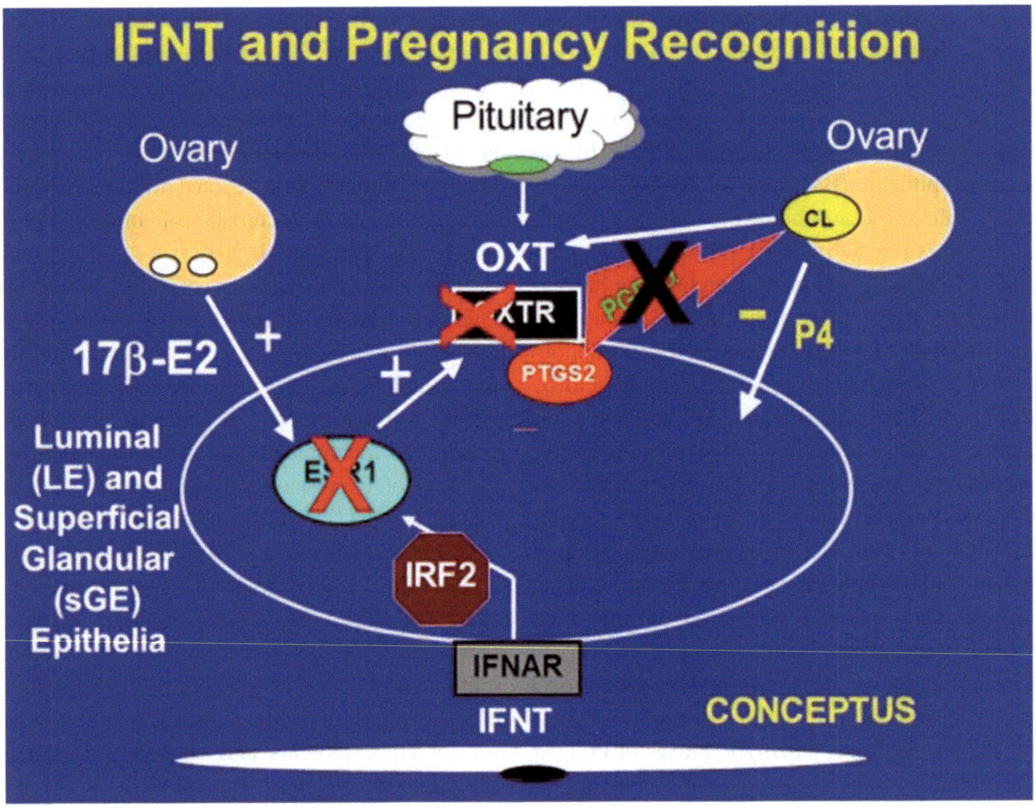

Fig. 2.1 Interferon tau (IFNT) is the pregnancy recognition hormone in sheep and other ruminants. It acts to silence expression of estrogen receptor alpha (ESR1) and, in turn, oxytocin receptor (OXTR) to prevent development of the luteolytic mechanism which requires oxytocin (OXT) from the corpus luteum (CL) and posterior pituitary to induce luteolytic pulses of prostaglandin $F_{2\alpha}$ (PGF). Thus, IFNT blocks the ability of the uterus to develop the luteolytic mechanism but does not inhibit prostaglandin synthase 2 (PTGS2) or the basal production of PGF during pregnancy. Adapted from the Open Access article of Bazer (2013)

signaling pathways that differentially effect gene expression in uterine LE, sGE, GE, and stromal cells [reviewed by (Bazer et al. 2015b)].

In sheep, IFNT up-regulates IRF2 in uterine LE and sGE and this inhibits expression of classical ISGs such as STAT1 and interferon regulatory factor 9 (IRF9) in those cells (Choi et al. 2001; Bazer et al. 2015b). However, there is a growing number of P4-induced and IFNT-stimulated genes being discovered to be expressed uterine LE/sGE that lack both PGR and STAT1 and are critical for implantation and conceptus development. Those genes include wingless-type MMTV integration site family member 7A (WNT7A), galectin 15 (LGALS15), cathepsin L (CTSL), cystatin C (CST3), solute carrier family 2 member 1 (SLC2A1), solute carrier family 7 member 1 (SLC7A1), solute carrier family 7 member 2 (SLC7A2), and hypoxia inducible factor 2 α (HIF2A).

2.3.5 Pregnancy Recognition Signaling in Pigs

After stimulation of the uterine endometrium by P4 for 10–12 days, luteolysis occurs during late diestrus and early proestrus, there is an accumulation of phospholipids and necessary enzymes for production of PGF in a pulsatile manner (Bazer 1989). It is not until Days 12–13 of the estrous cycle that porcine CL are sufficiently responsive to luteolytic PGF because until that time they express few receptors for PGF (PTGFR). The posterior pituitary and CL produce OXT which binds to uterine OXTR to elicit pulsatile release of PGF. In contrast to other species, the porcine CL has a very low abundance of both OXT and vasopressin; therefore, the role(s), if any, of these neuropeptides in luteolysis in pigs is not known. Further, the uterine endometrium is a source of OXT in pigs, although its potential roles in regulating the estrous cycle or aspects of pregnancy are not known. However, administration of exogenous OXT to gilts decreases the inter-estrous interval if administered between Days 10 and 16 post-estrus. This finding is not observed when OXT is administered to hysterectomized gilts with intact ovaries, suggesting that any effect of OXT on length of the estrous cycle is uterine-dependent. Both OXTR and lysine vasopressin receptors are expressed by cells of the porcine endometrium, but the endometrium only responds to OXT with increased secretion of PGF. Further, both OXT and vasopressin stimulate calcium-calmodulin kinase and protein kinase C signaling pathways in the porcine endometrium. Increased pulsatile secretion of PGF occurs between Days 14 and 18 of the estrous cycle as OXT increases the activity of phospholipase C and the hydrolysis of phosphatidylinositol (Bazer et al. 1984). Increases in intracellular calcium and diacylglycerol activate protein kinase C and calcium-calmodulin kinase which, in turn, activate phospholipase A_2 and the release of arachidonic acid, ultimately leading to the pulsatile production of PGF. During luteolysis, concentrations of OXT increase in maternal blood. Interestingly, the OXT-induced increases in circulating concentrations of prostaglandin F2α metabolite (PGFM) are lower in pregnant than cyclic gilts or gilts induced into pseudopregnancy by injection of exogenous E2 from Day 11 to Day 15 post-estrus. Concentrations of PGFM in blood of pregnant gilts increase beginning on Day 12. In addition, inhibition of prostaglandin synthesis results in pregnancy failure. Collectively these findings suggest an important role of prostaglandins in the establishment of pregnancy in pigs.

Porcine blastocysts hatch from the zona pellucida and expand before undergoing a rapid morphological transition to large spherical, tubular, and filamentous forms between Days 10 and 12 of pregnancy as they become conceptuses. These rapidly elongating conceptuses achieve a length of 800–1000 mm between Days 12 and 15 of pregnancy (Bazer and Johnson 2014), with the trophectoderm secreting many critical molecules for the establishment of pregnancy including estrogens, interleukin 1B, interferon gamma (IFNG), and interferon delta (IFND). In the absence of a conceptus, the uterine endometrium secretes PGF in an endocrine manner into its venous drainage to be transported through the maternal vasculature to the ovary where it can exert its luteolytic effect on the CL. In contrast, in pregnant gilts, the conceptus produces E2 from around Days 11–12 until Day 15 of pregnancy that is the maternal recognition signal. E2 acts on uterine epithelia to direct secretion of PGF away from the uterine vasculature and into the uterine lumen (exocrine secretion) where it is sequestered and metabolized, thereby preventing luteolysis (Fig. 2.2).

The conceptus estrogens not only act as the signal for maternal recognition in pigs, but also modulate uterine gene expression responsible for endometrial remodeling for implantation between Days 13 and 25 of gestation (Johnson et al. 2009). There is a report (Meyer et al. 2019) that silencing expression of the aromatase (CYP19A1) gene, utilizing CRISPR/Cas9 genome editing technology, did not prevent elongation and implantation of pig conceptuses or maintenance of CL through the peri-implantation period, but the pregnancies failed around Day 30 of gestation. Therefore, further investigation to ascertain the regulatory role of estrogens in the establishment of pregnancy in the pig should be performed.

The following critical events must be tightly regulated to allow the establishment and maintenance of pregnancy in the pig: (i) the conceptus must secrete estrogens to act as the maternal recognition of pregnancy signal; (ii) the uterine LE and GE must provide, through nutrient

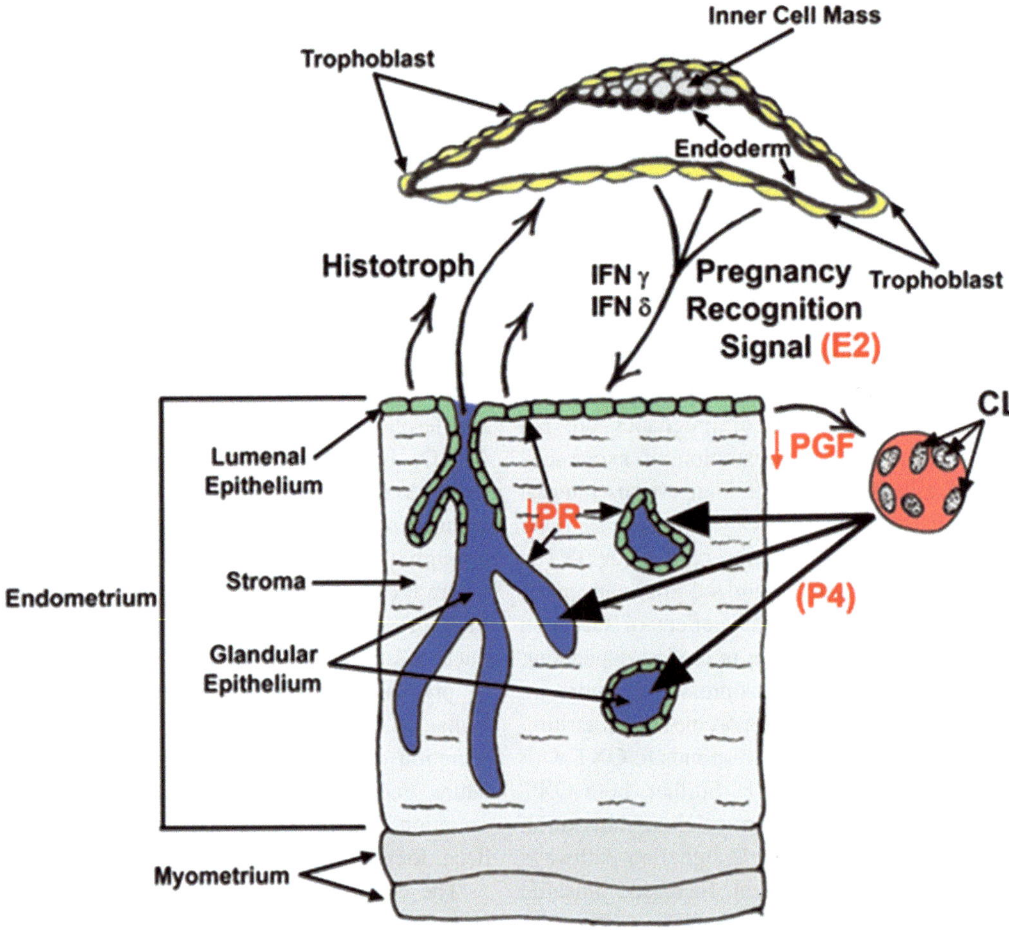

Fig. 2.2 The theory of pregnancy recognition in the pig is based on evidence that estradiol (E2), as the pregnancy recognition signal, which redirects secretion of prostaglandin $F_{2\alpha}$ (PGF) away from endocrine secretion into the uterine vasculature and into exocrine secretion into the uterine lumen, allowing metabolism to PGE or its inactive metabolite. The roles of interferon delta and gamma in early pregnancy are not yet fully understood. This figure also illustrates that there is down-regulation of expression of receptors for progesterone (PR) in uterine epithelia; therefore, progesterone (P4) acts on uterine stromal cells to regulate expression of genes associated with the secretion of histotroph into the uterine lumen. Adapted from Bazer and Johnson (2014)

transport and secretions, nutrient rich histotroph to support attachment, development, and growth of the conceptus; and (iii) cellular remodeling at the uterine LE:trophectoderm required for implantation and the initiation of placentation. In addition to the critical role of P4 and E2, the spatio-temporal changes in gene expression in both the trophectoderm and endometrium are regulated by interleukin-1 beta (IL1B), the interferons (IFND and IFNG), transforming growth factor beta (TGFB) and fibroblast growth factor 7 (FGF7). Interferons appear to have a crucial role in the establishment of pregnancy across mammalian species, with reports of up-regulation of interferon stimulated gene expression in response to conceptus secreted IFNs in many species. Significant antiviral activity is present in Day 14 uterine flushings from pregnancy pigs, with IFNG accounting for approximately 75% of this activity, suggesting a pivotal role for this interferon in the establishment of pregnancy. The pig is unique in that conceptus produced estrogens induce expression of IRF2 only in uterine LE (Joyce et al. 2007). Considering this, it could be speculated that IFND and IFNG work synergistically to induce classical interferon responsive genes only in uterine GE and stromal cells which do not express IRF2, while uterine LE in direct contact with conceptus trophectoderm is induced to express novel genes, e.g., nutrient transporters for glucose and amino acids that enhance conceptus development.

Uterine histotroph in pigs is composed of secretions from uterine epithelia and molecules that are selectively transported into the uterine lumen (Bazer et al. 2018). In addition to glucose, fructose, amino acids, and other micromolecules, it contains a very complex mixture of peptides that proteins that include: uteroferrin, now known as ACP5 (phosphatase, acid, type 5, tartrate resistant) which transports iron to the conceptus for erythropoiesis and stimulates hematopoiesis; retinol binding protein, plasmin/trypsin inhibitor, leucine aminopeptidase, glucose phosphate isomerase, serine protease inhibitors, lysozyme, various proteases, hexosaminadase, phospholipases, prostaglandin synthases, insulin-like growth factors 1 and 2, insulin-like growth factor binding proteins, high molecular weight glycoproteins, colony stimulating factor 1, secreted phosphoprotein 1 (SPP1), integrins, FGF7, FGF10, HGF, staniocalcins, and transforming growth factors beta-1, -2 and -3, nitric oxide synthase, GTP cyclohydrolase 1 (GCH1), tetrahydrobiopterin (BH4), and interleukins 1 and 4. The roles of many of these proteins remain to be determined. Nevertheless, they contribute to a uterine microenvironment that supports growth and development of the conceptus.

2.4 Implantation of Blastocysts in Humans, Sheep, Pigs, and Rodents

2.4.1 Implantation of Blastocysts in Mammals

Implantation of the blastocyst in mammals is the process of attachment of trophectoderm to uterine luminal epithelium (LE) and, in some species, invasion of the blastocyst into the uterine endometrium. Implantation is a prerequisite for placentation that begins later in gestation (McGowen et al. 2014). The two primary classifications of implantation in mammals are based on the extent to which the blastocyst invades into the uterine endometrium. Central-type or superficial implantation (pig, horse, sheep, and cow) does involve attachment of trophectoderm to the uterine LE, but not invasion into the endometrium. Interstitial attachment involves invasion and embedding of the blastocyst entirely within the uterine endometrium (rodents and primates). The implantation cascade has up to five stages: (1) shedding of the zona pellucida from the blastocyst; (2) pre-contact and orientation of the blastocyst; (3) apposition of trophectoderm and uterine LE; (4) adhesion of trophectoderm to uterine LE; and (5) invasion of the blastocyst into the endometrium in species with interstitial implantation.

2.4.2 Implantation in Humans

There is a well-defined "window of implantation" in women during which the uterus is receptive to the blastocyst for the initiation of implantation (Su and Fazleabas 2015; Kim and Kim 2017). That window of implantation is during the mid-secretory phase of the menstrual cycle, specifically cycle Days 20–24 (6–10 Days after ovulation). Markers of uterine receptivity to implantation in humans include: (1) pinopods or uterodomes that are hairlike microvilli of epithelial cells which transiently fuse to form a single flowerlike membrane projection only on the luminal surface of endometrial epithelial cells during the window of implantation; (2) epithelial plaques that are an endometrial response in primates to implantation of the blastocyst involving transformation of uterine LE and sGE epithelia as they undergo hypertrophy, hyperplasia, and form a rounded acinar multicellular pad; (3) down-regulation of expression of receptors for estradiol (ESR1) and progesterone (PGR) in uterine epithelia; (4) decrease in expression of MUC1 by uterine LE; (5) expression of the integrins $\alpha 1\beta 1$,

$\alpha4\beta1$, and $\alpha v\beta3$ by uterine LE/sGE; (6) increased expression and secretion of SPP1 by uterine epithelia; and (7) expression of heparin-binding epidermal growth factor (HBEGF) by uterine LE and the surface of pinopods. Implantation occurs in three stages. First, the blastocyst comes into apposition with the uterine LE, then trophectoderm attaches to the uterine LE, and finally the invasive trophoblast cells cross the endometrial epithelial basement membrane and invade into the uterine endometrium.

Trophoblast cells penetrate the uterine LE via gaps between cells to reach the basement membrane, but without destroying the uterine LE (Carson et al. 2000). Formation of thin folds of trophoblast cells between uterine LE is followed by degradation of the basement membrane and extracellular matrix (ECM), allowing trophoblast cells to be in close contact with the endometrial stromal cells. Activated matrix metalloproteinases (MMPs) play a primary role in degrading matrices during this process, and various integrins guide the invading trophoblast through different layers of cells and matrices within the endometrium. Next, trophoblast cells that migrate into the endometrium and continue to proliferate, differentiate, and fuse to become the multinucleated syncytiotrophoblasts (STB). The other trophoblast cells, those that surround the inner cell mass, are mononuclear cytotrophoblasts (CTB). STB cells guide invasion of the blastocyst into the endometrium until it is completely embedded within the endometrial stroma, 8 days after ovulation. The site of entry of the invading blastocyst is covered by fibrin that also supports growth of endometrial epithelial cells to cover the site of implantation.

The STB layer has fluid-filled spaces known as lacunae separated by trabeculae, resembling a sponge. As the STB contacts maternal blood vessels, maternal blood is trapped within the lacunae for transfer of oxygen and nutrients to the developing conceptus. CTB grow into the trabeculae formed by invading STB to form primary chorionic villi that initiate placentation. The uterine stroma at the site of implantation undergoes decidualization to form three areas of the decidua in humans during pregnancy. The area of decidua directly beneath the site of implantation is the decidua basalis, the region that overlies the developing conceptus and separates it from the uterine cavity is the decidua capsularis, and the remaining decidua is the decidua parietalis. The decidua basalis and the decidua capsularis are invaded by trophoblast cells and chorionic villi of the conceptus, but only the decidua basalis supports formation of the discoid placenta in mid- to late-pregnancy as the rest of the decidua degenerates later in pregnancy.

2.4.3 Implantation in Rodents

As for humans, there is a "window of implantation" in mice when the uterus is receptive to supporting growth, attachment, and implantation of the blastocyst. P4 and E2 are critical for implantation in mice as they regulate proliferation and/or differentiation of uterine cells in a time and cell-specific manner that is required for uterine receptivity to implantation of the blastocyst (Cockburn and Rossant 2010; McGowen et al. 2014; Aplin and Ruane 2017; Matsumoto 2017). E2 from preovulatory follicles induces proliferation of uterine epithelial cells and then, post-ovulation, P4 from the CL induces proliferation of stromal cells from Day 3. The uterus becomes receptive to implantation on Day 4 in response to an acute increase in E2. Attachment of blastocyst trophectoderm to uterine LE on Day 4 of pregnancy results in proliferation of stromal cells surrounding the implanting blastocyst and then the stromal cells undergo full decidualization. The receptive phase of the uterus for implantation is only approximately 24 h, therefore the processes governing implantation must be tightly regulated to ensure implantation occurs during this short time.

Blastocysts are found in uterine crypts on the anti-mesometrial side of the uterine lumen, and the uterine lumen then closes to confine the blastocyst to a limited space within the uterine lumen called the implantation chamber (Fig. 2.3). The tight space formed by the implantation chamber restricts blastocyst

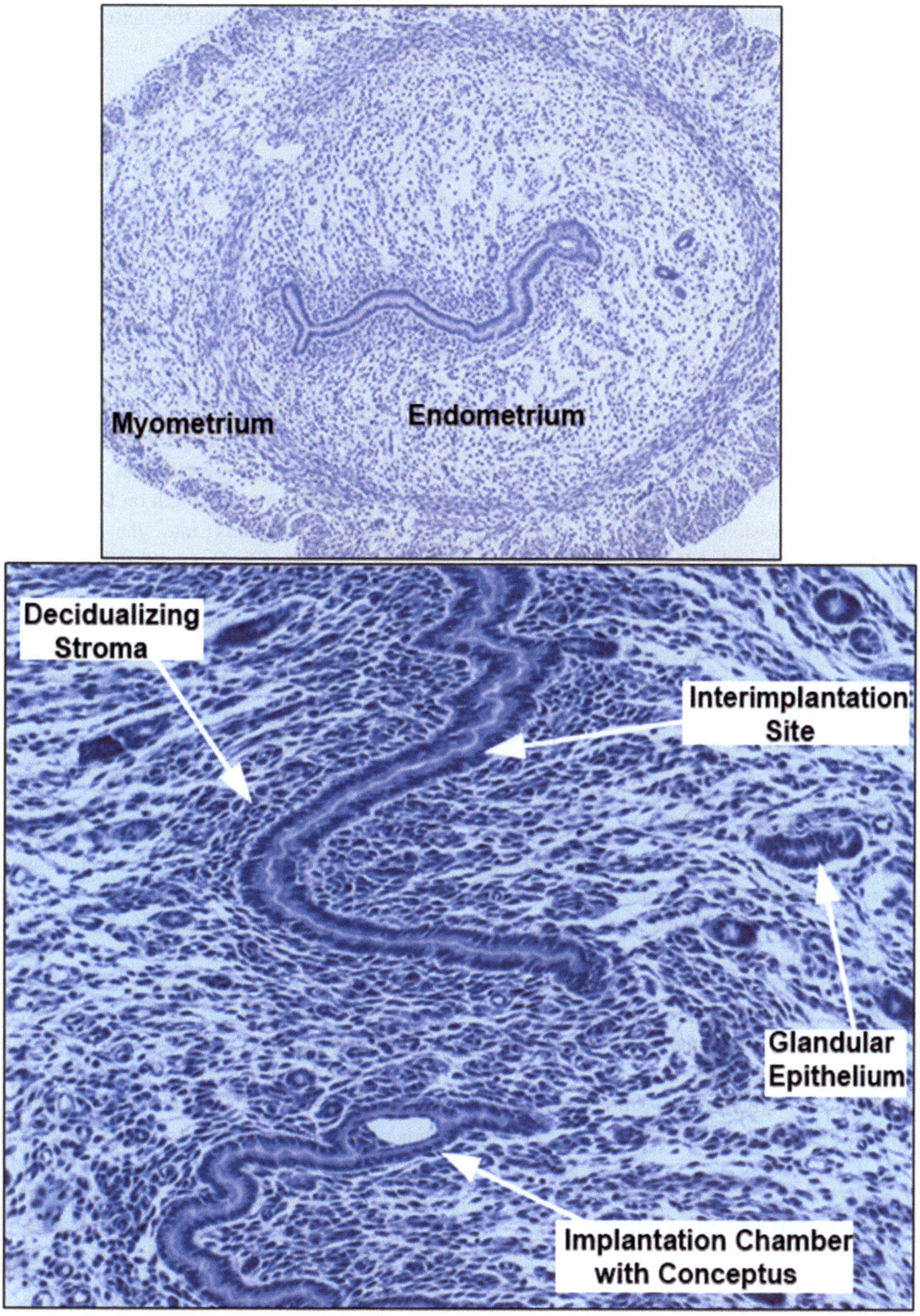

Fig. 2.3 Attachment of the mouse blastocyst to the uterine luminal epithelium (LE). Shown are mouse blastocysts within implantation chambers at the initiation of implantation. Closure of the uterine lumen at inter-implantation sites brings the blastocyst into close apposition to the uterine LE, and interaction of the blastocyst with the LE elicits the beginning of a decidual response within the uterine stroma. Decidualization begins following blastocyst adhesion, but prior to LE degradation for implantation into the uterine wall. Pregnancies were produced using delayed implantation so that the presence of implantation chambers could be accurately predicted. Unpublished work of Kramer AC, Erikson DW, McLendon BA, Seo H, Hyashi K, Spencer TE, Bazer FW, Burghardt RC, Johnson GA

movement and facilitates close apposition of the apical surfaces of trophoblast cells to endometrial LE cells. The integrity of the implantation chamber is maintained via closure of the lumen surrounding the chamber. The mechanisms regulating the closure of the uterine lumen at inter-implantation sites in mice are not completely understood, but likely involve absorption of fluid within the uterine lumen, mediated by aquaporins expressed by the endometrium (Richard et al. 2003; Beall et al. 2007; Chan et al. 2009; De Oliveira et al. 2020). The next phase involves invasion of the blastocyst into the uterine endometrium, which is mediated by expression of genes required for remodeling the endometrium. MMPs are zinc-dependent endopeptidases that breakdown ECM proteins. The MMPs most important to invasion are those with gelatinase activity (MMP2 and MMP9). MMP9 breaks down collagen type IV of the endometrial basal membrane. Serine proteases, including urokinase-type plasminogen activator (UPA) and tissue-type plasminogen activator (TPA) convert plasminogen to plasmin that accounts for proteolytic degradation of the ECM during implantation. During the invasion phase of blastocyst implantation, uterine decidual cells express transforming growth factor beta (TGFB) that plays a key regulatory role to limit the extent of invasion by increasing expression of tissue inhibitors of MMPs (TIMPs) and plasminogen activator inhibitor (PAI). Decorin, a TGFB binding proteoglycan, inhibits proliferation, migration, and invasiveness of human extravillous trophoblast cells and limits migration of mouse trophoblast cells via a mechanism that appears to be inhibitory to plasminogen activator activity (Strickland et al. 1976).

2.4.4 Implantation in Sheep

After fertilization of a sheep oocyte within the oviduct, the resulting one-cell zygote undergoes cleavage divisions to the 8–16 cell stage when activation of the embryonic transcriptome occurs (Johnson et al. 2018). The 32- to 64-cell morula remains enclosed in the zona pellucida and leaves the oviduct to enter the uterus on Day 3 or 4 of pregnancy to continue to develop to a blastocyst by Day 6 of gestation. The blastocyst includes the embryonic disc that will give rise to the embryo/fetus with ectoderm, endoderm, and mesoderm, trophectoderm that will form the chorion of the placenta, a blastocoel or primitive gut, and extra-embryonic endoderm and mesoderm. The sheep blastocyst hatches from the zona pellucida between Days 8 and 9 of gestation and expands to 400–900 μm in diameter (Spencer et al. 2004). The hatched blastocyst then undergoes a rapid morphological transition called elongation to first a tubular form (10–22 mm) on Day 12, followed by rapid growth and elongation to filamentous forms of 100 mm in length on Day 14, and 250 mm in length on Day 16 before trophectoderm extends into the contralateral uterine horn between Days 18 and 20 of pregnancy (Spencer et al. 2004). Prior to conceptus attachment, the conceptus is reliant entirely on the secretions of water, amino acids, hexose sugars, ions, growth factors, hormones, enzymes, cytokines, mitogens, and vitamins (collectively referred to as histotroph) by the uterine LE, sGE, and GE (Bazer et al. 2015a). Conceptus elongation rapidly increases the surface area of contact between trophectoderm and uterine LE for exchange of nutrients and gasses, maximizing paracrine effects of the conceptus to prevent regression of the CL (luteolysis), and signaling pregnancy recognition via IFNT.

Implantation in sheep occurs as the trophectoderm of the filamentous conceptus apposes and then adheres to uterine LE by Day 14 of pregnancy (Johnson et al. 2018). Apposition begins near the inner cell mass and moves toward the ends of the elongated conceptus. By Day 16, the conceptus trophectoderm is firmly attached to uterine LE with significant interdigitation between microvilli on uterine LE, and conceptus trophectoderm cells. There are also papillae of trophectoderm that extend into the mouths of uterine glands to take up nutrients and exchange gases, e.g., oxygen and carbon dioxide. Attachment of the conceptus to both the caruncular and intercaruncular regions of the endometrium is complete by Day 22 of pregnancy. Attachment

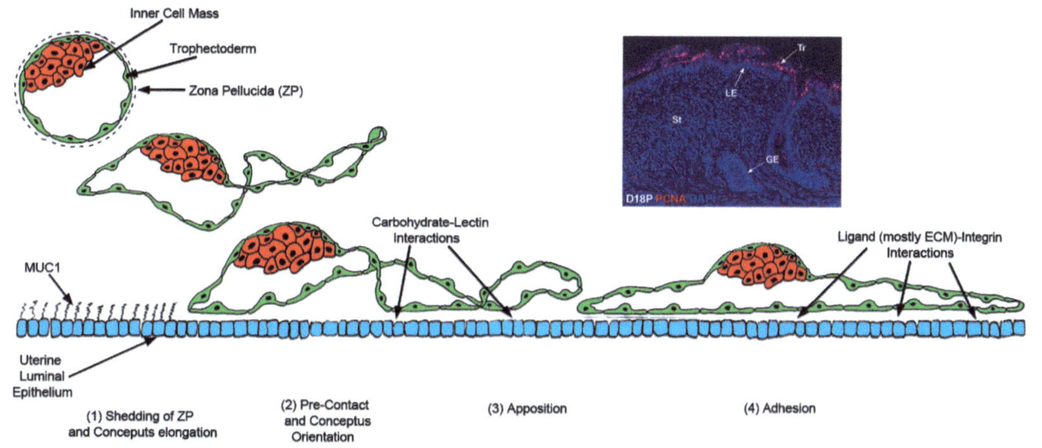

Fig. 2.4 The initial stages of implantation are common across species and are characterized as the "Adhesion Cascade for Implantation". The phases of this adhesion cascade in pigs include (1) elongation of the conceptus trophectoderm and shedding of the zona pellucida; (2) down-regulation of MUC1 at the apical surface of uterine LE to expose potential, but not yet identified, low affinity carbohydrate-lectin binding molecules to mediate pre-contact and conceptus trophectoderm orientation to the uterine LE; (3) low affinity contacts are then replaced by a more stable and extensive repertoire of adhesive interactions between integrins and maternal ECM to mediate apposition of trophectoderm to LE; and (4) integrin receptors expressed at the apical surface of uterine LE cells bind to Arg-Gly-Asp (RGD) and non-RGD amino acid sequence-containing ECM molecules and bridge to another complement of potential integrin receptors expressed at the apical surface of conceptus trophectoderm cells to mediate conceptus trophectoderm adhesion. Immunofluorescence staining for PCNA (red) illustrates that the conceptus trophectoderm (Tr) proliferates, but the uterine luminal epithelium (LE) does not proliferate during the peri-implantation period of sheep

and the adhesion cascade for implantation in sheep involves down-regulation of MUC1 across the entire endometrial surface to unmask glycosylation dependent cell adhesion molecule 1 (GLYCAM1), LGALS15, and SPP1 for interactions with lectins and integrins (Fig. 2.4). Initial attachment is likely mediated by GLYCAM1 and LGALS15, and firm attachment is likely mediated by SPP1. Integrins are constitutively present on uterine LE and conceptus trophectoderm during the peri-implantation period when expression of LGALS15 is induced by P4 and further increased by IFNT, and expression of SPP1 is induced by P4 (Johnson et al. 2000, 2014).

2.4.5 Implantation in Pigs

Pig embryos enter the uterus at the four-cell stage, reach the blastocyst stage by Day 5 of pregnancy, hatch from the zona pellucida between Days 6 and 7, and expand to 2–6 mm diameter spherical embryo by Day 10 of gestation (Fig. 2.5) (Bazer and Johnson 2014). Pig blastocysts then undergo rapid morphological transition from spherical to tubular and elongated filamentous forms between Days 10 and 12 of pregnancy at a rate of about 0.25 mm/h between the early spherical blastocyst stage and 4–9 mm diameter spherical blastocyst stage. The rate of conceptus elongation significantly increases to 30–45 mm/h from the 10 mm blastocyst to the 200 mm long filamentous conceptus due to increased cellular hypertrophy. As the mitotic index of spherical blastocysts is greater than for tubular blastocysts, but cellular hyperplasia does not account for initial elongation of pig blastocysts. Once the blastocyst reaches 10 mm in diameter, the conceptus begins to undergo the rapid morphological changes in both trophectoderm and extra-embryonic endoderm. A dense band of cells (the elongation zone) containing both endoderm and trophectoderm extends from the inner cell mass (ICM) to the tip of the ovoid

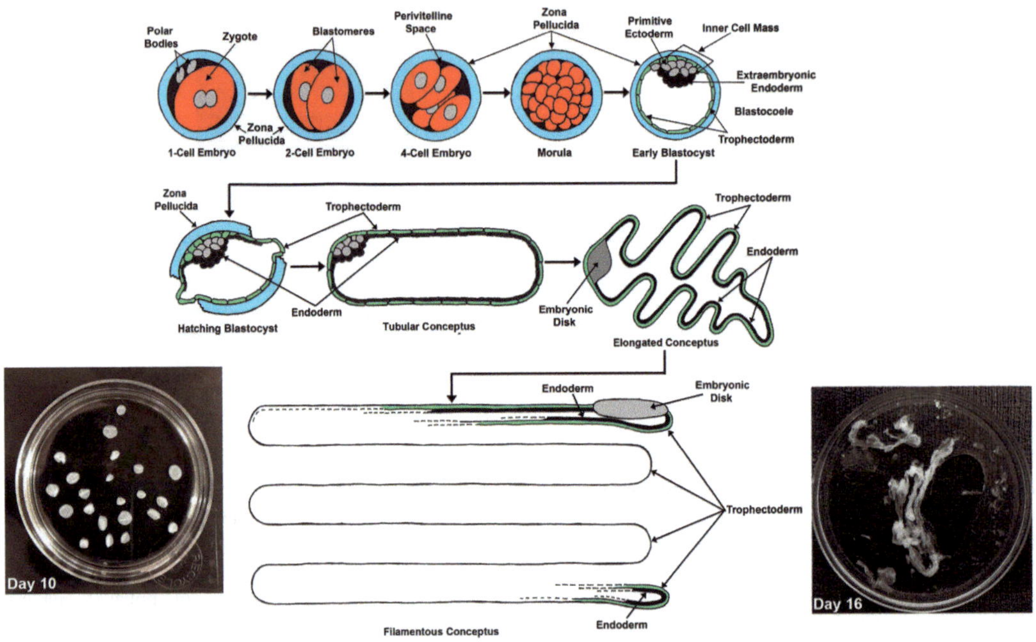

Fig. 2.5 Pig embryos undergo cell divisions, enter the uterus at about the 4-cell stage, hatch from the zona pellucida around Day 7, reach the expanded blastocyst stage around Day 10 and then change rapidly in morphology from spherical to tubular to filamentous forms to achieve maximum area of surface contact between the trophectoderm and uterine luminal epithelium. The insets are of spherical blastocysts from Day 10 and filamentous conceptuses on Day 16 of pregnancy. As spherical pig blastocysts expand there are increases in proliferation and migration of trophectoderm and extra-embryonic endoderm cells toward the inner cell mass (ICM). This process of elongation of the conceptus results in central-type implantation initially and then the development of true epitheliochorial placenta later in gestation. Reprinted from the freely available article of Johnson et al. (2018)

blastocyst. Following formation of the elongation zone, additional rapid elongation of the 100–200 mm long pig conceptus occurs to form a conceptus of 800–1,000 mm in length by Day 16 of pregnancy. These morphological changes primarily occur due to alterations in microfilaments and junctional complexes of trophectoderm cells and formation of filapodia by endodermal cells. The second period of elongation involves cellular hyperplasia and each conceptus within the litter achieves maximum surface area for contact between trophectoderm and uterine LE to facilitate uptake of nutrients from uterine LE and GE. Increasing surface area of the conceptus;uterine interface is a critical adaptation in the pig, which exhibits a minimally invasive epithelochorial placentation, to maximize nutrient and oxygen transport to the fetal-placental tissues.

Increased cellular density of both extra-embryonic endoderm and trophectoderm in tubular blastocysts results in the formation of a thin cell-dense band approximately 1–2 mm wide extending from and in the same plane as the embryonic disc to the end of the trophectoderm. The endoderm cells outside this thin band are sparsely populated and make cellular contact only through filapodia. The dense band of extra-embryonic endoderm and trophectoderm forms the elongation zone that decreases in width as elongation progresses but continues to extend along the entire length of the conceptus trophectoderm. Histologically, it has been demonstrated that the trophectoderm cells present in the elongation zone are columnar in shape as compared to cuboidal trophectoderm cells in areas peripheral to the elongation zone. This structural modification is associated with changes in length and orientation of microfilaments as early as the 10 mm stage of blastocyst development. Within the elongation zone, extension of filopodia from extra-embryonic endodermal cells in conjunction with alterations in microfilaments and junctional complexes of trophectoderm cells allows the movement and redistribution of cells toward the ends of tubular blastocysts. While the actin cytoskeleton initially exhibits a pericellular distribution, this later becomes a continuous actin-rich lateral border with stress fibers along the basal surface in the filamentous conceptus. The actin cytoskeleton, in association with myosin II, has a crucial function in generating the force required for conceptus elongation as constricted regions along the length of filamentous conceptuses contain polarized trophectoderm cells with a distinct F-actin array. The orientation of microfilaments within the trophectoderm changes from horizontal to parallel relative to the lateral cell borders likely due to a complex cellular response to torsional forces generated by the elongation process and mediated through transmembrane integrin receptors and the focal adhesions they assemble. Heterodimeric transmembrane integrin receptors [e.g., ITGAV: ITGB3 ($\alpha v\beta 3$), ITGA5:ITGB1 ($\alpha 5\beta 1$)] are the major components of focal adhesions. They transmit diverse signals between the ECM components, such as SPP1 and the actin cytoskeleton, to regulate cellular growth, proliferation, survival, and migration. Additionally, these focal adhesions alter gene expression and the morphology of trophectoderm and extra-embryonic endoderm in elongating pig conceptuses. The process of conceptus elongation likely involves several cell signaling pathways with serine-threonine kinases such as insulin growth factor 2 acting via receptor tyrosine kinases, integrin heterodimer-ECM complexes (e.g., SPP1-ITGAV:ITGB3 and/or SPP1-ITGA5:ITGB1), and arginine acting simultaneously and independently to stimulate mechanistic target of rapamycin 1 (mTORC1) and/or mTORC2) required for proliferation, migration, cytoskeletal reorganization, and adhesion of trophectoderm cells to uterine LE.

2.5 Overview of Placentation/Placental Growth and Function in Humans, Sheep, Pigs, and Rodents

The primary functions of the placenta are transplacental exchange of gases, micronutrients (amino acids, glucose) and macromolecules (proteins), production of hormones, and production of cytokines and other regulatory molecules that affect growth and development of the conceptus. Placental efficiency is achieved as maternal and fetal-placental vasculatures are brought into close apposition to allow for transplacental exchange of molecules while maintaining separation of the maternal and fetal circulatory systems. Endometrial and placental tissues are remodeled to achieve areas with reduced interhaemal distances regardless of whether the placenta is epitheliochorial, synepitheliochorial, endotheliochorial, or hemochorial to maximize transplacental exchange of gasses, micronutrients, and macronutrients.

2.5.1 Placentation in Humans

Following implantation of the human blastocyst, the primary syncytium invades into the uterine stroma and forms fluid-filled spaces called lacunae that enlarge and merge to form a system of trabeculae (Soares et al. 2018; Turco and Moffett 2019). The syncytium erodes into the uterine glands to become bathed in their secretions. Trophoblast cells beneath the syncytium are CTB cells that proliferate and form villi that penetrate through the STB to form primary villi with a CTB core and outer STB. The villi undergo further proliferation and branching, and the lacunae become the intervillous space. CTB cells penetrate through the primary syncytium and merge laterally to surround the conceptus in a continuous CTB shell between the villi and the decidua. The blastocyst then has three layers: inner chorionic plate in contact with the intervillous space; villi separated by the intervillous space; and the CTB shell in contact with the decidua. On Days 17–18 of pregnancy, extra-embryonic mesenchymal cells penetrate through the villous core to form secondary villi and soon thereafter fetal capillaries develop within the core of tertiary villi. The villous tree continues to enlarge through branching from the chorionic plate to form a system of vascularized villous trees. The CTB shell is in contact with uterine decidual cells and individual CTB cells invade into decidua as extravillous trophoblast via a process similar to that for an epithelial-mesenchymal transition.

The STB of placental villi are in direct contact with uterine gland secretions and maternal blood flowing into the intervillous space for maternal/fetal exchange of gases and nutrients supporting growth of the conceptus. The STB microvilli express receptors for growth factors and hormones and transporters for amino acids and glucose. They also secrete hormones and proteins into the maternal circulation to influence physiological and metabolic adaptations to pregnancy. The STB provides a protective immunological barrier as it does not express human leukocyte antigen (HLA) molecules that would otherwise subject it to immunological rejection.

With advancing development of the placenta, the CTB shell becomes discontinuous and CTB cells form columns that emerge from the anchoring villi in contact with the decidua. These extravillous trophoblast (EVT) cells migrate into the decidua along two differentiation pathways: the interstitial EVT (iEVT) cells migrate through the decidual stroma toward the maternal spiral arteries, while the endovascular trophoblast EVT (eEVT) cells migrate into the spiral arteries to displace endothelial cells and ensure maximum dilation and blood flow. The iEVT invade as far as the inner one-third of the myometrium and fuse to form a bed of placental giant cells. After the arterial transformation occurs, the eEVT move in a retrograde manner down the artery to form a plug that prevents blood flow into the intervillous space until the full hemochorial circulation is established. As a result of trophoblast plug, the placenta develops in a low oxygen environment during the first trimester. In humans, the yolk and allantoic sacs undergo regression. Therefore, the chorioamniotic placenta contains an abundance of amniotic fluid in which the fetus develops. The human placenta is hemochorial as the chorion is in direct contact with maternal blood for transplacental transport of nutrients and exchange of oxygen and carbon dioxide.

2.5.2 Placentation in Rodents

During implantation, trophectoderm attachment to uterine LE induces differentiation of uterine stromal cells into decidual cells followed by complete penetration of the blastocyst into the decidualized uterine stroma (Picut et al. 2009; Soares et al. 2018; Furukawa et al. 2019). As hemochorial placentation progresses in rodents, there is the maternal interface and the fetal interface determined by the extent to which extra-embryonic mesenchyme and associated vasculature penetrate into the trophoblast compartment. The region that includes trophoblast and extra-embryonic mesenchyme forms the labyrinth zone of the placenta in mice and rats. Labyrinth and villous trophoblast compartments include layers of STB. Trophoblast cells extend beyond the trophoblast-extra-embryonic mesenchyme admixture are at the maternal boundary and arranged into the junctional zone and extravillous trophoblast columns, respectively. As gestation advances, invasive trophoblast cells arise from the junctional zone to form extravillous trophoblast columns that migrate into the uterine parenchyma. There are two types of invasive trophoblast cells, interstitial and endovascular. Interstitial invasive trophoblast cells coalesce between the uterine vasculature, whereas endovascular invasive trophoblasts infiltrate uterine blood vessels, especially arteries/arterioles, and replace the vascular endothelium as in humans. During invasive trophoblast cell differentiation, there are changes in expression of integrins that alter interactions with surrounding ECM. As endovascular invasive trophoblast cells differentiate, they acquire an endothelial cell phenotype. For rats, there is deep intrauterine trophoblast cell invasion, but it is nominal for trophoblast cells migrating into the mouse uterine parenchyma. Cellular constituents of the maternal uterine interface, including decidual cells, endometrial glands, and immune/inflammatory cell populations, influence behavior of trophoblast cells.

Rodents have a chorioallantoic placenta and its formation is highly dependent of the ectoplacental cone that is a critical generative zone. The ectoplacental cone trophoblast is considered cytotrophoblastic and develops from the polar trophoblast covering the embryonic disc. Mesenchyme invades the ectoplacental cone leading to a division between the main functional exchange area, the labyrinth, and the superficial zone, the remnant of the ectoplacental cone, and the trophospongium or basal zone. There is maternal blood flow through the ectoplacental area before the labyrinth is formed to derive nutrition from the maternal circulation. During and after labyrinth formation, the trophospongium undergoes differentiation into secondary giant cell formation (primary giant cells are derived from the mural trophoblast developing in relation to the yolk sac) and are restricted to the zone immediately facing maternal tissues. Most of the trophospongium differentiates into islands of glycogen cells surrounded by basophilic trophoblast in contact with maternal blood.

The fully formed placenta includes the transient yolk sac, amnion, and chorioallantois with functions described previously. In addition, outside the placental membranes and between the chorion and metrial gland, there is the labyrinth, trophospongiosum or basal zone, and uterine decidua. The labyrinth is the largest layer of the placenta and the site for most, if not all, nutrient and gas exchange between the maternal and fetal-placental vasculatures. The basal zone forms just below the labyrinth zone and is composed of spongiotrophoblasts, glycogen cells, and (secondary) trophoblastic giant cells. The primary trophoblastic giant cells derive from the mural trophectoderm and are important for implantation. The secondary trophoblastic giant cells form from the ectoplacental cone and represent the main components of the basal zone. The basal zone is a site of production of steroids and peptide hormones required for the maintenance of pregnancy, storage of glycogen, and establishment of an immunological barrier to the maternal immune system. Metrial glands are located in the mesometrial triangle of the pregnant uterus from gestational Day 8 to parturition.

The uterine metrial gland is a distinct structure in the uterus that is composed of granulated metrial gland cells, endometrial stromal cells, trophoblast cells, blood vessels, and fibroblasts. Granulated metrial gland cells are hallmark cells of the metrial gland derived from bone marrow. They are perforin-positive, natural killer cells that proliferate in the pregnant uterus and encircle newly formed blood vessels in the mesometrial triangle, producing proteases that destroy basement membranes and vascular endothelial growth factor to stimulate endothelial cell proliferation. The metrial gland cells promote angiogenesis and remodel the uterine vasculature at the point of entry of blood vessels into the uterus.

2.5.3 Placentation in Sheep

Implantation is a prerequisite for placentation, and both are critical for a successful pregnancy. Conceptus attachment and adhesion to uterine LE/sGE first requires removal of large mucins from the glycocalyx that block direct physical interactions between carbohydrates and lectins at the apical surfaces of the opposing uterine LE and conceptus trophectoderm (Bazer et al. 2012; Johnson et al. 2018). These low affinity contacts are replaced by firm focal adhesions between integrins and ECM proteins like SPP1. Sheep have a synepitheliochorial placenta in which fusion of conceptus trophectoderm with uterine LE occurs and then the uterine LE is degraded. Both mononuclear trophectoderm cells and multinucleated trophoblast giant cells (TGCs) are present in the trophectoderm of ruminant placentae. The mononuclear cells constitute the majority of the trophectoderm cells and TGCs differentiate from the mononuclear trophectoderm cells in concert with trophectoderm outgrowth during conceptus elongation (Seo et al. 2019; Seo et al. 2020a). TGCs first appear between Days 14 and 16 of gestation in sheep conceptuses and comprise 15–20% of the trophectoderm during the apposition and attachment phases of implantation. TGCs migrate to LE and remove LE cells that are undergoing apoptosis to form multinucleated syncytia. The syncytia of sheep subsequently enlarge through continued TGC migration and fusion to form syncytial plaques. The syncytial plaques form the epithelial interface between endometrial caruncles and placental cotyledons that comprise the placentomes. Syncytial plaques are a consistent feature in placentomes throughout pregnancy in sheep.

Fetal fluids (allantoic and amniotic fluids) are of maternal origin via active transport of water, as well as other molecules, across the placenta and into the allantoic sac for distribution to other components of the conceptus including the fetus and amniotic sac. The driving force for expansion of the allantois, and in turn the chorioallantois, is the rapid accumulation of water in the allantoic sac from about 1 ml on Day 18 to 90 ml on Day 40 and then from Day 70 (32 ml) to Day 140 (438 ml) of the 147 day period of gestation. Similarly, amniotic fluid volume increases throughout gestation in sheep from 2 ml on Day 30 to over 700 ml on Day 140 of gestation. Amniotic fluid buoys the fetus to allow it to develop symmetrically, prevents fetal skin from adhering to the amnion and it is swallowed by the fetus in the last one-third of gestation to provide water, minerals, and other nutrients.

Development of placentomes begins to occur Days 25–30 of gestation (Fig. 2.6). The highly branched villous placental structures termed cotyledons protrude into crypts in the maternal endometrial caruncles (aglandular areas of endometrium consisting of stroma covered by a single layer of epithelium). As the cotyledonary chorioallantoic villi interdigitate extensively with endometrial caruncles there is, within the placentomes, opposing vascular beds that provide a large surface area for active transfer of nutrients and gases from maternal blood to the fetal-placental vasculature. Consequently, there is a high correlation between the placentomal mass and fetal weight at birth. In contrast, there is epitheliochorial attachment to uterine LE in inter-

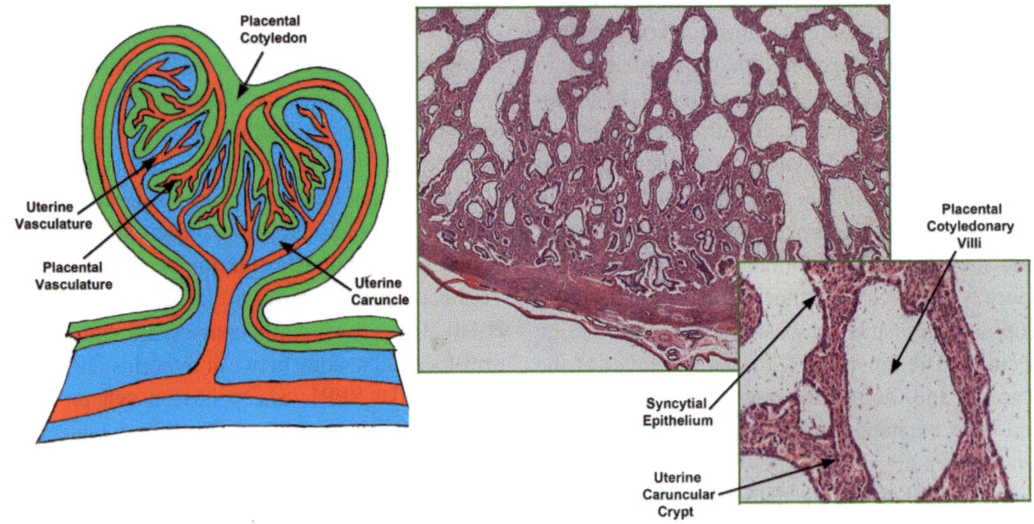

Fig. 2.6 Overview of placental development in the sheep. Illustration describing placentome structure that indicates the uterine caruncle in blue, placental cotyledon in green, and the network of blood vessels in red. Collectively, this is the placentome that provides sites for the exchange of gases and micronutrients such as glucose and amino acids between the vascular systems of the fetal-placental tissues and maternal vascular system. The figure on the right is a histological section of a placentome with the various cellular components of the placentome. Reprinted from the freely available article of Johnson et al. (2018)

placentomal regions of the placenta. The interplacentomal chorioallantois contains areolae that are in direct apposition to openings of the mouths of uterine glands for direct uptake of components of histotroph secreted by or transported by uterine GE. The components of histotroph are transported across the areolae and into the fetal-placental vasculature via fluid-phase pinocytosis.

2.5.4 Placentation in Pigs

The elongated conceptuses in pigs expand through the accumulation of water initially within the yolk sac and then the allantois of the chorioallantoic placenta as described for sheep (Knight et al. 1977; Bazer and Johnson 2014). The yolk sac derives from an evagination of the embryonic foregut and accumulates fluid that first brings the trophectoderm into apposition with the uterine wall (between Days 17 and 22 of gestation in pigs) for absorption of nutrients. After Day 22, the yolk sac becomes a vestigial structure. The allantoic sac forms as an evagination of the hindgut and expands rapidly as it fills with allantoic fluid between Days 18 (1 ml) and 30 (250 to 300 ml) of gestation to fill the extra-embryonic coelom and establish the chorioallantoic placenta. Allantoic fluid then decreases to about 50 ml on Day 40 of gestation and then increases again to around 450 ml on Day 55 of gestation. By Day 70 of gestation in pigs, development of the epitheliochorial placenta is considered complete based on placental weight, surface area, and numbers of placental areolae. Amniotic fluid serves the protective roles for the embryo/fetus in pigs as described for sheep. Amniotic fluid volume increases from around 2 ml on Day 20 to 200 ml on Day 70 of gestation and then decreases to term.

The chorioallantoic placenta attaches directly to uterine LE for hematrophic and histotrophic support of conceptus growth and development. Given the non-invasive nature of the pig placenta, it is critical to increase the surface area available at the uterine (endometrial)-placental (chorioallantoic) interface to minimize the distance between maternal and placental micro-vasculatures, thereby optimizing the transport of nutrients and gasses from maternal to placental blood vessels for eventual utilization by the embryo/fetus. To do this, extensive remodeling occurs at the uterine-placental interface by the formation of chorionic (placental) ridges that correspond with endometrial invaginations that result in extensive folding (Fig. 2.7). The interface between the endometrium and chorion in pigs begins to undergo folding between Days 20 and Day 25 of pregnancy. By Day 30 of pregnancy the chorioallantoic and endometrial surfaces interlock into folds composed of endometrial ridges and chorioallantoic troughs (Friess et al. 1980). These folds proceed to increase in length until Day 35 of gestation, followed by a second increase in length between Days 50 and 60 of gestation (Seo et al. 2020b). As the growth rate of the placenta decreases, the fetus undergoes a period of exponential growth (Marrable 1971). It is critical that the depth of the folds increases between Days 65 and 105 of gestation, to increase the surface area available for nutrient transport to accommodate the high nutritional demands of the exponentially growing fetus (Vallet and Freking 2007). The morphological folding characteristic of the epitheliochorial placentation in pigs is likely the result of mechanotransduction and mechanosensation at the interface between the endometrium and the chorion. It has been proposed that dilation of subepithelial uterine blood vessels delivers increased blood flow that pushes upward on the interface between the uterine LE and the placental chorioallantois. These protrusive forces from growing uterine blood vessels trigger integrin adhesion complex assembly and actin polymerization between the uterine LE and chorionic epithelium (CE) at the bottoms of the chorioallantoic troughs, and uterine fibroblasts differentiate into contractile myofibroblasts that pull the connective tissue downward and inward to sculpt folds at the uterine-placental interface (Seo et al. 2020b). The folding increases the surface area of the uterine-placental interface for each conceptus in the litter of piglets. Indentation of uterine LE and CE by underlying capillaries reduces the diffusion distance between the maternal and fetal-placental vasculatures. Indeed, placental and uterine capillaries lie immediately beneath the uterine LE and the chorionic epithelium, minimizing the distance between maternal and fetal blood vessels (Dantzer and Leiser 1994). In summary, the lateral sides and tops of the chorioallantoic ridges are designed for gaseous exchange, whereas the base of the chorioallantoic troughs is designed for the transport of blood-borne nutrients, i.e., hemotroph (Friess et al. 1980). The precise cell-specific spatio-temporal regulation of nutrient transporters is essential for the regulation of fetal growth and development. For example, at Day 60 of gestation the glucose transporter SLC2A1 is expressed by the uterine LE and in the CE but not by the tall columnar CE cells at the tips of the uterine-placental folds and the areolae (Kramer et al. 2020). In contrast, at Day 60 of gestation, the glucose transporter SLC2A3 is expressed by the CE of the areolae and the LE cells in close proximity to the tall columnar cells of the CE, and SLC2A8, a glucose and fructose transporter, is expressed by the tall columnar cells of the CE and by the areolae. Together, these findings suggest that differential expression of transporters for glucose across the uterine-chorionic folds is critical for the transportation of glucose from maternal circulation to the fetus.

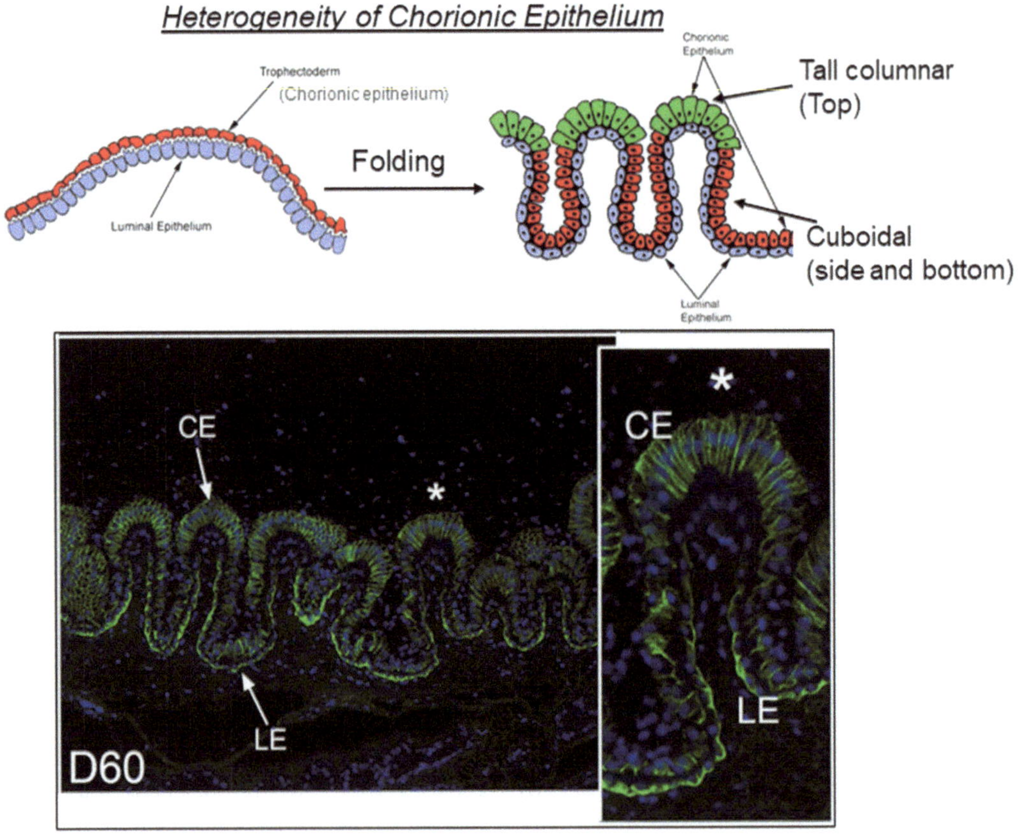

Fig. 2.7 Overview of placental development in the pig. An illustration depicting the uterine-placental interface during implantation (top left panel) and placentation (top right panel). Green and red colors indicate heterogeneity of gene expression within the chorionic epithelium. Immunofluorescence staining for $\alpha2\beta1$ integrin by uterine epithelia on Day 60 of pregnancy is shown (bottom panel). The asterisk indicates a fold or villus with the apposition of uterine luminal epithelium (LE) and chorionic epithelium (CE) that significantly increases the surface area for exchange of nutrients and gases between the fetal-placental and maternal vascular systems

Progressive interdigitation of microvilli on trophectoderm and uterine LE eventually occurs over the entire uterine-placental interface, except at the openings of uterine glands. At openings of the mouths of uterine glands, the CE forms areolae to transport components of histotroph into the fetal-placental vasculature via fluid-phase pinocytosis. Areolae are initially observed as small white circular discs with a prominent peripheral thickening of 1 mm in diameter (Friess et al. 1981), but quickly develop to cover the openings of the uterine gland(s). The cavity that forms collects the secretions of the uterine glands, and the columnar chorionic epithelial cells that line the placental border of this cavity form a seal between the uterine LE and the walls of the placental areola to prevent dissipation of histotroph into inter-areolar regions of the placenta (Fig. 2.8). The allantoic vasculature that receives the histotroph is clearly discernable from the vasculature that supplies inter-areolar regions of the placenta (Leiser and Dantzer 1994). The endometrial vasculature that supplies the areola develops more slowly than the endometrial vasculature of inter-areolar regions, presumably due to a less intimate association with the trophectoderm. This prevents direct physical interaction between the trophectoderm and endometrium and decreases the influence of paracrine products that are secreted by the trophectoderm. As the placenta grows, areolar diameter increases and a stretching of the areolar capillary network leads to a progressively widening size. During the early stages of placentation, the placental surface of the areolae is flat, but as placentation progresses the flat surface becomes more complex with formation of ridges and papilla-like structures lined by a columnar chorionic epithelium (Amoroso 1952). The balloon shape of the areola implies that there is an interior pressure against the chorioallantoic surface of the areola delivered by the continuous

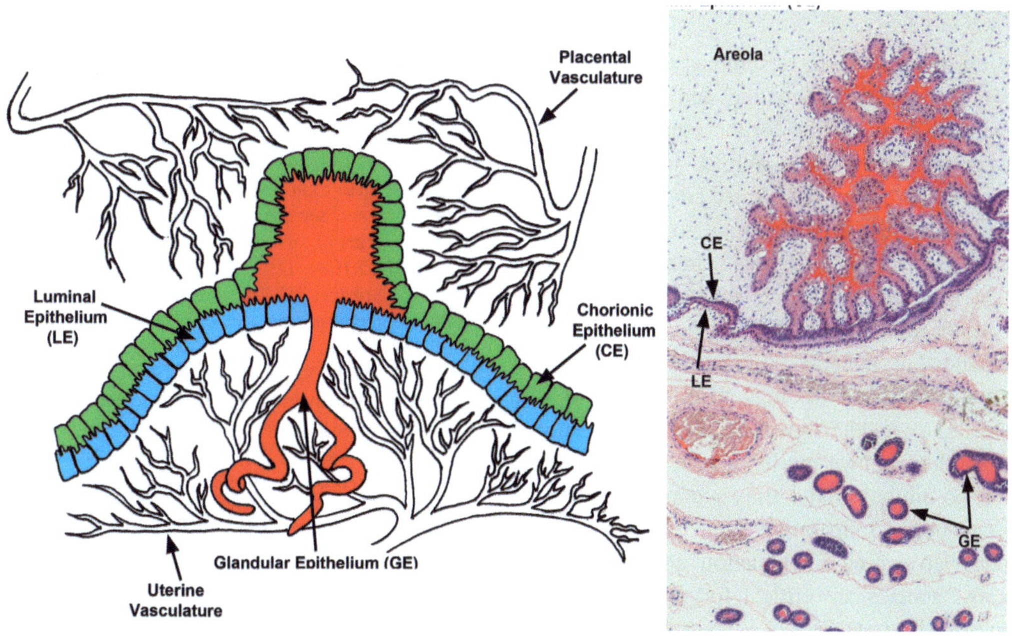

Fig. 2.8 Areolae structure in the pig. Illustration depicting an areola that exists in placentae of species such as pigs, horses, sheep, cattle, and goats for the transport of secretions from uterine glands into the fetal-placental vasculature via fluid-phase pinocytosis. Nutrients and gases are transported from the maternal capillaries into the placental capillaries. The hematoxylin and eosin stained image in the right panel illustrates the uterine-placental interface of mature placentation in the pig, with the areola having a critical function for histotrophic support of the fetus. The red staining indicates the synthesis, secretion, and transport of histotroph by the glands and into the lumen of the areola

accumulation of histotroph from the uterine glands. Indeed, the cavity of an areola is a small reservoir for the histotroph that is potentially secreted by the much larger uterine glands (Leiser and Dantzer 1994). There are some 2,500 areolae per placenta in pigs and a correlation between areolar number and fetal weight has been suggested.

Uteroferrin (UF, also known as acid phosphatase 5, tartrate resistant, ACP5) secreted by uterine GE is taken up by placental areolae by fluid-phase pinocytosis and released into the fetal-placental circulation. UF transports iron required for the synthesis of hemoglobin in the fetal liver, and it is a hematopoietic growth factor, regulating both the differentiation and proliferation of hematopoietic stem cells and their colonization in the yolk sac, liver, spleen, and bone marrow (Bazer et al. 1991; Ying et al. 2014).

2.6 Summary and Conclusions

In this review, we have summarized some of the critical molecular signaling and morphological events that are crucial for the establishment of a successful pregnancy. Whilst many of these processes are conserved, there are key differences across species. It is important to consider these species-specific differences when designing an experiment to investigate the mechanisms controlling implantation and placentation, placental transport of minerals and nutrients, and when extrapolating the findings from studies of one species to another species. While each animal model has its own merits, it is important to note that no animal model truly recapitulates human pregnancy. Comparative studies of the mechanisms governing implantation and placentation across species are critical for the discovery of improved strategies to enhance reproductive health and fertility in both humans and livestock species.

Conflicts of Interest The authors have no conflicts of interest to declare.

References

Amoroso E (1952) Placentation. In: Parkes A (ed) Marshall's physiology of reproduction. Little Brown & Co, Boston, MA, USA, pp 127–311

Aplin JD, Ruane PT (2017) Embryo—epithelium interactions during implantation at a glance. J Cell Sci 130:15–22

Bazer F (1989) Establishment of pregnancy in sheep and pigs. Reprod Fertil Dev 1:237–242

Bazer FW (2013) Pregnancy recognition signaling mechanisms in ruminants and pigs. J Anim Sci Biotechnol 4:1–10

Bazer FW, Johnson GA (2014) Pig blastocyst-uterine interactions. Differentiation 87:52–65

Bazer FW, Marengo SR, Geisert RD, Thatcher WW (1984) Exocrine versus endocrine secretion of prostaglandin f2a in the control of pregnancy in swine. Anim Reprod Sci 7:115–132

Bazer FW, Worthington-White D, Fliss MFV, Gross S (1991) Uteroferrin: a progesterone-induced hematopoietic growth factor of uterine origin. Exp Hematol 19:910–915

Bazer FW, Ott TL, Spencer TE (1994) Pregnancy recognition in ruminants, pigs and horses: signals from the trophoblast. Theriogenology 41:79–94

Bazer FW, Burghardt RC, Johnson GA, Spencer TE, Wu G (2008) Interferons and progesterone for establishment and maintenance of pregnancy: interactions among novel cell signaling pathways. Reprod Biol 8:179–211

Bazer FW, Spencer TE, Johnson GA (2009) Interferons and uterine receptivity. Semin Reprod Med 27:90–102

Bazer FW, Wu G, Johnson GA, Kim J, Song G (2011) Uterine histotroph and conceptus development: Select nutrients and secreted phosphoprotein 1 affect mechanistic target of rapamycin cell signaling in ewes. Biol Reprod 85:1094–1107

Bazer FW, Spencer TE, Thatcher WW (2012) Growth and development of the ovine conceptus. J Anim Sci 90:159–170

Bazer FW, Wang X, Johnson GA, Wu G (2015a) Select nutrients and their effects on conceptus development in mammals. Anim Nutr 1:85–95

Bazer FW, Ying W, Wang X, Dunlap KA, Zhou B, Johnson GA (2015b) The many faces of interferon tau. Amino Acids 47:449–460

Bazer FW, Burghardt RC, Johnson GA, Spencer TE, Wu G (2018) Mechanisms for the establishment and maintenance of pregnancy: Synergies from scientific collaborations. Biol Reprod 99:225–241

Beall MH, Wang S, Yang B, Chaudhri N, Amidi F, Ross MG (2007) Placental and membrane aquaporin water channels: correlation with amniotic fluid volume and composition. Placenta 28:421–428

Bonduelle ML, Dodd R, Liebaers I, Van Steirteghem A, Williamson R, Akhurst R (1988) Chorionic gonadotrophin-β mRNA, a trophoblast marker, is expressed in human 8-cell embryos derived from tripronucleate zygotes. Hum Reprod 3:909–914

Carson DD, Bagchi I, Dey SK, Enders AC, Fazleabas AT, Lessey BA, Yoshinaga K (2000) Embryo implantation. Dev Biol 223:217–237

Chan HC, Ruan YC, He Q, Chen MH, Chen H, Xu WM, Chen WY, Xie C, Zhang XH, Zhou Z (2009) The cystic fibrosis transmembrane conductance regulator in reproductive health and disease. J Physiol 587:2187–2195

Chen C, Spencer TE, Bazer FW (2000) Fibroblast growth factor-10: a stromal mediator of epithelial function in the ovine uterus. Biol Reprod 63:959–966

Chen C, Spencer TE, Bazer FW (2000) Expression of hepatocyte growth factor and its receptor c-met in the ovine uterus. Biol Reprod 62:1844–1850

Choi Y, Johnson GA, Burghardt RC, Berghman LR, Joyce MM, Taylor KM, Stewart MD, Bazer FW, Spencer TE (2001) Interferon regulatory factor-two restricts expression of interferon-stimulated genes to the endometrial Stroma and glandular epithelium of the ovine uterus. Biol Reprod 65:1038–1049

Cockburn K, Rossant J (2010) Making the blastocyst: lessons from the mouse. J Clin Invest 120:995–1003

Dantzer V, Leiser R (1994) Initial vascularisation in the pig placenta: I. Demonstration of nonglandular areas by histology and corrosion cases. Anat Rec 238:177–190

De Oliveira V, Schaefer J, Abu-Rafea B, Vilos GA, Vilos AG, Bhattacharya M, Radovick S, Babwah AV (2020) Uterine aquaporin expression is dynamically regulated by estradiol and progesterone and ovarian stimulation disrupts embryo implantation without affecting luminal closure. Mol Hum Reprod 26:154–166

Fazleabas AT, Kim JJ, Strakova Z (2004) Implantation: embryonic signals and the modulation of the uterine environment—a review. Placenta 25:S26–S31

Fleming JGW, Spencer TE, Safe SH, Bazer FW (2006) Estrogen regulates transcription of the ovine oxytocin receptor gene through GC-rich SP1 promoter elements. Endocrinology 147:899–911

Freeman ME, Smith MS, Nazian SJ, Neill JD (1974) Ovarian and hypothalamic control of the daily surges of prolactin secretion during Pseudopregnancy in the rat. Endocrinology 94:875–882

Friess AE, Sinowatz F, Skolek-Winnisch R, Träutner W (1981) The placenta of the pig II the ultrasound of the areolae. Anat Embryol (berl) 163:43–53

Friess AE, Sionwatz F, Skolek-Winnisch R, Trautner W (1980) The placenta of the Pig I. Finestructural changes of the placental barrier during pregnancy. Anat Embryol (berl) 158:179–191

Furukawa S, Tsuji N, Sugiyama A (2019) Morphology and physiology of rat placenta for toxicological evaluation. J Toxicol Pathol 32:1–17

Gunnet JW, Freeman ME (1983) The Mating-induced release of prolactin: a unique neuroendocrine response. Endocr Rev 4:44–61

Igarashi M, Finch PW, Aaronson SA (1998) Characterization of recombinant human fibroblast growth factor (FGF)-10 reveals functional similarities with keratinocyte growth factor (FGF-7). J Biol Chem 273:13230–13235

Johnson GA, Spencer TE, Burghardt RC, Taylor KM, Gray CA, Bazer FW (2000) Progesterone modulation of osteopontin gene expression in the ovine uterus. Biol Reprod 62:1315–1321

Johnson GA, Bazer FW, Burghardt RC, Spencer TE, Wu G, Bayless KJ (2009) Conceptus-uterus interactions in pigs: endometrial gene expression in response to estrogens and interferons from conceptuses. Soc Reprod Fertil Suppl 66:321–332

Johnson GA, Burghardt RC, Bazer FW (2014) Osteopontin: a leading candidate adhesion molecule for implantation in pigs and sheep. J Anim Sci Biotechnol 5:1–14

Johnson GA, Bazer FW, Burghardt RC, Wu G, Seo H, Kramer AC, McLendon BA (2018) Cellular events during ovine implantation and impact for gestation. Anim Reprod 15:843–855

Joyce MM, Burghardt JR, Burghardt RC, Hooper RN, Jaeger LA, Spencer TE, Bazer FW, Johnson GA (2007) Pig conceptuses increase uterine interferon-regulatory factor 1 (IRF1), but restrict expression to stroma through estrogen-induced IRF2 in luminal epithelium. Biol Reprod 77:292–302

Kim S, Kim J (2017) A review of mechanisms of implantation. Dev Reprod 21:351–359

Knight JW, Bazer FW, Thatcher WW, Franke DE, Wallace D (1977) Conceptus development in intact and unilaterally Hysterectomized-Ovariectomized gilts: interrelations among hormonal status, placental development, fetal fluids and fetal growth. J Anim Sci 44:620–637

Kramer AC, Steinhauser CB, Gao H, Seo H, Mclendon BA, Burghardt RC, Wu G, Bazer FW, Johnson GA (2020) Steroids regulate SLC2A1 and SLC2A3 to deliver glucose into Trophectoderm for metabolism via glycolysis. Endocrinology 161:1–19

Marrable AW (1971) The embryonic pig: a chronological account. Pitman Medical, London

Matsumoto H (2017) Molecular and cellular events during blastocyst implantation in the receptive uterus: clues from mouse models. J Reprod Dev 63:445–454

McGowen MR, Erez O, Romero R, Wildman DE (2014) The evolution of embryo implantation. Int J Dev Biol 58:155–161

Meyer AE, Pfeiffer CA, Brooks KE, Spate LD, Benne JA, Cecil R, Samuel MS, Murphy CN, Behura S, Mclean MK, Ciernia LA, Smith MF, Whitworth KM, Wells KD, Spencer TE, Prather RS, Geisert RD (2019) New perspective on conceptus estrogens in maternal recognition and pregnancy establishment in the pig. Biol Reprod 101:148–161

Picut CA, Swanson CL, Parker RF, Scully KL, Parker GA (2009) The metrial gland in the rat and its similarities to granular cell tumors. Toxicol Pathol 37:474–480

Richard C, Gao J, Brown N, Reese J (2003) Aquaporin water channel genes are differentially expressed and regulated by ovarian steroids during the periimplantation period in the mouse. Endocrinology 144:1533–1541

Seo H, Bazer FW, Burghardt RC, Johnson GA (2019) Immunohistochemical examination of trophoblast syncytialization during early placentation in sheep. Int J Mol Sci 20:1–13

Seo H, Frank JW, Burghardt RC, Bazer FW, Johnson GA (2020a) Integrins and OPN localize to adhesion complexes during placentation in sheep. Reproduction 160:521–532

Seo H, Li X, Wu G, Bazer FW, Burghardt RC, Bayless KJ, Johnson GA (2020b) Mechanotransduction drives morphogenesis to develop folding during placental development in pigs. Placenta 90:62–70

Soares MJ (2004) The prolactin and growth hormone families: pregnancy-specific hormones/cytokines at the maternal-fetal interface. Reprod Biol Endocrinol 2:1–15

Soares MJ, Alam SMK, Konno T, Ho-chen JK, Ain R (2006) The prolactin family and pregnancy-dependent adaptations. Anim Sci J 77:1–9

Soares MJ, Konno T, Alam SMK (2007) The prolactin family: effectors of pregnancy-dependent adaptations. Trends Endocrinol Metab 18:114–121

Soares MJ, Varberg KM, Iqbal K (2018) Hemochorial placentation: development, function, and adaptations. Biol Reprod 99:196–211

Spencer TE, Bazer FW (2004) Conceptus signals for establishing and maintenance of pregnancy. Reprod Biol Endocrinol 2:1–15

Spencer TE, Johnson GA, Bazer FW, Burghardt RC (2004) Focus on Implantation Implantation mechanisms: insights from the sheep. Reproduction 128:657–668

Srisuparp S, Strakova Z, Fazleabas AT (2001) The role of chorionic gonadotropin (CG) in blastocyst implantation. Arch Med Res 32:627–634

Stouffer RL (1988) Perspectives on the corpus luteum of the menstrual cycle and early pregnancy. Semin Reprod Endocrinol 6:103–113

Stouffer RL, Bishop CV, Bogan RL, Xu F, Hennebold JD (2014) Endocrine and local control of the primate corpus Luteum. Reprod Biol 13:259–271

Stouffer RL, Hearn JP (1998) Endocrinology of the transition from menstrual cyclicity to establishment of pregnancy in primates. In: Bazer FW (ed) Endocrinology of Pregnancy. Humana Press

Strickland S, Reich E, Sherman MI (1976) Plasminogen activator in early embryogenesis: enzyme production by trophoblast and parietal endoderm. Cell 9:231–240

Su R-W, Fazleabas AT (2015) Implantation and establishment of pregnancy in human and nonhuman primates. Adv Anatomy, Embryol Cell Biol 216:189–213

Syrkasheva AG, Dolgushina NV, Romanov AY, Burmenskaya OV, Makarova NP, Ibragimova EO, Kalinina EA, Sukhikh GT (2017) Cell and genetic predictors of human blastocyst hatching success in assisted reproduction. Zygote 25:631–636

Thatcher W, Meyer MD, Danet-Desnoyers G (1995) Maternal recognition of pregnancy. J Reprod Fertil Suppl 49:15–28

Turco MY, Moffett A (2019) Development of the Human Placenta. Development 146:1–14

Vallet JL, Freking BA (2007) Differences in placental structure during gestation associated with large and small pig fetuses. J Anim Sci 85:3267–3275

Wassarman PM (1988) Zona Pellucida Glycoproteins. Annu Rev Biochem 57:415–442

Wassarman PM, Litscher ES (2008) Mammalian fertilization: the egg's multifunctional zona pellucida. Int J Dev Biol 52:665–676

Wu G (2022) Nutritionand metabolism: Foundations for animal growth, development, reproduction, and health.Adv Exp Med Biol 1354:1-24

Ying W, Wang H, Bazer FW, Zhou B (2014) Pregnancy-secreted acid phosphatase, uteroferrin, enhances fetal erythropoiesis. Endocrinology 155:4521–4530

A Role for Fructose Metabolism in Development of Sheep and Pig Conceptuses

3

Robyn M. Moses, Avery C. Kramer, Heewon Seo, Guoyao Wu, Gregory A. Johnson, and Fuller W. Bazer

Abstract

The period of conceptus (embryo and extraembryonic membrane) development between fertilization and implantation in mammalian species is critical as it sets the stage for placental and fetal development. The trophectoderm and endoderm of pre-implantation ovine and porcine conceptuses undergo elongation, which requires rapid proliferation, migration, and morphological modification of the trophectoderm cells. These complex events occur in a hypoxic intrauterine environment and are supported through the transport of secretions from maternal endometrial glands to the conceptus required for the biochemical processes of cell proliferation, migration, and differentiation. The conceptus utilizes glucose provided by the mother to initiate metabolic pathways that provide energy and substrates for other metabolic pathways. Fructose, however, is in much greater abundance than glucose in amniotic and allantoic fluids, and fetal blood during pregnancy. Despite this, the role(s) of fructose is largely unknown even though a switch to fructosedriven metabolism in subterranean rodents and some cancers are key to their adaptation to hypoxic environments.

Keywords

Pregnancy · Glucose · Fructose · Conceptus Development

Abbreviations

BNC	Binucleated trophoblast giant cells
DHAP	Dihydroxyacetone phosphate
ESR1	Estrogen receptor α
F1P	Fructose-1-phosphate
F6P	Fructose-6-phosphate
G6P	Glucose-6-phosphate
G6PDH	Glucose-6-phosphate dehydrogenase
GAP	Glyceraldehyde-3-phosphate
HIF1A	Hypoxia inducible factor-1α
IFNT	Interferon tau
KHK	Ketohexokinase
LE	Luminal epithelia
NAD^+/ NADH	Nicotinamide adenine dinucleotide
$NADP^+$/ NADPH	Nicotinamide adenine dinucleotide phosphate
OXTR	Oxytocin receptor
PFK	Phosphofructokinase-1
PGF2α	Prostaglandin $F_{2\alpha}$
PHGDH	Phosphoglycerate dehydrogenase
PPP	Pentose phosphate pathway

R. M. Moses · A. C. Kramer · H. Seo · G. Wu · G. A. Johnson · F. W. Bazer (✉)
Departments of Animal Science and Veterinary Integrative Biosciences, Texas A&M University, College Station, TX 77843, USA
e-mail: fbazer@cvm.tamu.edu

G. Wu (ed.), *Recent Advances in Animal Nutrition and Metabolism*, Advances in Experimental Medicine and Biology 1354,
https://doi.org/10.1007/978-3-030-85686-1_3

PSPH	Phosphoserine phosphatase
SHMT	Serine hydroxymethyltransferase
SLC2A	Facilitative glucose transporter family
SLC5A	Sodium-dependent glucose transporter family
TCA	Tricarboxylic acid cycle
TGC	Multinucleated trophoblast giant cell
UDP-glcNAc	Uridine diphosphate N-acetylglucosamine
VEGF	Vascular endothelial growth factor

3.1 Introduction

Commercial sheep and pig operations have high incidences of prenatal loss, which negatively impacts their profitability and efficiency (Bazer et al. 2012b; Van der Lende et al. 2001). These prenatal losses are hypothesized to result from insufficient placental development and endometrial gland histotroph support because the developing ovine and porcine conceptuses (embryo and extraembryonic membranes) undergo significant morphological and metabolic changes as they prepare for attachment to the maternal endometrium for implantation. This early development, from fertilization to conceptus apposition and adhesion to uterine epithelium, is critical for the establishment of a successful and healthy pregnancy (Johnson 2018). Unfortunately, this is also the time during which most pregnancy losses occur (Spencer 2013). Two major causes of early pregnancy loss in the sheep are malnourishment of the conceptus and failure of the blastocyst to elongate (Gray et al. 2002; Spencer et al. 2004). Similar to sheep, pigs have two periods of high embryonic/fetal mortality (Bazer and Johnson 2014). The first period is during the peri-implantation period between Days 14 and 25 of pregnancy, when the free-floating conceptuses are undergoing elongation and implantation, and early stages of placentation. The second period of embryonic/fetal mortality is during mid-gestation from Days 50 to 70 of pregnancy when areolae are developing, and the uterus is folding in order to increase surface area to maximize nutrient transport required for rapid growth and development of the fetus (Seo et al. 2021).

The developing conceptus, like any other tissue in the body, requires nutrients (e.g., amino acids, carbohydrates, lipids, vitamins, minerals, and water) to sustain life and regulate cellular processes (Bazer et al. 2018; Wu 2018a, b). Glucose and its metabolite fructose are two major hexose sugars required for various metabolic pathways, and both are present in fetal blood and fetal fluids later in gestation, although fructose is much more abundant (Zavy et al. 1982). Despite this, the role of fructose during the peri-implantation period of pregnancy is often overlooked as fructose is not considered a substrate for glycolysis or the tricarboxylic acid (TCA) cycle during pregnancy (Battaglia and Meschia 1978, 1981; Bazer et al. 2012a; Kim et al. 2012).

This review serves to examine the current literature regarding conceptus development and metabolism of glucose and fructose in sheep and pigs. The metabolic pathways by which glucose and fructose may be utilized are those required for cellular proliferation, migration, and differentiation required for survival of the conceptus. Because the role of fructose is overlooked during pregnancy, this review will also explore potential metabolic pathways utilizing fructose under conditions of hypoxia and in highly prolific cancer cells.

3.2 Conceptus Development in Sheep

3.2.1 Sheep

Elongation of the sheep blastocyst occurs in the uterus after hatching from the zona pellucida on Day 8 or Day 9 of pregnancy (Seshagiri et al. 2009; Bazer et al. 2012b; Johnson et al. 2018). The blastocyst includes the embryonic disc, trophectoderm, and blastocoel, and later with further development of the extraembryonic membranes and the embryo, it is referred to as the conceptus. A key event in conceptus development is the transition of the blastocyst from a spherical to a tubular form, and then rapid elongation into a long filamentous form (Bazer and First 1983; Bazer et al. 2012b; Johnson et al. 2018). This elongation is required to achieve sufficient numbers of trophectoderm cells to produce enough interferon tau (IFNT) to signal the maternal recognition of pregnancy and to establish sufficient surface area for the exchange of nutrients and gases at the uterine–trophectoderm interface (Anthony et al. 1988; Bazer et al. 2012b). IFNT produced by the ovine trophectoderm silences the expression of receptors for estrogen (ESR1) and oxytocin (OXTR). This mechanism prevents the binding of oxytocin to its receptor and thus the oxytocin-dependent pulsatile release of prostaglandin $F_{2\alpha}$ (PGF2α) that would otherwise cause regression of the corpus luteum that produces progesterone (Fleming et al. 2006). Pregnancy in mammals requires sufficient levels of progesterone to successfully establish and maintain pregnancy (Spencer et al. 1995a, b; Spencer and Bazer 1996). At Day 16 of pregnancy, the sheep conceptus initiates implantation (attachment) to the uterine luminal epithelium (LE) (Guillomot et al. 1981) and it is fully adhered to the uterine LE by Day 22 of pregnancy (Spencer et al. 2017).

Proper elongation of the conceptus before implantation, as noted earlier, is also required to increase the surface area of the trophectoderm in contact with the uterine LE. As the conceptus makes contact with the uterine LE, adhesion molecules in the extra-cellular matrix and integrins on the trophectoderm and uterine LE provide for firm adhesion required for completion of implantation and then initiation of placentation (Johnson et al. 1999, 2018, 2018; Spencer et al. 1999, 2004; Gray et al. 2004). Sheep have a synepitheliochorial placenta, wherein a syncytial layer is formed at the uterine–conceptus interface, resulting in a non-invasive form of placentation. The syncytial layer has been known to be a hybrid of trophoblast and maternal composed of binucleated trophoblast giant cells (BNCs) and uterine LE cells (Wooding and Burton, 2008). However, the results of recent studies indicate that this syncytial layer is entirely of trophoblast origin. Mononuclear trophectoderm cells fuse with one another to form BNCs and multinucleated trophoblast giant cells (TGCs) within the trophoblast layer, which then migrate through uterine LE cells undergoing apoptosis being removed (Seo et al. 2019). The TGCs fuse with each other to form syncytial plaques that expand through continued migration and fusion to cover the entire surface of the endometrium at sites of implantation (Seo et al. 2019). As gestation continues, the caruncular regions of the endometrium interdigitate with specialized areas of the chorioallantois, known as the cotyledons. Collectively, the caruncle and cotyledon form the placentome, of which there are typically 50 to 70 for a sheep placenta. The placentomes are the primary sites for the exchange of nutrients and gasses between the fetal–placental and maternal vasculatures (Grazul-Bilska et al. 2010; Reynolds et al. 2010; Bazer et al. 2012c; Johnson et al. 2018).

Following hatching from the zona pellucida, the conceptus is capable of synthesizing its own proteins and enzymes for metabolic reactions (Crosby et al. 1988), but still requires nutrients provided by the mother. Glucose, amino acids, lipids, growth factors, and an array of proteins may be transported across the placenta or secreted by the uterine glands and transported across the placenta into the fetal–placental vasculature at specialized sites on the placenta termed areolae by a process known as fluid-phase pinocytosis. Secretions from the uterine glands are collectively known as histotroph. Histotrophis is required for the development of the conceptus in sheep beyond Day 14 of pregnancy (Brinsfield and Hawk 1973; Spencer and Bazer 2004; Bazer et al. 2012b). Interruptions in histotroph delivery to the conceptus due to the ablation of endometrial glands are embryonic lethal in sheep (Gray et al. 2002).

3.2.2 Pigs

Pig blastocysts hatch from the zona pellucida and undergo rapid elongation from spherical to tubular and filamentous forms between Days 10 and 12 of pregnancy, reaching a final length of 800 to 1000 mm by Day 15 of pregnancy (Bazer 2012b; Bazer and Johnson 2014). During this period of rapid elongation, pig conceptus trophectoderm secretes estrogen (E2) which is hypothesized to be the pregnancy recognition signal in pigs (Bazer and Thatcher 1977). However, when estrogen synthesis by the conceptus was recently ablated by targeting the aromatase (CYP19A1) gene utilizing CRISPR/Cas9 genome editing technology, estrogen was not found to be essential for pre-implantation conceptus development, conceptus elongation, or early CL maintenance, but pregnancies failed around Day 30 of gestation (Meyer et al. 2019). Pig conceptuses also secrete high amounts of interferon gamma (IFNG) and interferon delta (IFND), along with other proteinaceous paracrine factors (McLendon et al. 2020). These factors are likely important for the establishment of pregnancy and communication between the conceptus and uterine endometrium necessary for tissue remodeling required to support implantation and placentation, but they are not known to be pregnancy recognition signals.

The process of blastocyst elongation in the pig is similar to that described for sheep; including shedding of the zona pellucida, elongation of the conceptus trophectoderm, pre-contact and orientation of the conceptus, apposition, and adhesion. However, pigs have a true epitheliochorial placenta, in which the conceptus trophectoderm attaches to the uterine LE that remains intact throughout pregnancy. As a result, the uterine LE plays a significant role in the transport of nutrients across the uterine–placental interface (Seo et al. 2021). To enhance the efficiency of nutrient transport, the uterine–placental interface of pigs undergoes significant morphological changes to increase fold complexity and surface area. The early stages of fold formation begin by Day 25 of pregnancy and increase to Day 35 of pregnancy, with another increase in fold formation between Days 50 to 60 of pregnancy (see Bazer and Johnson 2014; Seo et al. 2020). It is hypothesized that as the size of the fetus, volumes of fetal fluids, and endometrial blood flow increase, mechanical forces increase on the uterine–placental interface that drives morphological changes that coordinate the development of folds at the uterine–placental interface. The resulting complex folds of the interface increase the surface area of contact between uterine and placental vasculatures, which subsequently increases the efficiency of the exchange of gases, nutrients, and waste at the uterine–placental interface (Seo et al. 2020; Vallet and Freking 2007).

3.3 Glucose and Fructose Profiles During Pregnancy in Sheep and PIGS

Glucose and fructose are present within the porcine and ovine endometria and conceptuses; however, fructose is the most abundant hexose sugar in both species (Bacon and Bell 1948; Goodwin 1956). In pigs, the diet is the major source of glucose in the blood in the fed state. In contrast, gluconeogenesis from propionate and glucogenic amino acids in the liver and kidneys is the source of glucose in blood under both fasting and fed conditions (Hou et al. 2020; Li et al. 2020). Thus, dietary intakes of carbohydrates (e.g., starch and fiber) and glucogenic amino acids (present in both plant and animal proteins; Hou et al. 2019; Li and Wu 2020; Li et al. 2021) may affect the availability of fructose in the conceptuses of ruminants. Of particular note, gluconeogenesis is absent from both ovine and porcine placentae, resulting in the need for glucose to be transported from maternal blood to the uterine lumen to be accessed and transported into the free-floating conceptus trophectoderm. Glucose and fructose can be transported by members of the facilitative diffusion transporters of the solute carrier family 2A (SLC2A) or sodium-dependent transports of the solute carrier 5A (SLC5A). To date, 14 members of the SLC2A family are known to be responsible for the transport of glucose and many other sugars (Augustin 2010).

Members of the facilitative glucose transport (SLC2A) and the sodium-dependent glucose transporter (SLC5A) families (Wood and Trayhurn 2003), particularly SLC2A1, SLC2A4, and

SLC5A1, are abundantly expressed by the uterine LE and superficial glandular epithelium (sGE) in pregnant ewes. These transporters are present in even greater abundance in trophectoderm and extraembryonic endoderm cells (Gao et al. 2009a). Interestingly, SLC2A1 and SLC5A1, as well as SLC2A5, a transporter for both glucose and fructose, are upregulated in endometria of ewes between Days 9 and 12 of pregnancy when treated with exogenous progesterone beginning around 30 h after the onset of estrus and mating (Satterfield et al. 2009; Hoskins et al. 2021) and by conceptus-derived prostaglandins at Day 14 of pregnancy (Spencer et al. 2013).

The facilitative diffusion glucose transporters SLC2A1, SLC2A2, SLC2A3, SLC2A4, and SLC2A8 are regulated in a spatial–temporal pattern at the uterine–placental interface of pigs and are present in all epithelial layers (Kramer et al. 2020; Steinhauser et al. 2016). This allows the hexose sugars, glucose and fructose, to be effectively transported from the maternal vasculature into the conceptus. SLC2A1, SLC2A4, and SLC2A8 are expressed by uterine LE, while SLC2A3 and SLC2A8 are the predominant transporters expressed by conceptus trophectoderm. SLC2A1 is expressed by the endometrial and allantoic endothelium. SLC2A3 and SLC2A8 are expressed by the allantoic epithelium and endothelium. In addition, SLC2A1 is upregulated in the uterine LE by exogenous estradiol and progesterone (Kramer et al. 2020).

Glucose metabolism by ovine and porcine embryos has been thoroughly investigated. Glucose uptake by ovine embryos is low during the early cleavage stages when lactate and pyruvate are the primary energy sources, but the uptake of glucose increases during the transition to the morula and blastocyst stages (Butler and Williams 1991; Gardner et al. 1993). However, there is very little incorporation of glucose into glycogen by the conceptus (Wales et al. 1989; Thompson et al. 1991). Interestingly, the oxidative metabolism of glucose increases between the 4- to 8-cell stage and the blastocyst stage (Wales et al. 1989), but after hatching from the zona pellucida, the oxidation of glucose decreases, as it may spare glucose for other pathways and use non-glucose molecules, such as amino acids, as a source for producing ATP (Gardner et al. 1993).

Unlike human and rodent placentae, the placentae of sheep and pigs are fructogenic, meaning they can synthesize fructose from other substrates, such as glucose (Alexander et al. 1955; White et al. 1979; Bazer et al. 2012d; Kramer et al. 2020). More specifically, glucose from the mother is converted into fructose via the polyol pathway in the trophectoderm cells and chorion of the placenta (Alexander et al. 1955; Teng et al. 2002; Regnault et al. 2010; Steinhauser et al. 2016; Bazer et al. 2018). Glucose is reduced to sorbitol via aldose reductase and further oxidized to fructose by sorbitol dehydrogenase (Wu 2018a). The resulting fructose may be metabolized by various cells of the conceptus or pass into the amniotic and allantoic fluids for later use, but fructose is unable to be transported back to the maternal vasculature (White et al. 1979; Bazer et al. 2012d; Wang et al. 2016). Throughout pregnancy in sheep and pigs, fructose is 10 to 30 times more abundant than glucose in fetal blood and fetal fluids (Bazer et al. 2012d).

3.4 Metabolism of Glucose and Fructose: Pathways of Interest for Development of the Conceptus

The elongation phase of both ovine and porcine conceptuses requires various nutrients for the synthesis of biomolecules required for proliferation, migration, and differentiation of cells, as well as the survival of the conceptus. Many of these nutrients, including glucose, are transported and/or secreted by the uterine epithelial cells into the uterine lumen to be components of histotroph (Spencer and Bazer 2004; Bazer et al. 2012b). Glucose and fructose are two major hexose sugars that, with other metabolic intermediates, are used in multiple metabolic pathways as precursors for the synthesis of other molecules (Fig. 3.1). The metabolism of glucose through glycolysis provides adenosine triphosphate (ATP; the major form of biological energy for

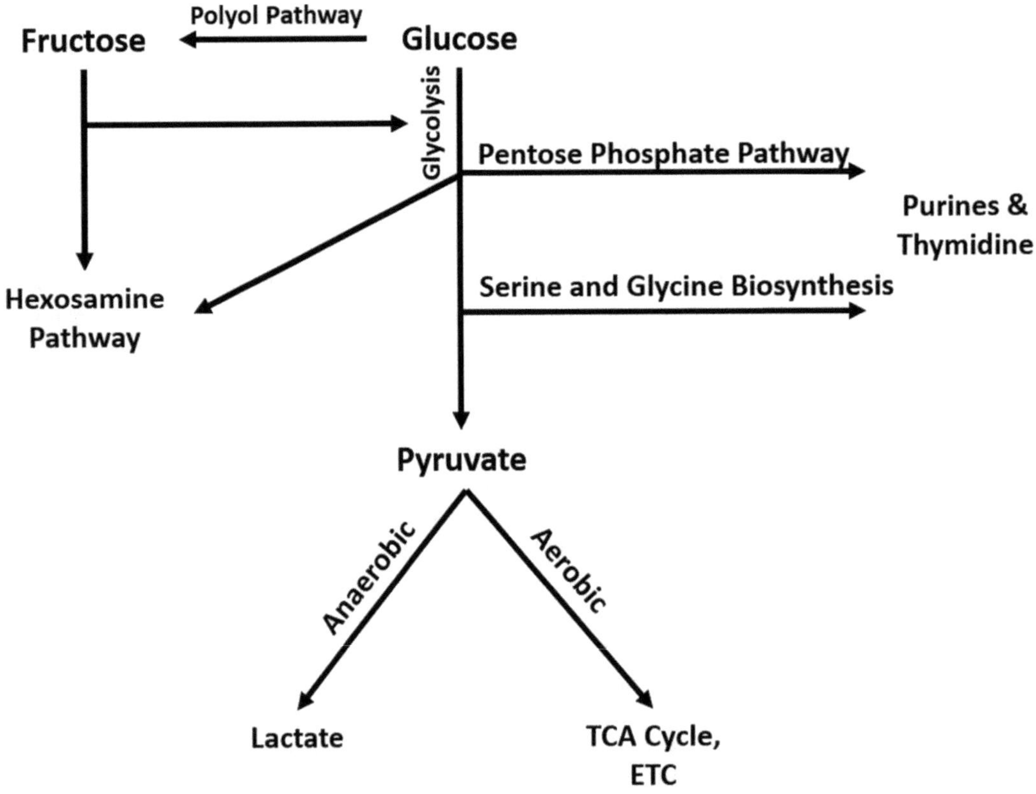

Fig. 3.1 Glucose and fructose are two molecules used in multiple metabolic pathways for producing ATP and other molecules required for proliferation, migration, and differentiation of cells. Pathways of interest regarding conceptus development are those for glycolysis, pentose phosphate pathway, one-carbon metabolism with serine and glycine biosynthesis, hexosamine biosynthesis pathway, and reduction of pyruvate to lactate

cell metabolism), as well as pyruvate which can be further metabolized via the tricarboxylic acid (TCA) cycle or converted to lactate (Wu 2018b). Recently, it was reported that once delivered to the conceptus, glucose and fructose are metabolized through glycolysis in support of placental and fetal development during pregnancy in pigs. When Day 16 conceptuses of pigs were incubated with either ^{14}C-glucose or ^{14}C-fructose and amounts of radiolabeled $^{14}CO_2$ released from the conceptuses were measured to determine rates of oxidation of glucose and fructose, both glucose and fructose were transported into conceptuses and metabolized (Kramer et al. 2020). Glucose was preferentially metabolized in the presence of fructose, while fructose was actively metabolized in the absence of glucose and to a lesser extent in the presence of glucose (Kramer et al. 2020).

The production of lactate from pyruvate allows for the regeneration of nicotinamide adenine dinucleotide (NAD^+) that is required for some enzymes in glycolysis, and lactate is often associated with anaerobic respiration in response to exposure to a hypoxic environment, as has been observed during early pregnancy. Lactic acid can decrease the pH of the immediate environment and, when produced by the conceptus, it can decrease the overall pH of the uterine lumen. This may be important during early pregnancy as the expression of some factors required for angiogenesis and vasculogenesis, such as vascular endothelial growth factor (VEGF) and hypoxia inducible factor-1a (HIF1A), are stimulated by lactate (Gardner 2015; Xiao et al. 2017). Lactate stimulation of HIF1A expression is achieved by inhibiting the enzyme prolyl hydroxylase (Polet and Feron 2013), which is responsible for regulating HIF1A in normoxic conditions (To and Huang 2005). Under hypoxic conditions, HIF1A will stimulate the expression of VEGF and other angiogenic factors (Semenza 2003) required for the development of the placental vasculature between the mother and conceptus to ensure proper blood flow between them for the transfer of nutrients and gases (Klagsbrun and D'Amore 1991; Cheung and Brace 1999; Reynolds and Redmer 2001; Grazul-Bilska et al. 2010). Of note, the mammalian conceptus is considered to develop in a hypoxic environment (Fischer and Bavister 1993) and conceptus trophectoderm of sheep expresses HIF1A and HIF2A to adapt to the hypoxic conditions of the uterine lumen (Song et al. 2008).

Glycolysis also provides intermediates used for DNA synthesis via the pathway for the synthesis of serine and glycine, as well as the pentose phosphate pathway (PPP; Wu 2022). Serine synthesis from glycolysis begins with the oxidation of 3-phosphoglycerate via phosphoglycerate dehydrogenase (PHGDH). Phosphoserine phosphatase (PSPH) catalyzes the final reaction to synthesize serine and it has been localized to uterine LE during the apposition phase of implantation (Seo et al. 2019). Glutamate (a product of glutamine hydrolysis) provides the amino group for the synthesis of serine, indicating an important link between glutaminolysis and fructose metabolism in the conceptus (Bazer et al. 2021). Serine hydroxymethyltransferase (SHMT) interconverts serine to glycine and this enzyme has been localized to conceptus trophectoderm of sheep during implantation (Seo et al. 2019). It should be noted that serine and glycine are both abundant in the uterine lumen of ewes during early pregnancy and in fetal fluids throughout gestation in ewes (Gao et al. 2009b; Kwon et al. 2003). Glycine, glutamine, aspartate, formate, and bicarbonate are required for the synthesis of purine nucleosides, whereas glutamine, aspartate, and bicarbonate are required for the synthesis of pyrimidine nucleosides (Lunt and Vander Heiden 2011; Wu 2013). This explains, in part, the nutritional and physiological importance of glycine, glutamine, and aspartate [all abundant in animal-sourced foodstuffs (Li and Wu 2020)] in swine (Zhang et al. 2021) and sheep (Cao et al. 2021; Gilbreath et al. 2021), as well as ruminant and nonruminant zoo animals (Herring et al. 2021).

Glucose is phosphorylated by hexokinase in the first step of glycolysis to yield glucose-6-phosphate (G6P). Conversion of G6P to 6-phosphoglucono-δ-lactone via G6P dehydrogenase (G6PDH) is the first committed step of the pentose phosphate pathway (PPP) and generates the reduced form of nicotinamide adenine dinucleotide phosphate (NADPH). Multiple metabolic pathways utilize NADPH as a cofactor, including lipid metabolism and the conversion of glucose to sorbitol via aldose reductase. De novo synthesis of lipids in the conceptus is fairly low as the mother provides the necessary lipids via histotroph (Brinsfield and Hawk 1973; Boshier et al. 1987; Ribeiro et al. 2016). The PPP produces ribose-5-phosphate that is the five-carbon sugar required for the synthesis of nucleosides for DNA and RNA. In vitro studies showed that contributions of glucose to the PPP increased from the 2-cell stage to the blastocyst stage of development in sheep, but then the contribution by glucose progressively decreased with advancing stages of conceptus development (Wales and Du 1993).

As previously mentioned, sheep and pig placentae convert glucose to fructose; however, the role of fructose has been largely overlooked, despite it being 10 to 30 times more abundant than glucose in fetal blood and fetal fluids (Edwards et al. 1997). Fructose itself can also be phosphorylated to fructose-6-phosphate (F6P) by hexokinase for use in glycolytic reactions or the hexosamine biosynthesis pathway. However, glucose is the preferred substrate for hexokinase, as it has a much higher affinity for glucose than fructose (Weinhouse 1976; Wu 2018b). In the hexosamine biosynthesis pathway, either F6P or G6P, which is converted to F6P by way of phosphoglucose isomerase, are the initial substrates to produce uridine diphosphate N-acetylglucosamine (UDP-GlcNAc), which is used to synthesize glycosaminoglycans and provide for post-translational glycosylation of proteins (Spiro 2002; Bazer et al. 2012c; Wang et al. 2016). UDP-GlcNAc is responsible for *O*-linked glycosylation of serine or threonine residues on proteins (Spiro 2002) and can stimulate the AKT (proto-oncogenic protein kinase Akt), TSC2 (tuberous sclerosis complex 2), and mTOR (mechanistic target of rapamycin; mTORC1) signaling pathways, all of which are important for proliferation of ovine trophectoderm cells (Wang et al. 2016). It should be noted that the first enzyme involved in the hexosamine biosynthesis pathway, glucosamine-F6P transaminase 1 (GFPT1), utilizes F6P and glutamine, a highly abundant amino acid in allantoic fluid of about 25 mM on Day 60 of pregnancy in ovine allantoic fluid (Kwon et al. 2003), to produce glucosamine-6-phosphate (GlcN-6-P). Thus, GFPT1is the key regulatory enzyme for this pathway (Broschat et al. 2002; Nagel and Ball 2015).

Fructose metabolism may also be important in allowing the conceptus to adapt to the hypoxic environment of the uterus during the peri-implantation period of pregnancy. A notable study performed by Park et al. (2017) provided evidence for a link between fructose metabolism and hypoxia in the naked mole-rat, which is a subterranean rodent that lives in low oxygen conditions and has adapted to hypoxic conditions. The study revealed that during extremely low oxygen conditions, and even anoxic conditions, the metabolism of the naked mole-rat switched to a fructose-driven anaerobic metabolism, with marked increases in expression of ketohexokinase (KHK), SLC2A1, and SLC2A5 (Park et al. 2017). KHK phosphorylates fructose at the first carbon position (fructose-1-phosphate, F1P), rather than at the six carbon position as for hexokinase. Thus, cleavage of F1P by aldolase B yields dihydroxyacetone (DHAP) and glyceraldehyde, both of which can be converted to glyceraldehyde-3-phosphate (GAP) for use in the ATP production phase of glycolysis (Fig. 3.2) (Wu 2018a). Fructose catabolism is primarily in the liver; however, KHK has been localized to porcine trophectoderm and placental chorion throughout pregnancy (Steinhauser et al. 2016; Bazer et al. 2018). Incorporation of fructose into the aerobic glycolytic pathway in this manner bypasses the regulatory mechanisms of phosphofructokinase-1 (PFK), which is inhibited by acid pH, as well as high levels of ATP and citrate (Fig. 3.2) (Adelman et al. 1967).

3.5 Similarities in Metabolism Between Conceptuses and Tumors

There are some intriguing similarities in the metabolic profiles of cancer cells and the developing pre-implantation conceptus (Krisher and Prather 2012). In the 1920s, Warburg et al. (1927) observed that cancer cells had a metabolic profile with a characteristic increase in glycolytic activity to produce lactate, despite oxygen being readily available for respiration. While glycolysis itself is considered to be an inefficient energy-generating process producing only a net of two ATP molecules per glucose molecule, increasing glucose uptake and the rate of glycolysis provides energy during hypoxic conditions and during periods of rapid cellular proliferation (Lunt and Vander Heiden 2011; Wu 2018b). Increasing glycolysis and performing aerobic respiration may also be a protective measure for cancer cells to decrease the amount of reactive oxygen species that would be produced by metabolism via the TCA cycle and electron transport chain (Brand and Hermfisse 1997). As previously mentioned, glycolysis also provides precursors for other pathways that are critical for the growth and survival of cells, especially those required by rapidly proliferating cancer cells and cells of the developing conceptus (Vander Heiden et al. 2009).

Fructose metabolism can contribute to, and even exacerbate, cellular proliferation and tumor growth (Nakagawa et al. 2020; Santhekadur 2020). Glioma (Gao et al. 2018) and intestinal (Goncalves et al. 2019) tumors exhibiting increased fructose metabolism express KHK which allows for fructose entry into glycolysis via a pathway that bypasses the upstream regulatory mechanisms of hexokinase and PFK. By doing so, the increased activity of the ATP

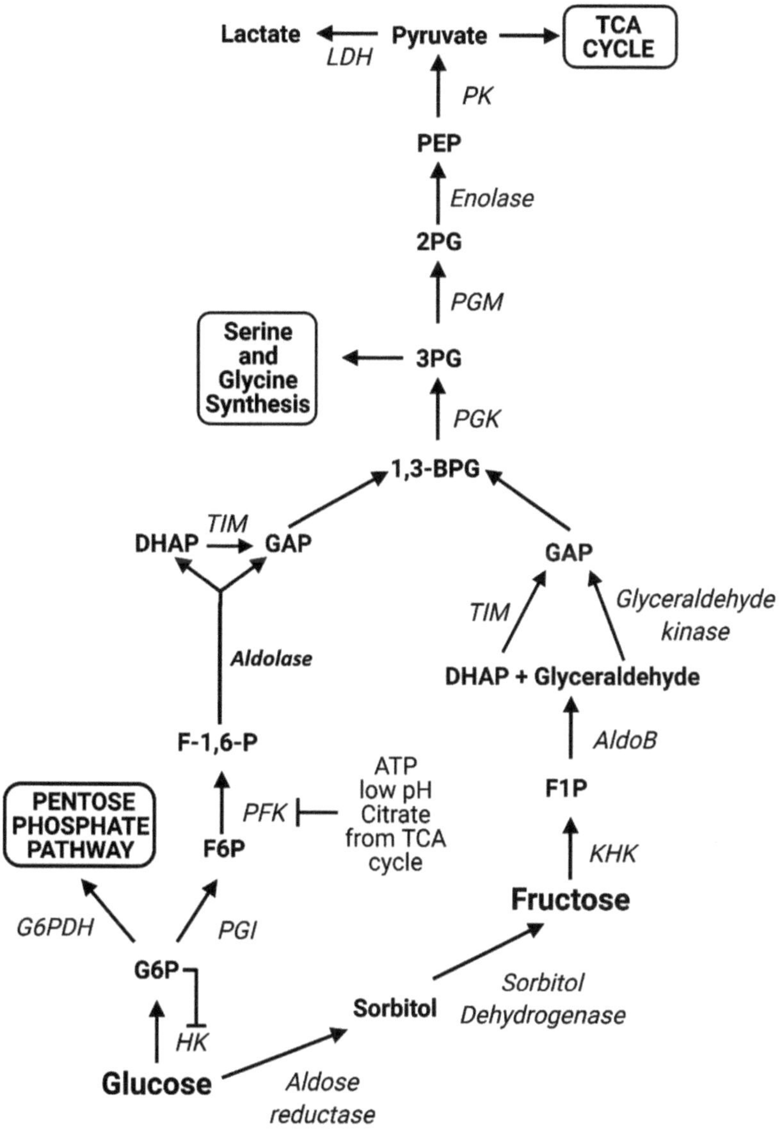

Fig. 3.2 Dihydroxyacetone phosphate (DHAP) is the common metabolite between glucose and fructose entry into glycolysis. Fructose can be phosphorylated via ketohexokinase (KHK) to fructose-1-phosphate (F1P), which can be cleaved by aldolase b (AldoB) to yield glyceraldehyde and DHAP. Glyceraldehyde can be converted to DHAP via glyceraldehyde kinase, providing two molecules of DHAP that can enter the ATP-producing phase of glycolysis. G6P: glucose-6-phosphate; HK: hexokinase; PGI: phosphoglucose isomerase; F6P: fructose-6-phosphate; PFK: phosphofructokinase-1; F-1,6-P: fructose-1–6-bisphosphate; GAP: glyceraldehyde-3-phosphate; TIM: triose phosphate isomerase; 1,3-BPG: 1,3-bisphosphoglycerate; PGK: phosphoglycerate kinase; 3PG: 3-phosphoglycerate; 2PG: 2-phosphoglycerate; PGM: phosphoglycerate mutase; PEP: phosphoenolpyruvate; PK: pyruvate kinase; LDH: lactate dehydrogenase. Created with BioRender.com

production phase of glycolysis generates more pyruvate and other metabolic intermediates required to sustain cellular processes (Vander Heiden et al. 2009; Abbaszadeh et al. 2020), as well as generate ATP at a faster rate than that achieved via metabolism through the TCA cycle and electron transport chain (Pfeiffer et al. 2001). Interestingly, this shift to fructose metabolism by cancer cells is similar to the strategy used by the subterranean naked mole-rats and Gansu zokor rats which used to overcome the hypoxic conditions in which they live. These rodents have greater expression of KHK and GLUT5 in vital organs to increase energy production in the absence of oxygen (Park et al. 2017; Lin et al. 2021). Since the elongating conceptus also develops in a hypoxic intrauterine environment prior to implantation, fructose may have a role in providing ATP via the KHK dependent entry of F1P into glycolysis.

Lactate production is also characteristic of developing conceptuses and cancer cells (Warburg et al. 1927; Du and Wales 1993; Gardner et al. 1993). Pyruvate produced through glycolysis can enter the TCA cycle via pyruvate dehydrogenase (PDH) or it can be reduced to lactate via lactate dehydrogenase (LDH). Pyruvate production by glycolysis that exceeds the metabolic capabilities of the cell may be converted to lactate via LDH, regenerating NAD^+ in the process that is required for glycolysis to continue (Curi et al. 1988). As previously mentioned, lactic acid accumulation decreases the pH of the surrounding environment, thus stimulating VEGF to facilitate angiogenesis, which is also the case for paracrine signaling between endothelial cells and tumor cells (Weis and Cheresh 2011).

Amino acids and their metabolic intermediates have roles in protein synthesis, DNA synthesis, and one-carbon metabolism, all of which are critical for the survival of rapidly proliferating cells. There are similarities between cancer cells and the elongating conceptuses of sheep and pigs regarding amino acid metabolism. In some estrogen negative breast cancers, PHGDH, the first committed step in serine biosynthesis from glycolysis is upregulated resulting in an increase in serine production (Possemato et al. 2011). Serine is the major one-carbon unit utilized for DNA synthesis in cancer cells (Labuschagne et al. 2014) and excessive serine metabolism to formate can promote metastasis and invasion of cancer cells (Meiser et al. 2018). The enzymes to convert 3-phosphoglycerate and glutamate to serine have been localized to ovine trophectoderm (Seo et al. 2019), and serine and glycine are among the most abundant amino acids in the uterine lumen during the peri-implantation period of pregnancy (Gao et al. 2009b), as well as allantoic and amniotic fluids in sheep during gestation (Kwon et al. 2003). Formate is greatly increased in some cancers (Meiser et al. 2018) and is present in greater abundance in fetal plasma than in maternal plasma in sheep during late gestation (Washburn et al. 2015). Understanding these changes during the peri-implantation period of pregnancy can further elucidate how glucose and fructose metabolism enhances or compromises these key metabolic processes.

3.6 Summary

The time between fertilization of the oocyte and implantation of the blastocyst/conceptus is critical in all mammalian species. Sheep and pig conceptuses must elongate after hatching from the zona pellucida to ensure proper implantation and subsequent placentation. The trophectoderm and endoderm cells of elongating conceptuses undergo structural and metabolic changes required for rapid proliferation, migration, and differentiation that require metabolic adaptation during a short period of time during pregnancy. These metabolic requirements mirror those of some cancer cell types. While this review focused on the roles of both glucose and fructose in various metabolic pathways for elongating sheep and pig conceptuses, there is evidence that fructose may have a larger role in supporting conceptus development. Fructose metabolism may allow the conceptus to adapt to the hypoxic environment of the uterus by providing energy or intermediates for other metabolic pathways, much as it does in the hypoxic environment for some cancers and subterranean rodents.

Acknowledgements Work in our laboratories was supported by Agriculture and Food Research Initiative Competitive Grant no. 2018-67015-28093 from the USDA National Institute of Food and Agriculture.

Conflicts of Interest The authors have no conflicts of interest to declare.

References

Abbaszadeh Z, Çeşmeli S, Biray Avcı Ç (2020) Crucial players in glycolysis: Cancer progress. Gene 726:144158

Adelman RC, Ballard FJ, Weinhouse S (1967) Purification and properties of rat liver fructokinase. J Biol Chem 242:3360–3365

Alexander DP, Andrews RD, Huggett ASG, Nixon DA, Widdas WF (1955) The placental transfer of sugars in the sheep: studies with radioactive sugar. J Physiol 129:352–366

Anthony RV, Helmer SD, Sharif SF, Roberts RM, Hansen PJ, Thatcher WW, Bazer FW (1988) Synthesis and processing of ovine trophoblast protein-1 and bovine trophoblast protein-1, conceptus secretory proteins involved in the maternal recognition of pregnancy. Endocrinology 123:1274–1280

Augustin R (2010) The protein family of glucose transport facilitators: it's not only about glucose after all. IUBMB Life 62:315–333

Bacon JS, Bell DJ (1948) Fructose and glucose in the blood of the foetal sheep. Biochem J 397–405

Battaglia FC, Meschia G (1978) Principal Substrates of Fetal Metabolism. Physiol Rev 58:499–527

Battaglia FC, Meschia G (1981) Foetal and placental metabolisms: their interrelationship and impact upon maternal metabolism. Proc Nutr Soc 40:99–113

Bazer FW, First NL (1983) Pregnancy and parturition. J Anim Sci 57:425–460

Bazer FW, Thatcher WW (1977) Theory of maternal recognition of pregnancy in swine based on estrogen controlled endocrine versus exocrine secretion of prostaglandin F2α by the uterine endometrium. Prostaglandins 14:397–400

Bazer FW, Johnson GA (2014) Pig blastocyst-uterine interactions. Differentiation 87:52–65

Bazer FW, Kim J, Song G, Ka H, Tekwe CD, Wu G (2012a) Select nutrients, progesterone, and interferon tau affect conceptus metabolism and development. Ann N Y Acad Sci 1271:88–96

Bazer FW, Song G, Kim J, Dunlap KA, Satterfield MC, Johnson GA, Burghardt RC, Wu G (2012b) Uterine biology in pigs and sheep. J Anim Sci Biotechnol 3:23

Bazer FW, Spencer TE, Thatcher WW (2012c) Growth and development of the ovine conceptus. J Anim Sci 90:159–170

Bazer FW, Burghardt RC, Johnson GA, Spencer TE, Wu G (2018) Mechanisms for the establishment and maintenance of pregnancy: Synergies from scientific collaborations. Biol Reprod 99:225–241

Bazer FW, Seo H, Johnson GA, Wu G (2021) One-carbon metabolism and development of the conceptus during pregnancy: Lessons from studies with sheep and pigs. Adv Exp Med Biol 1285:1–15

Boshier DP, Fairclough RJ, Holloway H (1987) Assessment of sheep blastocyst effects on neutral lipids in the uterine caruncular epithelium. J Reprod Fertil 79:569–573

Brand KA, Hermfisse U (1997) Aerobic glycolysis by proliferating cells: a protective strategy against reactive oxygen species. FASEB J 11:388–395

Brinsfield TH, Hawk HW (1973) Control by progesterone of the concentration of lipid droplets in epithelial cells of the sheep endometrium. J Anim Sci 36:919–922

Broschat KO, Gorka C, Page JD, Martin-Berger CL, Davies MS, Huang H, Gulve EA, Salsgiver WJ, Kasten TP (2002) Kinetic Characterization of Human Glutamine-fructose-6-phosphate Amidotransferase I. J Biol Chem 277:14764–14770

Butler JE, Williams JE (1991) Noninvasive measurement of pyruvate uptake by ovine preimplantation embryos and unfertilized ova. Theriogenology 36:1043–1048

Cao Y, Yao J, Sun X, Liu S, Martin GB (2021) Amino acids in the nutrition and production of sheep and goats. Adv Exp Med Biol 1285:63–79

Cheung CY, Brace RA (1999) Developmental expression of vascular endothelial growth factor and its receptors in ovine placenta and fetal membranes. J Soc Gynecol Invest 6:179–185

Crosby IM, Gandolfi F, Moor RM (1988) Control of protein synthesis during early cleavage of sheep embryos. J Reprod Fertil 82:769–775

Curi R, Newsholme P, Newsholme EA (1988) Metabolism of pyruvate by isolated rat mesenteric lymphocytes, lymphocyte mitochondria and isolated mouse macrophages. Biochem J 250:383–388

Du ZF, Wales RG (1993) Glycolysis and glucose oxidation by the sheep conceptus at different oxygen concentrations. Reprod Fertil Dev 5:383–393

Edwards EM, Rattenbury J, Varnam GC, Dhand UK, Jeacock MK, Shepherd DA (1997) Enzyme activities in the sheep placenta during the last three months of pregnancy. Biochim Biophys Acta 497:133–143

Fischer B, Bavister BD (1993) Oxygen tension in the oviduct and uterus of rhesus monkeys, hamsters and rabbits. J Reprod Fertil 99:673–679

Fleming JG, Spencer TE, Safe SH, Bazer FW (2006) Estrogen regulates transcription of the ovine oxytocin receptor gene through GC-rich SP1 promoter elements. Endocrinology 147:899–911

Gao H, Wu G, Spencer TE, Johnson GA, Bazer FW (2009a) Select nutrients in the ovine uterine lumen. II. Glucose transporters in the uterus and peri-implantation conceptuses. Biol Reprod 80:94–104

Gao H, Wu G, Spencer TE, Johnson GA, Li X, Bazer FW (2009b) Select nutrients in the ovine uterine lumen. I. Amino acids, glucose, and ions in uterine lumenal flushings of cyclic and pregnant ewes. Biol Reprod 80:86–93

Gao W, Li N, Li Z, Xu J, Su C (2018) Ketohexokinase is involved in fructose utilization and promotes tumor progression in glioma. Biochem Biophys Res Commun 503:1298–1306

Gardner DK (2015) Lactate production by the mammalian blastocyst: Manipulating the microenvironment for uterine implantation and invasion? BioEssays 37:364–371

Gardner DK, Lane M, Batt P (1993) Uptake and metabolism of pyruvate and glucose by individual sheep preattachment embryos developed in vivo. Mol Reprod Dev 36:313–319

Gilbreath KR, Bazer FW, Satterfield MC, Wu G (2021) Amino acid nutrition and reproductive performance in ruminants. Adv Exp Med Biol 1285:43–61

Goncalves MD, Lu C, Tutnauer J, Hartman TE, Hwang S, Murphy CJ, Pauli C, Morris R, Taylor S, Bosch K, Yang S, Wang Y, Van Riper J, Lekaye HC, Roper J, Kim Y, Chen Q, Gross SS, Rhee KY, Cantley LC, Yun J (2019) High-fructose corn syrup enhances intestinal tumor growth in mice. Science 363:1345–1349

Goodwin RF (1956) Division of the common mammals into two groups according to the concentration of fructose in the blood of the foetus. J Physiol 132:146–156

Gray CA, Adelson DL, Bazer FW, Burghardt RC, Meeusen ENT, Spencer TE (2004) Discovery and characterization of an epithelial-specific galectin in the endometrium that forms crystals in the trophectoderm. Proc Natl Acad Sci U S A 101:7982–7987

Gray CA, Burghardt RC, Johnson GA, Bazer FW, Spencer TE (2002) Evidence that absence of endometrial gland secretions in uterine gland knockout ewes compromises conceptus survival and elongation. Reproduction 124:289–300

Grazul-Bilska AT, Borowicz PP, Johnson ML, Minten MA, Bilski JJ, Wroblewski R, Redmer DA, Reynolds LP (2010) Placental development during early pregnancy in sheep: Vascular growth and expression of angiogenic factors in maternal placenta. Reproduction 140:165–174

Guillomot M, Fléchon JE, Wintenberger-Torres S (1981) Conceptus attachment in the ewe: An ultrastructural study. Placenta 2:169–181

Herring CM, Bazer FW, Wu G (2021) Amino acid nutrition for optimum growth, development, reproduction, and health of zoo animals. Adv Exp Med Biol 1285:233–253

Hoskins EC, Halloran KM, Stenhouse C, Moses RM, Dunlap KA, Satterfield MC, Seo H, Johnson GA, Wu G, Bazer FW (2021) Pre-implantation exogenous progesterone and pregnancy in sheep: I. polyamines, nutrient transport, and progestamedins. J Anim Sci Biotechnol 12:1–17

Hou YQ, He WL, Hu SD, Wu G (2019) Composition of polyamines and amino acids in plant-source foods for human consumption. Amino Acids 51:1153–1165

Hou YQ, Hu SD, Li XY, He WL, Wu G (2020) Amino acid metabolism in the liver: nutritional and physiological significance. Adv Exp Med Biol 1265:21–37

Johnson GA (2018) Domestic animal placentation. Encyclopedia of Reproduction, vol 2. Academic Press, Elsevier, pp 448–454

Johnson GA, Bazer FW, Burghardt RC, Wu G, Seo H, Kramer AC, McLendon BA (2018) Cellular events during ovine implantation and impact for gestation. Anim Reprod 15:843–855

Johnson GA, Burghardt RC, Spencer TE, Newton GR, Ott TL, Bazer FW (1999) Ovine osteopontin: II. Osteopontin and alpha-v beta-3 Integrin Expression in the Uterus and Conceptus During the Periimplantation Period. Biol Reprod 61:892–899

Kim J, Song G, Wu G, Bazer FW (2012) Functional roles of fructose. Proc Natl Acad Sci 109:E1619–E1628

Klagsbrun M, D'Amore PA (1991) Regulators of angiogenesis. Annu Rev Physiol 53:217–239

Kramer AC, Steinhauser CB, Seo GH, H, McLendon BA, Burghardt RC, Wu G, Bazer FW, Johnson GA, (2020) Steroids regulate SLC2A1 and SLC2A3 to deliver glucose into trophectoderm for metabolism via glycolysis. Endocrinology 161:1–19

Krisher RL, Prather RS (2012) A role for the Warburg effect in preimplantation embryo development: Metabolic modification to support rapid cell proliferation. Mol Reprod Dev 79:311–320

Kwon H, Spencer TE, Bazer FW, Wu G (2003) Developmental changes of amino acids in ovine fetal fluids. Biol Reprod 68:1813–1820

Labuschagne CF, van den Broek NJF, Mackay GM, Vousden KH, Maddocks ODK (2014) Serine, but not glycine, supports one-carbon metabolism and proliferation of cancer cells. Cell Rep 7:1248–1258

Li P, Wu G (2020) Composition of amino acids and related nitrogenous nutrients in feedstuffs for animal diets. Amino Acids 52:523–542

Li XY, Zheng SX, Wu G (2020) Amino acid metabolism in the kidneys: nutritional and physiological significance. Adv Exp Med Biol 1265:71–95

Li P, He WL, Wu G (2021) Composition of amino acids in foodstuffs for humans and animals. Adv Exp Med Biol 1332:189–209

Lin J, Fan L, Han Y, Guo J, Hao Z, Cao L, Kang J, Wang X, He J, Li J (2021) The mTORC1/eIF4E/HIF-1α Pathway Mediates Glycolysis to Support Brain Hypoxia Resistance in the Gansu Zokor, *Eospalax cansus*. Front Physiol 12:626240

Lunt SY, Vander Heiden MG (2011) Aerobic glycolysis: Meeting the metabolic requirements of cellular proliferation. Annu Rev Cell Dev Biol 27:441–464

McLendon BA, Seo H, Kramer AC, Burghardt RC, Bazer FW, Johnson GA (2020) Pig conceptuses secrete interferon gamma to recruit T cells to the endometrium during the peri-implantation period. Biol Reprod 103:1018–1029

Meiser J, Schuster A, Pietzke M, Voorde JV, Athineos D, Oizel K, Burgos-Barragan G, Wit N, Dhayade S, Morton JP, Dornier E, Sumpton D, Mackay GM, Blyth K, Patel KJ, Niclou SP, Vazquez A (2018) Increased formate overflow is a hallmark of oxidative cancer. Nat Commun 9:1368

Meyer AE, Pfeiffer CA, Brooks KE, Spate LD, Benne JA, Cecil R, Samuel MS, Murphy CN, Behura S, McLean MK, Ciernia LA, Smith MF, Whitworth KM, Wells KD, Spencer TE, Prather RS, Geisert RD (2019) New perspective on conceptus estrogens in maternal recognition and pregnancy establishment in the pig. Biol Reprod 101:148–161

Nagel AK, Ball LE (2015) Intracellular Protein O-GlcNAc Modification Integrates Nutrient Status with Transcriptional and Metabolic Regulation. In: Advances in Cancer Research, 1st edn. Elsevier, New York, pp 137–166

Nakagawa T, Lanaspa MA, San Millan I, Mehdi F, Rivard JC, Sanchez-Lozada LG, Andres-Hernando A, Tolan DR, Johnson RJ (2020) Fructose contributes to the Warburg effect for cancer growth. Cancer Metab 8:1–12

Park TJ, Reznick J, Peterson BL, Blass G, Omerbašić D, Bennett NC, Henning P, Kuich JL, Zasada C, Browe BM, Hamann W, Applegate DT, Radke MH, Kosten T, Lutermann H, Gavaghan V, Eigenbrod O, Bégay V, Amoroso VG, Govind V, Minshall RD, Smith ESJ, Larson J, Gotthardt M, Kempa S, Lewin GR (2017) Fructose-driven glycolysis supports anoxia resistance in the naked mole-rat. Science 356:307–311

Pfeiffer T, Schuster S, Bonhoeffer S (2001) Cooperation and competition in the evolution of ATP-producing pathways. Science 292:504–507

Polet F, Feron O (2013) Endothelial cell metabolism and tumour angiogenesis: Glucose and glutamine as essential fuels and lactate as the driving force. J Intern Med 273:156–165

Possemato R, Marks KM, Shaul YD, Pacold ME, Kim D, Birsoy K, Sethumadhavan S, Woo H, Jang HG, Jha AK, Chen WW, Barrett FG, Stransky N, Tsun Z, Cowley GS, Barretina J, Kalaany NY, Hsu PP, Ottina K, Chan AM, Yuan B, Garraway LA, Root DE, Mino-Kenudson M, Brachtel EF, Driggers EM, Sabatini DM (2011) Functional genomics reveal that the serine synthesis pathway is essential in breast cancer. Nature 476:346–350

Regnault TRH, Teng C, De Vrijer B, Galan HL, Wilkening RB, Battaglia FC (2010) The tissue and plasma concentration of polyols and sugars in sheep intrauterine growth retardation. Exp Biol Med 235:999–1006

Reynolds LP, Borowicz PP, Caton JS, Vonnahme KA, Luther JS, Buchanan DS, Hafez SA, Grazul-Bilska AT, Redmer DA (2010) Uteroplacental vascular development and placental function: An update. Int J Dev Biol 54:355–365

Reynolds LP, Redmer DA (2001) Angiogenesis in the placenta. Biol Reprod 64:1033–1040

Ribeiro ES, Santos JEP, Thatcher WW (2016) Role of lipids on elongation of the preimplantation conceptus in ruminants. Reproduction 152:R115–R126

Santhekadur PK (2020) The dark face of fructose as a tumor promoter. Genes Dis 7:163–165

Satterfield MC, Song G, Kochan KJ, Riggs PK, Simmons RM, Elsik CG, Adelson DL, Bazer FW, Zhou H, Spencer TE (2009) Discovery of candidate genes and pathways in the endometrium regulating ovine blastocyst growth and conceptus elongation. Physiol Genomics 39:85–99

Semenza GL (2003) Targeting HIF-1 for cancer therapy. Nat Rev Cancer 3:721–732

Seo H, Bazer FW, Burghardt RC, Johnson GA (2019) Immunohistochemical examination of trophoblast syncytialization during early placentation in sheep. Int J Mol Sci 20:4530

Seo H, Li X, Wu G, Bazer FW, Burghardt RC, Bayless JK, Johnson GA (2020) Mechanotransduction drives morphogenesis to develop folding during placental development in pigs. Placenta 90:62–70

Seo H, Johnson GA, Bazer FW, Wu G, McLendon BA, Kramer AC (2021) Cell-specific expression of enzymes for serine biosynthesis and glutaminolysis in farm animals. Adv Exp Med Biol 1285:17–28

Seshagiri PB, Sen Roy S, Sireesha G, Rao RP (2009) Cellular and molecular regulation of mammalian blastocyst hatching. J Reprod Immunol 83:79–84

Song G, Kim J, Bazer FW, Spencer TE (2008) Progesterone and interferon tau regulate hypoxia-inducible factors in the endometrium of the ovine uterus. Endocrinology 149:1926–1934

Spencer TE (2013) Early pregnancy: Concepts, challenges, and potential solutions. Anim Front 3:48–55

Spencer TE, Bartol FF, Bazer FW, Johnson GA, Joyce MM (1999) Identification and characterization of glycosylation-dependent cell adhesion molecule 1-like protein expression in the ovine uterus. Biol Reprod 60:241–250

Spencer TE, Bazer FW (1996) Ovine interferon tau suppresses transcription of the estrogen receptor and oxytocin receptor genes in the ovine endometrium. Endocrinology 137:1144–1147

Spencer TE, Bazer FW (2004) Uterine and placental factors regulating conceptus growth in domestic animals. J Anim Sci 82 E-Suppl:4–13

Spencer TE, Becker WC, George P, Mirando MA, Ogle TF, Bazer FW (1995a) Ovine interferon-τ regulates expression of endometrial receptors for estrogen and oxytocin but not progesterone. Biol Reprod 53:732–745

Spencer TE, Becker WC, George P, Mirando MA, Ogle TF, Bazer FW (1995b) Ovine Interferon-τ Inhibits Estrogen Receptor Up-Regulation and Estrogen-Induced Luteolysis in Cyclic Ewes. Endocrinology 136:4932–4944

Spencer TE, Forde N, Dorniak P, Hansen TR, Romero JJ, Lonergan P (2013) Conceptus-derived prostaglandins regulate gene expression in the endometrium prior to pregnancy recognition in ruminants. Reproduction 146:377–387

Spencer TE, Forde N, Lonergan P (2017) Insights into conceptus elongation and establishment of pregnancy in ruminants. Reprod Fertil Dev 29:84–100

Spencer TE, Johnson GA, Bazer FW, Burghardt RC (2004) Implantation mechanisms: Insights from the sheep. Reproduction 128:657–668

Spiro RG (2002) Protein glycosylation: Nature, distribution, enzymatic formation, and disease implications of glycopeptide bonds. Glycobiology 12:43R-56R

Steinhauser CB, Landers M, Myatt L, Burghardt RC, Vallet JL, Bazer FW, Johnson GA (2016) Fructose synthesis and transport at the uterine-placental

interface of pigs: Cell-specific localization of SLC2A5, SLC2A8, and components of the polyol pathway. Biol Reprod 95:1–14
Teng CC, Tjoa S, Fennessey PV, Wilkening RB, Battaglia FC (2002) Transplacental carbohydrate and sugar alcohol concentrations and their uptakes in ovine pregnancy. Exp Biol Med 227:189–195
Thompson JGE, Simpson AC, Pugh PA, Wright Jnr RW, Tervit HR (1991) Glucose utilization by sheep embryos derived in vivo and in vitro. Reprod Fertil Dev 3:571–576
To KKW, Huang LE (2005) Suppression of hypoxia-inducible factor 1α (HIF-1α) transcriptional activity by the HIF prolyl hydroxylase EGLN1. J Biol Chem 280:38102–38107
Vallet JL, Freking BA (2007) Differences in placental structure during gestation associated with large and small pig fetuses. J Anim Sci 85:3267–3275
Vander Heiden MG, Cantley LC, Thompson CB (2009) Understanding the Warburg Effect: the metabolic requirements of cell proliferation. Science 324:1029–1033
Van der Lende T, Knol EF, Leenhouwers JI (2001) Prenatal development as a predisposing factor for perinatal losses in pigs. Reprod Suppl 58:247–261
Wales RG, Cuneo CL, Waugh EE (1989) Incorporation of glucose by the sheep conceptus between days 13 and 19 of pregnancy. Reprod Fertil Dev 1:137–145
Wales RG, Du ZF (1993) Contribution of the pentose phosphate pathway to glucose utilization by preimplantation sheep embryos. Reprod Fertil Dev 5:329–340
Wang X, Li D, Wu G, Bazer FW (2016) Functional Roles of Fructose: Crosstalk between O-Linked Glycosylation and Phosphorylation of Akt-TSC2-MTOR Cell Signaling Cascade in Ovine Trophectoderm Cells. Biol Reprod 95:1–17
Warburg O, Wind F, Negelein E (1927) The Metabolism of Tumors in the Body. J Gen Physiol 8:519–530
Washburn SE, Caudill MA, Malysheva O, MacFarlane AJ, Behan NA, Harnett B, MacMillan L, Pongnopparat T, Brosnan JT, Brosnan ME (2015) Formate metabolism in fetal and neonatal sheep. Am J Physiol 308:E921–E927
Weinhouse S (1976) Regulation of glucokinase in liver. Curr Top Cell Regul 11:1–50
Weis SM, Cheresh DA (2011) Tumor angiogenesis: Molecular pathways and therapeutic targets. Nat Med 17:1359–1370
White CE, Piper EL, Noland PR (1979) Conversion of glucose to fructose in the fetal pig. J Anim Sci 48:585–590
Wood IS, Trayhurn P (2003) Glucose transporters (GLUT and SGLT): expanded families of sugar transport proteins. Br J Nutr 89:3–9
Wooding FPB, Burton G (2008) Synepitheliochorial placentation: Ruminants (ewe and cow). In: Comparative Placentation: Structures, Functions and Evolution. Springer, Berlin, Germany, pp 133–167
Wu G (2013) Amino Acids: Biochemistry and Nutrition. CRC Press, Boca Raton, FL
Wu G (2018a) Fructose metabolism in animal tissues. Principles of Animal Nutrition. CRC Press, Boca Raton, FL, pp 253–256
Wu G (2018b) Pathway of Glycolysis. Principles of Animal Nutrition. CRC Press, Boca Raton, FL, pp 215–218
Wu G (2022) Nutrition and metabolism: Foundations for animal growth, development, reproduction, and health. Adv Exp MedBiol 1354:1–24
Xiao S, Li R, El Zowalaty AE, Diao H, Zhao F, Choi Y, Ye X (2017) Acidification of uterine epithelium during embryo implantation in mice. Biol Reprod 96:232–243
Zavy MT, Clark WR, Sharp DC, Roberts RM, Bazer FW (1982) Comparison of glucose, fructose, ascorbic acid and glucosephosphate isomerase enzymatic activity in uterine flushings from nonpregnant and pregnant gilts and pony mares. Biol Reprod 27:1147–1158
Zhang Q, Hou YQ, Bazer FW, He WL, Posey EA, Wu G (2021) Amino acids in swine nutrition and production. Adv Exp Med Biol 1285:81–107

Nutritional Regulation of Embryonic Survival, Growth, and Development

4

Lawrence P. Reynolds, Kyle J. McLean, Kacie L. McCarthy, Wellison J. S. Diniz, Ana Clara B. Menezes, J. Chris Forcherio, Ronald R. Scott, Pawel P. Borowicz, Alison K. Ward, Carl R. Dahlen, and Joel S. Caton

Abstract

Maternal nutritional status affects conceptus development and, therefore, embryonic survival, growth, and development. These effects are apparent very early in pregnancy, which is when most embryonic losses occur. Maternal nutritional status has been shown to affect conceptus growth and gene expression throughout the periconceptual period of pregnancy (the period immediately before and after conception). Thus, the periconceptual period may be an important "window" during which the structure and function of the fetus and the placenta are "programmed" by stressors such as maternal malnutrition, which can have long-term consequences for the health and well-being of the offspring, a concept often referred to as Developmental Origins of Health and Disease (DOHaD) or simply developmental programming. In this review, we focus on recent studies, using primarily animal models, to examine the effects of various maternal "stressors," but especially maternal malnutrition and Assisted Reproductive Techniques (ART, including in vitro fertilization, cloning, and embryo transfer), during the periconceptual period of pregnancy on conceptus survival, growth, and development. We also examine the underlying mechanisms that have been uncovered in these recent studies, such as effects on the development of both the placenta and fetal organs. We conclude with our view of future research directions in this critical area of investigation.

Keywords

Maternal nutrition · Maternal stressors · Embryo · Fetus · Placenta · Survival · Growth · Development · Gene expression · Epigenetics · Developmental programming

L. P. Reynolds (✉) · K. J. McLean · K. L. McCarthy · W. J. S. Diniz · A. C. B. Menezes · P. P. Borowicz · A. K. Ward · C. R. Dahlen · J. S. Caton
Department of Animal Sciences, North Dakota State University, Fargo, ND 58108, USA
e-mail: larry.reynolds@ndsu.edu

L. P. Reynolds · W. J. S. Diniz · A. C. B. Menezes · P. P. Borowicz · A. K. Ward · C. R. Dahlen · J. S. Caton
Center for Nutrition and Pregnancy, North Dakota State University, Fargo, ND 58108, USA

K. J. McLean
Department of Animal Science, University of Tennessee, Knoxville, TN, USA

K. L. McCarthy
Department of Animal Sciences, University of Nebraska-Lincoln, Lincoln, NE, USA

J. C. Forcherio · R. R. Scott
Purina Animal Nutrition LLC, Gray Summit, MO 63039, USA

G. Wu (ed.), *Recent Advances in Animal Nutrition and Metabolism*, Advances in Experimental Medicine and Biology 1354,
https://doi.org/10.1007/978-3-030-85686-1_4

Abbreviations

ART	Assisted Reproductive Techniques
DOHaD	Developmental Origins of Health and Disease
ET	Embryo Transfer
IVP	In Vitro Embryo Production
IVF	In Vitro Fertilization
mTOR	Mechanistic Target of Rapamycin
NCDs	Non-communicable Diseases
PI3K	Phosphoinositide 3-kinase
PPAR	Peroxisome Proliferator-Activated Receptor

4.1 Developmental Programming

Developmental programming refers to factors that lead to altered pre- and (or) postnatal development, i.e., permanent changes in structure and (or) function of organs during the fetal period or postnatally, ultimately affecting the long-term health and productivity of the offspring. Factors associated with developmental programming include maternal "stressors" like poor maternal nutritional status or Assisted Reproductive Techniques (ART), but also factors such as metabolic syndrome diseases (obesity, diabetes, and cardiovascular disease), maternal age (both young and aged pregnancies; Borowicz 2008), multiple fetuses, preterm birth, maternal and fetal genetic background, environmental (e.g., heat stress, high altitude stress, and environmental contaminants) and other types of stress (e.g., relational stress), and, in humans, lifestyle factors (e.g., smoking, alcohol consumption, and drug use) and socioeconomic status (Table 4.1).

The consequences of developmental programming include not only preterm delivery, low birth weight, and poor survival of newborns, but also a two to tenfold increase in the risk of developing a host of "Non-communicable Diseases" (NCDs) in the offspring during infancy and as adults. Such NCDs include not only metabolic syndrome, growth abnormalities (including altered body composition), and cancer but also dysfunction of numerous organ systems including adipose, brain-neural (including behavioral and cognitive dysfunction), cardiovascular, endocrine, excretory, gastro-intestinal, immune, musculoskeletal, and reproductive systems (Reynolds and Caton 2012; Reynolds and Vonnahme 2016; Reynolds et al. 2019).

The initial observations that led to the concept of developmental programming (or Developmental Origins of Health and Disease, DOHaD) were based on epidemiological studies in humans. These studies suggested not only that an adverse intrauterine environment may lead to a greater incidence of NCDs in the offspring as adults, but also that poor maternal nutrition was "… an obvious suspect (Barker 1992, 2004)." The initial observations noted further that developmental programming was associated with "Discordance between placental and birth weights …." Key to our understanding of developmental programming was a period at the end of World War II known as the Dutch Hunger Winter, during which pregnant women in the Netherlands were limited to a dietary intake of 400–800 cal per day (Schulz 2010). An important observation from the Dutch Hunger Winter was that the consequences of low maternal caloric intake depended on the stage of pregnancy. For example, offspring of women receiving low caloric intake during early pregnancy experienced dyslipidemia and an increased incidence of obesity, cardiovascular disease, and age-associated decline in cognitive function as adults, despite the fact their birth weights were normal (Schulz 2010; de Rooij et al. 2010). In contrast, offspring of mothers exposed to low caloric intake during mid-gestation exhibited low birth weight and reduced renal function (Schulz 2010).

In the 30 or so years since David Barker and colleagues first articulated the concept of developmental programming (Barker 1992, 2004), many studies of animals, including livestock, have shown that developmental programming occurs in other mammals and affects fetal and

Table 4.1 Risk factors that may lead to developmental programming[c]

Lifestyle Choices[a]
Smoking
Alcohol consumption
Sedentary lifestyle
Maternal Factors[b]
Malnutrition
Assisted reproductive techniques
Metabolic syndrome
Age (young or aged)
Multiple fetuses
Preterm birth
Ethnicity/Breed (genetic background)
Stress (e.g., relational stress)
Social or economic status
Poor health or health care access
Marital Status[a]
Environmental Exposures[b]
Herbicides, pesticides, fungicides
Others (e.g., temperature and humidity, high altitude, human waste fertilizer, smoke, phytosteroids, drugs, etc.)

[a]Risk factors in humans only
[b]Risk factors in both humans and animals including livestock
[c]Fowden and Forhead (2009), Reynolds et al. (2010b), Bellingham et al. (2010), Hellemans et al. (2010), Dupont et al. (2012), Reynolds and Caton (2012), Vonnahme (2012), Vonnahme et al. (2013), Aizer and Currie (2014), Been et al. (2014), Galbally et al. (2014), Juul et al. (2014), Skinner (2014), Oostingh et al. (2019), Reynolds et al. (2019)
Modified and updated from Reynolds and Vonnahme (2017)

placental growth and development as well as the long-term health and productivity of the offspring (Wu et al. 2006; Reynolds and Caton 2012; Reynolds and Vonnahme 2016, 2017; Reynolds et al. 2017, 2019; Caton et al. 2019, 2020). These studies also have shown that developmental programming occurs most often during critical periods, or "windows" of development, including the periconceptual period (the period immediately before and after conception), pregnancy, and infancy (Fig. 4.1), although developmental programming can probably occur during childhood and even adulthood (Wu et al. 2006; Sinclair and Singh 2007; Reynolds and Caton 2012; Vonnahme 2012; Reynolds and Vonnahme 2016, 2017; Reynolds et al. 2017, 2019; Caton et al. 2019, 2020). In addition, confirming the suggestion by Barker and colleagues (Barker 1992, 2004), maternal malnutrition does indeed seem to be a major player in developmental programming (Wallace et al. 2006; Oliver et al. 2007; Sinclair and Singh 2007; Reynolds and Caton 2012; Steegers-Theunissen et al. 2013; Diskin and Kenny 2014; Reynolds and Vonnahme 2016, 2017; Reynolds et al. 2010b, 2017, 2019; Gu et al. 2015; Bairagi et al. 2016; Caton et al. 2019, 2020; McCarthy 2019; Gauvin et al. 2020; Diniz et al. 2021a).

4.2 Importance of the Periconceptual Period

The periconceptual period encompasses numerous critical reproductive events, including follicular development to the ovulatory stage,

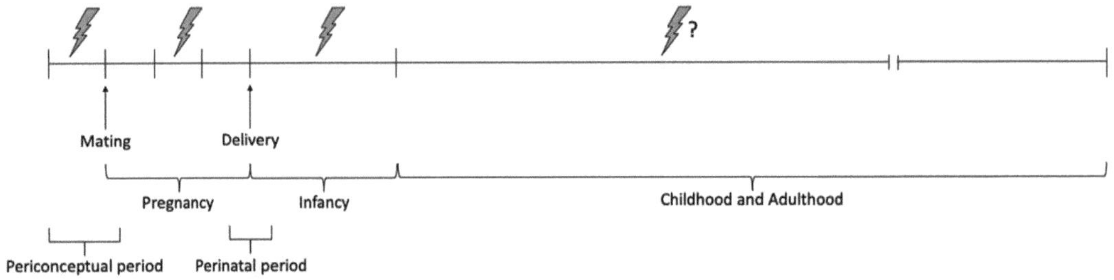

Fig. 4.1 Timeline of developmental programming. Insults (e.g., maternal malnutrition, etc.—see text) resulting in developmental programming ( ) occur primarily during the periconceptual period, pregnancy, and infancy, although they can probably occur during childhood and adulthood

ovulation, fertilization and early embryonic development (which occur in the oviduct), implantation, placentation, and embryonic/fetal organ development. Because of the many critical events occurring during early pregnancy (Fig. 4.2), it is also the period during which most conceptus (embryo/fetus plus placenta) loss (also referred to as spontaneous abortion or miscarriage) occurs (Reynolds et al. 2014; Bazer et al. 2015; Bairagi et al. 2016; Caton et al. 2020).

Human studies have shown that lifestyle factors (e.g., smoking, alcohol consumption) as well as maternal nutrition during the periconceptual period adversely affect not only fertility and pregnancy establishment but also are major contributors to developmental programming of embryo/fetal and placental growth and function, leading to altered birth weight and poor offspring health (Steegers-Theunissen et al. 2013; Oostingh et al. 2019). These observations in humans have been corroborated by studies of animals, including livestock (Sinclair et al. 2007; Sinclair and Singh 2007; Reynolds et al. 2014, 2019; Bazer et al. 2015; Bairagi et al. 2016; Caton et al. 2020).

Programming effects of maternal stressors can be observed as early as the oocyte stage, which may lead to the programming of the embryo/fetus and placenta throughout gestation (Grazul-Bilska et al. 2012; Kumarasamy et al. 2005). Once programmed, changes during early life are likely germline transmitted to subsequent generations, affecting the development of the offspring (Gu et al. 2015). For example, our group has shown that under- or over-feeding of ewes for 8 weeks before oocyte collection dramatically reduces the rate of In Vitro Fertilization (IVF) as well as the proportion of zygotes developing to blastocyst stage during in vitro development (also termed in vitro embryo production, IVP; Grazul-Bilska et al. 2012). Similarly, the studies of Jane Harding and colleagues have shown that underfeeding of ewes for 8 weeks before mating until 4 weeks after mating leads to premature delivery of dysmature offspring, most of whom die within the first week after birth (Kumarasamy et al. 2005).

Bazer et al. (2015) suggested the importance of amino acids in the maternal diet, especially "functional" amino acids such as arginine and ornithine, which are involved in the synthesis of both polyamines and nitric oxide, and serine and methionine, which are involved in 1-carbon metabolism. This suggestion agrees with observations in humans, in which levels of 1-carbon metabolites in maternal serum, including folate and vitamin B_{12}, are associated with gravid uterine blood flow and birth weights (Jonker et al. 2020; Lyon et al. 2020; Jankovic-Karasoulos et al. 2021). This suggestion also agrees with the studies of Sinclair and colleagues (2007), who showed that imposing a "methyl-deficient diet" (deficient in folate, vitamin B_{12}, and methionine) in embryo donor ewes during the periconceptual period led to the altered DNA methylation status of the fetal liver at day 90 of gestation, and subsequently altered body composition, altered immune function, insulin resistance, and hypertension in the offspring as adults.

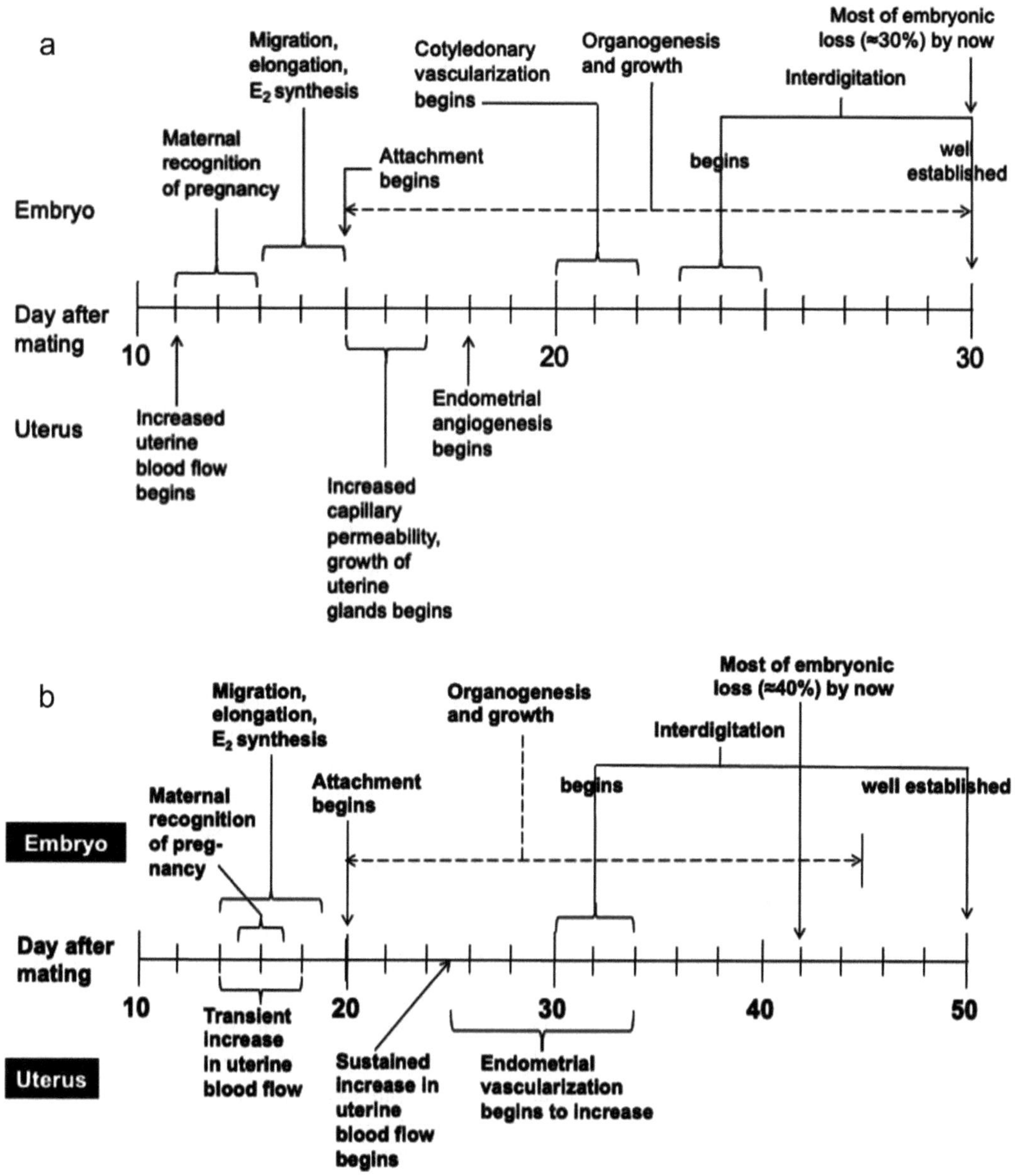

Fig. 4.2 Timelines of placental and embryonic/fetal development during early pregnancy in **a** sheep and **b** cows. E2 = estradiol-17beta, and "interdigitation" refers to interdigitation of the fetal cotyledonary and maternal caruncular tissues of the placentomes (the highly vascular sites that exhibit the most intimate contact between the fetal and maternal placental tissues). Taken from Reynolds et al. (2014) and Caton et al. (2020)

One of the best models for understanding the factors during early pregnancy that influence embryonic/fetal development and survival is the use of ART, including IVF, cloning, and Embryo Transfer (ET). This is because pregnancies established using ART result in high proportions of embryonic/fetal loss, especially early in pregnancy, as well as abnormal embryonic/fetal growth and development, which contributes to the high rates of postnatal death (Young and Fairburn 2000; Loi et al. 2006; Farin et al. 2006; Reynolds et al. 2014). Using ART (IVF and cloned embryos, as well as simple transfer of IVP embryos), our group and others showed increased global methylation of fetal (cotyledonary) and maternal (caruncular) placental DNA, altered trophoblast gene expression, decreased placental angiogenic factor expression and vascularization, and decreased fetal and placental growth in sheep during the first 3–4 weeks of pregnancy (Arnold et al. 2006; Grazul-Bilska et al. 2013, 2014; Fidanza et al. 2014; Reynolds et al. 2014). These observations agree with research in other species including humans (Beaujean et al. 2004; Loi et al. 2006; Farin et al. 2006; Palmieri et al. 2007, 2008; Canovas et al. 2017). Related to these observations, including oviductal fluid in the culture media improves the success of IVP and subsequent pregnancy rates and also reduces the effects of IVP on embryonic/fetal gene expression and DNA methylation status (Coy and Yanagimachi 2015; Canovas et al. 2017).

4.3 Underlying Mechanisms

4.3.1 Programming of the Placenta

As the organ of exchange between the maternal system and the gravid uterus, placental development and function are critical to successful pregnancy establishment and subsequent fetal growth and development (Burton and Fowden 2015; Woods et al. 2018). A particularly important aspect of placental development and function is vascularization, which is critical to the large increase in placental blood flow and transplacental exchange that occurs during pregnancy (Reynolds and Caton 1992; Reynolds and Redmer 1995, 2001; Reynolds et al. 2010a; Borowicz et al. 2007). Importantly, numerous investigators have shown that placental vascular development and thus blood flow are sensitive to many of the same stressors that affect embryonic survival and fetal growth and development, including maternal nutrient intake and ART (Table 4.2; Mayhew et al. 2003, 2004; Miles et al. 2004, 2005; Redmer et al. 2004; Cross and Mickelson 2006; Reynolds et al. 2006, 2010a, b, 2013, 2014; Wallace et al. 2006; Coan et al. 2010; Vonnahme 2012; Bairagi et al. 2016).

As mentioned in the previous paragraph, both maternal nutritional status (under- and overfeeding, altered dietary protein, or specific nutrients [e.g., vitamins and minerals]) as well as ART have been shown to affect placental angiogenesis late in pregnancy (Miles et al. 2004, 2005; Redmer et al. 2004; Cross and Mickelson 2006; Reynolds et al. 2006, 2010a, b, 2013, 2014; Wallace et al. 2006; Coan et al. 2010; Bairagi et al. 2016; McLean et al. 2017). One of the first indications that these effects may be "programmed" very early in pregnancy was from the study of Redmer et al. (2005), who showed that placental expression of angiogenic factors was reduced at mid-gestation before any effects on fetal or placental size were detected, but by late gestation, placental size and vascularity and fetal size were dramatically reduced. This suggestion has been confirmed by several subsequent studies (Grazul-Bilska et al. 2013, 2014; Reynolds et al. 2013, 2014; Bairagi et al. 2016; Quinn et al. 2016; Crouse et al. 2017, 2020; Greseth et al. 2017; Johnson et al. 2017; McLean et al. 2017, 2018; Diniz et al. 2021b), which have shown that both maternal nutritional status and ART affect placental growth and development, including vascular development, as well as placental function during the first third of pregnancy. More recently, we have utilized a model of moderate maternal nutrient restriction in cows during the first 50 days of pregnancy (approximately 0.2 of gestation) and showed effects on concentrations of hexoses (glucose and fructose) and amino acids in maternal serum, histotroph, and fetal fluids (allantoic and amniotic) (Crouse et al. 2019a).

A recent report (Diniz et al. 2021b) found that supplementing vitamins and minerals pre- and post-breeding to diets of beef heifers growing at a low or moderate rate of maternal weight gain led to placental adaptations in gene expression by the end of the first trimester of pregnancy. The

Table 4.2 Reduced placental (uterine and umbilical) vascular development and blood flows in several models of compromised pregnancy in sheep[a]

Model	Fetal weight (%)	Placental weight (%)	Placental vascular development	Maternal placental blood flow[b] (%)	Fetal placental blood flow[b]
Overfed adolescent	−20–40	−20–45	−31	−36	−37%
Underfed adolescent	−17	NSE	−20	ND	ND
Underfed adult	−12	ND	−14	−25	NSE
Heat-stressed adult	−42	−51	ND	−26	−60%
Multiple pregnancy[c]	−30	−37	−30	−23	ND
Adolescent versus adult[c]	−16	−26	−24	ND	ND
Maternal breed[c]	−44	−28	−33	ND	ND

[a]Percent reduction compared with non-compromised pregnancy, i.e., pregnancy with normal fetal and placental size. All observations are from late pregnancy (day 130–135, approximately 0.9 of gestation)
[b]Maternal placental blood flow = blood flow to the pregnant uterus; fetal placental blood flow = umbilical blood flow
[c]Multiple pregnancy compared singletons with triplets; Adolescent versus Adult were first pregnancies in the first season postnatally versus the second season postnatally; maternal breed compared Columbia with Romanov ewes
Table adapted from Reynolds et al. (2006)

authors identified 267 unique differentially expressed genes acting in biological processes and pathways, that include nutrient transporters, ion homeostasis, insulin secretion, Peroxisome Proliferator-Activated Receptor (PPAR) transcription factors, and amino acid biosynthesis. Likewise, Che et al. (2017) reported a differential abundance of placental proteins due to maternal diet. These authors suggested that differences in piglet birth weight were modulated by the over-represented processes of placental nutrient transport and lipid and energy metabolism. Together, these studies indicate that specific biological processes and pathways of the placenta are affected in response to maternal nutritional status.

In fact, as mentioned previously, maternal nutritional status even during the pre-conception period affects the rates of IVF as well as the rate of embryonic growth to the blastocyst stage during IVP (Grazul-Bilska et al. 2012), suggesting that even pre-fertilization maternal stressors may underlie programming of embryonic/fetal and placental growth and development (Caton et al. 2020). Similarly, as also mentioned previously, underfeeding of ewes for 8 weeks before mating until 4 weeks after mating leads to premature delivery of dysmature offspring, most of whom die within the first week after birth (Bloomfield et al. 2003; Kumarasamy et al. 2005; Oliver et al. 2007).

Thus, the periconceptual period may be an important "window" during which the structure and function of the placenta are programmed by stressors such as maternal malnutrition, having a powerful and potentially long-lasting effect on conceptus development.

4.3.2 Programming of the Fetus and Fetal Organ Systems

The periconceptual period is important in terms of not only conceptus growth and development but also programming of fetal organ development. It is now generally accepted that the genetic information of the developing embryo and embryonic stem cells once reprogramed is extended to all subsequent tissues generated that may lead to transcriptomic and phenotypic changes (Nilsson et al. 2019; Vickers 2014). We previously noted the very early effects of IVP and ART on conceptus development, including those of the embryo/fetus and placenta.

Utilizing our model of moderate maternal nutrient restriction in cows during the first 50 days of pregnancy, we discovered dramatic changes in gene expression of the fetal liver, muscle, and cerebrum, with more than three-fourths of the 291 differentially expressed genes upregulated in the nutrient-restricted group (Crouse et al. 2019b). In the same study, we also identified that although nutrient restriction led to differential tissue regulation, over-represented nutrient sensing pathways, such as the Mechanistic Target of Rapamycin (mTOR) and phosphoinositide 3-kinase (PI3K)/protein kinase B (Akt), were common among the fetal tissues, suggesting a coordinated adaptive response mediated by differential gene expression. Furthermore, maternal nutrient restriction led to a gain in connectivity of differentially expressed genes driven primarily by transcription factors in a tissue-specific fashion (Diniz et al. 2021a). In agreement with these observations, effects on fetal muscle development and downregulation of myogenic genes were reported in sheep in response to poor maternal nutrition (Gauvin et al. 2020).

Like the muscle, the liver is also responsive to maternal nutrition. Hyatt et al. (2007) reported that male sheep born to nutrient-restricted dams had smaller livers when they become adults likely due to changes in the expression of hepatic genes. The findings reported by Palombo et al. (2021) highlighted the effects of methionine supplementation during late pregnancy on hepatic function in the offspring. These authors found changes not only in hepatic gene expression but also in DNA methylation. Further, these authors suggested that transcription factors play a role in mediating the effects of maternal diet on energy metabolism and immune system processes.

Protein supplementation or restriction before or after conception also led to differential fetal development in cattle. Copping et al. (2014) reported that although there was no detectable dietary effect on birth weight at term, maternal protein intake during the periconceptual period and first trimester of pregnancy affected fetal development at days 36, 60, and 98 of gestation in a sex-specific manner. These authors also reported changes in fetal hepatic gene expression including those involved in regulating growth, glucose output, and lipid metabolism at day 98 of pregnancy (Copping et al. 2020).

Bioactive amino acids, such as L-arginine, also may improve pregnancy outcomes (Wu 2022). For example, intra-jugular injection of L-arginine from day 1 to day 15 of pregnancy in ewes reduced ovarian resistance index, increased systemic progesterone concentrations, and reduced embryo loss (Luther et al. 2008). In addition, daily dietary supplementation with 70 g of rumen-protected L-arginine to beef cows between days 1 and 60 after artificial insemination enhanced the birth rate of live-born calves from 22% in the control group (without arginine supplementation) to 36% (Gilbreath et al. 2021).

Preliminary data from our group demonstrated that maternal vitamin and mineral supplementation and two different rates of gain during the first 83 days of gestation modified the concentrations of key metabolic fuels available to the conceptus within the uterine environment (Menezes et al. 2021a) and affected the weight of fetal liver and intestines (McCarthy et al. 2020); trace mineral concentrations in fetal liver, muscle, and allantoic fluid (Menezes et al. 2021b); fetal liver concentrations of vitamins A and E (Crouse et al. 2021b); and maternal hormone and metabolic status (McCarthy et al. 2020). Additionally, Diniz et al. (2021a) found that maternal vitamin and mineral supplementation increased the expression of genes related to mineral homeostasis, lipid transport, and metabolism in the fetal liver. Jacometo et al. (2015) reported that maternal mineral supplementation 30 days prepartum led to changes in neonatal innate immune response in calves, at least in part via changes in gene and miRNA expression. As the maternal plane of nutrition is one of the main factors influencing the availability of trace minerals for the fetus, an inadequate supply of these critical nutrients may have a long-lasting impact on offspring growth and health (Hostetler et al. 2003; Van Emon et al. 2020).

Thus, maternal nutrition during the periconceptual period is important not only for placental development but also for fetal development, potentially contributing to the programming of offspring growth and metabolism, as well as health and, ultimately, productivity.

4.3.3 Common Mechanisms?

An insightful observation is that programming of fetal and placental growth and development may involve common mechanisms because of the "... close association between placental defects and abnormal cardiovascular and brain development ..." (Woods et al. 2018). This suggestion was based on the use of CRISPR-Cas9 technology to alter placental gene expression in pregnant mice. These authors also pointed out that the large proportion (approx. 70%) of mutant mice that exhibit poor pregnancy outcomes also show a placental phenotype; that is, of the approx. 5000 genes associated with embryonic lethality in mice, around two-thirds are associated with placental defects. These observations agree with the recent studies we discussed previously which show that maternal nutritional modulation results in altered expression of many genes and perhaps more importantly gene networks in both the placenta and fetal organs (Jacometo et al. 2015; Crouse et al. 2019b; Gauvin et al. 2020; Diniz et al. 2021a, b). Whether such altered gene expression can be regulated to improve reproductive development and function is an open question that needs to be addressed.

4.4 Summary and Future Directions

In this review, we focused on results from recent studies using primarily animal models that have examined the effects of various maternal stressors that affect conceptus survival, growth, and development during the periconceptual period, especially maternal malnutrition and ART. We also discussed some of the underlying mechanisms that have been uncovered in these recent studies, such as effects on the development of both placental and fetal organs, including effects on gene expression and gene networks. In addition, we briefly discussed some of the management and therapeutic strategies designed to overcome the negative consequences, or conversely, to promote positive effects on developmental programming, such as energy or vitamin-mineral supplementation of the mother, supplementation with specific nutrients or hormones, and use of oviductal or other reproductive fluids during IVP of embryos. While these approaches have shown promising results, they have not always been consistent and will likely require much more investigation including studies in more "real-world" settings with much large numbers of subjects.

Other areas we believe should receive much more attention in the future are strategies designed to take advantage of the positive effects of developmental programming. This idea is based on the supposition that developmental programming must ultimately be adaptive (Bateson et al. 2014; Nettle et al. 2013; Mueller et al. 2015). For example, our studies of maternal nutritional status during the first 50 days of pregnancy in cattle showed that the vast majority of genes in fetal liver, muscle, and brain were upregulated in fetuses from nutrient-restricted dams. Based on these observations, we suggested that the upregulation of genes may represent an adaptive response to the maternal nutrient restriction (Crouse et al. 2019a). This suggestion is consistent with the observation that cattle and other grazers (deer, goats, sheep, etc.) in an "extensive," pasture-based system are usually severely nutrient restricted during early pregnancy and are thus well adapted not only to yearly cycles of low availability and quality of forages but also to short-term fluctuations in body weight and body condition (Krysl et al. 1987; Johnson et al. 1998; Cline et al. 2009, 2010).

This idea of adaptive versus maladaptive responses to maternal malnutrition also has relevance to humans, as low caloric intake and malnutrition are often experienced by pregnant women, with severe consequences for pregnancy outcomes including very low birth weight, high rates of postnatal morbidity and mortality, and long-term consequences for health and productivity of the offspring, including stunting and wasting during infancy and early childhood (Latham 1997; Bhutta and Salam 2012; WHO 2016; UNICEF 2019; UNSCN 2021). Thus, we need a better understanding of whether changes in expression of fetal and placental genes are adaptive, and therefore positive, or maladaptive, and therefore detrimental in both the short and long term.

Acknowledgements We would like to acknowledge the support for our studies described in this paper, including the Agriculture and Food Research Initiative of the National Institute of Food and Agriculture, U.S. Department of Agriculture; the Eunice Kennedy Shriver National Institute of Child Health and Human Development and the National Heart, Lung and Blood Institute, both of the U.S. National Institutes of Health, U.S. Department of Health and Human Services; Purina Animal Nutrition LLC; the North Dakota State Board of Agricultural Research and Education; and the North Dakota Agricultural Experiment Station.

References

Aizer A, Currie J (2014) The intergenerational transmission of inequality: maternal disadvantage and health at birth. Science 344:856–861

Arnold DR, Bordignon V, Rj L, Murphy BD, Smith LC (2006) Somatic cell nuclear transfer alters peri-implantation trophoblast differentiation in bovine embryos. Reproduction 132:279–290

Bairagi S, Quinn KE, Crane AR, Ashley RL, Borowicz PP, Caton JS, Redden RR, Grazul-Bilska AT, Reynolds LP (2016) Maternal environment and placental vascularization in small ruminants. Theriogenology 86:288–305

Barker DJP (1992) Fetal and infant origins of adult disease. BMJ Publishing Group, London, England

Barker DJ (2004) The developmental origins of well-being. Philos Trans R Soc Lond B Biol Sci 359:1359–1366

Bateson P, Gluckman P, Hanson M (2014) The biology of developmental plasticity and the predictive adaptive response hypothesis. J Physiol 592:2357–2368

Bazer F, Johnson G, Wu G (2015) Amino acids and conceptus development during the peri-implantation period of pregnancy. Adv Exp Med Biol 843:23–52

Beaujean N, Taylor J, Gardner J, Wilmut I, Meehan R, Young L (2004) Effect of limited DNA methylation reprogramming in the normal sheep embryo on somatic cell nuclear transfer. Biol Reprod 71:185–193

Been JV, Nurmatov UB, Cox B, Nawrot TS, van Schayck CP, Sheikh A (2014) Effect of smoke-free legislation on perinatal and child health: a systematic review and meta-analysis. Lancet 383:1549–1560

Bellingham M, Fowler PA, Amezaga MR, Whitelaw CM, Rhind SM, Cotinot C, Mandon-Pepin B, Sharpe RM, Evans NP (2010) Foetal hypothalamic and pituitary expression of gonadotrophin-releasing hormone and galanin systems is disturbed by exposure to sewage sludge chemicals via maternal ingestion. J Neuroendocrinol 22:527–533

Bhutta ZA, Salam RA (2012) Global nutrition epidemiology and trends. Ann Nutr Metab 61(Suppl 1):19–27

Bloomfield FH, Oliver MH, Hawkins P, Campbell M, Phillips DJ, Gluckman PD, Challis JRG, Harding JE (2003) A periconceptional nutritional origin for noninfectious preterm birth. Science 300:606

Borowicz PP, Arnold DR, Johnson ML, Grazul-Bilska AT, Redmer DA, Reynolds LP (2007) Placental growth throughout the last two thirds of pregnancy in sheep: Vascular development and angiogenic factor expression. Biol Reprod 76:259–267

Borowicz PP (2008) Maternal factors affecting placental vascular development in sheep. Dissertation, North Dakota State University, Fargo, ND, USA

Burton GJ, Fowden AL (2015) The placenta: a multifaceted, transient organ. Philos Trans R Soc Lond B Biol Sci 370:20140066

Canovas S, Ivanova E, Romar R, García-Martínez S, Soriano-Úbeda C, García-Vázquez FA, Saadeh H, Andrews S, Kelsey G, Coy P (2017) DNA methylation and gene expression changes derived from assisted reproductive technologies can be decreased by reproductive fluids. eLife 6:e23670

Caton JS, Crouse MS, Reynolds LP, Neville TL, Dahlen CR, Ward AK, Swanson KC (2019) Board invited review: maternal nutrition and programming of offspring energy requirements. Transl Anim Sci 3:976–990

Caton JS, Crouse MS, McLean KJ, Dahlen CR, Ward AK, Cushman RA, Grazul-Bilska AT, Neville BW, Borowicz PP, Reynolds LP (2020) Maternal periconceptual nutrition, early pregnancy, and developmental outcomes in beef cattle. J Anim Sci 98:skaa358

Che L, Yang Z, Xu M, Xu S, Che L, Lin Y, Fang Z, Feng B, Li J, Chen D, Wu D (2017) Maternal nutrition modulates fetal development by inducing placental efficiency changes in gilts. BMC Genomics 18:213

Cline HJ, Neville BW, Lardy GP, Caton JS (2009) Influence of advancing season on dietary composition, intake, site of digestion, and microbial efficiency in

beef steers grazing a native range in western North Dakota. J Anim Sci 87:375–383
Cline HJ, Neville BW, Lardy GP, Caton JS (2010) Influence of advancing season on dietary composition, intake, site of digestion, and microbial efficiency in beef steers grazing season-long or twice-over rotation native range pastures in western North Dakota. J Anim Sci 88:2812–2824
Coan PM, Vaughan OR, Sekita Y, Finn SL, Burton GJ, Constancia M, Fowden AL (2010) Adaptations in placental phenotype support fetal growth during undernutrition of pregnant mice. J Physiol 588:527–538
Copping KJ, Hoare A, Callaghan M, McMillen IC, Rodgers RJ, Perry VEA (2014) Fetal programming in 2-year-old calving heifers: peri-conception and first trimester protein restriction alters fetal growth in a gender-specific manner. Anim Prod Sci 54:1333–1337
Copping KJ, Hernandez-Medrano J, Hoare A, Hummitzsch K, McMillen IC, Morrison JL, Rodgers RJ, Perry VEA (2020) Maternal periconceptional and first trimester protein restriction in beef heifers: Effects on placental parameters and fetal and neonatal calf development. Reprod Fertil Dev 32:495–507
Coy P, Yanagimachi R (2015) The common and species-specific roles of oviductal proteins in mammalian fertilization and embryo development. Bioscience 65:973–984
Cross JC, Mickelson L (2006) Nutritional influences on implantation and placental development. Nutr Rev 64 (suppl 2):S12–S18
Crouse MS, McLean KJ, Greseth NP, Crosswhite MR, Pereira NN, Ward AK, Reynolds LP, Dahlen CR, Neville BW, Borowicz PP, Caton JS (2017) Maternal nutrition and stage of early pregnancy in beef heifers: impacts on expression of glucose, fructose, and cationic amino acid transporters in utero-placental tissues. J Anim Sci 95:5563–5572
Crouse MS, Caton JS, Cushman RA, McLean KJ, Dahlen CR, Borowicz PP, Reynolds LP, Ward AK (2019a) Moderate nutrient restriction of beef heifers alters expression of genes associated with tissue metabolism, accretion, and function in fetal liver, muscle, and cerebrum by day 50 of gestation. Transl Anim Sci 3:855–866
Crouse MS, Greseth NP, McLean KJ, Crosswhite MR, Pereira NN, Ward AK, Reynolds LP, Dahlen CR, Neville BW, Borowicz PP, Caton JS (2019b) Maternal nutrition and stage of early pregnancy in beef heifers: impacts on hexose and AA concentrations in maternal and fetal fluids. J Anim Sci 97:1296–1316
Crouse MS, McLean KJ, Greseth NP, Ward AK, Reynolds LP, Dahlen CR, Neville BW, Borowicz PP, Caton JS (2020) The effects of maternal nutrient restriction and day of early pregnancy on the location and abundance of neutral amino acid transporters in beef heifer utero-placental tissues. J Anim Sci 98: skaa197
Crouse MS, McLean KJ, Dwamena J, Neville TL, Menezes ACB, Ward AK, Reynolds LP, Dahlen CR, Neville BW, Borowicz PP, Caton JS (2021a) The effects of maternal nutrition during the first 50 days of gestation on the location and abundance of hexose and cationic amino acid transporters in beef heifer utero-placental tissues. J Anim Sci 99:skaa386
Crouse MS, McCarthy KL, Menezes ACB, Kasseta CJ, Baumgaertner F, Kirsch JD, Dorsam S, Neville TL, Ward AK, Borowicz PP, Reynolds LP, Sedivec KK, Forcherio JC, Scott R, Caton JS, Dahlen CR (2021b). Vitamin and mineral supplementation and rate of gain during the first trimester of gestation affect the abundance of fat-soluble vitamins in fetal liver at d 83 of gestation. J Anim Sci Abstr (in press)
de Rooij SR, Wouters H, Yonker JE, Painter RC, Roseboom TJ (2010) Prenatal undernutrition and cognitive function in late adulthood. Proc Natl Acad Sci USA 107:16881–16886
Diniz WJS, Crouse MS, Cushman RA, McLean KJ, Caton JS, Dahlen CR, Reynolds LP, Ward AK (2021a) Cerebrum, liver, and muscle regulatory networks uncover maternal nutrition effects in developmental programming of beef cattle during early pregnancy. Sci Rep 11:2771
Diniz WJS, Reynolds LP, Borowicz PP, Ward AK, Sedivec KK, McCarthy KL, Kassetas CJ, Baumgaertner F, Kirsch JD, Dorsam ST, Neville TL, Forcherio JC, Scott RR, Caton JS, Dahlen CR (2021b) Maternal vitamin and mineral supplementation and rate of maternal weight gain affects placental expression of energy metabolism and transport-related genes. Genes 12:385
Diskin MK, Kenny DA (2014) Optimising reproductive performance of beef cows and replacement heifers. Animal 8(s1):27–39
Dupont C, Cordier AG, Junien C, Mandon-Pépin B, Levy R, Chavatte-Palmer P (2012) Maternal environment and the reproductive function of the offspring. Theriogenology 78:1405–1414
Farin PW, Piedrahita JA, Farin CE (2006) Errors in development of fetuses and placentas from in vitro-produced bovine embryos. Theriogenology 65:178–191
Fidanza A, Toschi P, Zacchini F, Czernik M, Palmieri C, Scapolo P, Modlinski JA, Loi P, Ptak GE (2014) Impaired placental vasculogenesis compromises the growth of sheep embryos developed in vitro. Biol Reprod 91:21
Fowden AL, Forhead AJ (2009) Hormones as epigenetic signals in developmental programming. Exp Physiol 94:607–625
Galbally M, Snellen M, Power J (2014) Antipsychotic drugs in pregnancy: a review of their maternal and fetal effects. Ther Adv Drug Saf 5:100–109
Gauvin MC, Pillai SM, Reed SA et al (2020) Poor maternal nutrition during gestation in sheep alters prenatal muscle growth and development in offspring. J Anim Sci 98:1–15
Gilbreath KR, Bazer FW, Satterfield MC, Wu G (2021) Amino acid nutrition and reproductive performance in ruminants. Adv Exp Med Biol 1285:43–61

Grazul-Bilska AT, Borowczyk E, Bilski JJ, Reynolds LP, Redmer DA, Caton JS, Vonnahme KA (2012) Overfeeding and underfeeding have detrimental effects on oocyte quality measured by in vitro fertilization and early embryonic development in sheep. Domest Anim Endocrinol 43:289–298

Grazul-Bilska AT, Johnson ML, Borowicz PP, Baranko L, Redmer DA, Reynolds LP (2013) Placental development during early pregnancy in sheep: effects of embryo origin on fetal and placental growth and global methylation. Theriogenology 79:94–102

Grazul-Bilska AT, Johnson ML, Borowicz PP, Bilski JJ, Cymbaluk T, Norberg S, Redmer DA, Reynolds LP (2014) Placental development during early pregnancy in sheep: effects of embryo origin on vascularization. Reproduction 147:639–648

Greseth NP, Crouse MS, McLean KJ, Crosswhite MR, Negrin Pereira N, Dahlen CR, Borowicz PP, Reynolds LP, Ward AK, Neville BW, Caton JS (2017) The effects of maternal nutrition on the messenger ribonucleic acid expression of neutral and acidic amino acid transporters in bovine uteroplacental tissues from day sixteen to fifty of gestation. J Anim Sci 95:4668–4676

Gu L, Liu H, Gu X, Boots C, Moley KH, Wang Q (2015) Metabolic control of oocyte development: linking maternal nutrition and reproductive outcomes. Cell Mol Life Sci 72:251–271

Hellemans KGC, Sliwowska JH, Verma P, Weinberg J (2010) Prenatal alcohol exposure: fetal programming and later life vulnerability to stress, depression and anxiety disorders. Neurosci Biobehav Rev 34:791–807

Hostetler CE, Kincaid RL, Mirando MA (2003) The role of essential trace elements in embryonic and fetal development in livestock. Vet J 166:125–139

Hyatt MA, Gopalakrishnan GS, Bispham J, Gentili S, McMillen IC, Rhind SM, Rae MT, Kyle CE, Brooks AN, Jones C, Budge H, Walker D, Stephenson T, Symonds ME (2007) Maternal nutrient restriction in early pregnancy programs hepatic mRNA expression of growth-related genes and liver size in adult male sheep. J Endocrinol 192:87–97

Jacometo CB, Osorio JS, Socha M, Corrêa MN, Piccioli-Cappelli F, Trevisi E, Loor JJ (2015) Maternal consumption of organic trace minerals alters calf systemic and neutrophil mRNA and microRNA indicators of inflammation and oxidative stress. J Dairy Sci 98:7717–7729

Jankovic-Karasoulos T, Furness DL, Leemaqz SY, Dekker GA, Grzeskowiak LE, Grieger JA, Andraweera PH, McCullough D, McAninch D, McCowan LM, Bianco-Miotto T, Roberts CT (2021) Maternal folate, one-carbon metabolism and pregnancy outcomes. Matern Child Nutr 17:e13064

Johnson JA, Caton JS, Poland W, Kirby DR, Dhuyvetter DV (1998) Influence of season on dietary composition, intake, and digestion by beef steers grazing mixed-grass prairie in the Northern Great Plains. J Anim Sci 76:1682–1690

Johnson ML, Redmer DA, Reynolds LP, Grazul-Bilska AT (2017) Gap junctional connexin messenger RNA expression in the ovine uterus and placenta: effects of estradiol-17β-treatment, early pregnancy stages, and embryo origin. Domest Anim Endocrinol 58:104–112

Jonker H, Capelle N, Lanes A, Wen SW, Walker M, Corsi DJ (2020) Maternal folic acid supplementation and infant birthweight in low- and middle-income countries: a systematic review. Matern Child Nutr 16: e12895

Juul A, Almstrup K, Andersson A-M, Jensen TK, Jørgensen N, Main KM, Rajpert-De Meyts E, Toppari J, Skakkebæk NE (2014) Possible fetal determinants of male infertility. Nat Rev Endocrinol 10:553–562

Krysl LJ, Galyean ML, Wallace JD, McCollum FT, Judkins MB, Branine ME, Caton JS (1987) Cattle nutrition on blue grama rangeland in New Mexico Bulletin 727. Las Cruces, NM, USA

Kumarasamy V, Mitchell MD, Bloomfield FH, Oliver MH, Campbell ME, Challis JRG, Harding JE (2005) Effects of periconceptional undernutrition on the initiation of parturition in sheep. Am J Physiol 288:R67–R72

Latham MC (1997) Human nutrition in the developing world. Food and Nutrition Series No. 29 Food and Agriculture Organization of the United Nations, Rome

Loi P, Clinton M, Vackova I, Fulka J, Feil R, Palmieri C, Salda LD, Ptak G (2006) Placental abnormalities associated with post-natal mortality in sheep somatic cell clones. Theriogenology 65:1110–1121

Luther JS, Windorski EJ, Schauer CS, Kirsch JD, Vonnahme KA, Reynolds LP, Caton JS, Wu G (2008) Impacts of L-arginine on ovarian function and reproductive performance in ewes. J Anim Sci 86(E-Suppl 2)/J Dairy Sci 9(E-Suppl 1): ii (Abstr LB5)

Lyon P, Strippoli V, Fang B, Cimmino L (2020) B Vitamins and one-carbon metabolism: implications in human health and disease. Nutrients 12:2867

Mayhew TM, Ohadike C, Baker PN, Crocker IP, Mitchell C, Ong SS (2003) Stereological investigation of placental morphology in pregnancies complicated by pre-eclampsia with and without intrauterine growth restriction. Placenta 24:219–226

Mayhew TM, Wijesekara J, Baker PN, Ong SS (2004) Morphometric evidence that villous development and fetoplacental angiogenesis are compromised by intrauterine growth restriction but not by pre-eclampsia. Placenta 25:829–833

McCarthy KL (2019) Energy and mineral supplementation strategies for beef cattle grazing the northern great plains. Dissertation, North Dakota State University, Fargo, USA

McCarthy KL, Nestande J, Kassetas CJ, Baumgaertner F, Kirsch JD, Dorsam ST, Neville TL, Ward AK, Borowicz PP, Reynolds LP, Sedivec KK, Forcherio JC, Scott R, Caton J, Dahlen CR (2020) Effects of feeding vitamin and mineral and (or) energy supplements to beef heifers during the first 83 days of gestation on progesterone concentrations, corpus

luteum size, and fetal body measurements. J Anim Sci 98(Supplement 4):161–162
McLean KJ, Crouse MS, Crosswhite MR, Negrin Pereira N, Dahlen CR, Borowicz PP, Reynolds LP, Ward AK, Neville BW, Caton JS (2017) Impacts of maternal nutrition on uterine and placental vascularity and mRNA expression of angiogenic factors during the establishment of pregnancy in beef heifers. Transl Anim Sci 1:160–167
McLean KJ, Crouse MS, Crosswhite MR, Negrin Pereira N, Dahlen CR, Borowicz PP, Reynolds LP, Ward AK, Neville BW, Caton JS (2018) The effects of nutrient restriction on mRNA expression of endogenous retroviruses, interferon-tau, and pregnancy-specific protein-B during the establishment of pregnancy in beef heifers. J Anim Sci 96:950–963
Menezes ACB, McCarthy KL, Kassetas CJ, Baumgaertner F, Kirsch JD, Dorsam S, Neville TL, Ward AK, Borowicz PP, Reynolds LP, Sedivec KK, Forcherio JC, Scott R, Caton JS, Dahlen CR (2021a) Vitamin and mineral supplementation and rate of gain during the first trimester of gestation affect concentrations of amino acids in maternal serum and allantoic fluid of beef heifers. J Anim Sci 99:skab024
Menezes ACB, McCarthy KL, Kassetas CJ, Baumgaertner F, Kirsch JD, Dorsam ST, Neville TL, Ward AK, Borowicz PP, Reynolds LP, Sedivec KK, Forcherio JC, Scott R, Caton JS, and Dahlen CR (2021b) Vitamin and mineral supplementation and rate of gain in beef heifers: Effects on fetal trace mineral reserves at day 83 of gestation. J Anim Sci Abstr (in Press)
Miles JR, Farin CE, Rodriguez KF, Alexander JE, Farin PW (2004) Angiogenesis and morphometry of bovine placentas in late gestation from embryos produced in vivo or in vitro. Biol Reprod 71:1919–1926
Miles JR, Farin CE, Rodriguez KF, Alexander JE, Farin PW (2005) Effects of embryo culture on angiogenesis and morphometry of bovine placentas during early gestation. Biol Reprod 73:663–671
Mueller CA, Eme J, Burggren WW, Roghair RD, Rundle SD (2015) Challenges and opportunities in developmental integrative physiology. Comp Biochem Physiol A 184:113–124
Nettle D, Frankenhuis WE, Rickard IJ (2013) The evolution of predictive adaptive responses in human life history. Proc Royal Soc b: Biological Sciences 280:20131343
Nilsson E, Ben Maamar M, Skinner MK (2019) Definition of epigenetic transgenerational inheritance and biological impacts. In: Tollefsbol TO (ed) Transgenerational epigenetics, vol 13. 2nd ed. Elsevier, Cambridge, MA, pp 13–24
Oliver MH, Jaquiery AL, Bloomfield FH, Harding JE (2007) The effects of maternal nutrition around the time of conception on the health of the offspring. Soc Reprod Fertil Suppl 64:397–410
Oostingh EC, Hall J, Koster MPH, Grace B, Jauniaux E, Steegers-Theunissen RPM (2019) The impact of maternal lifestyle factors on periconception outcomes: a systematic review of observational studies. Reprod BioMed Online 38:77–94
Palmieri C, Loi P, Reynolds LP, Ptak G, Della Salda L (2007) Placental abnormalities in ovine somatic cell clones at term: a light and electron microscopic investigation. Placenta 28:577–584
Palmieri C, Loi P, Ptak G, Della Salda L (2008) A review of the pathology of abnormal placentae of somatic cell nuclear transfer clone pregnancies in cattle, sheep, and mice. Vet Pathol 45:865–880
Palombo V, Alharthi A, Batistel F, Parys C, Guyader J, Trevisi E, D'Andrea M, Loor JJ (2021) Unique adaptations in neonatal hepatic transcriptome, nutrient signaling, and one-carbon metabolism in response to feeding ethyl cellulose rumen-protected methionine during late-gestation in Holstein cows. BMC Genomics 22:280
Quinn KE, Reynolds LP, Grazul-Bilska AT, Borowicz PP, Ashley RL (2016) Placental development during early pregnancy: effects of embryo origin on expression of chemokine ligand twelve (CXCL12). Placenta 43:77–80
Redmer DA, Wallace JM, Reynolds LP (2004) Effect of nutrient intake during pregnancy on fetal and placental growth and vascular development. Domest Anim Endocrinol 27:199–217
Redmer DA, Aitken RP, Milne JS, Reynolds LP, Wallace JM (2005) Influence of maternal nutrition on messenger RNA expression of placental angiogenic factors and their receptors at midgestation in adolescent sheep. Biol Reprod 72:1004–1009
Reynolds LP, Caton JS (2012) Role of the pre- and post-natal environment in developmental programming of health and productivity. Mol Cell Endocrinol 354:54–59
Reynolds LP, Redmer DA (1995) Utero-placental vascular development and placental function. J Anim Sci 73:1839–1851
Reynolds LP, Redmer DA (2001) Angiogenesis in the placenta. Biol Reprod 64:1033–1040
Reynolds LP, Vonnahme KA (2016) Triennial reproduction symposium: developmental programming of fertility. J Anim Sci 94:2699–2704
Reynolds LP, Vonnahme KA (2017) Livestock as models for developmental programming. Anim Front 7:12–17
Reynolds LP, Caton JS, Redmer DA, Grazul-Bilska AT, Vonnahme KA, Borowicz PP, Luther JS, Wallace JM, Wu G, Spencer TE (2006) Evidence for altered placental blood flow and vascularity in compromised pregnancies. J Physiol 572:51–58
Reynolds LP, Borowicz PP, Caton JS, Vonnahme KA, Luther JS, Buchanan DS, Hafez SA, Grazul-Bilska AT, Redmer DA (2010a) Uteroplacental vascular development and placental function: an update. Int J Dev Biol 54:355–366
Reynolds LP, Borowicz PP, Caton JS, Vonnahme KA, Luther JS, Hammer CJ, Maddock Carlin KR, Grazul-Bilska AT, Redmer DA (2010b) Developmental programming: the concept, large animal models, and

the key role of uteroplacental vascular development. J Anim Sci 88 (E. Suppl):E61–E72
Reynolds LP, Vonnahme KA, Lemley CO, Redmer DA, Grazul-Bilska AT, Borowicz PP, Caton JS (2013) Maternal stress and placental vascular function and remodeling. Curr Vasc Pharmacol 11:564–593
Reynolds LP, Borowicz PP, Palmieri C, Grazul-Bilska AT (2014) Placental vascular defects in compromised pregnancies. Adv Exp Med Biol 814:193–204
Reynolds LP, Ward AK, Caton JS (2017) Epigenetics and developmental programming in ruminants: Long-term impacts on growth and development. In: Scanes CF, Hill R (eds) Biol domest anim. CRC Press/Taylor & Francis Group, Milton Park, UK, pp 85–121
Reynolds LP, Borowicz PP, Caton JS, Crouse MS, Dahlen CR, Ward AK (2019) Developmental programming of fetal growth and development. Vet Clin North Am Food Anim Pract 35:229–247
Schulz LC (2010) The Dutch Hunger Winter and the developmental origins of health and disease. Proc Natl Acad Sci USA 107:16757–16758
Sinclair KD, Singh R (2007) Modelling the developmental origins of health and disease in the early embryo. Theriogenology 67:43–53
Sinclair KD, Allegrucci C, Singh R, Gardner DS, Sebastian S, Bispham J, Thurston A, Huntley JF, Rees WD, Maloney CA, Lea RG, Craigon J, McEvoy TG, Young LE (2007) DNA methylation, insulin resistance, and blood pressure in offspring determined by maternal periconceptional B vitamin and methionine status. Proc Natl Acad Sci USA 104:19351–19356
Skinner MK (2014) Endocrine disruptor induction of epigenetic transgenerational inheritance of disease. Mol Cell Endocrinol 398:4–12
Steegers-Theunissen RPM, Twigt J, Pestinger V, Sinclair KD (2013) The periconceptional period, reproduction and long-term health of offspring: the importance of one-carbon metabolism. Hum Reprod Update 19:640–655
UNICEF (2019) The State of the World's Children 2019. Children, food and nutrition: growing well in a changing world. United Nations Children's Fund, New York. https://www.unicef.org/media/60806/file/SOWC-2019.pdf
UNSCN (2021) The UN Decade of Action on Nutrition 2016–2025. United Nations System Standing Committee on Nutrition. https://www.unscn.org/en/topics/un-decade-of-action-on-nutrition
Van Emon M, Sanford C, McCoski S (2020) Impacts of bovine trace mineral supplementation on maternal and offspring production and health. Animals 10:2404
Vickers MH (2014) Developmental programming and transgenerational transmission of obesity. Ann Nutr Metab 64(Suppl 1):26–34
Vonnahme KA (2012) How the maternal environment impacts fetal and placental development: implications for livestock production. Anim Reprod 9:789–797
Vonnahme KA, Lemley CO, Shukla P, O'Rourke ST (2013) 2011 and 2012 early careers achievement awards: placental programming: how the maternal environment can impact placental function. J Anim Sci 91:2467–2480
Wallace JM, Luther JS, Milne JS, Aitken RP, Redmer DA, Reynolds LP, Hay WW (2006) Nutritional modulation of adolescent pregnancy outcome—a review. Placenta 27:61–68
WHO (2016) Good maternal nutrition: the best start in life. WHO Regional Office for Europe, Copenhagen, Denmark. https://www.euro.who.int/__data/assets/pdf_file/0008/313667/Good-maternal-nutrition-The-best-start-in-life.pdf
Woods L, Perez-Garcia V, Hemberger M (2018) Regulation of placental development and its impact on fetal growth—new insights from mouse models. Front Endocrinol 9:570
Wu G (2022) Nutritionand metabolism: Foundations for animal growth, development, reproduction, and health.Adv Exp Med Biol 1354:1–24
Wu G, Bazer FW, Wallace JM, Spencer TE (2006) BOARD-INVITED REVIEW: Intrauterine growth retardation: implications for the animal sciences. J Anim Sci 84:2316–2337
Young LE, Fairburn HR (2000) Improving the safety of embryo technologies: possible role of genomic imprinting. Theriogenology 53:627–648

Phosphate, Calcium, and Vitamin D: Key Regulators of Fetal and Placental Development in Mammals

5

Claire Stenhouse, Larry J. Suva, Dana Gaddy, Guoyao Wu, and Fuller W. Bazer

Abstract

Normal calcium and bone homeostasis in the adult is virtually fully explained by the interactions of several key regulatory hormones, including parathyroid hormone, 1,25 dihydroxy vitamin D3, fibroblast growth factor-23, calcitonin, and sex steroids (estradiol and testosterone). In utero, bone and mineral metabolism is regulated differently from the adult. During development, it is the placenta and not the fetal kidneys, intestines, or skeleton that is the primary source of minerals for the fetus. The placenta is able to meet the almost inexhaustible needs of the fetus for minerals by actively driving the transport of calcium and phosphorus from the maternal circulation to the growing fetus. These fundamentally important minerals are maintained in the fetal circulation at higher concentrations than those in maternal blood. Maintenance of these inordinately higher fetal levels is necessary for the developing skeleton to accrue sufficient minerals by term. Importantly, in livestock species, prenatal mineralization of the skeleton is crucial for the high levels of offspring activity soon after birth. Calcium is required for mineralization, as well as a plethora of other physiological functions. Placental calcium and phosphate transport are regulated by several mechanisms that are discussed in this review. It is clear that phosphate and calcium metabolism is intimately interrelated and, therefore, placental transport of these minerals cannot be considered in isolation.

Keywords

Phosphate · Calcium · Vitamin D · Placenta · Endometrium · Pregnancy

List of Abbreviations

1,25(OH)2D3	1,25 Dihydroxyvitamin D3
ADAM	A disintegrin and metalloprotease domain
cAMP	Cyclic adenosine monophosphate
CTB	Cytotrophoblasts
CYPs	Cytochrome P450 mixed-function oxidases

C. Stenhouse · G. Wu
Departments of Animal Science, Texas A&M University, College Station, TX 77843, USA

L. J. Suva
Veterinary Physiology and Pharmacology, Texas A&M University, College Station, TX 77843, USA

D. Gaddy
Veterinary Integrative Biosciences, Texas A&M University, College Station, TX 77843, USA

F. W. Bazer (✉)
Department of Animal Science, Kleberg Center, Texas A&M University, College Station, TX 77843-2471, USA
e-mail: fbazer@exchange.tamu.edu

G. Wu (ed.), *Recent Advances in Animal Nutrition and Metabolism*, Advances in Experimental Medicine and Biology 1354, https://doi.org/10.1007/978-3-030-85686-1_5

E2	Estradiol
ECM	Extracellular Matrix
FGF	Fibroblast growth factor
FGFR	Fibroblast growth factor receptor
GE	Glandular epithelium
IFNT	Interferon tau
KL	Klotho
LE	Luminal epithelium
P21	Cyclin-dependent kinase inhibitor
P4	Progesterone
P53	Tumor protein 53
PGR	Progesterone receptor
PKC	Protein kinase C
PMCA	Plasma membrane Ca^{2+}ATPase
PTH	Parathyroid hormone
PTHR	Parathyroid hormone receptor
PTHrP	Parathyroid hormone-related protein
S100	S100 calcium binding proteins
sFRP4	Secreted frizzled-related protein 4
sGE	Superficial glandular epithelium
SLC	Solute carrier
SLC20	Solute carrier family 20
SLC34	Solute carrier family 34
SPP1	Secreted phosphoprotein 1
STB	Syncytiotrophoblast
STC	Stanniocalcin
TRPV	Transient receptor potential cation channel
VDR	Vitamin D receptor
WNT	Wingless-related integration site

5.1 Introduction

The strength and durability of the adult vertebrate skeleton come not from the construction of a monolithic static structure but from building a dynamic and renewable biologically active scaffold (Suva et al. 2005). This process requires an enormous supply of minerals, principally calcium and phosphate, along with the meticulous regulation of mineral storage, availability, and utility by a plethora of hormones acting in concert on kidney, intestine, and bone. Indeed, normal adult calcium and bone homeostasis can be almost fully explained by the interactions of several regulatory hormones, including parathyroid hormone (PTH), PTH-related protein (PTHrP), 1,25 dihydroxyvitamin D3 (1,25(OH)2D3), fibroblast growth factor-23 (FGF23), calcitonin, and sex steroids (estradiol, progesterone, and testosterone).

During normal fetal development minerals such as calcium, phosphate and magnesium are maintained at significantly higher concentrations in utero to achieve adequate bone accretion (Taylor-Miller and Allgrove 2021). The availability of these critical minerals is an integral component of fetal development that facilitates the safe neonatal transition to post-natal life in all mammalian species (Taylor-Miller and Allgrove 2021). It is perhaps not surprising that many of the same mineral transport regulators active throughout adult life are associated with mineral transport across the placenta to the developing fetus; a concept that will be discussed throughout this review.

With regard to mineral ion action, inorganic phosphate (Pi) is critical for a wide variety of metabolic pathways including synthesis of DNA and RNA, formation of phospholipids and membranes, the generation of ATP/GTP/UTP, cellular signaling (phosphorylation and dephosphorylation), the buffering of pH in intracellular fluids and urine, as well as in the extracellular matrix (ECM) of bone and teeth in the forms of hydroxyapatite ($Ca_{10}[PO_4]_6[OH]_2$) (Voelkl et al. 2021) and glucosamine-phosphate-derived glycoproteins (Wu 2021).

During embryonic development after the early chondrocyte anlage pattern for the endochondral skeleton is laid down, rapid bone formation and mineralization of the developing fetal skeleton creates a significant demand for minerals. In this scenario, kidneys, intestines, and skeleton (active in adults to control the extent of mineral ion metabolism) do not supply minerals to the normal fetus. Rather, the placenta meets the extensive fetal need for minerals by actively transporting calcium, phosphorus, and magnesium from the maternal circulation (Kovacs 2014). The placenta provides these essential minerals even when confronted with their reduced concentrations in the maternal circulation. Indeed, the fetus develops in the face of a significant hypercalcemia yet maintains even higher mineral concentrations than in the mother or a normal adult (Kovacs 2014). These significantly elevated levels are required for the skeleton to form and mineralize normally (Care 1991). In this review, the regulatory mechanisms responsible for the transport of minerals across the placenta to the developing fetus are described in relation to the well-characterized mechanisms in play in the adult.

5.2 Roles of Phosphate, Calcium, and Vitamin D Postnatally in Bone and Kidney

5.2.1 Phosphate

Several factors and endocrine hormones modulate homeostasis of renal calcium and phosphate. Among those, (1,25(OH)2D3), PTH, and FGF23 have gained the most attention as very important regulators of phosphate homeostasis (Suva and Friedman 2020). These three hormones regulate each other, thereby forming a classic regulatory endocrine loop (Kovacs 2014).

An increase in dietary phosphate intake or plasma phosphate levels stimulates PTH secretion, which enhances bone-derived FGF23 synthesis and release as well as the synthesis of 1,25 (OH)2D3 by kidneys (Fig. 5.1). Similarly, modest decreases in serum calcium also produce profound and rapid increases in PTH secretion that enhance bone-derived FGF23 formation and release, as well as the synthesis of 1,25(OH)2D3 by the kidney. The release of FGF23 suppresses PTH and 1,25(OH)2D3 levels (PTH inhibits 1α hydroxylase action in the kidney), whereas 1,25 (OH)2D3 stimulates FGF23 release and inhibits PTH synthesis and secretion. In addition to these positive and negative feedback loops, all three hormones are regulated by a plethora of other factors.

Blood in the vascular circulation and extracellular fluid are responsible for the transport of phosphate to and from the organs involved in phosphate metabolism. It is the skeleton that provides the largest reservoir of phosphate, and it is also the source of a major hormone regulating phosphate transport, namely FGF23 (Dias et al. 2006; Peacock 2021). Bone mineral phosphate is exchangeable with extracellular fluid phosphate, but this flux does not substantially contribute to extracellular fluid phosphate homeostasis (Dias et al. 2006; Peacock 2021). Fundamentally, two transport pathways for phosphate exist in bone. One involves the transport of extracellular fluid phosphate into and out of bone whereas the second is the transport of phosphate within bone, directly from the bone-forming osteoblast as a component of the mineralizing apatite crystal (Peacock 2021). Daily net transport of phosphate from extracellular fluid to new bone formation and back to the extracellular fluid by bone resorption is less than that managed by the kidney and the gut (Peacock 2021).

It is the kidney that handles the major fraction of daily phosphate transport, as well as being the primary source for the generation of the active form of vitamin D (1,25(OH)2D3), through the action of the PTH-regulated 1α hydroxylase (Li et al. 2020). 1,25(OH)2D3 is a critically important hormone for the regulation of phosphate transport. The process of glomerular filtration transports more than 5,000 mg phosphate every 24 h to the proximal renal tubule that reabsorbs and transfers the majority (>80%) back to extracellular fluid. As such, the kidney is a primary organ responsible for controlling the circulating concentrations of phosphate (Peacock 2018).

Given the fundamental role of the kidney, the impact of the gut on the supply of phosphate to the circulation is also important to consider. The amount of dietary phosphate is inadequate for herbivorous mammals without receiving supplementation (Wu 2018) and humans consuming little or no animal products. It is possible that

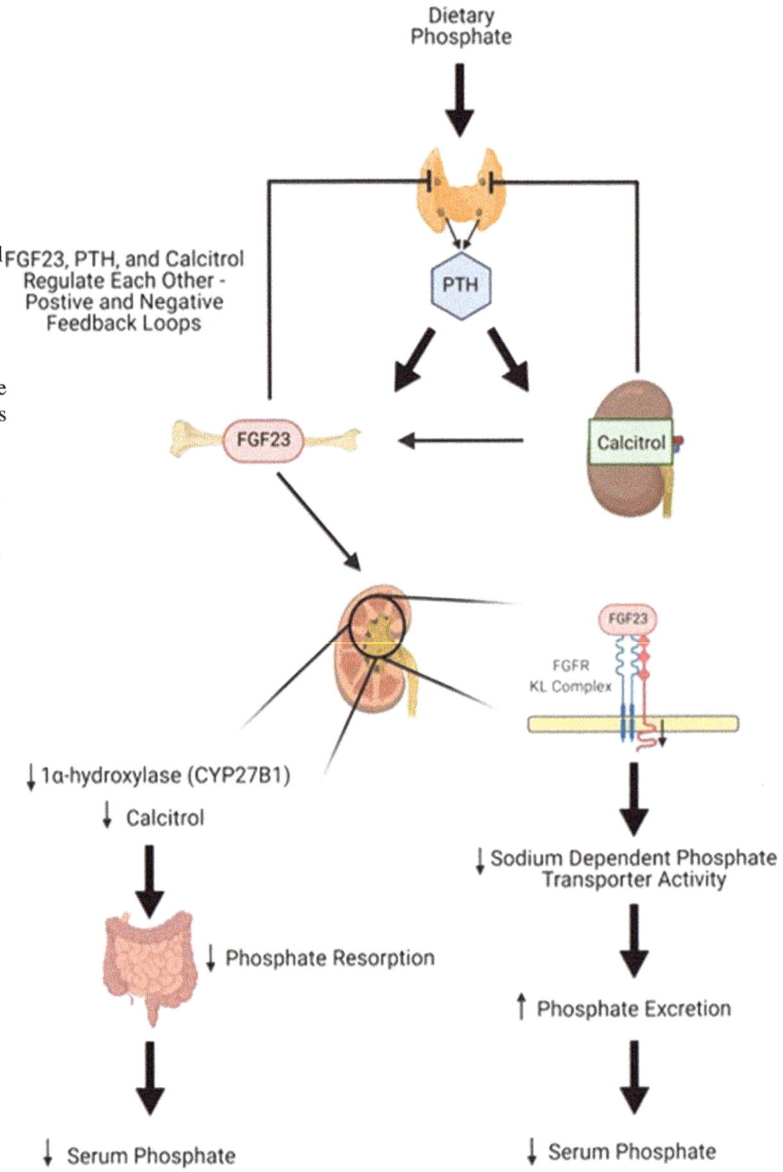

Fig. 5.1 PTH, FGF23, and 1,25(OH)2D3 are essential regulators of phosphate **homeostasis** postnatally. An increase in dietary phosphate intake or plasma phosphate levels stimulates PTH secretion, which enhances bone-derived FGF23 formation and release as well as the synthesis of 1,25(OH)2D3 by kidneys. The release of FGF23 suppresses PTH and 1,25(OH)2D3 levels (PTH inhibits 1α hydroxylase action in the kidney), whereas 1,25(OH)2D3 stimulates FGF23 release and inhibits PTH synthesis and secretion. In addition to these positive and negative feedback loops, all three hormones are regulated by a plethora of other factors.

dietary phosphate is surplus over the daily requirements for phosphate in carnivores. Paracellular diffusion is an important mechanism for phosphate uptake in humans and other animals, but the mechanistic detail(s) and its relative magnitude remain to be fully elucidated. The absorption of phosphate occurs by both paracellular diffusion and active, saturable, transcellular mechanisms that act at low phosphate intakes (Peacock 2021). In addition, the small intestine is an important site for phosphate transport, and it is directly responsible for the bulk of dietary phosphate absorption (Hernando and Wagner 2018). In contrast to animal products, the bioavailability of phosphate in plant-sourced foods is relatively low (Wu 2018). The efficiency of absorption is estimated at around 80% with some 20% of the absorbed phosphate returning to the lumen of the gut from the extracellular fluid as endogenous secretions.

With this in mind, it is clear that phosphate transport is both passively and actively performed and regulated by a number of transport mechanisms distributed across many tissues, but with a particularly important focus on the kidney, gut and bone (Peacock 2021).

5.2.2 Calcium

Similar to phosphate, the minute to minute regulation of serum calcium is also intimately and elegantly controlled by the interplay between calcium and phosphate as well as the level of the major calcium regulating hormones in the adult, PTH and 1,25(OH)2D3 (Kovacs 2014). Calcium is involved in many physiological and biochemical processes as it is an essential element for cardiac function, the structural integrity of bone, muscle contraction, and it also acts as a second messenger in cell signaling and a cofactor for many enzymes in a multitude of biochemical pathways. Serum calcium can be measured in venous blood samples, with normal physiologic levels ranging from 8.8 to 10.4 mg/dL for total calcium, and 4.7–5.2 mg/dL for free calcium (Kavsak 2017). Total calcium values should be corrected for current concentrations of albumin as their interaction can affect reported levels.

A consistent finding throughout fetal development is the high demand for calcium. Indeed, concentrations of calcium in fetal serum are significantly greater than the simultaneous concentrations of calcium in maternal serum in many species (Kovacs 2014). In fetal mice, free calcium is 0.25–0.50 mM greater than maternal values (Kovacs et al. 1996), and the values can be even higher in relation to maternal levels in a wide variety of species including pigs (Care et al. 1986), lambs (Delivoria-Papadopoulos et al. 1967), foals (Garel 1972), and perinatal primates (Fleischman et al. 1975). These important measurements indicate that the fetus is indeed hypercalcemic relative to both the maternal and normal adult serum calcium values. Interestingly, an explanation continues to elude us regarding how early in gestation do concentrations of calcium in fetal serum exceed those in maternal serum.

Given the profound hypercalcemia of the fetus, an obvious question is why? What purpose is served by the fetus maintaining a high concentration of calcium in its blood? The conservation of this fetal hypercalcemia across mammals suggests some fundamentally important physiological function. Several rationales exist including the obvious need for high fetal calcium to ensure normal skeletal mineralization (especially in large animals) and the possibility that elevated blood calcium in utero provides a survival advantage after birth. Whatever the physiological reason, the details of why are unclear, especially since moderate fetal hypocalcemia does not appear to impair survival to the end of gestation as shown in murine studies (Kovacs et al. 2001a; Suzuki et al. 2008).

During endochondral bone formation in the fetus, the regulation of fetal mineral homeostasis is of utmost importance. In addition to the requirement for calcium, phosphorus also plays a key role. Since the mineralization of osteoid requires the incorporation of phosphate prior to calcium binding (Zhang et al. 2011), dietary phosphorus is an important determinant of bone mineralization during fetal development. In sum, concentrations of calcium, ionized calcium, and phosphorus in serum are significantly higher in fetuses than in the mother and can generally be maintained despite abnormal concentrations in the maternal circulation (Kovacs 2014).

As discussed previously, and throughout this review, the transport of both phosphate and calcium is regulated by a complex hormonal axis comprised of FGF23, PTH, and 1,25(OH)2D3. Perhaps most importantly, the metabolism of phosphate and calcium does not occur in isolation from each other. They closely interact at transport mechanisms in extracellular fluid, gut, bone, and kidney. Changes in the extracellular fluid concentrations of phosphate and calcium independently regulate secretion of the hormones, FGF23, PTH, and 1,25(OH)2D3 that maintain phosphate and calcium homeostasis. The regulation of phosphate and calcium metabolism is intimately inter-related and neither should be considered in isolation.

5.3 Animal Models for the Study of Placental Mineral Transport (see Enders and Carter 2004; Carter 2007; Barry and Anthony 2008; Grigsby 2016)

The placenta is an underappreciated organ, which is critical for the exchange of minerals, gases, amino acids, sugars, and proteins, while producing regulatory molecules such as cytokines, growth factors, and hormones that are crucial for the development and growth of the conceptus (Enders and Blankenship 1999). Given the ethical concerns and the significant limitations associated with performing human placental research, comparative physiology is essential to improve understanding of the mechanisms governing placental nutrient and mineral transport, and conceptus development and growth. As there are often confounding factors in human pregnancy studies, animal models can provide valuable data from a controlled experimental design to allow further investigation into the importance of mineral transport in pregnancy. Despite the striking similarities that exist between mammalian species during pregnancy, placentation varies substantially, which can lead to differences in mineral and nutrient transport to the fetus (Enders and Blankenship 1999; Enders and Carter 2004; Montiel et al. 2013).

There are many published reports on placental transport of minerals with rodents, sheep, and pigs commonly used as animal models. Rodents are an extensively utilized as an inexpensive animal model for studies of conceptus growth and development. The gestation period for rats, mice, and hamsters is 20–22 days, allowing generation of data from large numbers of animals in a timely manner. Rodents undergo invasive implantation and have hemochorial placentae. While there are several similarities between human and rodent placentae in structure and function, there are several considerations that must be made when interpreting results from studies with rodents (Wu 2022). Considering the significant interspecies differences in bone structure, composition, and density (Aerssens et al. 1998) and physiological maturity at birth, it is unsurprising that there are significant variations in the timings of ossification and skeletal mineralization. Compared with humans, rodents are relatively immature at birth, and their small size makes both surgical and non-surgical procedures challenging (Swanson and David 2015). In contrast to humans, rodents are a litter-bearing species, with individual feto-placental units allowing for direct comparison among samples from different fetuses within the same uterus, without the confounding factors of maternal genotype, nutrition, or husbandry practices. However, as they are a litter bearing species, there are much greater demands being placed upon the mother to ensure that adequate mineral transfer occurs to allow appropriate fetal growth and development. Data suggest that the fetal mineral demand is much greater in rodents than in other commonly utilized animal models, with approximately 80% of the calcium in maternal blood transported to the fetuses in late gestation in rats, compared with 5–10% of calcium in maternal blood of women (Comar 1956; Widdowson 1962). This makes rodents highly susceptible to secondary hyperparathyroidism and hypocalcemia compared with other species. While this allows for unique opportunities to study the effects of these conditions on fetal growth and development, it does raise the question of how comparable placental mineral transport is in rodents compared with other species.

In contrast to rodents, sheep are much larger in size, making them easier to handle and perform experimental manipulations. Importantly, the sheep fetus has a similar physiological maturity at birth to human infants. Furthermore, like humans, sheep usually have singleton or twin pregnancies. Sheep have a much longer gestation length than rodents, which allows investigations to focus on comparable stages of gestation to those for human pregnancies. While the sheep placenta is viewed by some as being anatomically distinct from the human placenta, there are many important similarities. The sheep have a multi-cotyledonary placenta, consisting of placentomes composed of maternal caruncular tissue and fetal cotyledonary tissue. Large variations in cotyledonary numbers exist and it is known that breed, litter size, and parity can influence cotyledonary number and size (Dwyer et al. 2005). Despite the gross anatomical differences, the human placenta also has cotyledonary structures, making them more similar in structure to those for sheep than may be appreciated by some scientists. Additionally, while the fetal vascular trees of sheep and human placentae differ in size, the architecture of the stem, intermediate, and terminal villi is comparable (Leiser et al. 1997).

Similar to rodents, pigs are a litter-bearing species with individual feto-placental units, so comparisons can be made among different fetuses within the same uterus, with minimal confounding effects. Pigs are much larger in size than rodents, making them easier to handle and perform experimental manipulations. Furthermore, piglets have a similar physiological maturity at birth to human infants and a longer gestation length than rodents, which allows investigations to occur at desired, comparable stages of gestation to human pregnancies. Pigs have a true epitheliochorial placenta, wherein the uterine luminal epithelium (LE) remains intact throughout pregnancy and the trophectoderm directly attaches to the LE. This type of conceptus attachment is much less invasive than for other types of placental types; therefore, extensive remodeling must occur to maximize placental surface area available for nutrient, mineral, and gas exchange.

It is important to consider these species-specific differences when designing an experiment to investigate placental transport of minerals and nutrients and when extrapolating the findings from studies of one species to another species. While each animal model has its own merits, it is important to note that no animal model fully recapitulates human pregnancy.

5.4 Maternal Mineral Adaptations

Calcium and phosphorous are two of the most abundant minerals in the body, with approximately 99 and 85% of this calcium and phosphorous, respectively, stored in bone in the form of hydroxyapatite ($Ca_{10}[PO_4]_6[OH]_2$) (Mitchell et al. 1945; Penido and Alon 2012). In addition to their critical roles in the skeletal and renal systems outlined earlier, calcium and phosphate are regulators of many processes including cellular proliferation, protein synthesis, and cellular metabolism (Chin et al. 1987; Santella 1998; Brostrom and Brostrom 2003; Jeon 2008; Glancy and Balaban 2012; Penido and Alon 2012; Kovacs 2014). During pregnancy, the mother must adapt to meet the nutritional and mineral requirements of the developing fetus, without compromising the maternal mineral reserve. Studies investigating placental transport of minerals have demonstrated that the rate of calcium and phosphate transport significantly increases in the last trimester of pregnancy to allow for fetal skeletal mineralization. In fact, it is estimated that around 80% of the minerals present in a fetus at birth are transported during the last trimester of pregnancy. In humans, this translates to approximately 20 g of phosphate and 30 g of calcium (Givens and Macy 1933; Widdowson and McCance 1965; Ziegler et al. 1976). Unsurprisingly, this is a significant burden on the mother's mineral reserve and the mechanisms regulating maternal mineral homeostasis must be altered to avoid the mother becoming hypocalcemic or hypophosphatemic.

To accommodate the increasing mineral demands, the maternal intestine must double the absorption of phosphate and calcium (Kent et al. 1991; Ritchie et al. 1998; Kovacs 2016) (Fig. 5.2). Increasing intestinal mineral absorption is a crucial gestational adaptation to provide sufficient minerals for transport across the placenta to meet the demands of the developing fetus. In humans, this begins after approximately

12 weeks of pregnancy and is maintained until parturition (Heaney and Skillman 1971; Kent et al. 1991; Cross et al. 1995). This strategy increases the maternal mineral reserve so that in the latter stages of gestation, during the period of exponential fetal growth and skeletal mineralization, there are sufficient amounts of calcium and phosphate available for the fetus. There are conflicting theories regarding what regulates this change in maternal mineral absorption. Some consider calcitrol to be the central regulator of this adaptation as it is known to increase intestinal expression of molecules involved in calcium transport including calcium-binding proteins, transient receptor potential cation channel family members, and Ca^{2+} ATPases [reviewed by (Kovacs 2016)]. While this is plausible, the increase in intestinal calcium absorption begins early in gestation, prior to the increase in maternal calcitrol. Additionally, recent evidence from research with mice indicates that calcitrol is not the key regulator of this increase in intestinal calcium absorption during pregnancy (Fudge and Kovacs 2010; Gillies et al. 2018). Evidence from animal models suggests that prolactin and placental lactogen can regulate intestinal calcium absorption; therefore, the potential role of these molecules in the regulation of increased intestinal calcium absorption during the first trimester warrants further investigation. Despite this crucial and significant increase in intestinal calcium absorption, the maternal kidneys do not alter their resorption of minerals, ultimately leading to significant increases in calcium excretion in urine.

Considering the extensive transport of calcium that must occur across the placenta to the fetus, it is unsurprising that calcium in maternal serum decreases in late gestation in many species including rats, sheep, goats, and cows [reviewed by Kovacs (2016)]. Results of several studies of women suggest that calcium in maternal serum decreases in pregnancy; however, results of other studies suggest that the abundance of free calcium in maternal serum is not affected by stage of pregnancy [reviewed by Kovacs (2016)].

5.5 Gestational Changes in Phosphate and Calcium in Fetal Fluids

Prior to the establishment of a functional placenta for transplacental exchange of gases, micronutrients (amino acids, glucose), and macromolecules (proteins), the conceptus is entirely reliant upon the secretion and transport of nutrients into the uterine lumen by the maternal uterine glandular epithelium (GE) and LE. A simple and effective manner to assess the nutrient and mineral composition within the uterine luminal environment is to flush the uterus with a buffer solution and compare the nutrient and mineral composition across different days of gestation. Data from pigs and sheep indicate that calcium in uterine flushings increases during the peri-implantation period and is more abundant in uterine flushings from pregnant animals when compared with cyclic animals during the mid-luteal phase of the estrous cycle (Gao et al. 2009; Choi et al. 2019; Stenhouse et al. 2021a). Furthermore, it has been demonstrated in mice that this increase in calcium is essential for the regulation of conceptus development and implantation, with infusion of calcium channel blockers into the mouse uterine lumen causing failure of blastocysts to implant (Banerjee et al. 2011).

As the major components of the skeleton, calcium, and phosphate must be transported across the placenta to the fetus for adequate mineralization of the skeleton (Kovacs 2014, 2016). Studies investigating placental transport of minerals have demonstrated that the rate of calcium and phosphate transport increases significantly during the last trimester of pregnancy to allow mineralization of the fetal skeletal system to occur, as noted previously. In humans, >300 mg of calcium must be transported per day to the fetus between weeks 35 and 38 of gestation. It has been suggested that pregnant ewes in late gestation must transport 1.25 g of phosphate and 2.7 g of calcium to the ovine fetus each day (Grace et al. 1986).

Fetal fluids (allantoic and amniotic fluids) are of maternal origin via active transport of water,

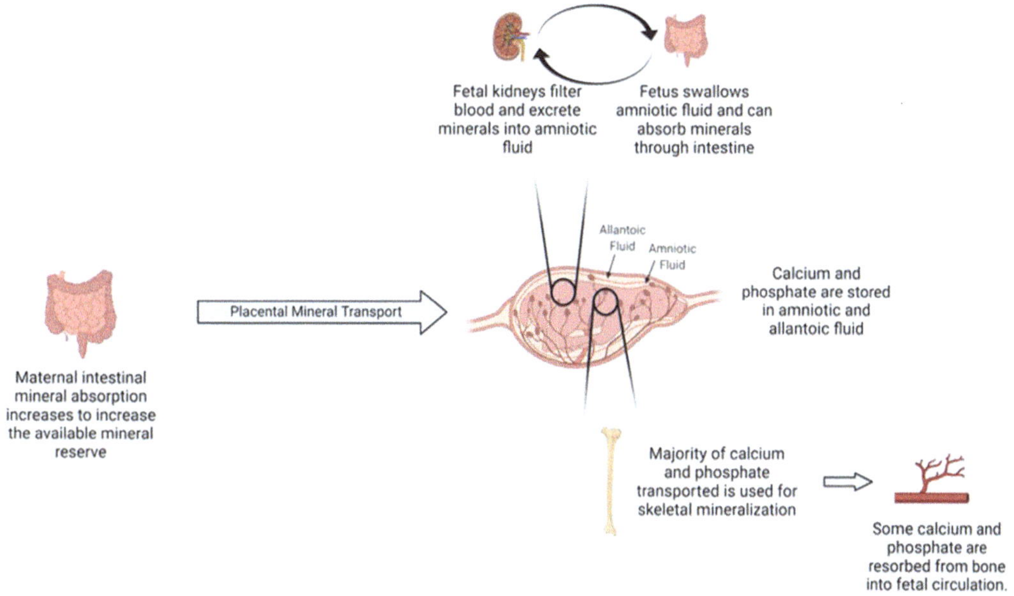

Fig. 5.2 Maternal mineral absorption increases during gestation from an increased mineral reserve for fetal skeletal mineralization. During pregnancy, the mother must adapt to meet the nutritional and mineral requirements of the developing fetus, without compromising the maternal mineral reserve. Studies investigating placental mineral transport have demonstrated that the rate of calcium and phosphate transport significantly increases in the last trimester of pregnancy to allow for fetal skeletal mineralization, with the majority of transported calcium and phosphate forming the fetal skeleton. To accommodate the increasing mineral demands, the maternal intestine must double the absorption of phosphate and calcium. Following placental transport of these minerals, they can be stored in amniotic and allantoic fluid. The fetus can drink amniotic fluid, which is an important additional source of minerals for the fetus as the fetal intestines can absorb minerals present in amniotic fluid. The fetal kidneys then filter the blood and excrete minerals into urine, which in turn makes up much of the volume of amniotic fluid; thereby forming an intestinal-renal-amniotic fluid loop

as well as other molecules, across the placenta and into the allantoic sac for distribution to other components of the conceptus including the fetus and amniotic sac. The driving force for expansion of the allantois, and in turn the chorioallantois, is the rapid accumulation of water in the allantoic sac. In the pregnant ewe, this volume increases from about 1 ml on Day 18 to 90 ml on Day 40 and then from Day 70 (32 ml) to Day 140 (438 ml) of the 147-day period of gestation. The rapid changes in allantoic fluid volume allow expansion of the chorioallantoic membranes and ensure their intimate apposition across the maximum surface area for attachment to the maternal endometrium. Allantoic fluid is a nutrient reservoir that is rich in electrolytes, sugars, amino acids, minerals, and other nutrients, as well as proteins. Importantly, the allantoic epithelium derives from the hindgut and functions to transport nutrients from allantoic fluid into the fetal-placental vasculature. Amniotic fluid is isosmotic to fetal serum, and its volume increases throughout gestation in sheep from 2 ml on Day 30 to over 700 ml on Day 140 of gestation. Amniotic fluid has several important roles. First, it buoys the fetus to allow it to develop symmetrically in three dimensions. Second, it prevents fetal skin from adhering to the amnion. Third, the fetus may drink up to 1 L of amniotic fluid in the last one-third of gestation to gain water and other nutrients, including proteins secreted by the lungs, salivary glands, and amniotic membranes. This act of drinking amniotic fluid is an important additional source of minerals for the fetus as the fetal intestines can absorb minerals present in amniotic fluid.

Phosphate and calcium are both very abundant in ovine amniotic and allantoic fluids across gestation (Stenhouse et al. 2021a; b), indicative of extensive mineral transport across the placenta. Total phosphate and calcium in ovine allantoic fluid increase significantly with advancing gestational day, highlighting that in late gestation when increased placental mineral transport occurs, these minerals are stored in the allantoic fluid for the fetus to utilize for skeletal mineralization. However, total calcium and phosphate are relatively constant in ovine amniotic fluid across gestation, despite significant changes in fluid volume across pregnancy (Stenhouse et al. 2021a; b). In goats, the concentrations of phosphorous and calcium in allantoic fluid increase with advancing gestational day, whereas in amniotic fluid, the concentrations of both calcium and phosphorous decrease with advancing gestational stage (Tabatabaei 2012). Similarly, it has been suggested that the concentrations of phosphate in human amniotic fluid decrease with advancing gestational day (Fotiou et al. 2015; Correia-Branco et al. 2020). It is important to note that differences in placentation type, stage of gestation, and fetal number may all influence the abundance of these minerals in fetal fluids. Additionally, is known that the volume of fetal fluids fluctuates significantly across gestation (Bazer et al. 2012; Dubil and Magann 2013). Therefore, care should be taken when comparing fetal fluid data presented as concentration as compared with total abundance (concentration X fluid volume).

While it is understood that mineral transport at the maternal–conceptus interface increases in late gestation, it is important to ascertain the spatiotemporal profile of molecules that regulate calcium and phosphate transport and vitamin D metabolism at the maternal–conceptus interface across species. Many studies have investigated calcium, phosphate, and vitamin D signaling in pregnancy. While these are all important signaling pathways individually, we must ascertain how these pathways work in concert at the maternal–conceptus interface to fully understand the importance of these minerals during the course of gestation.

5.6 Impact of Maternal Calcium and Vitamin D Deficiencies on Fetal Development and Pregnancy Outcomes

Hypocalcemia in pregnancy occurs when the fetal demands for calcium exceed the availability of calcium in maternal serum. Deficiency in maternal calcium is a significant problem in pregnancy and has been associated with several adverse pregnancy outcomes including preeclampsia, preterm birth, and intrauterine growth restriction (Chhabra and Singh 2017; Wilson et al. 2020). Thus, calcium has an important role in the regulation of conceptus development and growth. In mothers with severe hypocalcemia, the increase in intestinal calcium absorption that occurs in a normal pregnancy is not sufficient to increase the available calcium reserve in the mother. In these instances, calcium must be resorbed from maternal bone, which can decrease maternal bone density and increase the risk of the mother developing osteoporosis during pregnancy (Kovacs and Ralston 2015). Despite these adaptations in mineral status of the mother, it is necessary to monitor mineral status and ensure that there is sufficient mineral transport across the placenta because severe calcium deficiencies impact placental transport of calcium to the fetus which, in severe cases, can lead to demineralization of the fetal skeleton and development of secondary hyperparathyroidism [reviewed by Kovacs (2014)]. Similar findings have been reported for pregnant rats fed a calcium-deficient diet that led to maternal hypocalcemia and secondary hyperparathyroidism in dams, and resorption of minerals from the fetal skeletal system and secondary hyperparathyroidism in the fetuses [reviewed by Kovacs (2014)].

In dairy cows, the periparturient period (4 weeks before to 4 weeks after calving) is often associated with the development of 'milk fever' (hypocalcemia). Reports vary substantially on the incidence of this condition (DeGaris and Lean 2009); however, a meta-analysis of 137 controlled trials estimated that the mean incidence of the condition is 21% (range 0–83%) (Lean et al. 2006). In this condition, calcium in maternal serum can reach < 50% of the recommended normal concentrations of calcium in serum (Mayer et al. 1969; Allen and Sansom 1986), which can have severe consequences for both the mother and fetus. Hypocalcemia in dairy cows is associated with an increased risk for many undesirable conditions including parturition recumbency, dystocia, still born caves, ketosis, displaced abomasum, and even death. Uterine involution is less efficient in cows with hypocalcemia, therefore, there is a higher incidence of uterine prolapse and retained fetal membranes in hypocalcemic cows (Risco et al. 1994). Additionally, cows with hypocalcemia are more susceptible to infections, including mastitis and metritis after calving (Ducusin et al. 2003). This condition is not restricted to dairy cows as a similar condition occurs in ewes during the periparturient period that is estimated to account for between 4 and 6% of the mortality rate in sheep (Hindson and Winter 2002). Collectively, these findings indicate that hypocalcemia in large animals during the periparturient period has significant adverse impacts on pregnancy outcomes and maternal and offspring health, which can be costly for the livestock industry.

Feeding low amounts of calcium to pregnant ewes for 2 months from Day 60 of gestation significantly decreased maternal calcium while increasing circulating concentrations of hydroxyproline, parathyroid hormone, and $1,25(OH)_2D$ (Lima et al. 1993). Interestingly, maternal hypocalcemia in ewes does not influence calcium, hydroxyproline, parathyroid hormone, or $1,25(OH)_2D$ in fetal plasma at Day 120 of gestation. Despite the lack of differences in calcium in fetal plasma, higher concentrations of phosphate in fetal plasma result in delayed ossification of the skeletal system, a lower proportion of bone compared with cartilage, lower specific gravity, and lower ash content in fetuses from hypocalcemic mothers (Lima et al. 1993).

Maternal vitamin D deficiency is also a significant problem in pregnancy and has been linked to several adverse pregnancy outcomes including pre-eclampsia, preterm birth, and intrauterine growth restriction (Bodnar et al. 2007; Ma et al. 2012; Gernand et al. 2014; Achkar et al. 2015; Chen et al. 2015; Qin et al. 2016; Zhou et al. 2017; Alimohamadi et al. 2020; Wilson et al. 2020). Thus, vitamin D has an important role in the regulation of conceptus development and growth. In addition to impacts on pregnancy outcome, it has been suggested that maternal hypovitaminosis D (defined as maternal 25(OH)D] levels <20 ng/ml or <50 nmol/l) is a significant risk factor for adverse neonatal outcomes in humans (Karras et al. 2016). There are large variations in the reported incidence of maternal hypovitaminosis D, presumably due to seasonal and locational variations in maternal vitamin D status [reviewed by Karras et al. (2016)], but this problem does affect the global population. Despite the clear associations between maternal calcium status and fetal skeletal development, there are contradicting reports on the association between maternal vitamin D status and fetal skeletal development in humans [reviewed by Karras et al. (2016)]. However, there is some evidence suggesting a link between maternal vitamin D status and fetal bone development, with maternal hypovitaminosis D impacting fetal bone development from as early as 19 weeks of gestation (Mahon et al. 2010; Ioannou et al. 2012; Young et al. 2012).

5.7 Regulators of Placental Calcium and Phosphate Transport

Given the importance of phosphate, calcium, and vitamin D for many critical processes postnatally, it is unsurprising that many regulatory mechanisms exist (Fig. 5.3). In this review, we will briefly summarize some of the available evidence for the roles of several of the regulatory processes present at the maternal–conceptus interface across species affecting phosphate, calcium, and vitamin D.

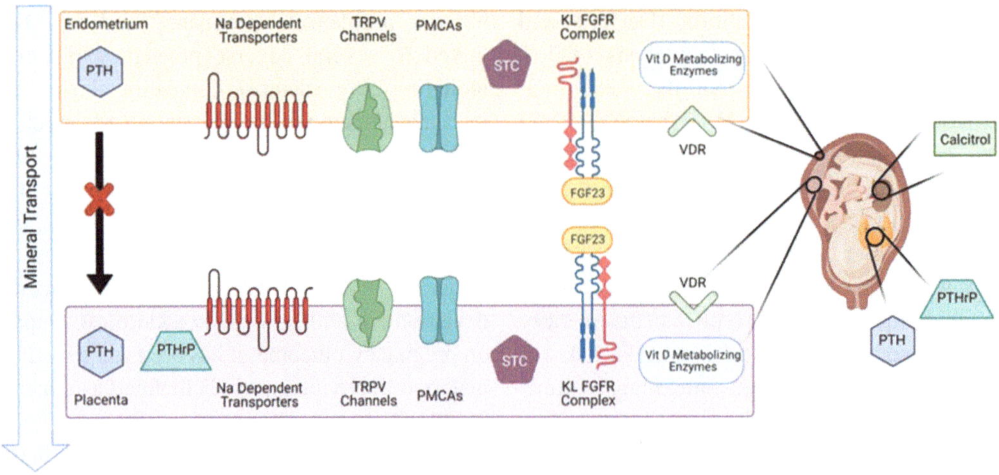

Fig. 5.3 Many signaling pathways are present at the maternal–conceptus interface across species for the regulation of phosphate, calcium, and vitamin D transport and metabolism. As the major components of the skeleton, calcium and phosphate must be transported across the placenta to the fetus, where they must be present in excess abundance compared with that in the mother to allow adequate mineralization of the skeleton. Considering this, it is perhaps unsurprising that many pathways are expressed at the maternal–conceptus interface for the regulation of the transport of these molecules. These pathways are intimately related to one another and therefore must not be considered in isolation from one another

5.7.1 PTH and PTHrP

Parathyroid hormone (PTH), PTH-related protein (PTHrP), and the cognate G protein-coupled PTH receptor (PTHR) play defining roles in the regulation of extracellular calcium and phosphate metabolism and in controlling skeletal growth and repair. Through a series of complex signaling mechanisms that, in many instances proceed in a tissue-specific manner, precise control of these processes is achieved and maintained.

The shared receptor (PTHR) mediates all known biological actions of PTH and PTHrP (Suva and Friedman 2020). Full length PTH is a peptide of 84 amino acids whereas PTHrP exists as peptide products of 139, 141, or 173 amino acid residues derived from a single alternatively spliced gene distinct from the PTH gene (Suva and Friedman 2020). The fundamental role of PTH in phosphate and calcium metabolism in adults is well defined as it has been the focus of intense investigation for many years (Quarles 2008).

Within the fetal circulation, PTH is less abundant when compared with concentrations in maternal blood of several species such as lambs and rodents (Thomas et al. 1981; Collignon et al. 1996; Simmonds et al. 2010). These low values for PTH in the fetus are from sources within the fetus since intact PTH does not cross the placenta (Garel 1972; Northrop et al. 1977; Günther et al. 2000). During development, fetal parathyroid glands begin to express PTH by mid-gestation (Tucci et al. 1996). The fetal parathyroid gland appears to be the primary source since genetic ablation of the fetal parathyroids using Hoxa3 null mice results in undetectable PTH in serum, measured using a rodent PTH immunoradiometric assay (Kovacs et al. 2001b). An additional source of PTH is likely the placenta since the PTH gene is expressed by placentae of mice. As such, the placenta may contribute a small amount of PTH to fetal circulation (Simmonds and Kovacs 2010). The normally low concentration of PTH in fetal blood is critical for the maintenance of calcium concentrations in fetal blood since fetal mice lacking PTH or the PTHR are hypocalcemic compared with WT littermates (Kovacs et al. 1996).

Only since the discovery and cloning of PTHrP (Suva et al. 1987) have details of the movement of calcium across the placenta begun to be elucidated (Abbas et al. 1989; MacIsaac et al. 1991). Since fetal blood contains high PTH-like bioactivity, but low levels of immunoreactive PTH (Allgrove et al. 1985; Loveridge et al. 1988; Rodda et al. 1988; Bourdeau et al. 1990), the finding of elevated PTHrP in fetal blood (Khosla et al. 1990; Seki et al. 1994) explained earlier reports of PTH-like bioactivity. These results led to the overarching hypothesis that PTHrP is the primary calcium-regulating hormone in fetuses, and that the suppressed PTH levels suggested that PTH is largely unimportant until after birth. Furthermore, studies in chronically cannulated sheep fetuses demonstrated a PTHrP-specific action (Abbas et al. 1989) and investigation of PTHrP null mice fetuses compared with their WT and PTHrP ± littermates demonstrated the importance of PTHrP in regulating mineral and bone metabolism in the fetus (Kovacs 2014).

5.7.2 Calcitrol

Postnatally, vitamin D has a central role in the regulation of mineral homeostasis (Bouillon et al. 2019). Vitamin D is acquired through the diet, in the form of fish liver and fortified dairy products, and vitamin D3 (cholecalciferol) can be synthesized from cholesterol in the skin following exposure to ultraviolet irradiation (Schmid and Walther 2013). Cytochrome P450 mixed-function oxidases (CYPs) are the regulators of vitamin D metabolism (Jones et al. 2014). Vitamin D enters the systemic circulation bound to vitamin D binding protein and is transported to the liver, where it is hydroxylated by CYP2R1 to 25-hydroxyvitamin D (25[OH]D3) (Bikle 2014). 25[OH]D3 is then transported to the kidney, bone, and placenta where it is further hydroxylated to the active hormonal form 1,25-dihydroxyvitamin D (1,25[OH]2D3) by 1α-hydroxylase CYP27B1 (Bikle 2014; Wu 2018). The activity of calcitrol is regulated by the catabolic activity of CYP24A1 (24-hydroxylase) (Bikle 2014), which inactivates 1,25[OH]2D3 via conversion to 1,24,25[OH]3D3. The biologically active form of vitamin D (1,25(OH)2D3 or calcitrol) binds to the vitamin D receptor (VDR). The VDR (also known as NR1I1 (nuclear receptor subfamily 1, group I, member 1) is a member of the nuclear receptor family of transcription factors. 1,25(OH)2D3 binds to the VDR, which then heterodimerizes with the retinoid-X receptor to elicit transcriptional effects. Vitamin D2 from sunlight-dried plants is almost as effective as vitamin D3 in most mammals, including humans, cattle, sheep, and pigs (Wu 2018).

As gestation progresses, the abundance of calcitrol in maternal blood increases significantly in many species including humans, rabbits, rodents, and pigs (Pike et al. 1979; Kubota et al. 1982; Ardawi et al. 1997; Boass et al. 1997; Kovacs et al. 2005; Kirby et al. 2013; Jang et al. 2017). This increase in maternal calcitrol occurs due to increased synthesis of calcitrol and not due to alterations in the metabolic clearance of calcitrol (Ross et al. 1989; Paulson et al. 1990). Species specific differences in the abundance of calcitrol in fetal blood versus maternal blood do exist. In rodents (Lester et al. 1978; Verhaeghe et al. 1988; Kovacs et al. 2005) and pigs (Lachenmaier-Currle et al. 1989; Lachenmaier-Currle and Harmeyer 1989), the abundance of calcitrol in fetal blood is <50% of that in maternal blood. It has been suggested that calcitrol cannot cross the placenta in rats (Noff and Edelstein 1978); therefore, the conceptus itself must be responsible for the generation of calcitrol from 25-hydroxyvitamin D that can be transported across the placenta in rats (Haddad et al. 1971). In contrast, calcitrol abundance is significantly greater in fetal blood than maternal blood in pregnant ewes (Abbas et al. 1987; Paulson et al. 1987) and calcitrol can pass between maternal and fetal blood in pregnant ewes (Devaskar et al. 1984). Furthermore, given the rapid metabolism of calcitrol in sheep fetuses, it is thought that calcitrol production is significantly upregulated in the ovine conceptus when compared with maternal production of calcitrol (Ross et al. 1989).

While it is well established that vitamin D is an important regulator of mineral homeostasis postnatally, contradicting data from animal models exists regarding whether vitamin D is required to maintain normal concentrations of minerals in fetal serum (Kovacs 2014). Loss of 1α-hydroxylase in fetal pigs, vitamin D deficiencies in rats, and some data from VDR null mice suggest that abundances of calcium, phosphorous, and PTH in fetal serum are not regulated by calcitrol, vitamin D, or VDR [reviewed by Kovacs (2014)]. However, other data from studies of other VDR null mice lines are inconsistent with these reports [reviewed by Kovacs (2014)].

In humans, rodents, and pigs, vitamin D metabolizing enzymes and the production of vitamin D at the maternal–conceptus interface have been demonstrated (Shahbazi et al. 2011; Bergada et al. 2014; O'Brien et al. 2014; Jang et al. 2017), suggesting potentially important roles of vitamin D in the establishment and maintenance of pregnancy, and in the regulation of conceptus growth and development. The expression of CYP27B1 mRNA and protein in placentae is positively correlated to the abundance of vitamin D in blood of humans (O'Brien et al. 2014). VDR is more abundant in the endometrium of pregnant mice compared with cyclic mice, and VDR expression increases in both the placenta and decidua with advancing days of gestation (Shahbazi et al. 2011). Additionally, the expression of CYP2R1, CYP27B1, CYP24A1, GC, and VDR varies significantly across gestation at the porcine maternal–conceptus interface (Jang et al. 2017), suggesting that vitamin D metabolism at the maternal–conceptus interface is regulated locally and varies across gestation.

Recently, we demonstrated the expression of vitamin D regulatory molecules at the ovine maternal–conceptus interface across gestation (Stenhouse et al. 2021a). VDR protein is expressed by both the trophectoderm and endoderm of the Day 17 conceptus, with more intense staining in the endoderm. Furthermore, VDR protein is expressed by the myometrium, blood vessels, uterine LE, uterine superficial glandular epithelium (sGE), uterine caruncles, and chorioallantois throughout pregnancy. The expression of VDR protein at the ovine maternal–conceptus interface throughout pregnancy demonstrates that the synthesis of active vitamin D may have broad functional consequences locally throughout pregnancy. Molecules involved in the metabolism of vitamin D (*CYP2R1*, *CYP11A1*, *CYP24*, and *CYP27B1*, and *VDR*) are expressed at the mRNA level in both endometria and placentae across gestation in sheep (Stenhouse et al. 2021a). Interestingly, *CYP2R1* and *CYP27B1* mRNAs have stable expression in ovine endometria and placentae across days of gestation (Stenhouse et al. 2021a). Since CYP2R1 and CYP27B1 are essential for the production of calcitrol, stable expression across gestation suggests an essential role for vitamin D throughout pregnancy across the maternal–conceptus interface. The expression of *CYP24*, *VDR*, and *CYP11A1* mRNAs at the ovine maternal–conceptus interface varies depending upon the gestational day investigated, with high expression in early pregnancy that decreases with advancing days of gestation (Stenhouse et al. 2021a). The activity of calcitrol is regulated by the catabolic activity of CYP24A1 (24-hydroxylase) that inactivates 1,25 [OH]2D3 via conversion to 1,24,25[OH]3D3 (Bikle 2014). The high expression of *CYP24* mRNA in ovine endometria and placentae in early pregnancy compared with mid- and late-gestation could be indicative of high requirements for these tissues to catabolize calcitrol during implantation and placentation, which may not be required as extensively in mid- to late-gestation when the placenta is established and readily transporting minerals and nutrients to the fetus. Similar findings have been reported for porcine endometria, with high expression of *CYP24* mRNA on Days 12 and 15 of gestation (Jang et al. 2017). Thus, it could be postulated that a localized negative feedback mechanism is present to regulate the abundance of calcitrol at the maternal–conceptus interface. Further investigations into the spatiotemporal expression and enzymatic activity of these enzymes would provide valuable insights into the importance of vitamin D in the establishment and maintenance of pregnancy in mammals.

Using porcine endometrial explant cultures, it has been demonstrated that calcitriol regulates the expression of several genes with essential roles in the establishment and maintenance of pregnancy including fibroblast growth factor 7 (FGF7), and secreted phosphoprotein 1 (SPP1) (Jang et al. 2017). In addition to the essential roles of vitamin D in the regulation of phosphate and calcium homeostasis, there are several 'non-classical' extra-skeletal functions of vitamin D. Vitamin D is a known regulator of both the adaptive and innate immune systems (Adams and Hewison 2008), targeting multiple immune cell types including T- and B-lymphocytes, monocytes, macrophages, and dendritic cells [reviewed by Baeke et al. (2010)]. Importantly, vitamin D modulates cytokine production by T-lymphocytes and antigen-presenting cells, and enhances anti-microbial activity of macrophages and monocytes, [reviewed by Baeke et al. (2010)]. During pregnancy, the uterine immune system must be tightly regulated to ensure that required anti-microbial protection occurs while ensuring that the conceptus is not rejected by the maternal immune system (Hunt 2006). Therefore, an immunomodulatory role of vitamin D at the maternal–conceptus interface has been suggested [reviewed by Tamblyn et al. (2015)]. Metabolism of vitamin D stimulates both anti-bacterial and anti-inflammatory responses by human decidual and trophoblast cells (Evans et al. 2006; Díaz et al. 2009; Liu et al. 2009). Additionally, trophoblast-derived vitamin D has been suggested to be a regulator of placental inflammation in mice (Liu et al. 2011). Given the importance of immune modulation during pregnancy, and the described spatiotemporal expression profile of vitamin D regulatory molecules at the maternal–conceptus interface in multiple species, it could be hypothesized that vitamin D not only modulates phosphate and calcium transport but plays an important role in immunomodulation at the maternal–conceptus interface during implantation and placentation, which warrants further investigation.

5.7.3 Phosphatonins

Fibroblast growth factor 23 (FGF23), secreted frizzled-related protein 4 (sFRP4), matrix extracellular phosphoglycoprotein, and fibroblast growth factor 7 (FGF7) are members of the phosphatonin family of circulating molecules with suggested roles in the maintenance of phosphate homeostasis (Berndt and Kumar 2007; Shaikh et al. 2008). Of the phosphatonins, FGF23 has been the most extensively investigated and is considered a significant regulator of phosphate transport through its interactions with Klotho (KL). KL acts as a coreceptor with members of the fibroblast growth factor receptor family (FGFR), and FGF23 binds to the FGFR-KL complex. Binding to this complex can regulate WNT (wingless-related integration site), PKC (protein kinase C), cAMP (cyclic adenosine monophosphate), p53 (tumor protein 53), and p21 (cyclin-dependent kinase inhibitor) signaling cascades to maintain phosphate homeostasis (Wang and Sun 2009). The metalloproteinases ADAM10 and ADAM17 can cleave KL from the plasma membrane to form a secreted form of KL (Chen et al. 2007). KL can mediate phosphate absorption by the intestine, phosphate excretion in urine, and phosphate distribution in bone in both an FGF23-dependent and -independent manner. Furthermore, the expression of type II and type III sodium-dependent phosphate transporters is tightly regulated via KL and FGF23 (Bon et al. 2018; Hu et al. 2019). In addition, other FGF family members such as FGF19 and FGF21 can utilize the KL-FGFR complex to enhance their cell signaling (Dolegowska et al. 2019). In addition to the role of KL signaling in the regulation of phosphate homeostasis, KL likely has essential roles in the regulation of calcium (Nabeshima and Imura 2008), acting as a glucuronidase to activate transient receptor potential cation channel 5 (TRPV5) (Chang et al.

2005). Furthermore, KL is a mediator of calcitrol synthesis (Yoshida et al. 2002; Haussler et al. 2013).

In mice, the fetal heart, liver, and somites express FGF23 from embryonic day 12.5 (Sitara et al. 2004). Despite low expression of *FGF23* mRNA by the murine placenta, several genes involved in FGF23-KL signaling are expressed including KL, SLC34A1, SLC34A2, SLC34A3, CYP27B1, CYP24A1, FGFR1, FGFR2, FGFR3, and FGFR4 (Ma et al. 2014). Transgenic mouse lines have provided interesting insights into the role of FGF23 in the regulation of placental phosphate transport. While FGF23 null mice are fertile, they have a growth restriction phenotype accompanied by shorter longevity (Sitara et al. 2004). Interestingly, these studies revealed that intact FGF23 cannot cross the murine placenta despite fetal levels of intact FGF23 being comparable to intact FGF23 levels in wild-type pregnant mice (Ma et al. 2014).

A role of KL in the maintenance of calcium and phosphate transport in pigs has also been suggested (Choi et al. 2014a). *KL* mRNA is expressed in both the porcine endometria and chorioallantois and analysis of porcine endometrial *KL* mRNA expression demonstrated that *KL* mRNA decreased between Days 12–15 of the estrous cycle. Interestingly, porcine endometrial *KL* mRNA has a biphasic expression profile, with the highest expression observed on Days 12 and 90 of gestation. A role of KL-FGF23 signaling in the regulation of phosphate transport during pregnancy in ewes has also been suggested (Stenhouse et al. 2021b). FGF23 and KL proteins are localized to both the endoderm and trophectoderm of Day 17 ovine conceptuses, and both were abundantly expressed in the myometrium, uterine LE, sGE, and GE, and stratum compactum stroma, as well as blood vessels throughout gestation. FGF23 protein is expressed by the mononuclear and binucleate cells at the interface of the chorioallantois and caruncular tissue, and in areas undergoing syncytialization within the placentome. As pregnancy progresses, FGF23 protein is expressed in greater abundance in caruncular tissue than cotyledonary tissue of placentomes. KL protein is expressed in caruncular tissue, the chorioallantois and blood vessels throughout gestation. There is expression of FGF23-KL signaling molecules (FGF23, FGFR1-4, KL, KLB, ADAM10, and ADAM17) at the mRNA level at the ovine maternal–conceptus interface throughout gestation. The spatiotemporal expression profiles suggest potential roles of KL-FGF23 signaling in the regulation of conceptus implantation in sheep, which warrants further investigation.

KL is very abundant in human cord blood (Ohata et al. 2011) and it is expressed by the syncytiotrophoblast (STB) cells of the chorionic and basal plates of the human placenta (Iñiguez et al. 2019). Data from studies with humans suggest that KL regulates placental transport of phosphate and fetal growth, with alterations in the expression of KL, ADAM17, and FGFR1 associated with pre-eclampsia and small for gestational age term placentae (Miranda et al. 2014; Loichinger et al. 2016; Iñiguez et al. 2019).

In addition to the role of KL in the regulation of pathways associated with cellular proliferation and apoptosis, KL can regulate angiogenesis, nitric oxide production by endothelial cells, and protect against endothelial dysfunction, all of which are important in the establishment and maintenance of pregnancy (Saito et al. 1998; Fukino et al. 2002; Yamamoto et al. 2005; Kusaba et al. 2010). It is also known that vascular smooth muscle cells have high expression of KL and that knockdown of KL expression in human vascular cells revealed that they are a KL-dependent target tissue for FGF23 (Lim et al. 2012). Extensive angiogenesis must occur in the uterus and placentae to allow adequate nutrient transport from the mother to the fetus. Future studies to ascertain whether KL-FGF signaling also acts as an essential regulator of angiogenesis at the ovine maternal–conceptus interface should be performed.

Despite extensive research into the role of FGF23 in the regulation of phosphate transport, few studies have investigated the mechanisms whereby the phosphatonin FGF7 regulates phosphate transport. FGF7 inhibits phosphate uptake in tumor cells (Carpenter et al. 2005), and its expression may be regulated by vitamin D in breast cancer cells (Lyakhovich et al. 2000). FGF7 may also have an essential role in the female reproductive tract as it is classified as a progestamedin. FGF7 is expressed by uterine stromal cells in response to progesterone (P4) and acts in a paracrine manner via $FGFR_{2IIIb}$ to regulate the function of uterine epithelial cells and conceptus trophectoderm (Chen et al. 2000). Ovine *FGF7* mRNA localizes to the tunica muscularis of blood vessels in the endometrium and myometrium (Chen et al. 2000). Given the essential roles of the progestamedins in the establishment and maintenance of pregnancy, and the limited studies that have implicated FGF7 in phosphate transport, future studies should determine whether progestamedins act as regulators of phosphate and calcium transport during pregnancy.

5.7.4 Calcium-Binding Proteins

The S100 family of proteins are cytosolic calcium-binding proteins that have central roles in the regulation of many cellular processes including cellular proliferation, differentiation, migration, metabolism, apoptosis, protein phosphorylation, and the maintenance of calcium homeostasis (Hermann et al. 2012).

Calbindin-D9K, also known as S100G, is a vitamin D-regulated calcium binding protein expressed by epithelial cells of the intestine for the modulation of calcium transport (Darwish and DeLuca 1992). While S100G is a known regulator of calcium absorption in the kidney and intestine, and multiple studies provide evidence for a critical role of this protein in the establishment of pregnancy and regulation of conceptus growth and development. In the mouse uterus, *S100G* mRNA is localized to uterine LE and GE when the endometrium is receptive to implantation of the blastocyst (Nie et al. 2000). Furthermore, *S100G* mRNA expression is down-regulated at the implantation site in mice (Nie et al. 2000). Similarly, S100G protein is localized to the uterine stroma and myometrium in the rat, with decreased uterine expression during the period of blastocyst implantation (An et al. 2003). In contrast to findings from rodents, S100G protein is expressed only by uterine LE and GE of the non-pregnant bovine endometrium, and not by myometrial or uterine stromal cells (Inpanbutr et al. 1994). S100G mRNA and protein expression are greatest during the luteal phase of the estrous cycle, suggesting that S100G may be important in the regulation of uterine gland secretions during this phase of the estrous cycle in cattle. Furthermore, S100G is highly expressed by the porcine uterus on Day 12 of pregnancy, and its expression may be regulated by P4 (Choi et al. 2009, 2012). In sheep, *S100G* mRNA is expressed in endometria across gestation (Stenhouse et al. 2021a), with high expression at Day 30 of gestation, which could indicate a vital role for S100G in placental development and function. Both the cytotrophoblast (CTB) and STB cells of human term placentae express S100G mRNA, with greater expression by STB than CTB cells (Belkacemi et al. 2004), highlighting a role of S100G in the mediation of calcium flux at the maternal–conceptus interface. S100G is expressed by the yolk sac and placenta in multiple species (reviewed by Choi and Jeung (2008), and it is hypothesized that this protein has a critical role in regulation of the accumulation of excess calcium required in late gestation for mineralization of the fetal skeletal system.

While S100G is expressed at the conceptus–maternal interface in multiple species, with differing spatiotemporal expression profiles across species, the functional significance of S100G in pregnancy remains poorly understood. It is proposed that other calcium regulatory molecules may undergo compensatory increases in the kidney and intestines of S100G null mice, such as TRPV6 and plasma membrane Ca^{2+}-ATPase 1 (PMCA1) (reviewed by (Choi and Jeung 2008). S100G null mice are fertile but it could be hypothesized that in the absence of S100G, other calcium regulatory molecules undergo compensatory increases in expression at the maternal–conceptus interface to ensure that calcium transport is appropriately regulated.

Additional S100 family members, while not as extensively investigated, have been suggested to play critical roles in pregnancy. S100A8 is associated with early recurrent pregnancy loss in humans (Nair et al. 2013) and S100A8 null mice (Passey et al. 1999; Baker et al. 2011). S100A7, S100A8, S100A9, and S100A12 have been suggested to have a role in the establishment of pregnancy in pigs (Zeng et al. 2018, 2019). Similarly, roles for S100A9 and S100A12 in the establishment of pregnancy in sheep have been suggested because of high endometrial expression early in gestation (Days 9, 12, and 17) (Stenhouse et al. 2021a). Interestingly, the spatiotemporal profile of S100A9 protein in the ovine uterus varies significantly across gestation. S100A9 protein is localized to uterine LE, sGE, and GE on Days 9, 12, and 17 of gestation, suggesting a pivotal role of this molecule in the regulation of calcium trafficking at the maternal–conceptus interface during the peri-implantation period of pregnancy. In contrast, S100A9 protein is only detected in uterine GE on Days 30 and 90 of gestation, and uterine LE at Day 125 of gestation. This striking change in the localization of this protein across gestation in uteri of sheep suggests alterations in the functional importance of S100A9 during gestation that warrants further investigation. S100A11 may have a role in the regulation of endometrial receptivity to implantation of blastocysts and immunotolerance through regulation of epidermal growth factor-stimulated adhesion through its roles in regulating intracellular calcium uptake and release (Liu et al. 2012; Poeter et al. 2013). Knockdown of S100A11 in mice reduced the expression of genes associated with uterine receptivity to implantation and reduced rates of implantation of blastocysts (Liu et al. 2012). Additionally, S100A11 has been detected in uterine flushings from ewes on Days 14 and 16 of pregnancy, suggesting a potential role of S100A11 during the peri-implantation period of pregnancy (Romero et al. 2017). The results from studies with these animal models translate directly to those from humans in which S100A11 is a factor associated with pregnancy failure, and low endometrial expression of S100A11 is associated with adverse immune responses (Liu et al. 2012).

While S100 family members are known to regulate calcium flux, many of the S100 proteins are also mediators of inflammation (Donato et al. 2013; Xia et al. 2018). Available data suggest that the S100 family of calcium-binding proteins plays a critical role in early pregnancy when the uterine immune environment must be tightly regulated to ensure that anti-microbial protection occurs and the conceptus is not rejected by the maternal immune system (Hunt 2006). It could be speculated that in addition to ensuring that sufficient calcium is available for the rapidly dividing and differentiating conceptus tissues in early gestation, members of the S100 family play an important role in the establishment of immunotolerance at the maternal–conceptus interface.

5.7.5 Transient Receptor Potential Cation Channel (TRPV) Family

The TRPV family members, TRPV5 and TRPV6, have a critical role in the regulation of calcium transport by epithelial cells in mammals (Islam 2011). The TRPV channels, such as TRPV5 and TRPV6, have important functions in the regulation of calcium transport at the maternal–conceptus interface to regulate calcium transport in a manner similar to that for transport of calcium by intestinal epithelia. It is speculated that TRPV channels are important for the transfer of calcium in maternal blood to cells on the maternal side of the maternal–conceptus interface, and that Ca^{2+}-ATPases are present on the basement membranes of cells on the fetal side of the maternal–conceptus interface that allow that calcium to enter the fetal circulation.

TRPV6 mRNA and protein are expressed by the murine yolk sac (Suzuki et al. 2008) with expression of TRPV6 increasing 14-fold in the placentae of mice during the last 4 days of gestation (Suzuki et al. 2008) to ensure that required calcium is available for mineralization of the fetal skeletal system. TRPV6 null mice have a 40% reduction in placental calcium transport, accompanied by severe hypocalcemia and a 50% decrease in weight of fetal skeletal ash, indicating that placental TRPV6 is critical for the regulation of placental transport of calcium (Suzuki et al. 2008). TRPV6 is highly expressed by the porcine uterus on Day 12 of pregnancy (Choi et al. 2009, 2012), and a role for TRPV channels in early pregnancy has also been suggested for sheep (Stenhouse et al. 2021a). Day 17 ovine conceptus tissue expresses *TRPV5* mRNA, and there is also high expression of *TRPV6* mRNA in Day 17 endometria and Day 30 placentae. Interestingly, endometria and placentae supplying twin fetuses have greater expression of *TRPV6* mRNA compared with endometria from ewes with singleton pregnancies at Day 125 of gestation. Thus, TRPV6 may be critical in regulating the increase in calcium transport required to meet the increased mineral demands in twin pregnancies.

5.7.6 Sodium-Dependent Phosphate Transporters

Solute carrier family members (SLCs) have central roles in the regulation of solute transport across the plasma membrane, which is critical for the regulation of intracellular contents of metabolites and ions, cellular volume, and the removal of waste products from cells (Pizzagalli et al. 2020). Many SLC transporter family members exist, allowing for passive facilitative transport or secondary active transport of molecules such as sugars, amino acids, polyamines, vitamins, nucleotides, metals, and inorganic and organic ions (Pizzagalli et al. 2020). The SLC34 and SLC20 families of type II and type III sodium-dependent phosphate transporters play critical roles in bone, kidney, choroid plexus, and vascular physiology and pathophysiology (Lederer 2014; Segawa et al. 2015).

Interestingly, evidence suggests that members of the SLC34 and SLC20 families of type II and type III sodium-dependent phosphate transporters have a critical role in the regulation of growth and development of mammalian conceptuses. In vitro placental perfusion studies have provided interesting insights into the mechanisms of mineral transport in many species including humans, guinea pigs, rats, mice, and sheep [reviewed by Kovacs (2014)]. Manipulation of Na^+ concentrations in these models demonstrated that placental phosphate transport is sodium-dependent (Husain and Mughal 1992; Stulc and Stulcova 1996). Given that sodium-dependent phosphate transporters are regulators of post-natal phosphate homeostasis, it could be hypothesized that they are regulators of phosphate transport at the maternal–fetal interface. SLC20A1 is expressed by STB cells of the human placenta (Yang et al. 2014) and defects in vascular remodeling of the yolk sac in SLC20A1-deficient mice is embryonic lethal (Wallingford and Giachelli 2014). Similarly, knockout of the type II sodium-dependent transporter SLC34A2 is embryonic lethal in mice (Shibasaki et al. 2009). SLC20A2 is localized to intravillous connective tissues in placentae of humans (Yang et al. 2014). A role for sodium-dependent phosphate transporters in the development of preeclampsia is based on evidence that a reduction in placental expression of SLC20A1 and SLC20A2 is associated with development of preeclampsia and, in late gestation, there is a significant increase in expression of SLC20A2 in preeclamptic pregnancies. SLC20A2 deficient mice are fertile, but some 25% of the mice have pregnancy complications including abnormal placental vascular remodeling, fetal growth restriction, and placental calcification and about 50% of the offspring of those mice do not survive until weaning (Wallingford et al. 2016).

A role for sodium-dependent phosphate transporters SLC20A1 and SLC20A2 in the regulation of phosphate transport at the maternal–conceptus interface in sheep has also been suggested (Stenhouse et al. 2021b). Both *SLC20A1* and *SLC20A2* mRNAs are expressed by ovine placentae and endometria throughout gestation. Interestingly, expression of *SLC20A1* mRNA decreased in ovine endometria at Day 17 of gestation, corresponding to the period of conceptus attachment to uterine LE. There was a stable expression of *SLC20A2* mRNA across gestation in both ovine endometria and placentae, highlighting a potentially crucial role for this transporter in the regulation of placental phosphate transport across gestation.

5.7.7 Plasma Membrane Ca^{2+}-ATPases

Plasma membrane Ca^{2+}-ATPases (PMCAs) are Ca^{2+} extrusion pumps that function to maintain intracellular concentrations of calcium within a tightly regulated range. There are four main isoforms of PMCA (PMCA1-4) expressed in adult tissues. PMCA expression doubles during the last week of gestation in rodents (Tuan and Bigioni 1990; Glazier et al. 1992). Similarly, PMCA 1–4 is expressed by the STB of human placentae and ATP-dependent Ca^{2+} transport increases linearly during the third trimester of pregnancy (Strid and Powell 2000). A critical role for Ca^{2+}-ATPase activity in the regulation of placental transport of calcium was suggested based on results obtained following inhibition of Ca^{2+}-ATPase activity in the rat placentae (Care 1991; Strid and Powell 2000). Homozygous PMCA1 knockout mice have an embryonic lethal phenotype prior to implantation of the blastocyst (Okunade et al. 2004; Prasad et al. 2007). In contrast, PMCA2 and PMCA4 null mice have mild phenotypes although further analyses should be performed to investigate effects on the development of the fetal skeletal system and on placental development and structure (Okunade et al. 2004; Prasad et al. 2007).

The *PMCA1-4* mRNAs are expressed by the uterine endometrium of pigs during the estrous cycle and throughout gestation, in a pregnancy status and stage of pregnancy-specific manner (Choi et al. 2014b), suggesting a role of Ca^{2+}-ATPases in the regulation of calcium ion abundance in uterine endometria. A role for plasma membrane Ca^{2+}-ATPases in early pregnancy in sheep has also been suggested (Stenhouse et al. 2021a). Day 17 ovine conceptus tissue expresses *PMCA4* mRNA which, in combination with high expression of *PMCA3* and *PMCA4* mRNA in endometria from Day 17 of pregnancy, and high expression of *PMCA4* mRNA in placentae on Day 30 of pregnancy, suggests a role in the regulation of implantation and placental development. Collectively, these findings suggest a role for *PMCAs* in the regulation of calcium at the maternal–conceptus interface in pregnancy that warrants further investigation, particularly during the critical periods of implantation and placentation.

5.7.8 Stanniocalcin

The homodimeric phosphoglycoprotein stanniocalcin (STC) is a central regulator of calcium levels in fish (Wagner et al. 1986; Lu et al. 1994; Wagner 1994; Wagner and Jaworski 1994). STC decreases the influx of calcium from the aquatic environment through the gills and promotes absorption of phosphate in the kidney, while chelating excess calcium, and inhibiting calcium uptake in the intestine. STC1 and STC2 are mammalian orthologs of STC. STC1 has relatively high amino acid sequence identity (approximately 50%) to fish STC and is expressed in many tissues including brain, lung, and heart. While it is known that mammalian STC1 is a regulator of calcium and phosphate transport in the kidney and intestine, in part through the regulation of sodium-phosphate co-transporters, the role of STC2 remains poorly understood.

In addition to the role of STCs in the kidney and intestine, STC may have an important role in female reproduction. During pregnancy, the abundance of *STC1* mRNA in the murine ovary increases substantially and STC1 protein is present in maternal blood during pregnancy despite not normally being found in blood (Deol et al. 2000). The localization of *STC1* mRNA in the mouse uterus varies substantially depending upon pregnancy status and stage. In cyclic mice, the uterine LE expresses *STC1* mRNA (Stasko et al. 2001). Following implantation, *STC1* mRNA localizes to the mesometrial stromal cells bordering the uterine lumen until Days 6.5–8.5 when expression is in the mesometrial lateral sinusoids, after which time expression decreases. STC1 protein, however, is expressed by the uterine epithelia, stromal cells and decidual cells, as well as trophoblast giant cells. Similarly, STC1 and STC2 may have roles in the regulation of implantation and decidualization in the rat (Xiao et al. 2006). *STC1* and *STC2* mRNA expressions are induced in stromal cells at implantation sites on Day 6 of pregnancy, with additional expression in the uterine LE. Furthermore, STC2 protein is immunolocalized to uterine GE from Day 2 of gestation, with expression peaking on Day 6. STC1 mRNA and protein are expressed by decidualized cells from Days 7–9 of gestation. Interestingly, while *STC1* mRNA is expressed by the whole decidua from Days 7–9 of gestation, *STC2* mRNA is only expressed in the decidua in close proximity to the implantation site on Days 7 and 8, with weak staining throughout the entire decidua by Day 9.

A role for STC has also been suggested for the female reproductive tract of large animals. In the pig, *STC1* mRNA is localized to uterine LE between Days 12 and 15 of the estrous cycle and, in pregnant pigs, it increases between Days 15 and 20 of gestation, suggesting that its expression may be regulated by P4 (Song et al. 2009). After Day 20, expression of *STC1* mRNA in the uterus decreases to minimal expression on Day 25 and no expression at Day 30, suggesting that STC1 may be important for conceptus attachment and implantation. Expression of *STC1* mRNA by the endometrial glands of the ovine uterus is evident from Day 18 of pregnancy and increased further to Day 80 of gestation (Song et al. 2006). STC1 protein is localized predominantly to the apical surface of uterine GE after Day 16 of pregnancy and in the placental areolae that form opposite the opening of uterine glands to transport histotroph across the placenta and into the fetal-placental blood. Ovine endometrial stroma and glands have low expression of *STC2* mRNA in both cyclic and early pregnant uteri (Song et al. 2006). The abundance of *STC2* mRNA significantly increases between Days 20 and 40 of gestation, with high expression in both placentomes and endometria until Day 120 of gestation. In late gestation, abundant expression of *STC2* mRNA is present in uterine LE and GE, trophectoderm, and uterine caruncules.

In addition to demonstrating that STC1 and STC2 mRNAs and proteins are expressed by ovine uterine and placental tissues, Song et al. established that STC1 protein is present in ovine and porcine uterine luminal fluid (Song et al. 2006, 2009) and ovine allantoic fluid (Song et al. 2006). Thus, STC1 can be secreted by endometrial glands and enter allantoic fluid and the fetal circulation. Further investigations are required to ascertain the role of secreted STC1 in the pregnant uterus, but it may act as a regulator of conceptus growth and development.

Given the spatiotemporal profiles of STC1 expression across gestation, its expression may be hormonally regulated in the female reproductive tract. While interferon tau (IFNT; signal for maternal recognition of pregnancy in ruminants) does not appear to be a regulator of STC1 expression in the ovine uterus, treatment of ewes with P4 does induce expression of *STC1* mRNA in the ovine endometria (Song et al. 2006). Intrauterine infusions of ovine placental lactogen increased the abundance of STC1 in progestinized ewes; demonstrating an important role of P4 and placental lactogen in the regulation of calcium transport by STC1 by the ovine uterus (Song et al. 2006). Furthermore, the expression of *STC1* mRNA by uterine LE in pigs is stimulated by P4, and this expression is enhanced by estrogen (Song et al. 2009). Given the pattern of expression of STC1 in the uterine LE of pigs, with complete downregulation occurring after conceptus attachment and implantation, the downregulation in expression may be in response to the progestinized uterus being exposed to estrogen, making STC1 a useful marker of implantation in pigs. Furthermore, in that study, treatment of ewes with estradiol (E2) increased the expression of progesterone receptor (PGR) by uterine GE and this was associated with decreased expression of STC1, which could indicate that downregulation of PGR is critical for the expression of STC1 and other genes expressed by uterine GE.

Collectively, the spatiotemporal expression patterns suggest an important role for the calcium and phosphate regulator STC in the female reproductive tract. Despite differences in expression profiles for STC1 among species, it is evident that STC1 plays a role in both the establishment and maintenance of pregnancy in species with differing types of placentation to influence growth and development of the conceptus. The intriguing finding that STC1 and STC2 have opposing expression profiles in the ovine uterus and placentomes indicate that they may work together to regulate calcium and phosphate transport at the ovine maternal–conceptus interface in different ways. The concept warrants further investigation to improve our understanding of these molecules.

5.7.9 Sex Steroids

Sex steroids, specifically estradiol and testosterone, are critical for the regulation of postnatal skeletal development and bone turnover throughout adulthood, and therefore are considered calciotropic hormones. Placental-derived hCG and the fetal pituitary gonadotropins can stimulate sex steroid production by developing fetal ovaries and testes. Additionally, the fetal adrenals synthesize low amounts of sex steroids. During pregnancy, concentrations of both estradiol (Habert and Picon 1984) and progesterone (Ward and Weisz 1984) are similar between male and female fetuses in rats. In contrast, concentrations of testosterone in fetuses are at least four times greater for male than female fetuses in rats (Habert and Picon 1984). In addition, concentrations of testosterone are equal to (Houtsmuller et al. 1995) or higher than those in maternal blood in rats [reviewed in Kovacs (2014)]. Similar findings have been reported for sheep, with higher concentrations of testosterone in plasma of male than female fetuses. Interestingly, there is a dramatic decrease in concentrations of testosterone in fetal plasma immediately prior to parturition, and concentrations of testosterone remain greater than those in maternal plasma. Concentrations of progesterone are less in male and female sheep fetuses compared with much higher concentrations in maternal plasma until just prior to parturition when there is a dramatic decline in progesterone in maternal plasma (Strott et al. 1974). Concentrations of estradiol in fetal plasma are slightly greater than those in maternal plasma throughout most of pregnancy (Strott et al. 1974). Thus, circulating levels of sex steroids in fetal lambs exceed maternal concentrations at all gestational time points investigated [reviewed in Kovacs (2014)]. Finally, the testes in human fetuses produce testosterone in response to gonadotropins and hCG, whereas negligible amounts of testosterone are produced by the ovaries or adrenals in female fetuses (Reyes et al. 1973). Concentrations of estradiol in human fetuses are quite low and similar between males and females, and much lower than those in maternal plasma at term [reviewed in Kovacs (2014)].

The levels of sex steroids in the placenta can be quite high during pregnancy, with concentrations of testosterone, estradiol, or both in umbilical cord blood being equal to or greater than those in maternal plasma [reviewed in Kovacs (2014)]. This appears to result from aromatase enzymatic activity in the placenta converting maternal androgens into estrogens. In the third trimester of pregnancy in humans, the placenta is the source of nearly all of the estradiol, estrone, and estriol present in maternal blood (Menon and Sperling 1998).

Although estrogen and testosterone are considered calciotropic hormones because of their effects on development of the post-natal skeleton, evidence that they play a role in mineral transport and homeostasis during pregnancy is less clear. Studies of calcium and phosphate in serum of murine fetuses that lack sex steroids, or their receptors have not been performed, apart from describing them as being normal at birth. This lack of evidence for mineralization deficiencies suggests that development of fetal bone and mineral homeostasis in fetuses is less dependent upon sex steroids (Kovacs 2015). In the rat, it has been suggested that fluctuations in concentrations of Ca^{2+} may result from activities of different calcium transporters in uteri that respond differentially to progesterone and estradiol during the estrous cycle (Kim et al. 2006; Yang et al. 2011). Additionally, in the pig, it has been suggested that estrogen increases the release and or transport of calcium into the uterine lumen during the peri-implantation period of gestation (Geisert et al. 1982).

Estrogen is a known regulator of calcium transport in the intestine (Arjmandi et al. 1993) and kidney (Bronner 1989). Estrogen stimulates expression of renal CYP27B1 that increases calcitriol secretion by the kidneys in multiple animal models and humans (Reddy et al. 1983; Kovacs 2016). Estradiol also reduces serum FGF23 and increases PTH in blood, which also may contribute to increase in calcitriol (Saki et al. 2020), although the requirement for estrogen to stimulate the 2–3-fold increase in calcitriol in maternal blood during pregnancy has not been tested directly (Ryan and Kovacs 2021).

In addition to the indirect regulation of calcitriol by estrogen via CYP27B1 expression and decreases in FGF23 and PTH, estrogen may also indirectly regulate calcium transport during pregnancy via regulation of TRPV6, S100G, and vitamin D metabolism. Estrogen is a key regulator of expression of TRPV6 during pregnancy (Lee and Jeung 2007), and TRPV6 is critical for the regulation of placental transport of calcium (Suzuki et al. 2008). TRPV6 is highly expressed in the porcine uterus on Day 12 of pregnancy (Choi et al. 2009, 2012) and in uteri of pregnant sheep (Stenhouse et al. 2021a). Thus, taken together, current evidence supports an indirect role for sex steroids in the regulation of mineral homeostasis during pregnancy, primarily via upregulation of expression of calcitriol.

5.8 Summary and Conclusions

Appropriate transport of nutrients and minerals at the maternal–conceptus interface is essential not only for the establishment and maintenance of pregnancy but also the appropriate regulation of conceptus development and growth. Despite our understanding that phosphate, calcium, and vitamin D are essential for conceptus development and mineralization of the fetal skeletal system, little is known regarding mechanisms of transport and metabolism of these molecules at the maternal–conceptus interface. Current research defining the spatiotemporal expression of the multiple pathways for calcium and phosphate transport and vitamin D metabolism at the maternal–conceptus interface across gestation and species is vital to improving understanding of these processes. Furthermore, understanding how IFNT, E2, P4, and nutrients regulate phosphate and calcium transporters, and enzymes involved in vitamin D metabolism will further understanding of the roles of these molecules in the nutrition and physiology of reproduction under normal and pathological conditions.

Acknowledgements This work was supported, in part, by Texas A&M University Presidential Impact Fellow Funds (to FWB) and funds from Texas A&M AgriLife Research (to FWB and GW), and the National Institutes of Health # 1R21DE028076-01 (to DG).

Conflicts of Interest The authors have no conflicts of interest to declare.

References

Abbas SK, Care AD, Van Baelen H, Bouillon R (1987) Plasma vitamin D-binding protein and free 1,25-dihydroxyvitamin D3 index in pregnant ewes and their fetuses in the last month of gestation. J Endocrinol 115:7–12

Abbas S, Pickard D, Rodda C, Heath J, Hammonds R, Wood W, Caple I, Martin T, Care A (1989) Stimulation of ovine placental calcium transport by purified natural and recombinant parathryoid hormone-related protein (PTHrP) preparations. Q J Exp Physiol 74:549–552

Achkar M, Dodds L, Giguère Y, Forest JC, Armson BA, Woolcott C, Agellon S, Spencer A, Weiler HA (2015) Vitamin D status in early pregnancy and risk of preeclampsia. Am J Obstet Gynecol 212:e1–e7

Adams JS, Hewison M (2008) Unexpected actions of vitamin D: new perspectives on the regulation of innate and adaptive immunity. Nat Clin Pract Endocrinol Metab 4:80–90

Aerssens J, Boonen S, Lowet G, Dequeker J (1998) Interspecies differences in bone composition, density, and quality: potential implications for in vivo bone reseach. Endocrinology 139:663–670

Alimohamadi S, Esna-Ashari F, Rooy RSB (2020) Relationship of vitamin D serum level with intrauterine growth retardation in pregnant women. Int J Women's Heal Reprod Sci 8:221–226

Allen WM, Sansom BF (1986) Parturient paresis (milk fever) and hypocalcemia (cows, ewes, and goats). In: Howard JL (ed) Current veterinary therapy: food animal practice, Philadelphia, vol 2, pp 311–320

Allgrove J, Adami S, Manning RM, O'Riordan JL (1985) Cytochemical bioassay of parathyroid hormone in maternal and cord blood. Arch Dis Child 60:110–115

An B, Choi K, Kang SK, Hwang WS, Jeung E (2003) Novel Calbindin-D 9k protein as a useful biomarker for environmental estrogenic compounds in the uterus of immature rats. Reprod Toxicol 17:311–319

Ardawi MSM, Nasrat HAN, Aqueel HSBA (1997) Calcium-regulating hormones and parathyroid hormone-related peptide in normal human pregnancy and postpartum: a longitudinal study. Eur J Endocrinol 137:402–409

Arjmandi BH, Salih MA, Herbert DC, Sims SH, Kalu DN (1993) Evidence for estrogen receptor-linked calcium transport in the intestine. Bone Miner 21:63–74

Baeke F, Takiishi T, Korf H, Gysemans C, Mathieu C (2010) Vitamin D: modulator of the immune system. Curr Opin Pharmacol 10:482–496

Baker JR, Jeffery R, May RD, Mathies M, Spencer-Dene B, Poulsom R, Hogg N (2011) Distinct roles for S100a8 in early embryo development and in the maternal deciduum. Dev Dyn 240:2194–2203

Banerjee A, Padh H, Nivsarkar M (2011) Hormonal crosstalk with calcium channel blocker during implantation. Syst Reprod Med 57:186–189

Barry JS, Anthony RV (2008) The pregnant sheep as a model for human pregnancy. Theriogenology 69:55–67

Bazer FW, Spencer TE, Thatcher WW (2012) Growth and development of the ovine conceptus. J Anim Sci 90:159–170

Belkacemi L, Gariépy G, Mounier C, Simoneau L, Lafond J (2004) Calbindin-D9k (CaBP9k) localization and levels of expression in trophoblast cells from human term placenta. Cell Tissue Res 315:107–117

Bergada L, Pallares J, Arcidiacono M, Cardus A, Santacan M, Valls J, Cao G, Fernandez E, Dolcet X, Dusso A, Matias-Guiu X (2014) Role of local bioactivation of vitamin D by CYP27A1 and CYP2R1 in the control of cell growth in normal endometrium and endometrial carcinoma. Lab Invest 94:608–622

Berndt T, Kumar R (2007) Phosphatonins and the regulation of phosphate homeostasis. Annu Rev Physiol 69:341–359

Bikle D (2014) Vitamin D metabolism, mechanism of action, and clinical applications. Chem Biol 21:319–329

Boass A, Garner SC, Schultz VL, Toverud SU (1997) Regulation of serum calcitriol by serum ionized calcium in rats during pregnancy and lactation. J Bone Miner Res 12:909–914

Bodnar LM, Catov JM, Simhan HN, Holick MF, Powers RW, Roberts JM (2007) Maternal vitamin D deficiency increases the risk of preeclampsia. J Clin Endocrinol Metab 92:3517–3522

Bon N, Frangi G, Sourice S, Guicheux J, Beck-cormier S, Beck L (2018) Phosphate-dependent FGF23 secretion is modulated by PiT2/Slc20a2. Mol Metab 11:197–204

Bouillon R, Marcocci C, Carmeliet G, Bikle D, White JH, Dawson-hughes B, Lips P, Munns CF, Lazaretti-castro M, Giustina A, Bilezikian J (2019) Skeletal and extraskeletal actions of Vitamin D: current evidence and outstanding questions. Endocr Rev 40:1109–1151

Bourdeau A, Manganella G, Thil-Trubert CL, Sachs C, Cournot G (1990) Bioactive parathyroid hormone in pregnant rats and fetuses. Am J Physiol 258:E549–E554

Bronner F (1989) Renal calcium transport: mechanisms and regulation-an overview. Am J Physiol 257:F707–F711

Brostrom MA, Brostrom CO (2003) Calcium dynamics and endoplasmic reticular function in the regulation of protein synthesis: implications for cell growth and adaptability. Cell Calcium 34:345–363

Care AD (1991) The placental transfer of calcium. J Dev Physiol 15:253–257

Care AD, Caple IW, Singh R, Peddie M (1986) Studies on calcium homeostasis in the fetal Yucatan miniature pig. Lab Anim Sci 36:389–392

Carpenter TO, Ellis BK, Insogna KL, Philbrick WM, Sterpka J, Shimkets R (2005) Fibroblast growth factor 7: an inhibitor of phosphate transport derived from oncogenic osteomalacia-causing tumors. J Clin Endocrinol Metab 90:1012–1020

Carter AM (2007) Animal models of human placentation —a review. Placenta 28:S41–S47

Chang Q, Hoefs S, Van Der Kemp AW, Topala CN, Bindels RJ, Hoenderop JG (2005) Cell signalling: the β-glucuronidase klotho hydrolyzes and activates the TRPV5 channel. Science 310:490–493

Chen C, Spencer TE, Bazer FW (2000) Fibroblast growth factor-10: a stromal mediator of epithelial function in the ovine uterus. Biol Reprod 63:959–966

Chen C, Podvin S, Gillespie E, Leeman SE, Abraham CR (2007) Insulin stimulates the cleavage and release of the extracellular domain of Klotho by ADAM10 and ADAM17. Proc Natl Acad Sci USA 104:19796–19801

Chen YH, Fu L, Hao JH, Yu Z, Zhu P, Wang H, Xu YY, Zhang C, Tao FB, Xu DX (2015) Maternal vitamin D deficiency during pregnancy elevates the risks of small for gestational age and low birth weight infants in Chinese population. J Clin Endocrinol Metab 100:1912–1919

Chhabra S, Singh A (2017) Role of calcium in hypertensive disorders of pregnancy current status of research a mini review. J Nutr Disord Ther 7:1–5

Chin KV, Cade C, Brostrom CO, Galuska EM, Brostrom MA (1987) Calcium-dependent regulation of protein synthesis at translational initiation in eukaryotic cells. J Biol Chem 262:16509–16514

Choi KC, Jeung EB (2008) Molecular mechanism of regulation of the calcium-binding protein calbindin-D9k, and its physiological role(s) in mammals: a

review of current research: molecular medicine. J Cell Mol Med 12:409–420
Choi Y, Seo H, Kim M, Ka H (2009) Dynamic expression of calcium-regulatory molecules, TRPV6 and S100G, in the uterine endometrium during pregnancy in pigs. Biol Reprod 81:1122–1130
Choi Y, Seo H, Shim J, Kim M, Ka H (2012) Regulation of S100G expression in the uterine endometrium during early pregnancy in pigs. Asian-Australasian J Anim Sci 25:44–51
Choi Y, Seo H, Shim J, Hyun S-H, Lee E, Ka H (2014a) Klotho: expression and regulation at the maternal-conceptus interface in pigs. J Anim Reprod Biotechnol 29:375–383
Choi Y, Seo H, Shim J, Yoo I, Ka H (2014b) Calcium extrusion regulatory molecules: differential expression during pregnancy in the porcine uterus. Domest Anim Endocrinol 47:1–10
Choi Y, Jang H, Seo H, Yoo I, Han J, Kim M, Lee S, Ka H (2019) Changes in calcium levels in the endometrium throughout pregnancy and the role of calcium on endometrial gene expression at the time of conceptus implantation in pigs. Mol Reprod Dev 86:883–895
Collignon H, Davicco MJ, Barlet JP (1996) Calcitonin mRNA expression and plasma calciotropic hormones in fetal lambs. Domestric Anim Endocrinol 13:269–276
Comar CL (1956) Radiocalcium studies in pregnancy. Ann NY Acad Sci 64:281–298
Correia-Branco A, Rincon MP, Pereira LM, Wallingford MC (2020) Inorganic phosphate in the pathogenesis of pregnancy-related complications. Int J Mol Sci 21:1–11
Cross N, Hillman L, Allen S, Krause G, Vieira N (1995) Calcium homeostasis pregnancy, lactation, and bone metabolism during and postweaning: a longitudinal study. Am J Clin Nutr 61:514–523
Darwish HM, DeLuca HF (1992) Identification of a 1,25-dihydroxyvitamin D3-response element in the 5′-flanking region of the rat calbindin D-9k gene. Proc Natl Acad Sci USA 89:603–607
DeGaris PJ, Lean IJ (2009) Milk fever in dairy cows: a review of pathophysiology and control principles. Vet J 176:58–69
Delivoria-Papadopoulos M, Battaglia FC, Bruns PD, Meschia G (1967) Total, protein-bound, and ultrafilterable calcium in maternal and fetal plasmas. Am J Physiol 213:363–366
Deol HK, Varghese R, Wagner GF, Dimattia GE (2000) Dynamic regulation of mouse ovarian stanniocalcin expression during gestation and lactation. Endocrinology 141:3412–3421
Devaskar UP, Ho M, Devaskar SU, Tsang RC (1984) 25-hydroxy- and 1 alpha,25-dihydroxyvitamin D. Maternal-fetal relationship and the transfer of 1,25-dihydroxyvitamin D3 across the placenta in an ovine model. Dev Pharmacol Ther 7:213–220
Dias RS, Kebreab E, Vitti DMSS, Roque AP, Bueno ICS, France J (2006) A revised model for studying phosphorus and calcium kinetics in growing sheep. J Anim Sci 84:2787–2794
Díaz L, Noyola-martínez N, Barrera D, Hernández G, Avila E, Halhali A, Larrea F (2009) Calcitriol inhibits TNFa-induced inflammatory cytokines in human trophoblasts. J Reprod Immunol 81:17–24
Dolegowska K, Marchelek-mysliwiec M, Nowosiad-magda M, Slawinski M, Dolegowska B (2019) FGF19 subfamily members: FGF19 and FGF21. J Physiol Biochem 75:229–240
Donato R, Cannon BR, Sorci G, Riuzzi F, Hsu K, Weber DJ, Geczy CL (2013) Functions of S100 Proteins. Curr Mol Med 13:24–57
Dubil EA, Magann EF (2013) Amniotic fluid as a vital sign for fetal wellbeing. Australas J Ultrasound Med 16:62–70
Ducusin RJT, Uzuka Y, Satoh E, Otani M, Nishimura M, Tanabe S, Sarashina T (2003) Effects of extracellular Ca^{2+} on phagocytosis and intracellular Ca^{2+} concentrations in polymorphonuclear leukocytes of postpartum dairy cows. Res Vet Sci 75:27–32
Dwyer CM, Calvert SK, Farish M, Donbavand J, Pickup HE (2005) Breed, litter and parity effects on placental weight and placentome number, and consequences for the neonatal behaviour of the lamb. Theriogenology 63:1092–1110
Enders AC, Blankenship TN (1999) Comparative placental structure. Adv Drug Deliv Rev 38:3–15
Enders AC, Carter AM (2004) What can comparative studies of placental structure tell us?—a review. Placenta 25:S3–S9
Evans KN, Nguyen L, Chan J, Innes BA, Bulmer JN, Kilby MD, Hewison M (2006) Effects of 25-hydroxyvitamin D 3 and 1, 25-dihydroxyvitamin D 3 on cytokine production by human decidual cells. Biol Reprod 75:816–822
Fleischman AR, Lerman S, Oakes GK, Epstein MF, Chez RA, Mintz DH (1975) Perinatal primate parathyroid hormone metabolism. Biol Neonate 21:40–49
Fotiou M, Michaelidou AM, Athanasiadis AP, Menexes G, Symeonidou M, Koulourida V, Ganidou M, Theodoridis TD, Tarlatzis BC (2015) Second trimester amniotic fluid glucose, uric acid, phosphate, potassium, and sodium concentrations in relation to maternal pre-pregnancy BMI and birth weight centiles. J Matern Neonatal Med 28:910–915
Fudge NJ, Kovacs CS (2010) Pregnancy up-regulates intestinal calcium absorption and skeletal mineralization independently of the vitamin D receptor. Endocrinology 151:886–895
Fukino K, Suzuki T, Saito Y, Shindo T, Amaki T, Kurabayashi M, Nagai R (2002) Regulation of angiogenesis by the aging suppressor gene klotho. Biochem Biophys Res Commun 293:332–337
Gao H, Wu G, Spencer TE, Johnson GA, Li X, Bazer FW (2009) Select nutrients in the ovine uterine lumen. I. Amino acids, glucose, and ions in uterine lumenal flushings of cyclic and pregnant ewes. Biol Reprod 80:86–93

Garel JM (1972) Distribution of labeled parathyroid hormone in rat fetus. Horm Metab Res 4:131–132

Geisert R, Thatcher W, Roberts R, Bazer F (1982) Establishment of pregnancy in the pig: III. Endometrial secretory response to estradiol administered on day 11 of the estrous cycle. Biol Reprod 27:957–965

Gernand AD, Simhan HN, Caritis S, Bodnar LM (2014) Maternal vitamin D status and small-for-gestational-age offspring in women at high risk for preeclampsia. Obstet Gynecol 123:40–48

Gillies BR, Ryna BA, Tonkin BA, Poulton IJ, Ma Y, Kirby BJ, St-Arnaud R, Sims NA, Kovacs CS (2018) Absence of calcitrol causes increased lactational bone loss and lower milk calcium but does not impair post-lactation bone recovery in Cyp27b1 null mice. J Bone Miner Res 33:16–26

Givens MH, Macy IG (1933) Chemical composition of human fetus. J Biol Chem 102:7–17

Glancy B, Balaban RS (2012) Role of mitochondrial Ca^{2+} in the regulation of cellular energetics. Biochemistry 51:2959–2973

Glazier JD, Sharpe PT, Atkinson DE, Thornburg KL, Edwards D, Boyd RDH, Sibley CP, Boyd RDH (1992) Gestational changes in Ca^{2+} transport across rat placenta and mRNA for calbinding K and Ca^{2+}-ATPase. Am J Physiol 263:R930–R935

Grace ND, Watkinson JH, Martinson PL (1986) Accumulation of minerals by the foetus(es) and conceptus of single- and twin-bearing ewes. New Zeal J Agric Res 29:207–222

Grigsby PL (2016) Animal models to study placental development and function throughout normal and dysfunctional human pregnancy. Semin Reprod Med 34:11–16

Günther T, Chen ZF, Kim J, Priemel M, Rueger JM, Amling M, Moseley JM, Martin TJ, Anderson DJ, Karsenty G (2000) Genetic ablation of parathyroid glands reveals another source of parathyroid hormone. Nature 406:199–203

Habert R, Picon R (1984) Testosterone, dihydrotestosterone, and estradiol-17b levels in maternal and fetal plasma and in fetal testes in the rat. J Steroid Biochem 21:193–198

Haddad JGJ, Boisseau V, Avioli LV (1971) Placental transfer of vitamin D3 and 25-hydroxycholecalciferol in the rat. J Lab Clin Med 77:908–915

Haussler MR, Whitfield GK, Kaneko I, Forster R, Saini R, Hsieh C, Haussler CA, Jurutka PW (2013) The role of Vitamin D in the FGF23, Klotho, and phosphate bone-kidney endocrine axis. Rev Endocr Metab Disord 13:57–69

Heaney RP, Skillman TG (1971) Calcium metabolism in normal human pregnancy. J Clin Endocrinol 33:661–670

Hermann A, Donato R, Weiger TM, Chazin WJ (2012) S100 calcium binding proteins and ion channels. Front Pharmacol 3:1–10

Hernando N, Wagner CA (2018) Mechanisms and regulation of intestinal phosphate absorption. Compr Physiol 8:1065–1090

Hindson J, Winter A (2002) Appendix 3 nutrition. In: Manual of sheep diseases. Wiley-Blackwell Publisher, Hoboken, New Jersey, 304 pp.

Houtsmuller EJ, De Jong FH, Rowland DL, Slob AK (1995) Plasma testosterone in fetal rats and their mothers on day 19 of gestation. Physiol Behav 57:495–499

Hu MC, Shi M, Moe OW (2019) Role of αKlotho and FGF23 in regulation of type II Na-dependent phosphate co-transporters. Pflugers Arch Eur J Physiol 471:99–108

Hunt JS (2006) Stranger in a strange land. Immunol Rev 213:36–47

Husain S, Mughal M (1992) Mineral transport across the placenta. Arch Dis Child 67:874–878

Iñiguez G, Gallardo P, Castro JJ, Gonzalez R, Garcia M, Kakarieka E, San Martin S, Johnson MC, Mericq V, Cassorla F (2019) Klotho gene and protein in human placentas according to birth weight and gestational age. Front Endocrinol (lausanne) 9:1–8

Inpanbutr N, Miller EK, Petroff BK, Iacopino AM (1994) CaBP(9K) levels during the luteal and follicular phases of the estrous cycle in the bovine uterus. Biol Reprod 50:561–571

Ioannou C, Javaid MK, Mahon P, Yaqub MK, Harvey NC, Godfrey KM, Noble JA, Cooper C, Papageorghiou AT (2012) The effect of maternal vitamin D concentration on fetal bone. J Clin Endocrinol Metab 97:2070–2077

Islam MS (ed) (2011) Transient receptor potential channels. Adv Exp Med Biol 704:1–1095

Jang H, Choi Y, Yoo I, Han J, Hong JS, Kim YY, Ka H (2017) Vitamin D-metabolic enzymes and related molecules: expression at the maternal-conceptus interface and the role of vitamin D in endometrial gene expression in pigs. PLoS ONE 12:1–20

Jeon US (2008) Kidney and calcium homeostasis. Electrolyte Blood Press 6:68–76

Jones G, Prosser DE, Kaufmann M (2014) Cytochrome P450-mediated metabolism of vitamin D. J Lipid Res 55:13–31

Karras SN, Fakhoury H, Muscogiuri G, Grant WB, van den Ouweland JM, Colao AM, Kotsa K (2016) Maternal vitamin D levels during pregnancy and neonatal health: evidence to date and clinical implications. Ther Adv Musculoskelet Dis 8:124–135

Kavsak P (2017) Textbook of clinical chemistry and molecular diagnostics, 6th ed. Elsevier, NewYork

Kent G, Price R, Gutteridge D, Rosman K, Smith M, Allen J, Hickling C, Blakeman S (1991) The efficiency of intestinal calcium absorption is increased in late pregnancy but not in established lactation. Calcif Tissue Int Tissue 48:293–295

Khosla S, Johansen KL, Ory SJ, O'Brien PC, Kao PC (1990) Parathyroid hormone-related peptide in lactation and in umbilical cord blood. Mayo Clin Proc 65:1407–1414

Kim HJ, Lee GS, Ji YK, Choi KC, Jeung EB (2006) Differential expression of uterine calcium transporter 1 and plasma membrane Ca^{2+} ATPase 1b during rat estrous cycle. Am J Physiol 291:234–241

Kirby BJ, Ma Y, Martin HM, Favaro KLB, Karaplis AC, Kovacs CS (2013) Upregulation of calcitriol during pregnancy and skeletal recovery after lactation do not require parathyroid hormone. J Bone Miner Res 28:1987–2000

Kovacs CS (2014) Bone development and mineral homeostasis in the fetus and neonate: roles of the calciotropic and phosphotropic hormones. Physiol Rev 94:1143–1218

Kovacs CS (2015) Early human development calcium, phosphorus, and bone metabolism in the fetus and newborn. Early Hum Dev 91:623–628

Kovacs CS (2016) Maternal mineral and bone metabolism during pregnancy, lactation, and post-weaning recovery. Physiol Rev 96:449–547

Kovacs CS, Ralston SH (2015) Presentation and management of osteoporosis presenting in association with pregnancy or lactation. Osteoporos Int 26:2223–2241

Kovacs CS, Lanske B, Hunzelman JL, Guo J, Karaplis AC, Kronenberg HM (1996) Parathyroid hormone-related peptide (PTHrP) regulates fetal-placental calcium transport through a receptor distinct from the PTH/PTHrP receptor. Proc Natl Acad Sci USA 93:15233–15238

Kovacs CS, Chafe LL, Fudge NJ, Friel JK, Manley NR (2001a) PTH regulates fetal blood calcium and skeletal mineralization independently of PTHrP. Endocrinology 142:4983–4993

Kovacs CS, Manley NR, Moseley JM, John Martin T, Kronenberg HM (2001b) Fetal parathyroids are not required to maintain placental calcium transport. J Clin Invest 107:1007–1015

Kovacs CS, Woodland ML, Fudge NJ, Friel JK, Christopher S, Woodland ML, Fudge NJ (2005) The vitamin D receptor is not required for fetal mineral homeostasis or for the regulation of placental calcium transfer in mice. Am J Physiol 289:E133–E144

Kubota M, Ohno J, Shiina Y, Suda T (1982) Vitamin D metabolism in pregnant rabbits: differences between the maternal and fetal response to administration of large amounts of vitamin D3. Endocrinology 110:1950–1956

Kusaba T, Okigaki M, Matui A, Murakami M, Ishikawa K, Kimuraa T, Sonomura K, Adachi Y, Shibuya M, Shirayama T, Tanda S, Hatta T, Sasaki S, Mori Y, Matsubara H (2010) Klotho is associated with VEGF receptor-2 and the transient receptor potential canonical-1 Ca^{2+} channel to maintain endothelial integrity. Proc Natl Acad Sci USA 107:19308–19313

Lachenmaier-Currle U, Harmeyer J (1989) Placental transport of calcium and phosphorus in pigs. J Perinat Med 17:127–136

Lachenmaier-Currle U, Breves G, Harmeyer J (1989) Role of 1,25-(OH)2D3 during pregnancy: studies with pigs suffering from pseudo-vitamin D-deficiency rickets, type 1. Q J Exp Physiol 74:875–881

Lean IJ, Degaris PJ, Mcneil DM, Block E (2006) Hypocalcemia in dairy cows: meta-analysis and dietary cation anion difference theory revisited. J Dairy Sci 89:669–684

Lederer E (2014) Renal phosphate transporters. Curr Opin Nephrol Hypertens 23:502–506

Lee GS, Jeung EB (2007) Uterine TRPV6 expression during the estrous cycle and pregnancy in a mouse model. Am J Physiol 293:132–138

Leiser R, Krebs C, Ebert B, Dantzer V (1997) Placental vascular corrosion cast studies: a comparison between ruminants and humans. Microsc Res Tech 38:76–87

Lester GE, Gray TK, Lorenc RS (1978) Evidence for maternal and fetal differences in vitamin D metabolism. Proc Soc Exp Biol Med 159:303–307

Li XY, Zheng SX, Wu G (2020) Amino acid metabolism in the kidneys: nutritional and physiological significance. Adv Exp Med Biol 1265:71–95

Lim K, Lu TS, Molostvov G, Lee C, Lam FT, Zehnder D, Hsiao LL (2012) Vascular klotho deficiency potentiates the development of human artery calcification and mediates resistance to fibroblast growth factor 23. Circulation 125:2243–2255

Lima MS, Kallfelz F, Krook L, Nathanielsz PW (1993) Humeral skeletal development and plasma constituent changes in fetuses of ewes maintained on a low calcium diet from 60 days of gestation. Calcif Tissue Int 52:283–290

Liu N, Kaplan AT, Low J, Nguyen L, Liu GY, Equils O, Hewison M (2009) Vitamin D induces innate antibacterial responses in human trophoblasts via an intracrine pathway. Biol Reprod 80:398–406

Liu NQ, Kaplan AT, Lagishetty V, Yuxin B, Ouyang Y, Simmons CF, Equils O, Hewison M (2011) Vitamin D and the regulation of placental inflammation. J Immunol 186:5974–5986

Liu XM, Ding GL, Jiang Y, Pan HJ, Zhang D, Wang TT, Zhang RJ, Shu J, Sheng JZ, Huang HF (2012) Down-regulation of S100A11, a calcium-binding protein, in human endometrium may cause reproductive failure. J Clin Endocrinol Metab 97:3672–3683

Loichinger M, Towner D, Thompson K, Jun Ahn H, Bryant-Greenwood G (2016) Systemic and placental α-klotho: effects of preeclampsia in the last trimester of gestation. Placenta 41:53–61

Loveridge N, Caple IW, Rodda C, Martin TJ, Care AD (1988) Further evidence for a parathryoid hormone-related protein in fetal parathyroid glands of sheep. Arthritis Rheum 73:781–784

Lu M, Wagner GF, Renfro JL (1994) Stanniocalcin stimulates phosphate reabsorption by flounder renal proximal tubule in primary culture. Am J Physiol 267: R1356–R1362

Lyakhovich A, Aksenov N, Pennanen P, Miettinen S, Ahonen MH, Syvälä H, Ylikomi T, Tuohimaa P (2000) Vitamin D induced up-regulation of keratinocyte growth factor (FGF-7/KGF) in MCF-7 human breast cancer cells. Biochem Biophys Res Commun 273:675–680

Ma R, Gu Y, Zhao S, Sun J, Groome LJ, Wang Y (2012) Expressions of vitamin D metabolic components VDBP, CYP2R1, CYP27B1, CYP24A1, and VDR in placentas from normal and preeclamptic pregnancies. Am J Physiol 303:E928–E935

Ma Y, Samaraweera M, Cooke-hubley S, Kirby BJ, Karaplis AC, Lanske B, Kovacs CS (2014) Neither absence nor excess of FGF23 disturbs murine fetal-placental phosphorus homeostasis or prenatal. Endocrinology 155:1596–1605

MacIsaac RJ, Heath JA, Rodda CP, Moseley JM, Care AD, Martin TJ, Caple IW (1991) Role of the fetal parathyroid glands and parathyroid hormone-related protein in the regulation of placental transport of calcium, magnesium and inorganic phosphate. Reprod Fertil Dev 3:447–457

Mahon P, Harvey N, Crozier S, Inskip H, Robinson S, Arden N, Swaminathan R, Cooper C, Godfrey K (2010) Low maternal vitamin D status and fetal bone development: cohort study. J Bone Miner Res 25:14–19

Mayer GP, Ramberg CFJ, Kronfeld DS, Buckle RM, Sherwood LM, Aurbach GD, Potts JTJ (1969) Plasma parathyroid hormone concentration in hypocalcemic parturient cows. Am J Vet Res 30:1587–1597

Menon RK, Sperling MA (1998) Developmental endocrinology in the fetal-placental unit. In: Cowett RM (ed) Principles of perinatal-neonatal metabolism, 2nd edn. Springer, New York, pp 425–436

Miranda J, Romero R, Korzeniewski SJ, Schwartz AG, Chaemsaithong P, Stampalija T, Yeo L, Dong Z, Hassan SS, Chrousos GP, Gold P, Chaiworapongsa T (2014) The anti-aging factor α-klotho during human pregnancy and its expression in pregnancies complicated by small-for-gestational-age neonates and/or preeclampsia. J Matern Neonatal Med 27:449–457

Mitchell H, Hamilton T, Steggerda F, Bean H (1945) The chemical composition of the adult human body and its bearing on the biochemistry of growth. J Biol Chem 168:625–637

Montiel JF, Kaune H, Maliqueo M (2013) Maternal-fetal unit interactions and eutherian neocortical development and evolution. Front Neuroanat 7:1–14

Nabeshima YI, Imura H (2008) a-Klotho: a regulator that integrates calcium homeostasis. Am J Nephrol 28:455–464

Nair RR, Khanna A, Singh K (2013) Role of inflammatory proteins S100A8 and S100A9 in pathophysiology of recurrent early pregnancy loss. Placenta 34:824–827

Nie GY, Li Y, Wang J, Minoura H, Findlay JK, Salamonsen LA (2000) Complex regulation of calcium-binding protein D9k (calbindin-D(9k)) in the mouse uterus during early pregnancy and at the site of embryo implantation. Biol Reprod 62:27–36

Noff D, Edelstein S (1978) Vitamin D and its hydroxylated metabolites in the rat. Placental and lacteal transport, subsequent metabolic pathways and tissue distribution. Horm Res 9:292–300

Northrop G, Misenhimer HR, Becker FO (1977) Failure of parathyroid hormone to cross the nonhuman primate placenta. Am J Obstet Gynecol 129:449–453

O'Brien KO, Li S, Cao C, Kent T, Young BV, Queenan RA, Pressman EK, Cooper EM (2014) Placental CYP27B1 and CYP24A1 expression in human placental tissue and their association with maternal and neonatal calcitropic hormones. Endocr Res 99:1348–1356

Ohata Y, Arahori H, Namba N, Kitaoka T, Hirai H, Wada K, Nakayama M, Michigami T, Imura A, Nabeshima YI, Yamazaki Y, Ozono K (2011) Circulating levels of soluble α-klotho are markedly elevated in human umbilical cord blood. J Clin Endocrinol Metab 96:943–947

Okunade GW, Miller ML, Pyne GJ, Sutliff RL, O'Connor KT, Neumann JC, Andringa A, Miller DA, Prasad V, Doetschman T, Paul RJ, Shull GE (2004) Targeted ablation of plasma membrane Ca^{2+}-ATPase (PMCA) 1 and 4 indicates a major housekeeping function for PMCA1 and a critical role in hyperactivated sperm motility and male fertility for PMCA4. J Biol Chem 279:33742–33750

Passey RJ, Williams E, Lichanska AM, Wells C, Hu S, Geczy CL, Little MH, Hume DA (1999) A null mutation in the inflammation-associated S100 protein S100A8 causes early resorption of the mouse embryo. J Immunol 163:2209–2216

Paulson SK, DeLuca HF, Battaglia F (1987) Plasma levels of vitamin D metabolites in fetal and pregnant ewes. Proc Soc Exp Biol Med 185:267–271

Paulson SK, Ford KK, Langman CB (1990) Pregnancy does not alter the metabolic clearance of 1,25-dihydroxyvitamin D in rats. Am J Physiol 258: E158–E162

Peacock M (2018) Hypoparathyroidism and the kidney. Endocrinol Metab Clin North Am 47:839–853

Peacock M (2021) Phosphate metabolism in health and disease. Calcif Tissue Int 108:3–15

Penido MGMG, Alon US (2012) Phosphate homeostasis and its role in bone health. Pediatr Nephrol 27:2039–2048

Pike JW, Parker JB, Haussler MR, Boass A, Toverud SU (1979) Dynamic changes in circulating 1,25-dihydroxyvitamin D during reproduction in rats. Science 204:1427–1429

Pizzagalli MD, Bensimon A, Superti-Furga G (2020) A guide to plasma membrane solute carrier proteins. FEBS J 1–52

Poeter M, Radke S, Koese M, Hessner F, Hegemann A, Musiol A, Gerke V, Grewal T, Rescher U (2013) Disruption of the annexin A1/S100A11 complex increases the migration and clonogenic growth by dysregulating epithelial growth factor (EGF) signaling. Biochim Biophys Acta 1833:1700–1711

Prasad V, Okunade G, Liu L, Paul RJ, Shull GE (2007) Distinct phenotypes among plasma membrane Ca^{2+}-ATPase knockout mice. Ann NY Acad Sci 1099:276–286

Qin LL, Lu FG, Yang SH, Xu HL, Luo BA (2016) Does maternal Vitamin D deficiency increase the risk of preterm birth: a meta-analysis of observational studies. Nutrients 8:1–10

Quarles LD (2008) Endocrine functions of bone in mineral metabolism regulation. J Clin Invest 118:3820–3828

Reddy GS, Willis DM, Goltzman D, Guyda H, Solomon S, Philips DR, Bishop JE, Mayer E (1983) Regulation of vitamin D metabolism in normal human pregnancy. J Clin Endocrinol Metab 56:363–370

Reyes FI, Winter JSD, Faiman C (1973) Studies on human sexual development. I. Fetal gonadal and adrenal sex steroids. J Clin Endocrinol Metab 37:74–78

Risco C, Drost M, Thatcher WW, Savio J, Thatcher MJ (1994) Effects of calving-related disorders on prostaglandin, calcium, ovarian activity and uterine involution in postpartum dairy cows. Theriogenology 42:183–203

Ritchie LD, Fung EB, Halloran BP, Turnlund JR, Van LMD, Cann CE, King JC (1998) A longitudinal study of calcium homeostasis during human pregnancy and lactation and after resumption of menses. Am J Clin Nutr 67:693–701

Rodda CP, Kubota M, Heath JA, Ebeling PR, Moseley JM, Care AD, Caple IW, Martin TJ (1988) Evidence for a novel parathyroid hormone-related protein in fetal lamb parathyroid glands and sheep placenta: comparisons with a similar protein implicated in humoral hypercalcaemia of malignancy. J Endocrinol 117:261–271

Romero JJ, Liebig BE, Broeckling CD, Prenni JE, Hansen TR (2017) Pregnancy-induced changes in metabolome and proteome in ovine uterine flushings. Biol Reprod 97:273–287

Ross R, Halbert K, Tsang RC (1989) Determination of the production and metabolic clearance rates of 1,25-dihydroxyvitamin D3 in the pregnant sheep and its chronically catheterized fetus by primed infusion technique. Pediatr Res 26:633–634

Ryan BA, Kovacs CS (2021) Maternal and fetal vitamin D and their roles in mineral homeostasis and fetal bone development. J Endocrinol Invest 44:643–659

Saito Y, Yamagishi T, Nakamura T, Ohyama Y, Aizawa H, Suga T, Matsumura Y, Masuda H, Kurabayashi M, Kuro-O M, Nabeshima YI, Nagai R (1998) Klotho protein protects against endothelial dysfunction. Biochem Biophys Res Commun 248:324–329

Saki F, Sadeghian F, Kasaee SR, Koohpeyma F, Omrani GHR (2020) Effect of prolactin and estrogen on the serum level of 1,25-dihydroxy vitamin D and FGF23 in female rats. Arch Gynecol Obstet 302:265–271

Santella L (1998) The role of calcium in the cell cycle: facts and hypotheses. Biochem Biophys Res Commun 244:317–324

Schmid A, Walther B (2013) Natural vitamin D content in animal products. Adv Nutr 4:453–462

Segawa H, Shiozaki Y, Kaneko I, Miyamoto KI (2015) The role of sodium-dependent phosphate transporter in phosphate homeostasis. J Nutr Sci Vitaminol (tokyo) 61:S119–S121

Seki K, Wada S, Nagata N, Nagata I (1994) Parathyroid hormone-related protein during pregnancy and the perinatal period. Gynecol Obstet Invest 37:83–86

Shahbazi M, Jeddi-tehrani M, Zareie M, Salek-moghaddam A, Akhondi MM, Bahmanpoor M, Sadeghi MR, Zarnani AH (2011) Expression profiling of vitamin D receptor in placenta, decidua and ovary of pregnant mice. Placenta 32:657–664

Shaikh A, Berndt T, Kumar R (2008) Regulation of phosphate homeostasis by the phosphatonins and other novel mediators. Pediatr Nephrol 23:1203–1210

Shibasaki Y, Etoh N, Hayasaka M, Takahashi MO, Kakitani M, Yamashita T, Tomizuka K, Hanaoka K (2009) Targeted deletion of the tybe IIb Na^+-dependent Pi-co-transporter, NaPi-IIb, results in early embryonic lethality. Biochem Biophys Res Commun 381:482–486

Simmonds CS, Kovacs CS (2010) Role of parathyroid hormone (PTH) and PTH-related protein (PTHrP) in regulating mineral homeostasis during fetal development. Crit Rev Eukaryot Gene Expr 20:235–273

Simmonds CS, Karsenty G, Karaplis AC, Kovacs CS (2010) Parathyroid hormone regulates fetal-placental mineral homeostasis. J Bone Miner Res 25:594–605

Sitara D, Razzaque MS, Hesse M, Yoganathan S, Taguchi T, Erben RG, Harald J, Lanske B (2004) Homozygous ablation of fibroblast growth factor-23 results in hyperphosphatemia and impaired skeletogenesis, and reverses hypophosphatemia in Phex- deficient mice. Matrix Biol 421–432

Song G, Bazer FW, Wagner GF, Spencer TE (2006) Stanniocalcin (STC) in the endometrial glands of the ovine uterus: regulation by progesterone and placental hormones 1. Biol Reprod 74:913–922

Song G, Dunlap KA, Kim J, Bailey DW, Spencer TE, Burghardt RC, Wagner GF, Johnson GA, Bazer FW (2009) Stanniocalcin 1. Is a luminal epithelial marker for implantation in pigs regulated by progesterone and estradiol. Endocrinology 150:936–945

Stasko SE, DiMattia GE, Wagner GF (2001) Dynamic changes in stanniocalcin gene expression in the mouse uterus during early implantation. Mol Cell Endocrinol 174:145–149

Stenhouse C, Halloran KM, Newton MG, Gaddy D, Suva LJ, Bazer FW (2021a) Novel mineral regulatory pathways in ovine pregnancy: II. Calcium binding proteins, calcium transporters, and vitamin D signaling. Biol Reprod 105:232–243

Stenhouse C, Halloran KM, Newton MG, Gaddy D, Suva LJ, Bazer FW (2021b) Novel mineral regulatory pathways in ovine pregnancy: 1. Phosphate, klotho signaling, and sodium dependent phosphate transporters. Biol Reprod 104:1084–1096

Strid H, Powell TL (2000) ATP-dependent Ca^{2+} transport is up-regulated during third trimester in human

syncytiotrophoblast basal membranes. Pediatr Res 48:58–63
Strott CA, Sundel H, Stahlman MT (1974) Maternal and fetal plasma progesterone, cortisol, testosterone and 17b-estradiol in preparturient sheep: response to fetal ACTH infusion. Endocrinology 95:1327–1339
Stulc J, Stulcova B (1996) Placental transfer of phosphate in anaesthetized rats. Placenta 17:487–493
Suva LJ, Winslow GA, Wettenhall RE, Hammonds RG, Moseley JM, Diefenbach-Jagger H, Rodda CP, Kemp BE, Rodriguez H, Chen EY (1987) A parathyroid hormone-related protein implicated in malignant hypercalcemia: cloning and expression. Science 237:893–896
Suva LJ, Gaddy D, Perrien DS, Thomas RL, Findlay DM (2005) Regulation of bone mass by mechanical loading: microarchitecture and genetics. Curr Osteoporos Rep 3:46–51
Suva L, Friedman P (2020) PTH and PTHrP actions on bone. In: Handbook of experimental pharmacology, vol 262. Springer, Berlin, pp 27–45
Suzuki Y, Kovacs CS, Takanaga H, Peng J, Landowski CP, Hediger MA (2008) Calcium channel TRPV6 is involved in murine maternal-fetal calcium transport. J Bone Miner Res 23:1249–1256
Swanson AM, David AL (2015) Animal models of fetal growth restriction: considerations for translational medicine. Placenta 36:623–630
Tabatabaei S (2012) Gestational variations in the biochemical composition of the fetal fluids and maternal blood serum in goat. Comp Clin Path 21:1305–1312
Tamblyn JA, Hewison M, Wagner CL, Bulmer JN, Kilby MD (2015) Immunological role of vitamin D at the maternal—fetal interface. J Endocrinol 224:R107–R121
Taylor-Miller T, Allgrove J (2021) Endocrine diseases of newborn: epidemiology, pathogenesis, therapeutic options, and outcome "Current Insights Into Disorders of Calcium and Phosphate in the Newborn." Front Pediatr 9:1–8
Thomas ML, Anast CS, Forte LR (1981) Regulation of calcium homeostasis in the fetal and neonatal rat. Am J Physiol 240:E367–E372
Tuan RS, Bigioni N (1990) Ca-activated ATPase of the mouse chorioailantoic placenta: developmental expression, characterization and cytohistochemical localization. Development 110:505–513
Tucci J, Russell A, Senior PV, Fernley R, Ferraro T, Beck F (1996) The expression of parathyroid hormone and parathyroid hormone-related protein in developing rat parathyroid glands. J Mol Endocrinol 17:149–157
Verhaeghe J, Thomasset M, Brehier A, Assche FAV, Bouillon R (1988) 1,25(OH)2D3 and Ca-binding protein in fetal rats: relationship to maternal vitamin D status. Am J Physiol 254:E505–E512
Voelkl J, Egli-Spichtig D, Alesutan I, Wagner CA (2021) Inflammation: a putative link between phosphate metabolism and cardiovascular disease. Clin Sci 135:201–227
Wagner GF (1994) The molecular biology of the corpuscles of stannius and regulation of stanniocalcin gene expression. Fish Physiol 13:273–306
Wagner GF, Jaworski E (1994) Calcium regulates stanniocalcin mRNA levels in primary cultured rainbow trout corpuscles of Stannius. Mol Cell Endocrinol 99:315–322
Wagner GF, Hampong M, Park CM, Copp DH (1986) Purification, characterization, and bioassay of teleocalcin, glycoprotein from salmon corpuscles of stannius. Gen Comp Endocrinol 63:481–491
Wallingford MC, Giachelli CM (2014) Loss of PiT-1 results in abnormal endocytosis in the yolk sac visceral endoderm. Mech Dev 133:189–202
Wallingford M, Gammill H, Giachelli C (2016) SLC20A2 deficiency results in fetal growth restriction and placental calcification associated with thickened basement membranes and novel CD13 and lamininα1 expressing cells. Reprod Biol 16:13–26
Wang Y, Sun Z (2009) Current understanding of klotho. Ageing Res Rev 8:43–51
Ward IL, Weisz J (1984) Differential effects of maternal stress of circulating levels of corticosterone, progesterone, and testosterone in male and female rat fetuses and their mothers. Endocrinology 114:1635–1644
Widdowson EM (1962) Metabolic relationship of calcium, magnesium and phosphorus in the foetus and newly born. Voeding 23:62–71
Widdowson EM, McCance RA (1965) The metabolism of calcium, phosphorus, magnesium and strontium. Pediatr Clin North Am 12:595–614
Wilson RL, Phillips JA, Bianco-Miotto T, McAninch D, Goh Z, Anderson PH, Roberts CT (2020) Reduced dietary calcium and vitamin D results in preterm birth and altered placental morphogenesis in mice during pregnancy. Reprod Sci 27:1330–1339
Wu G (2018) Principles of animal nutrition. CRC Press, Boca Raton, FL
Wu G (2021) Amino acids: biochemistry and nutrition. CRC Press, Boca Raton, FL
Wu G (2022) Nutritionand metabolism: Foundations for animal growth, development, reproduction, and health.Adv Exp Med Biol 1354:1–24
Xia C, Braunstein Z, Toomey AC, Zhong J, Rao X (2018) S100 proteins as an important regulator of macrophage inflammation. Front Immunol 8:1–11
Xiao L, Yuan J, Song X, Li Y, Hu Z, Liu Y (2006) Expression and regulation of stanniocalcin 1 and 2 in rat uterus during embryo implantation and decidualization. Reproduction 131:1137–1149
Yamamoto M, Clark JD, Pastor JV, Gurnani P, Nandi A, Kurosu H, Miyoshi M, Ogawa Y, Castrillon DH, Rosenblatt KP, Kuro-O M (2005) Regulation of oxidative stress by the anti-aging hormone klotho. J Biol Chem 280:38029–38034
Yang H, Choi KC, Hyun SH, Jeung EB (2011) Coexpression and estrogen-mediated regulation of TRPV6 and PMCA1 in the human endometrium during the menstrual cycle. Mol Reprod Dev 78:274–282

Yang H, Kim TH, Lee GS, Hong EJ, Jeung EB (2014) Comparing the expression patterns of placental magnesium/phosphorus-transporting channels between healthy and preeclamptic pregnancies. Mol Reprod Dev 81:851–860

Yoshida T, Fujimori T, Nabeshima Y (2002) Mediation of unusually high concentrations of 1,25-Dihydroxyvitamin D in homozygous klotho mutant mice by increased expression of renal 1a-hydroxylase gene. Endocrinology 143:683–689

Young BE, McNanley TJ, Cooper EM, McIntyre AW, Witter F, Harris ZL, O'Brien KO (2012) Maternal vitamin D status and calcium intake interact to affect fetal skeletal growth in utero in pregnant adolescents. Am J Clin Nutr 95:1103–1112

Zeng S, Bick J, Ulbrich SE, Bauersachs S (2018) Cell type-specific analysis of transcriptome changes in the porcine endometrium on day 12 of pregnancy. BMC Genomics 19:1–19

Zeng S, Ulbrich SE, Bauersachs S (2019) Spatial organization of endometrial gene expression at the onset of embryo attachment in pigs. BMC Genomics 20:1–19

Zhang R, Lu Y, Ye L, Yuan B, Yu S, Qin C, Xie Y, Gao T, Drezner MK, Bonewald LF, Feng JQ (2011) Unique roles of phosphorus in endochondral bone formation and osteocyte maturation. J Bone Miner Res 26:1047–1056

Zhou SS, Tao YH, Huang K, Zhu BB, Tao FB (2017) Vitamin D and risk of preterm birth: up-to-date meta-analysis of randomized controlled trials and observational studies. J Obstet Gynaecol Res 43:247–256

Ziegler E, O'Donnell A, Nelson S, Fomon S (1976) Body composition of the reference fetus. Growth 40:329–341

6 Nutritional and Physiological Regulation of Water Transport in the Conceptus

Cui Zhu, Zongyong Jiang, Gregory A. Johnson, Robert C. Burghardt, Fuller W. Bazer, and Guoyao Wu

Abstract

Water transport during pregnancy is essential for maintaining normal growth and development of conceptuses (embryo/fetus and associated membranes). Aquaporins (AQPs) are a family of small integral plasma membrane proteins that primarily transport water across the plasma membrane. At least 11 isoforms of AQPs (AQPs 1–9, 11, and 12) are differentially expressed in the mammalian placenta (amnion, allantois, and chorion), and organs (kidney, lung, brain, heart, and skin) of embryos/fetuses during prenatal development. Available evidence suggests that the presence of AQPs in the conceptus mediates water movement across the placenta to support the placentation, the homeostasis of amniotic and allantoic fluid volumes, as well as embryonic and fetal survival, growth and development. Abundances of AQPs in the conceptus can be modulated by nutritional status and physiological factors affecting the pregnant female. Here, we summarize the effects of maternal dietary factors (such as intakes of protein, arginine, lipids, all-trans retinoic acid, copper, zinc, and mercury) on the expression of AQPs in the conceptus. We also discuss the physiological changes in hormones (e.g., progesterone and estrogen), oxygen supply, nitric oxide, pH, and osmotic pressure associated with the regulation of fluid exchange between mother and fetus. These findings may help to improve the survival, growth, and development of embryo/fetus in livestock species and other mammals (including humans).

Keywords

Water transport · Aquaporin · Placenta · Fetal membrane · Conceptus · Nutrition

Abbreviations

AQP	Aquaporin
CFTR	Cystic fibrosis transmembrane conductance regulator
E2	Estrogen
hCG	Human chorionic gonadotropin
HIF	Hypoxia inducible factor
NHE	Na^+/H^+ exchanger
PI3K	Phosphatidylinositol 3-kinase
P4	Progesterone
pTr2	Porcine trophectoderm cells

C. Zhu
School of Life Science and Engineering, Foshan University, Foshan 528225, China

Z. Jiang
Institute of Animal Science, Guangdong Academy of Agricultural Sciences, Guangzhou 510640, China

G. A. Johnson · R. C. Burghardt
Veterinary Integrative Biosciences, Texas A&M University, College Station, TX 77843, USA

F. W. Bazer · G. Wu (✉)
Department of Animal Science, Texas A&M University, College Station, TX 77843, USA
e-mail: g-wu@exchange.tamu.edu

G. Wu (ed.), *Recent Advances in Animal Nutrition and Metabolism*, Advances in Experimental Medicine and Biology 1354,
https://doi.org/10.1007/978-3-030-85686-1_6

6.1 Introduction

The fetal body consists of 70–90% water (Pérez-Pérez et al. 2020b) and is the most abundant nutrient across the placenta from the maternal to the fetal-placental vascular system (Wilbur et al. 1978). Fetal requirements for water increase directly with fetal weight gain during gestation (Stulc 1997). Water transport and homeostasis in the conceptus play a critical role in the normal growth and development of the conceptus. Thus, intrauterine growth restriction is associated with a reduced volume of water in amniotic and allantoic fluids (Bazer et al. 2009; Wu et al. 2017, 2018).

The placenta is a key organ providing the interface for the exchanges of gases, nutrients (including water), and fetal wastes (e.g., bilirubin, ammonia, urea, and carbon dioxide) between mother and fetus, and also for the production of hormones (Martínez et al. 2017; Pérez-Pérez et al. 2020b; Schneider 1991; Sibley et al. 2018). The flow of water from mother to fetus is tightly regulated to control the expansion of the fetal vascular system and fetal fluid volumes (Wu et al. 2006; Faber and Anderson 2010). The transport of water across the placenta to the fetus occurs primarily through transcellular pathways mediated by aquaporins (AQPs) expressed on cell membranes and, to a lesser extent, simple paracellular diffusion (Faber and Anderson 2010; Sibley et al. 2018).

Aquaporins are water channel proteins in a family of small integral plasma membrane proteins that primarily transport water across the plasma membrane (Agre and Kozono 2003; Ishibashi et al. 2009). At least 13 isoforms of the AQP are differentially expressed in the female reproductive tract of mammals (Zhu et al. 2015), including AQPs 1–9, 11, and 12) in the fetal membranes (amnion, allantois, and chorion), and the developing embryo/fetus (Kordowitzki et al. 2020; Liu et al. 2008). Cell-specific expression of AQP9 at the uterine–placental interface of swine on different days of gestation is shown in Fig. 6.1. These water channels may play important roles in the regulation of implantation and placentation, the homeostasis of amniotic and allantoic fluids, and placental transport functions during pregnancy to support the development of embryo/fetus (Ducza et al. 2017). The abnormal expression of AQPs is associated with pregnancy complications and reproductive dysfunctions in humans and other mammals (including ruminants and swine), such as preeclampsia, fetal growth restriction or overgrowth, gestational diabetes mellitus, preterm birth, polyhydramnios, and oligohydramnios (de Oliveira et al. 2020; Shao et al. 2021).

There is increasing evidence that the maternal–fetal fluid balance and the associated water channel proteins can be modulated by various factors, including nutrients, hormones, oxygen supply, nitric oxide, pH, osmotic pressure, and other physiological factors (Sha et al. 2011a). The expression of key AQPs that mediate water transport in the conceptus is regulated by many signaling pathways at both the transcriptional and post-transcriptional levels (Zhu et al. 2015; Meli et al. 2018). This review summarizes the recent progress regarding the nutritional and physiological regulation of water transport from the maternal system to the conceptus.

6.2 Water Transport and Expression of AQPs in the Conceptus During Pregnancy

6.2.1 Blastocyst Formation and Implantation

Before implantation, the conceptus develops into a blastocyst, which consists of an inner cell mass and a fluid-filled cavity surrounded by trophectoderm cells (Damiano 2020). Blastocyst formation is essential for implantation and subsequent development of the conceptus (Offenberg et al. 2000). Implantation of the blastocyst/conceptus in the uterus takes place via multiple stages, including orientation, apposition, adhesion, attachment, and penetration in primates, but does not include penetration through the uterine epithelia into the uterine stroma in other species (Bazer et al. 2009).

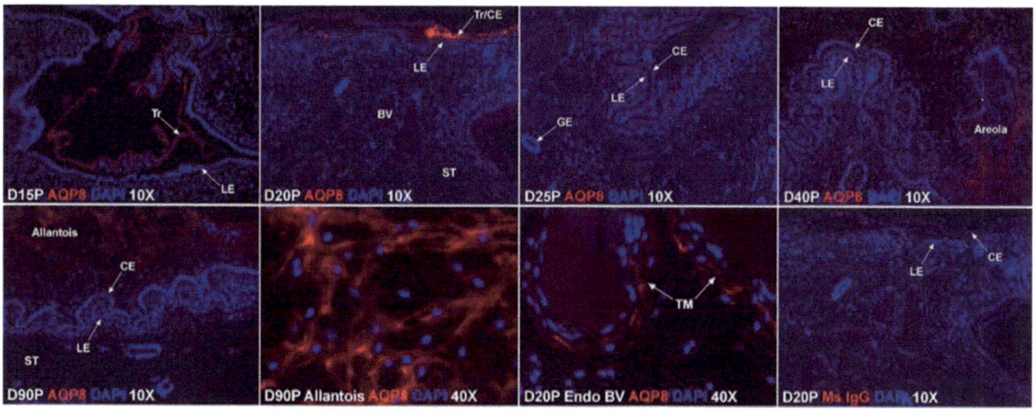

Fig. 6.1 Immunofluorescence microscopy for aquaporin 8 (AQP8; red) at the uterine–placental interface of gilts on Days 15 (D15), 20, 25, 40, and 90 of pregnancy. Immunofluorescence microscopy was performed as described by Seo et al. (2020). Paraffin-embedded sections (5 μm) of the uterine–placental interface were adhered to slides, deparaffinized and rehydrated in CitriSolv, ethanol, and water. For antigen unmasking, the sections were brought to a boil in a 10 mM Sodium Citrate Buffer solution. The sections were then washed in PBS 3 times for 5 min each, blocked with 10% normal goat serum, and then the sections were incubated with either rabbit anti-AQP1 (2.5 μg/ml), mouse anti-AQP8 (5 μg/ml), or rabbit anti-AQP9 (2.5 μg/ml) overnight at 4 °C in a humidified chamber. Normal rabbit or mouse IgG was substituted for a primary antibody and served as a negative control. Expression was detected with either fluorescein-conjugated goat anti-rabbit IgG or goat anti-mouse IgG (1:250) for 1 hour (Chemicon International). Slides were then overlaid with Prolong Gold Anti-fade mounting reagent containing DAPI (Molecular Probes) and a coverslip. Images were taken using an Axioplan 2 microscope and a Zeiss Imager.M2, (Carl Zeiss, Thornwood, NY) interfaced with an Axioplan HR and an AxioCam HRm digital camera, respectively. Photographic plates were assembled using Adobe Photoshop (version 6.0, Adobe Systems Inc., San Jose, CA). AQP8 protein is localized to the trophectoderm on Days 15 and 20, the luminal epithelium (LE) on Days 25 and 30, the tunica media of blood vessels within both uterine and placental tissues, to placental areolae, and cells within the allantois on Days 40 and 90 of gestation. Nuclei were stained with DAPI for histological reference. The Day 20 mouse IgG (Ms IgG) panel served as the negative control. The width of fields for microscopic images captured at 10X is 940 μm. The width of fields for microscopic images captured at 40X is 230 μm. Legend: D, day; P, pregnancy; GE, glandular epithelium; ST, stroma; BV, blood vessel; Tr, trophectoderm; CE, chorionic epithelium; TM, tunica media. (Unpublished work of McLendon B, Zhu C, Bazer FW, Johnson GA, Burghardt RC, Wu G at Texas A&M University, College Station, TX, USA)

These processes may require dynamic changes in fluid transport via ion/water channel proteins (Liu et al. 2014). Many ion/water channel proteins play critical roles in the process of blastocyst/conceptus implantation (Liu et al. 2014). Of note, AQPs 1, 3, 5–7, and 9 are expressed by embryos/blastocysts during the preimplantation period of pregnancy of mice from the one-cell stage to the blastocyst stage and they may account for fluid transport by trophectoderm cells required for the formation of the blastocyst (Offenberg et al. 2000). Moreover, mRNA transcripts of AQPs 1–5, 7, 9, 11, and 12 have been detected in human embryos during the preimplantation period of pregnancy (Xiong et al. 2013). Thus, knockdown of AQP3 or AQP7 expression in mouse embryos at the 2-cell stage by specific siRNAs significantly inhibited embryonic development (Xiong et al. 2013). Other than water channel proteins, the tight junctional permeability of trophectoderm cells contributes to the regulation of the leakage of fluid from the blastocoel (Watson 1992).

The pig is an excellent animal model for studying the placental transport of water, because dynamic changes in both amniotic and allantoic fluid volumes occur with the growth and development of the conceptus (Bazer 1989; Knight et al. 1977). On Days 12–13 of pregnancy, the elongating porcine conceptuses have completed migration and spacing throughout the uterus and start to attach to the uterine luminal epithelium (LE) for implantation (Geisert et al. 1982).

Interestingly, the development of porcine blastocysts can be enhanced by maternal transcription coactivator yes-associated protein (YAP) through the transcriptional modulation of key genes involved in lineage commitment (CDX2, TEAD4, OCT4, and SOX2), tight junction assembly (OCLN, CLDN4, CLDN6, CDH1, TJP1, and TJP2), and fluid accumulation (ATP1B1 and AQP3) (Cao et al. 2019).

6.2.2 Early and Late Pregnancy

Several AQPs have been found in the placenta and fetal membrane as well as the organs of the embryo/fetus in different mammalian species, including humans, mice, rats, sheep, pigs, dogs, giraffe, and macaques (Table 6.1). Water channel protein AQPs in the fetal membranes play important roles in the homeostasis of maternal–fetal fluid transport and exchange during pregnancy (Zhu et al. 2015; Sha et al. 2011a; Liu et al. 2008). For example, AQPs 1, 3–5, 8, 9, and 11 mRNAs are expressed in the placenta and chorionic villi between the 10th and 14th weeks of gestation in humans (Escobar et al. 2012). We have previously reported the expression of AQPs 1, 3–9, and 11 mRNAs in placentae of gilts at Day 25 of gestation (Zhu et al. 2015). In addition, AQP1 has been detected in placental blood vessels (Mann et al. 2002), and AQPs 3, 4, 8, and 9 have been detected in human syncytiotrophoblast at term (Damiano et al. 2001; De Falco et al. 2007; Wang et al. 2001). Similarly, AQP8 and AQP9 are expressed in the amnionic epithelium, cytotrophoblasts, and trophectoderm (Zhu et al. 2010). Thus, AQP8 and AQP9 may regulate the placental transport of water from mother to fetus, as well as the flow of water and solutes within the conceptus (Damiano 2011). The maternal–fetal fluid balance is also essential for regulating amniotic fluid volume and composition for fetal growth and development during pregnancy (Sha et al. 2011b; Pérez-Pérez et al. 2020b). In support of this view, decreases in AQPs 1, 3, 4, 8, and 9 expression in human amnion and chorion are closely associated with abnormal amniotic fluid volumes such as polyhydramnios and oligohydramnios (Damiano et al. 2011; De Falco et al. 2007; Jiang et al. 2012; Zhu et al. 2009), as well as increases in fetal morbidity and mortality (Ducza et al. 2017). Furthermore, five AQPs (AQPs 1, 3, 8, 9, and 11) are differentially expressed in ovine amnion, with AQP1 protein expression being positively associated with intramembranous absorption rates that affect amniotic fluid volume (Cheung et al. 2016). Therefore, alterations in the expression and localization of these AQPs in a species- and site-specific manner during pregnancy may be involved in the regulation of amniotic fluid homeostasis and fluid transfer in the conceptus (Cheung et al. 2018; Vilariño-García et al. 2016). These results further indicate a crucial role of AQPs in maintaining normal pregnancies as well as the proper growth and development of embryo/fetus.

There is also evidence for the roles of AQPs in the diverse functions of placentae, including the regulation of cell volume, migration, proliferation, and adhesion beyond water transport (Kitchen et al. 2015). As summarized in Table 6.2, deficiencies of several AQPs (AQPs 1, 3, 4, 7, and 8) in conceptuses may be associated with abnormal outcomes of pregnancy in mammals. For example, deletion of the AQP1 gene in mice resulted in greater amniotic fluid volume, while decreasing the osmolality and concentrations of calcium in amniotic fluid, as well as upregulating AQP8 expression and downregulating AQP9 expression in the fetal membranes (Luo et al. 2018). Research with AQP1-null mice demonstrated a critical role for AQP1 in placental and fetal growth as well as maternal–fetal fluid homeostasis (Mann et al. 2005; Zheng et al. 2014; Sha et al. 2015; Luo et al. 2018). Likewise, a deficiency of AQP3 in placentae induced significant impairments in fetal growth through changes in amniotic fluid volume and reductions in concentrations of metabolites in amniotic fluid (Seo et al. 2018). Moreover, AQP4-null mice were subfertile based on low pregnancy rates and litter size (Sun et al. 2009), but AQP8-deficient pregnant mice exhibited increases in the numbers of embryos (Su et al. 2010), weights of fetuses and neonatal pups, placental weight, and amniotic fluid volume when compared with wild-type mice (Sha et al. 2011b). To date, the underlying mechanisms for those findings are unknown.

Table 6.1 Expression of aquaporins in conceptuses of different species during pregnancy

Item	Blastocyst	Placenta and fetal membranes			Fetal organs				
		Placenta	Amnion	Chorion	Brain	Heart	Lung	Kidney	Skin
AQP1	Human, mouse	Human, mouse, rat, pig, sheep, giraffe, dog	Human, mouse, macaques, dog, sheep	Human, dog	Human	Sheep	Rat, sheep	Human, rat, sheep	
AQP2	Human	Human	Human	Human				Human, rat, sheep	
AQP3	Human, mouse	Human, mouse, pig, sheep, giraffe, dog	Human, mouse, macaques, sheep, dog	Human, sheep, dog		Sheep	Sheep	Rat	Rat, mouse
AQP4	Human	Human, pig	Human	Human	Human, rat	Sheep	Rat, sheep	Rat	
AQP5	Human, mouse	Human, pig	Dog	Human, dog			Sheep, mouse		
AQP6	Mouse	Pig							
AQP7	Human	human, pig							
AQP8		Human, mouse, rat, pig, sheep	Human, mouse, macaques, sheep, dog	Human, mouse, dog		Sheep			
AQP9	Human, mouse	Human, mouse, rat, pig, sheep	Human, mouse, macaques, sheep	Human, mouse					
AQP11	Human	Human, pig	Human, macaques, sheep	Human, sheep					
AQP12	Human								
References	Xiong et al. (2013), Offenberg et al. (2000)	Damiano et al. (2001), Wang et al. (2004), De Falco et al. (2007), Prat et al. (2012), Escobar et al. (2012), Belkacemi et al. (2011), Ma et al. (1997), Beall et al. (2007), Zhu et al. (2015), Johnston et al. (2000), Liu et al. (2004), Wooding et al. (2015), Aralla et al. (2012)	Mann et al. (2002), Wang et al. (2001), Wang et al. (2004), Mann et al. (2002), Beall et al. (2007), Cheung et al. (2018), Johnston et al. (2000), Cheung et al. (2016), Aralla et al. (2012)	Escobar et al. (2012), Mann et al. (2002), Wang et al. (2001), Wang et al. (2004), Johnston et al. (2000), Aralla et al. (2012)	Wen et al. (1999), Nico et al. (2001), Gömöri et al. (2006)	Wintour et al. (2004)	Liu et al. (2002, 2003), Torday and Rehan (2003), King et al. (1997)	Baum et al. (1998), Yamamoto et al. (1997), Butkus et al. (1997, 1999), Devuyst et al. (1996)	Agren et al. (1998, 2003)

Table 6.2 The deficiency of AQPs and pregnancy outcomes in mice

Item	AQPs deficiency model	Pregnancy outcome	References
AQP1	AQP1 $^{(-/-)}$ mice	Amniotic fluid volume ↑, amniotic fluid osmolality ↓	Mann et al. (2005)
	AQP1 $^{(-/-)}$ mice	Embryo number ↓, amniotic fluid volume ↑, fetal weight ↓	Zheng et al. (2014)
	AQP1 $^{(-/-)}$ mice	Water permeability ↓	Sha et al. (2015)
	AQP1 $^{(-/-)}$ mice	Amniotic fluid volume ↑, decreased osmolality ↓, calcium in amniotic fluid ↓	Luo et al. (2018)
AQP3	AQP3 siRNA in mouse embryo	Preimplantation embryo development ↓	Xiong et al. (2013)
	AQP3 $^{(-/-)}$ mice	Amniotic fluid volume ↓, fatty acid and triglycerides in amniotic fluid ↓, embryo numbers ↓, fetal growth ↓	Seo et al. (2018)
AQP4	AQP4 $^{(-/-)}$ mice	Pregnancy rate ↓, litter size ↓	Sun et al. (2009)
AQP7	AQP7 siRNA in mouse embryo	Preimplantation embryo development ↓	Xiong et al. (2013)
AQP8	AQP8 $^{(-/-)}$ mice	Offspring number ↑	Su et al. (2010)
	AQP8 $^{(-/-)}$ mice	Embryo number ↑, amniotic fluid volume ↑, placental weight ↑, fetal weight ↑	Sha et al. (2011a, b)

↓, decrease; ↑, increase

6.2.3 Development of Fetal Organs

The balance between the production of fetal fluids (including fetal urine and lung liquid) and its resorption through fetal swallowing and intramembranous flow is critical for maintaining adequate amniotic fluid volume (Modena and Fieni 2004). Thus, the AQPs expressed in the fetal organs have important roles in fetal fluid exchange and amniotic fluid homeostasis (Martínez and Damiano 2017). There is evidence that expression of AQPs 1–4 in the fetal kidneys for the production of fetal urine contributes to aminotic fluid volume (Baum et al. 1998; Yamamoto et al. 1997; Butkus et al. 1997, 1999; Devuyst et al. 1996). Furthermore, the fetal lung produces fluid that contributes to amniotic fluid volume and that is mediated by AQPs 1, 3, 4, and 5 in many species, including rats, sheep, and mice (Table 6.1) (Liu et al. 2002, 2003; Torday and Rehan 2003; King et al. 1997). Also, the expression of AQP1 and AQP4 in the fetal brain, AQPs 1, 3, 4, and 8 in the fetal heart, and AQP3 in the fetal skin (Table 6.1) may also contribute to fluid homeostasis and transcellular water transport in the fetus, but their functions are not confirmed.

6.3 Nutritional Regulation of Water Transport in the Conceptus

6.3.1 Dietary Protein

The nutritional regulation of AQP expression in the conceptus can modulate water transport through physiological or pathological alterations during pregnancy. Regulation of maternal dietary protein intake during gestation can affect embryonic survival, growth, and development, while inadequate or excessive intake of protein contributes to embryonic loss and impaired growth and development of conceptus due to the imbalance in amino acids (Herring et al. 2018a; Wu 2022). A low-protein diet during gestation decreases the renal expression of AQP2 in female rats (Cornock et al. 2010). Likewise, maternal undernutrition decreased AQP1 expression, while increasing AQP8 and AQP9 expression in the placenta that was accompanied with decreases in fetal and placental weights and amniotic fluid volume in rats (Belkacemi et al. 2011). Because dietary protein is terminally hydrolyzed to free amino acids in the small intestine via the combined actions of its luminal (extracellular) proteases and peptidases as well as intracellular (enterocyte) dipeptidases and tripeptidases, amino acids mediate the effects of dietary protein on AQP expression in both maternal and fetal tissues (Hu et al. 2021; Wu 2021).

6.3.2 Amino Acids

Adequate intakes of amino acids are essential for embryonic/fetal survival, growth, and development (Wu 2009; Wu et al. 2017). Earlier research has shown that water transfer is most sensitive to changes in passive cations and chloride, plasma lactate and glucose, bicarbonate, and amino acids (Wilbur et al. 1978). Selected nutrients including arginine (Arg), leucine (Leu), and glutamine (Gln) are critical for supporting the survival, growth, and development of conceptuses based on their abundant concentrations in the conceptus (Bazer et al. 2015; Gao 2020; Wu et al. 2011). Among these nutrients, Arg enhances placental growth (Gao et al. 2012; Li et al. 2014; Zhang et al. 2021) and promotes embryonic and fetal development in mammals as a source of nitric oxide and polyamines by increasing placental angiogenesis, blood flow, and protein synthesis (Wu et al. 2013; Bazer et al. 2015). Most mammals, including humans, pigs, rats, mice, cattle and sheep, synthesize Arg from glutamine, glutamate and proline via the intestinal-renal axis (Dillon and Wu 2021; Wu and Morris 1998; Wu et al. 2022). Because there are high rates of Arg utilization via multiple metabolic pathways (such as the syntheses of proteins, creatine, nitric oxide, and polyamines) in both mother and fetus, the ingestion of Arg by pregnant mammals particularly those (e.g., pigs and cattle) that typically have a restricted food intake or consume low-protein (e.g., 10–12% crude protein) diets during gestation, may be inadequate (Gilbreath et al. 2021; Wu et al. 2018).

We recently reported that dietary supplementation with Arg during Days 14–25 of gestation increased the abundance of AQP 1, 5 and 9 proteins and water transport [measured in Ussing chambers (Fig. 6.2)] in the placenta of gilts (Herring et al. 2018b; Zhu et al. 2016, 2021). An underlying mechanism may involve nitric oxide synthesis and the cGMP cell signaling (Fig. 6.3). This supports previous results that Arg enhances the volume of amniotic and allantoic fluids, as well as embryonic survival in pregnant gilts (Li et al. 2014). Hence, physiological concentrations of Arg upregulate the expression of AQP3 and enhance water transport by porcine trophectederm cells (pTr2), which is essential for blastocyst development and implantation (Zhu et al. 2018a). Therefore, Arg is a functional amino acid that improves embryonic survival and ameliorates intrauterine growth restriction (IUGR) in swine (Wu et al. 2018, 2021; Zhang et al. 2021) and other mammals (Gilbreath et al. 2021; Wu et al. 2013).

6.3.3 High-Fat Diet

Consumption of a high-fat diet for 3 weeks increases body weight in both wild-type and AQP8-knockout mice (Yang et al. 2005). Bivariate regression analysis revealed that the amniotic fluid index was positively associated with AQP11 expression in Japanese macaque fed a high-fat diet (Cheung et al. 2018). In contrast, maternal high-fat diets do not appear to alter the expression of AQPs 1, 3, 8, 9, and 11 in the amnion of non-obese Japanese macaques (Cheung et al. 2018).

6.3.4 Vitamins

All-trans retinoic acid (a bioactive metabolite of vitamin A) directly regulates the expression of AQP3 and amniotic fluid volume through binding retinoic acid receptor alpha (RARA) and DR5 retinoic acid receptor element (DR5-RARE) (Prat et al. 2015). Similarly, the retinoid pathway may play an essential role in regulating intramembranous transport across the ovine amnion into the fetal vasculature through effects on the expression of vascular endothelial growth factor (VEGF) (Cheung et al. 2019). Dietary vitamins are essential to embryonic/fetal survival and growth (Wu 2018) and are expected to influence water transport in both the mother and conceptus. More research is warranted in this area of nutrition and metabolism.

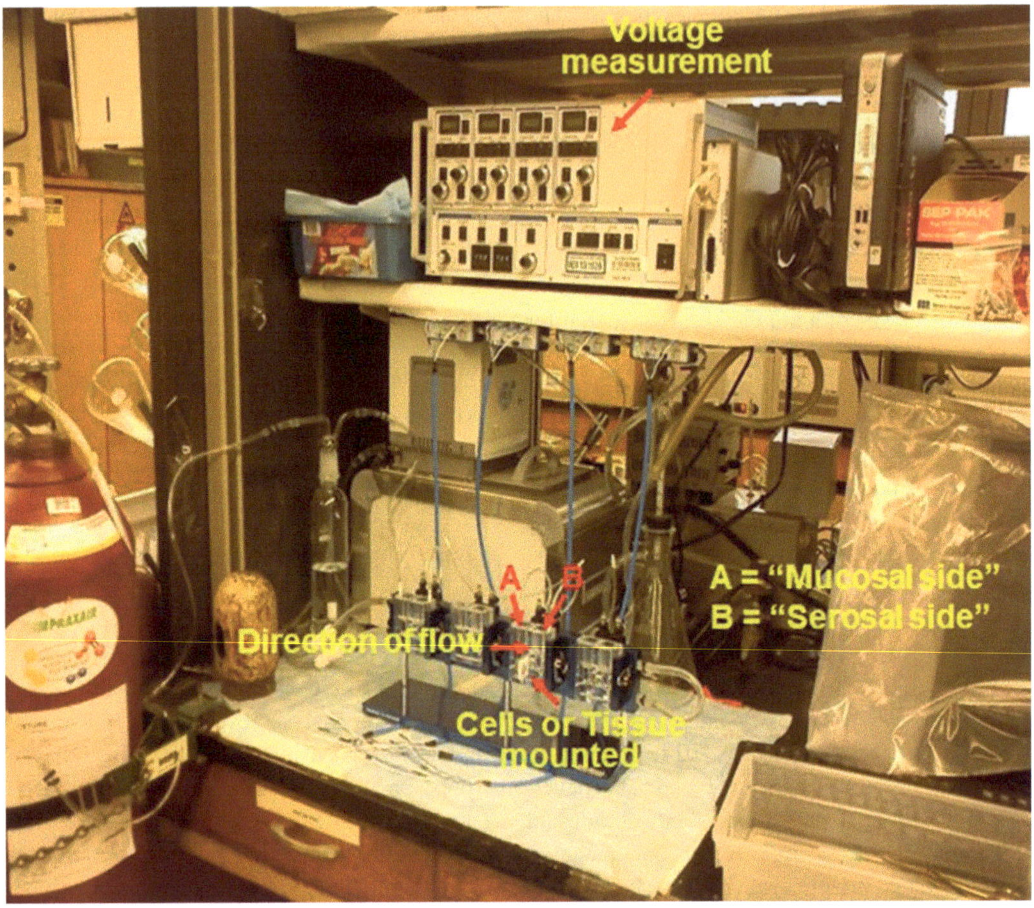

Fig. 6.2 Ussing chambers for measuring water and ion transport by epithelial tissues (e.g., placenta). Both chambers contain the same volume (e.g., 5 mL) of Krebs bicarbonate buffer (37 °C, pH 7.4) that is continuously gassed with 95% O_2/5% CO_2. Pieces of a physiologically viable and intact tissue (e.g., placental tissue; 1 cm^2) were mounted into the cassettes of Ussing chambers, followed by the addition of 20 μL of ^3H_2O to the "mucosal" side of each chamber. A 20 μL aliquot of solution is obtained from the "serosal" side of the chamber at 0, 5, 10, and 15 min for the measurement of the net transport of ^3H_2O or a ^{14}C-labeled amino acid. Radioactivity is determined using a liquid scintillation counter (Packard, Shelton, CT) (Zhang et al. 2019). A = Addition of a tested substance to the "mucosal side" of the chamber (i.e., allantoic membrane side). B = Sampling, from the "serosal side" of the chamber (i.e., chorion side), of a solution containing the tested substance transported by the mounted tissue (e.g., placenta). Water transport and an amino acid by the placenta are measured using ^3H_2O and a ^{14}C-labeled amino acid, respectively, whereas ion transport by the placenta is measured simultaneously by an ohmmeter as transepithelial voltage changes

6.3.5 Minerals

Several minerals, including copper, zinc, nickel, and mercury, may be involved in the regulation of water transport and permeability of AQPs (Zelenina et al. 2004; Németh-Cahalan et al. 2007; Szpilbarg et al. 2016). For instance, elevated concentrations of copper inhibit the transport of water and glycerol by AQP3 in humans (Zelenina et al. 2004). Hence, $CuSO_4$ could be used as an inhibitor of AQP3 in human trophoblast cells (Szpilbarg et al. 2016). Mercury chloride ($HgCl_2$) is recognized as a general inhibitor of most AQPs (Ishibashi 2009) and can rapidly increase the transport of water and anions by AQP6 (Yasui et al. 1999). This may explain, in part, the toxicity of mercury in animals. In addition, zinc or nickel increases the transport of water by AQP0 in Xenopus laevis oocytes but did not affect the activities of AQP1 and AQP4 (Németh-Cahalan et al. 2007). It remains largely unknown as to how

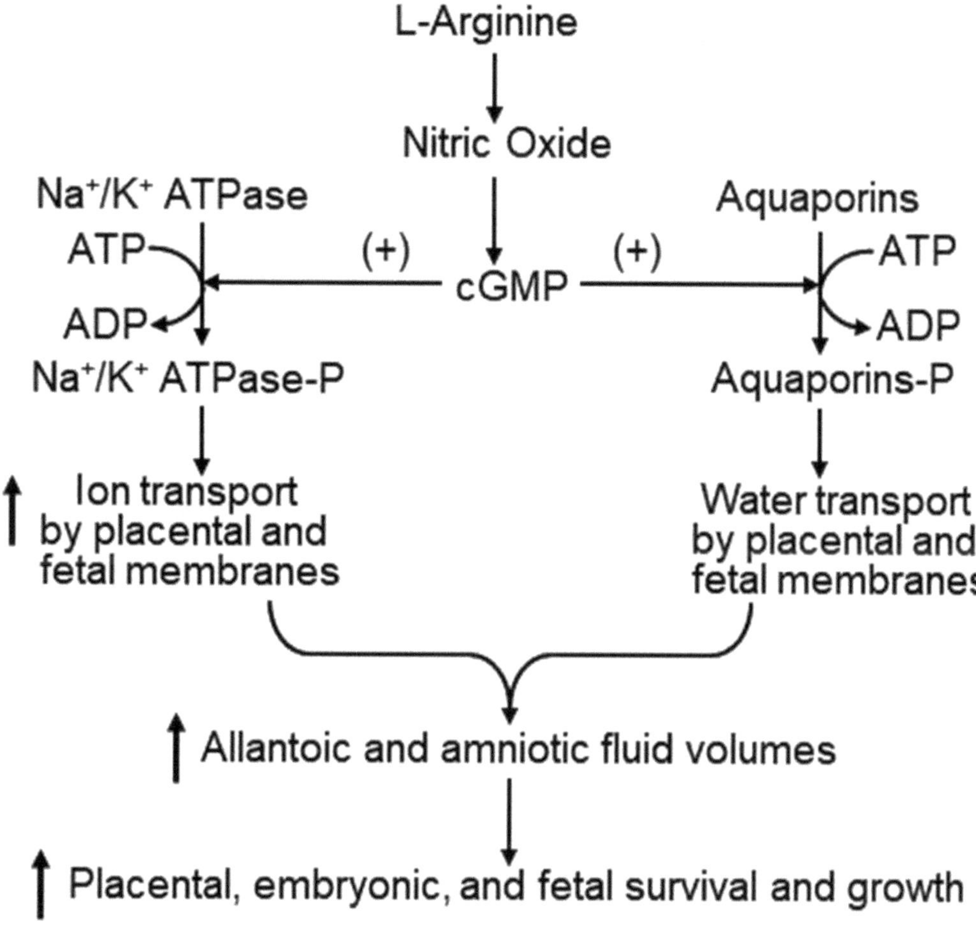

Fig. 6.3 A possible mechanism for L-arginine to increase water and ion transport in the porcine placenta. Nitric oxide (NO) synthesized from L-arginine stimulates the production of cGMP which then activates cGMP-dependent protein kinases to phosphorylate aquaporins and Na^+/K^+-ATPase. Phosphorylated forms of aquaporins and Na^+/K^+-ATPase promote the transport of water and ions by the placenta of the conceptus to increase the volumes of both allantoic and amniotic fluids. These effects of NO provide a uterine environment that beneficially enhances placental development and embryonic/fetal survival ↑, Increase

maternal mineral nutrition affects water transport in mammalian conceptuses.

6.4 Physiological Regulation of Water Transport in the Conceptus

6.4.1 Hormones

6.4.1.1 Estradiol (E2)

The expression of most ion/water channel proteins is regulated by steroid hormones including estradiol (E2), progesterone (P4), insulin, leptin, human chorionic gonadotropin (hCG), and relaxin (Liu et al. 2014; Damiano 2020). After fertilization, pig conceptuses secrete large amounts of E2, which plays a crucial role in maintaining pregnancy and conceptus development during the periods of implantation and placentation required for successful pregnancies (Geisert et al. 1982). The distribution of AQP1 in progesterone-primed uteri of mice shifts from the myometrium to the uterine stromal vasculature in response to exogenous E2 administration (Richard et al. 2013). The increased expression of AQP2 and/or the abundance of AQP3 in the uterus in response to E2 was associated with a greater viscosity of uterine luminal fluid that facilitated the movement of water into the uterine lumen of mice (Jablonski et al. 2003). The E2 also induced the expression of AQP5 in the uterus (Kobayashi et al. 2006), which may be via the presence of estrogen response elements in the promoter region of the AQP5 gene (Jiang et al. 2015). A recent study found that the uterine expression of AQPs 3, 4, 5, and 8 is induced by E2 in mice during pregnancy (de Oliveira et al. 2020). Similarly, E2 increased both mRNA and protein levels of AQPs 1, 2, 5, and 7 in uteri of rats (Chinigarzadeh et al. 2017). Available results indicate that the expression and cell-specific localization of specific AQPs in the uterus are regulated by E2. Of note, we reported that the addition of 0.025–0.1 ng E2/mL to culture medium increased the abundance of the AQP3 protein in pTr2 cells (Zhu et al. 2018b).

6.4.1.2 Progesterone (P4)

Progesterone is the hormone of pregnancy required for the establishment and maintenance of successful pregnancies in mammals (including pigs) (Spencer et al. 2004). Progesterone effectively induces the uterine expression of AQP1 and AQP11 but inhibits the E2 induction of the expression of AQP3 and AQP4 in uteri of mice (de Oliveira et al. 2020). In addition, the up-regulation of expression of AQP1 in the myometrium and AQP5 in uterine epithelial cells of rats is P4-dependent (Lindsay and Murphy 2006). The expression of AQP9 protein increases in response to P4 in rat oviductal epithelium (Brañes et al. 2005). Likewise, we reported that the addition of 20 ng P4/mL to culture medium increased the abundances of AQP3, AQP5, and AQP9 proteins in pTr2 cells (Zhu et al. 2018b).

6.4.1.3 Insulin

Insulin can suppress the transcriptional expression of AQP3 and AQP9 in the amnion of parturient women with type 2 diabetes or gestational diabetes in a phosphatidylinositol 3-kinase (PI3K)-dependent manner (Bouvier et al. 2015). Similarly, the expression of AQP9 in trophoblast is upregulated in patients with gestational diabetes with higher requirements for the transport of water associated with this pathology of pregnancy (Vilariño-García et al. 2016). The negative insulin response element in the promoter region of AQP9 may explain the downregulation of AQP9 expression in explants of normal placentae treated with insulin (Castro-Parodi et al. 2008). Furthermore, the regulation of the expression of AQPs in the fetal membranes by insulin may be responsible for the imbalance in amniotic fluid volume during pregnancies in diabetic women (Bouvier et al. 2015). However, the functional transport of water in normal placental explants was not affected by insulin (Castro-Parodi et al. 2011). It is unknown whether the effect of insulin on AQP expression may be dependent on the actions of other hormones in vivo.

6.4.1.4 Leptin

Leptin is expressed abundantly by trophoblast cells. An increase in concentrations of leptin in the plasma of patients with gestational diabetes and obesity has been reported (Pérez-Pérez et al. 2020a). AQP9 is over-expressed in placentae of pregnant women with gestational diabetes mellitus possibly due to alterations in the leptin/leptin receptor system (Pérez-Pérez et al. 2013). Similarly, the abundances of the AQP9 mRNA and protein in human trophoblast explants increase in response to leptin (Vilariño-García et al. 2018). Importantly, AQP9 facilitates not only the transport of water but also the transport of solutes such as glycerol that likely contribute to energy metabolism and gluconeogenesis during the development of the conceptus (Damiano 2011).

6.4.1.5 Relaxin

Relaxin is involved in the remodeling of the extracellular matrix of the cervix and vagina during pregnancy and the rupture of fetal membranes at term (Parry and Vodstrcil 2007). In the uterus, relaxin enhances angiogenesis to increase blood flow during early pregnancy (Jauniaux et al. 1994) and augments the water volume, glycoproteins, and weight of the uterus (Breenan and Zarrow 1959). Relaxin is also a growth factor for fetal membranes because the proliferation of amniotic epithelial and cytotrophoblast cells is increased by relaxin (Millar et al. 2003). We also found that the addition of 25–50 ng relaxin/mL to culture medium increased the abundances of AQP1 and AQP3 proteins in pTr2 cells (Zhu et al. 2018b), and the expression of AQP3 mRNA is upregulated by relaxin in the cervix of mice in late pregnancy (Soh et al. 2012).

6.4.1.6 Human Chorionic Gonadotropin

Concentrations of hCG in the serum of pregnant women with preeclampsia are positively correlated with the abundances of AQP2 (Zhao et al. 2018) and AQP9 (Damiano et al. 2006) proteins in the placenta. Expression of AQP2 is upregulated in placentae of pregnant women with preeclampsia (Zhao et al. 2018). Likewise, both water transport and AQP9 expression are greater in placental explants from women with preeclampsia by hCG in a cAMP-dependent manner (Marino et al. 2010).

6.4.2 Oxygen Supply

Due to low levels of oxygen in the lumen of uteri, hypoxia occurs naturally in developing embryos, which rely on direct diffusion for environmental oxygen and on transmembrane transporters for nutrients (Stenhouse et al. 2022). A possible relationship between oxygen homeostasis and preeclampsia in women was identified by Hung and Burton (2006) and they were associated with changes in the expression of AQPs and their distribution in placentae. Oxygen regulates the placental expression of AQP4 through the hypoxia inducible factor 1 alpha (HIF1A)-dependent pathway. Especially, hypoxia results in the upregulation of the expression of AQP4, but the downregulation of the expression of AQP4 in human placentae (Szpilbarg et al. 2018). Moreover, AQP9 expression and water transport are reduced in human placental explants when HIF1A is overexpressed under hypoxic conditions (Castro-Parodi et al. 2013). In contrast, reoxygenation of placental explants following hypoxia increases the expression of AQP9 and water uptake by the tissue (Castro-Parodi et al. 2013). Similarly, hypoxia decreases AQP3 expression and changes its distribution in the cytosol of syncytiotrophoblasts, but its expression can be partially restored after reoxygenation (Szpilbarg et al. 2016). Inhibition of AQP3 expression can attenuate placental apoptosis (Szpilbarg et al. 2016) and impair the migration and differentiation of extravillous trophoblast (Alejandra et al. 2018; Reppetti et al. 2020).

6.4.3 pH

Disturbances in pH homeostasis can alter water transport and cell volume in human placentae (Dietrich and Damiano 2015). Maintenance of intracellular pH of the syncytiotrophoblast is regulated through the inhibition of Na^+/H^+ exchangers (NHEs) (Johansson et al. 2002; Sibley et al. 2002). The expression of NHE1 is reduced in syncytiotrophoblast cells isolated from placentae of preterm women with an IUGR pregnancy (Johnston et al. 2000). Although water uptake was not affected by intracellular pH changes in syncytiotrophoblast, NHE can alter water transport mediated by placental AQPs during acidotic states (Dietrich and Damiano 2015). Thus, pH changes induced by NHE may be associated with the alterations in the placental permeability of AQPs for the homeostasis of amniotic fluid volume and normal conceptus development.

6.4.4 Osmotic Pressure

Transplacental transport of water is driven, in part, by hydrostatic and osmotic pressures (Faber and Anderson 2010). The lower osmolality and concentrations of sodium in the maternal blood than in the fetal blood create an osmotic gradient that favors water transport from mother to fetus (Moen et al. 2018). The trophoblast cells from AQP1-deficient pregnant mice have lower water permeability than those from wild-type mice, indicating the important role of AQP1 in mediating the maternal–fetal fluid balance (Sha et al. 2015). In addition, reductions in the expression of AQP1 in the amnion, as well as AQP3 in the amnion and chorion occur in women with oligohydramnios when compared with those with normal volumes of amniotic fluid (Zhu et al. 2009). Moreover, the expression of both AQP8 and AQP9 in the amnion as well as AQP9 in the chorion is less for pregnant women with oligohydramnios. Interestingly, the expression of AQP3 (Zhu et al. 2009) as well as AQP8 and AQP9 (Jiang et al. 2012) is greater in placentae from women with oligohydramnios, when compared with normal women. Furthermore, when subjecting amnionic epithelial cells to culture media with different osmolalities, AQP8 expression is greater in hypotonic than hypertonic medium (Qi et al. 2009). On the other hand, for human term pregnancies with idiopathic polyhydramnios, there is an increase in the expression of AQP8 in the amnion and AQP9 in the amnion and chorion, but a decrease in the expression of AQP9 in the placenta (Zhu et al. 2010). The basis for the differential regulation of the expression of AQPs in placentae from women with polyhydroamnios is not known.

6.4.5 Cystic Fibrosis Transmembrane Conductance Regulator (CFTR)

Double immunofluorescence labeling revealed that the expression of CFTR is co-localized with AQP9 in the apical membrane of syncytiotrophoblast cells of normal placentae (Castro-Parodi et al. 2009). In addition, the expression of AQP9 is less in placentae from women with preeclampsia than placentae from healthy women, whereas CFTR is not colocalized with AQ9 in the apical membrane of syncytiotrophoblast cells of placentae from women with preeclampsia (Castro-Parodi et al. 2009). This decrease in CFTR expression may affect the placental transport of water by AQPs.

6.5 Summary

The functionality of maternal–fetal water transport is critical for enabling the successful outcome of pregnancy and supporting normal growth and development of the conceptus. AQPs are primarily responsible for a rapid transfer of water between mother and fetus during this process. To date, at least 11 isoforms of AQPs (AQPs 1–9, 11, and 12) are known to be expressed in conceptuses of mammals in a species- and tissue/cell-specific manner. Normal expression and localization of AQPs in the conceptus play important roles in the maintenance of blastocyst formation and implantation, embryonic and fetal fluid balances, as well as embryonic and fetal survival and development. Studies with knockout mice models have shown that a reduction in abundances of AQPs can lead to pregnancy complications and impair the development of conceptuses. Maternal undernutrition has been shown to decrease amniotic fluid volume and fetal weight by downregulating AQP1 expression and upregulating AQP8 and AQP9 expression in the placenta. There is a reduction in AQP2 expression associated with females that consume a low-protein or Arg-inadequate diet during gestation. As a functional amino acid, Arg increases water transport by porcine trophectoderm cells and placentae. High-fat diets increase the amniotic fluid index independent of AQP expression in the amnion. Oxygen levels, nitric oxide, pH, osmotic pressure, E2, P4, insulin, relaxin, leptin, and hCG have been found to regulate the distribution and functional expression of AQPs in female reproductive tissues. The ion transporters, NHEs, and CFTR also influence water transport by AQPs in the placenta. The regulation of AQP expression by nutritional and physiological factors may be mediated by genes at the transcriptional level through their binding to specific response elements in the promoter regions of AQP genes. Further investigations are needed to identify potential mechanisms responsible for the nutritional and physiological regulation of water transport in the conceptus.

Acknowledgements The authors gratefully acknowledge the financial supports from the Guangdong Basic and Applied Basic Research Foundation (Grant Number 2020A1515010018 to CZ), and the Discipline Construction Program of Foshan University (Grant Number CGZ0400162 to CZ), Agriculture and Food Research Initiative Competitive Grants Number 2015-67015-23276 (to GW, FWB, and GAJ) and Number 2018-67015-28093 (to FWB, GAJ, and GW) from the USDA National Institute of Food and Agriculture, and Texas A&M AgriLife Research H-8200 (to GW).

References

Agre P, Kozono D (2003) Aquaporin water channels: molecular mechanisms for human diseases. FEBS Lett 555:72–78

Agren J, Sjors G, Sedin G (1998) Transepidermal water loss in infants born at 24 and 25 weeks of gestation. Acta Paediatr 87:1185–1190

Agren J, Zelenin S, Hakansson M, Eklof AC, Aperia A, Nejsum LN, Nielsen S, Sedin G (2003) Transepidermal water loss in developing rats: role of aquaporins in the immature skin. Pediatr Res 53:558–565

Alejandra R, Natalia S, Alicia ED (2018) The blocking of aquaporin-3 (AQP3) impairs extravillous trophoblast cell migration. Biochem Biophys Res Commun 499:227–232

Aralla M, Mobasheri A, Groppetti D, Cremonesi F, Arrighi S (2012) Expression of aquaporin water channels in canine fetal adnexa in respect to the regulation of amniotic fluid production and absorption. Placenta 33:502–510

Baum MA, Ruddy MK, Hosselet CA, Harris HW (1998) The perinatal expression of aquaporin-2 and aquaporin-3 in developing kidney. Pediatr Res 43:783–790

Bazer FW (1989) Allantoic fluid: regulation of volume and composition. In: Brace RA (ed) Fetal and neonatal body fluids. Perinatology Press, Cornell, NY, pp 135–157

Bazer FW, Spencer TE, Johnson GA, Burghardt RC, Wu G (2009) Comparative aspects of implantation. Reproduction 138:195–209

Bazer FW, Johnson GA, Wu G (2015) Amino acids and conceptus development during the peri-implantation period of pregnancy. Adv Exp Med Biol 843:23–52

Beall MH, Wang S, Yang B, Chaudhri N, Amidi F, Ross MG (2007) Placental and membrane aquaporin water channels: correlation with amniotic fluid volume and composition. Placenta 28:421–428

Belkacemi L, Desai M, Beall MH, Liu Q, Lin JT, Nelson DM, Ross MG (2011) Early compensatory adaptations in maternal undernourished pregnancies in rats: role of the aquaporins. J Matern Fetal Neonatal Med 24:752–759

Bouvier D, Rouzaire M, Marceau G, Prat C, Pereira B, Lemarié R, Deruelle P, Fajardy I, Gallot D, Blanchon L, Vambergue A, Sapin V (2015) Aquaporins and fetal Membranes from diabetic parturient women: expression abnormalities and regulation by insulin. J Clin Endocrinol Metab 100:E1270–1279

Brañes MC, Morales B, Ríos M, Villalón MJ (2005) Regulation of the immunoexpression of aquaporin 9 by ovarian hormones in the rat oviductal epithelium. Am J Physiol 288:C1048–1057

Breenan DM, Zarrow MX (1959) Water and electrolyte content of the uterus of the intact and adrenalectomized rat treated with relaxin and various steroid hormones. Endocrinology 64:907–913

Butkus A, Alcorn D, Earnest L, Moritz K, Giles M, Wintour EM (1997) Expression of aquaporin-1 (AQP1) in the adult and developing sheep kidney. Biol Cell 89:313–320

Butkus A, Earnest L, Jeyaseelan K, Moritz K, Johnston H, Tenis N, Wintour EM (1999) Ovine aquaporin-2: cDNA cloning, ontogeny and control of renal gene expression. Pediatr Nephrol 13:379–390

Cao Z, Xu T, Tong X, Wang Y, Zhang D, Gao D, Zhang L, Ning W, Qi X, Ma Y, Yu T, Knott JG, Zhang Y (2019) Maternal yes-associated protein participates in porcine blastocyst development via modulation of trophectoderm epithelium barrier function. Cells 8:1606

Castro Parodi M, Farina M, Dietrich V, Abán C, Szpilbarg N, Zotta E, Damiano AE (2011) Evidence for insulin-mediated control of AQP9 expression in human placenta. Placenta 32:1050–1056

Castro-Parodi M, Farina M, Dietrich V, Levi LN, Ibarra C, Damiano AE (2008) Dose dependent insulin-mediated regulation of AQP9 in human placenta. Placenta 29:120

Castro-Parodi M, Levi L, Dietrich V, Zotta E, Damiano AE (2009) CFTR may modulate AQP9 functionality in preeclamptic placentas. Placenta 30:642–648

Castro-Parodi M, Szpilbarg N, Dietrich V, Sordelli M, Reca A, Abán C, Maskin B, Farina MG, Damiano AE (2013) Oxygen tension modulates AQP9 expression in human placenta. Placenta 34:690–698

Cheung CY, Anderson DF, Brace RA (2016) Aquaporins in ovine amnion: responses to altered amniotic fluid volumes and intramembranous absorption rates. Physiol Rep 4:e12868

Cheung CY, Roberts VHJ, Frias AE, Brace RA (2018) High-fat diet effects on amniotic fluid volume and amnion aquaporin expression in non-human primates. Physiol Rep 6:e13792

Cheung CY, Anderson DF, Rouzaire M, Blanchon L, Sapin V, Brace RA (2019) Retinoic acid pathway regulation of vascular endothelial growth factor in ovine amnion. Reprod Sci 26:1351–1359

Chinigarzadeh A, Muniandy S, Salleh N (2017) Combinatorial effect of genistein and female sex-steroids on uterine fluid volume and secretion rate and aquaporin (AQP)-1, 2, 5, and 7 expression in the uterus in rats. Environ Toxicol 32:832–844

Cornock R, Langley-Evans SC, Mobasheri A, McMullen S (2010) The impact of maternal protein restriction

during rat pregnancy upon renal expression of angiotensin receptors and vasopressin-related aquaporins. Reprod Biol Endocrinol 8:105
Damiano AE (2011) Review: water channel proteins in the human placenta and fetal membranes. Placenta 32 (Suppl. 2):S207–S211
Damiano AE (2020) Chapter fifteen—aquaporins during pregnancy. Vitam Horm 112:327–355
Damiano AE, Zotta E, Ibarra C (2006) Functional and molecular expression of AQP9 channel and UT-A transporter in normal and preeclamptic human placentas. Placenta 27:1073–1081
Damiano AE, Zotta E, Goldstein J, Reisin I, Ibarra C (2001) Water channel proteins AQP3 and AQP9 are present in syncytiotrophoblast of human term placenta. Placenta 22:776–781
De Falco M, Cobellis L, Torella M, Acone G, Varano L, Sellitti A, Ragucci A, Coppola G, Cassandro R, Laforgia V, Varano L, De Luca A (2007) Downregulation of aquaporin 4 in human placenta throughout pregnancy. In Vivo 21:813–817
de Oliveira V, Schaefer J, Abu-Rafea B, Vilos GA, Vilos AG, Bhattacharya M, Radovick S, Babwah AV (2020) Uterine aquaporin expression is dynamically regulated by estradiol and progesterone and ovarian stimulation disrupts embryo implantation without affecting luminal closure. Mol Hum Reprod 26:154–166
Devuyst O, Burrow CR, Smith BL, Agre P, Knepper MA, Wilson PD (1996) Expression of aquaporins-1 and -2 during nephrogenesis and in autosomal dominant polycystic kidney disease. Am J Physiol 271:F169–F183
Dietrich V, Damiano AE (2015) Activity of Na^+/H^+ exchangers alters aquaporin-mediated water transport in human placenta. Placenta 36:1487–1489
Dillon EL, Wu G (2021) Cortisol enhances ctrulline synthesis from proline in enterocytes of suckling piglets. Amino Acids. http://doi.org/10.1007/s00726-021-03039-y
Ducza E, Csányi A, Gáspár R (2017) Aquaporins during pregnancy: their function and significance. Int J Mol Sci 18:2593
Escobar J, Gormaz M, Arduini A, Gosens K, Martinez A, Perales A, Escrig R, Tormos E, Roselló M, Orellana C, Vento M (2012) Expression of aquaporins early in human pregnancy. Early Hum Dev 88:589–594
Faber JJ, Anderson DF (2010) The placenta in the integrated physiology of fetal volume control. Int J Dev Biol 54:391–396
Gao H (2020) Amino acids in reproductive nutrition and health. Adv Exp Med Biol 1265:111–131
Gao K, Jiang Z, Lin Y, Zheng C, Zhou G, Chen F, Yang L, Wu G (2012) Dietary L-arginine supplementation enhances placental growth and reproductive performance in sows. Amino Acids 42:2207–2214
Geisert RD, Thatcher WW, Roberts RM, Bazer FW (1982) Establishment of pregnancy in the pig: III. Endometrial secretory response to estradiol valerate administered on day 11 of the estrous cycle. Biol Reprod 27:957–965
Gilbreath KR, Bazer FW, Satterfield MC, Wu G (2021) Amino acid nutrition and reproductive performance in ruminants. Adv Exp Med Biol 1285:43–61
Gömöri E, Pál J, Abrahám H, Vajda Z, Sulyok E, Seress L, Dóczi T (2006) Fetal development of membrane water channel proteins aquaporin-1 and aquaporin-4 in the human brain. Int J Dev Neurosci 24:295–305
Herring CM, Bazer FW, Johnson GA, Wu G (2018a) Impacts of maternal dietary protein intake on fetal survival, growth, and development. Exp Biol Med (Maywood) 243:525–533
Herring CM, Hu S, Bazer FW, Johnson GA, Seo H, Kramer A, McLendon B, Elmetwally M, Wu G (2018b) Dietary supplementation of 0.4% L-arginine between days 14 and 30 of gestation stimulates placental transport of water in gilts. In: Society of the study of reproduction annual meeting, July 10–13, 2018. New Orleans, LA
Hu SD, He WL, Wu G (2021) Hydroxyproline in animal metabolism, nutrition, and cell signaling. Amino Acids. http://doi.org/10.1007/s00726-021-03056-x
Hung TH, Burton GJ (2006) Hypoxia and reoxygenation: a possible mechanism for placental oxidative stress in preeclampsia. Taiwan J Obstet Gynecol 45:189–200
Ishibashi K (2009) New members of mammalian aquaporins: AQP10-AQP12. Handbook Exp Pharmacol 190:251–262
Ishibashi K, Hara S, Kondo S (2009) Aquaporin water channels in mammals. Clin Exp Nephrol 13:107–117
Jablonski EM, McConnell NA, Hughes FM Jr, Huet-Hudson YM (2003) Estrogen regulation of aquaporins in the mouse uterus: potential roles in uterine water movement. Biol Reprod 69:1481–1487
Jauniaux E, Johnson MR, Jurkovic D, Ramsay B, Campbell S, Meuris S (1994) The role of relaxin in the development of the uteroplacental circulation in early pregnancy. Obstetr Gynecol 84:338–342
Jiang SS, Zhu XJ, Ding SD, Wang JJ, Jiang LL, Jiang WX, Zhu XQ (2012) Expression and localization of aquaporins 8 and 9 in term placenta with oligohydramnios. Reprod Sci 19:1276–1284
Jiang XX, Wei FX, Zhao L, Ye XL, Xin LB, Qu Y, Wu RJ, Lin J (2015) Aquaporin 5 plays a role in estrogen-induced ectopic implantation of endometrial stromal cells in endometriosis. PLoS One 10: e0145290
Johansson M, Glazier JD, Sibley CP, Jansson T, Powell TL (2002) Activity and protein expression of the Na^+/H^+ exchanger is reduced in syncytiotrophoblast microvillous plasma membranes isolated from preterm intrauterine growth restriction pregnancies. J Clin Endocrinol Metab 87:5686–5694
Johnston H, Koukoulas I, Jeyaseelan K, Armugam A, Earnest L, Baird R, Dawson N, Ferraro T, Wintour EM (2000) Ontogeny of aquaporins 1 and 3 in ovine placenta and fetal membranes. Placenta 21:88–99

King LS, Nielsen S, Agre P (1997) Aquaporins in complex tissues. I. developmental patterns in respiratory and glandular tissues of rat. Am J Phys 273: C1541–C1548

Kitchen P, Day RE, Salman MM, Conner MT, Bill RM, Conner AC (2015) Beyond water homeostasis: diverse functional roles of mammalian aquaporins. Biochim Biophys Acta 1850:2410–2421

Knight JW, Bazer FW, Thatcher WW, Franke DE, Wallace HD (1977) Conceptus development in intact and unilaterally hysterectomized-ovariectomized gilts: interrelationships among hormonal status, placental development, fetal fluids and fetal growth. J Anim Sci 44:620–637

Kobayashi M, Takahashi E, Miyagawa S, Watanabe H, Iguchi T (2006) Chromatin immunoprecipitation-mediated target identification proved aquaporin 5 is regulated directly by estrogen in the uterus. Genes Cells:1133–1143

Kordowitzki P, Kranc W, Bryl R, Kempisty B, Skowronska A, Skowronski MT (2020) The relevance of aquaporins for the physiology, pathology, and aging of the female reproductive system in mammals. Cells 9:2570

Li XL, Bazer FW, Johnson GA, Burghardt RC, Frank JW, Dai ZL, Wang JJ, Wu ZL, Shinzato I, Wu GY (2014) Dietary supplementation with L-arginine between days 14 and 25 of gestation enhances embryonic development and survival in gilts. Amino Acids 46:375–384

Lindsay LA, Murphy C (2006) Redistribution of aquaporins 1 and 5 in the rat uterus is dependent on progesterone: a study with light and electron microscopy. Reproduction 131:369–378

Liu C, Morrisey EE, Whitsett JA (2002) GATA-6 is required for maturation of the lung in late gestation. Am J Physiol 283:L468–L475

Liu H, Hooper SB, Armugam A, Dawson N, Ferraro T, Jeyaseelan K, Thiel A, Koukoulas I, Wintour EM (2003) Aquaporin gene expression and regulation in the ovine fetal lung. J Physiol 551:503–514

Liu H, Koukoulas I, Ross MC, Wang S, Wintour EM (2004) Quantitative comparison of placental expression of three aquaporin genes. Placenta 25:475–478

Liu H, Zheng Z, Wintour EM (2008) Aquaporins and fetal fluid balance. Placenta 29:840–847

Liu XM, Zhang D, Wang TT, Sheng JZ, Huang HF (2014) Ion/water channels for embryo implantation barrier. Physiology (Bethesda) 29:186–195

Luo H, Xie A, Hua Y, Wang J, Liu Y, Zhu X (2018) Aquaporin 1 gene deletion affects the amniotic fluid volume and composition as well as the expression of other aquaporin water channels in placenta and fetal membranes. Clin Chim Acta 482:161–165

Ma T, Yang B, Verkman AS (1997) Cloning of a novel water and urea-permeable aquaporin from mouse expressed strongly in colon, placenta, liver, and heart. Biochem Biophys Res Commun 240:324–328

Mann SE, Ricke EA, Yang BA, Verkman AS, Taylor RN (2002) Expression and localization of aquaporin 1 and 3 in human fetal membranes. Am J Obstet Gynecol 187:902–907

Mann SE, Ricke EA, Torres EA, Taylor RN (2005) A novel model of polyhydramnios: amniotic fluid volume is increased in aquaporin 1 knockout mice. Am J Obstet Gynecol 192:2041–2046

Marino G, Castro-Parodi M, Dietrich V, Damiano AE (2010) High levels of human chorionic gonadotropin (HCG) may increase aquaporin-9 (AQP9) expression in human preeclamptic placenta. Reprod Sci 17:444–453

Martínez N, Damiano AE (2017) Aquaporins in fetal development. Adv Exp Med Biol 969:199–212

Meli R, Pirozzi C, Pelagalli A (2018) New perspectives on the potential role of aquaporins (AQPs) in the physiology of inflammation. Front Physiol 16(9):101

Millar LK, Reiny R, Yamamoto SY, Okazaki K, Webster L, Bryant-Greenwood GD (2003) Relaxin causes proliferation of human amniotic epithelium by stimulation of insulin-like growth factor-II. Am J Obstetr Gynecol 188:234–241

Modena AB, Fieni S (2004) Amniotic fluid dynamics. Acta Biomed 75(Suppl 1):11–13

Moen V, Brudin L, Tjernberg I, Rundgren M, Irestedt L (2018) Feto-maternal osmotic balance at term. A prospective observational study. J Perinat Med 46:183–189

Németh-Cahalan KL, Kalman K, Froger A, Hall JE (2007) Zinc modulation of water permeability reveals that aquaporin 0 functions as a cooperative tetramer. J Gen Physiol 130:457–464

Nico B, Frigeri A, Nicchia GP, Quondamatteo F, Herken R, Errede M, Ribatti D, Svelto M, Roncali L (2001) Role of aquaporin-4 water channel in the development and integrity of the blood-brain barrier. J Cell Sci 114:1297–1307

Offenberg H, Barcroft LC, Caveney A, Viuff D, Thomsen PD, Watson AJ (2000) mRNAs encoding aquaporins are present during murine preimplantation development. Mol Reprod Dev 57:323–330

Parry LJ, Vodstrcil LA (2007) Relaxin physiology in the female reproductive tract during pregnancy. Adv Exp Med Biol 612:34–48

Pérez-Pérez A, Maymó JL, Gambino YP, Guadix P, Dueñas JL, Varone CL, Sánchez-Margalet V (2013) Activated translation signaling in placenta from pregnant women with gestational diabetes mellitus: possible role of leptin. Horm Metab Res 45:436–442

Pérez-Pérez A, Vilariño-García T, Guadix P, Dueñas JL, Sánchez-Margalet V (2020a) Leptin and nutrition in gestational diabetes. Nutrients 12:1970

Pérez-Péreza A, Vilariño-Garcíaa T, Dietrich V, Guadixc P, Dueñasc JL, Varone CL, Damiano AE, Sánchez-Margalet V (2020b) Aquaporins and placenta. Vitam Horm 112:311–326

Prat C, Blanchon L, Borel V, Gallot D, Herbet A, Bouvier D, Marceau G, Sapin V (2012) Ontogeny of aquaporins in human fetal membranes. Biol Reprod 86:48

Prat C, Bouvier D, Comptour A, Marceau G, Belville C, Clairefond G, Blanc P, Gallot D, Blanchon L, Sapin V

(2015) All-trans-retinoic acid regulates aquaporin-3 expression and related cellular membrane permeability in the human amniotic environment. Placenta 36:881–887

Qi HB, Li L, Zong WJ, Hyer BJ, Huang J (2009) Expression of aquaporin 8 is diversely regulated by osmotic stress in amnion epithelial cells. J Obstet Gynaecol 35:1019–1025

Reppetti J, Reca A, Seyahian E, Medina Y, Martínez N, Szpilbarg N, Damiano AE (2020) Intact caveolae are required for proper extravillous trophoblast migration and differentiation. J Cell Physiol 235:3382–3392

Richard C, Gao J, Brown N, Reese J (2013) Aquaporin water channel genes are differentially expressed and regulated by ovarian steroids during the periimplantation period in the mouse. Endocrinology 144:1533–1541

Schneider H (1991) Placental transport function. Reprod Fertil Dev 3:345–353

Seo MJ, Lim JH, Kim DH, Bae HR (2018) Loss of aquaporin-3 in placenta and fetal membranes induces growth restriction in mice. Dev Reprod 22:263–273

Seo H, Li XL, Wu G, Bazer FW, Burghardt RC, Bayless KJ, Johnson GA (2020) Mechanotransduction drives morphogenesis to develop folding during placental development in pigs. Placenta 90:62–70

Sha XY, Xiong ZF, Liu HS, Di XD, Ma TH (2011a) Maternal-fetal fluid balance and aquaporins: from molecule to physiology. Acta Pharmacol Sin 32:716–720

Sha XY, Xiong ZF, Liu HS (2011b) Pregnant phenotype in aquaporin 8-deficient mice. Acta Pharmacol Sin 32:840–844

Sha XY, Liu HS, Ma TH (2015) Osmotic water permeability diversification in primary trophoblast cultures from aquaporin 1-deficient pregnant mice. J Obstet Gynaecol Res 41:1399–1405

Shao H, Gao S, Ying X, Zhu X, Hua Y (2021) Expression and regulation of aquaporins in pregnancy complications and reproductive dysfunctions. DNA Cell Biol 40:116–125

Sibley CP, Brownbill P, Glazier JD, Greenwood SL (2018) Knowledge needed about the exchange physiology of the placenta. Placenta 64(Suppl 1):S9–S15

Sibley CP, Glazier JD, Greenwood SL, Lacey H, Mynett K, Speake P, Jansson T, Johansson M, Powell TL (2002) Regulation of placental transfer: the Na^+/H^+ exchanger—a review. Placenta 23(Suppl A):S39–S46

Soh YM, Tiwari A, Mahendroo M, Conrad KP, Parry LJ (2012) Relaxin regulates hyaluronan synthesis and aquaporins in the cervix of late pregnant mice. Endocrinology 153:6054–6064

Spencer TE, Johnson GA, Burghardt RC, Bazer FW (2004) Progesterone and placental hormone actions on the uterus: insights from domestic animals. Biol Reprod 71:2–10

Stenhouse C, Seo H, Wu G, Johnson GA, Bazer FW (2022) Insights into the regulation of implantation and placentation in humans, rodents, sheep, and pigs. Adv Exp Med Biol 1354:25–48

Stulc J (1997) Placental transfer of inorganic ions and water. Physiol Rev 77:805–836

Su WH, Qiao Y, Yi F, Guan XG, Zhang D (2010) Increased female fertility in aquaporin 8-deficient mice. IUBMB Life 62:852–857

Sun XL, Zhang J, Fan Y, Ding JH (2009) Aquaporin-4 deficiency induces subfertility in female mice. Fertil Steril 92:1736–1743

Szpilbarg N, Castro-Parodi M, Reppetti J, Repetto M, Maskin B, Martinez N, Damiano AE (2016) Placental programmed cell death: insights into the role of aquaporins. Mol Hum Reprod 22:46–56

Szpilbarg N, Seyahian A, Di Paola M, Castro-Parodi M, Martinez N, Farina M, Damiano AE (2018) Oxygen regulation of aquaporin-4 in human placenta. Reprod BioMed Online 37:601–612

Torday JS, Rehan VK (2003) Testing for fetal lung maturation: a biochemical "window" to the developing fetus. Clin Lab Med 23:361–383

Vilariño-García T, Pérez-Pérez A, Dietrich V, Fernández-Sánchez M, Guadix P, Dueñas JL, Varone CL, Damiano AE, Sánchez-Margalet V (2016) Increased expression of aquaporin 9 in trophoblast from gestational diabetic patients. Horm Metab Res 48:535–539

Vilariño-García T, Pérez-Pérez A, Dietrich V, Guadix P, Dueñas JL, Varone CL, Damiano AE, Sánchez-Margalet V (2018) Leptin up-regulates aquaporin 9 expression in human placenta in vitro. Gynecol Endocrinol 34:75–77

Wang S, Kallichanda N, Song W, Ramirez BA, Ross MG (2001) Expression of aquaporin-8 in human placenta and chorioamniotic membranes: evidence of molecular mechanism for intramembranous amniotic fluid resorption. Am J Obstet Gynecol 185:1226–1231

Wang S, Chen J, Beall M, Zhou W, Ross MG (2004) Expression of aquaporin 9 in human chorioamniotic membranes and placenta. Am J Obstet Gynecol 191:2160–2167

Watson AJ (1992) The cell biology of blastocyst development. Mol Reprod Dev 33:492–504

Wen H, Nagelhus EA, Amiry-Moghaddam M, Agre P, Ottersen OP, Nielsen S (1999) Ontogeny of water transport in rat brain: postnatal expression of the aquaporin-4 water channel. Eur J Neurosci 11:935–945

Wilbur WJ, Power GG, Longo LD (1978) Water exchange in the placenta: a mathematical model. Am J Physiol 235:R181–199

Wintour EM, Liu H, Dodic M, Moritz K (2004) Differential renal and cardiac gene expression in ovine fetuses programmed to become hypertersive adults by early glucocorticoid treatment. In: Proceedings of the 12th International Congress of Endocrinology, Lisbon, Portugal. pp 501–505

Wooding FB, Wilsher S, Benirschke K, Jones CJ, Allen WR (2015) Immunocytochemistry of the placentas of giraffe (Giraffa cameleopardalis giraffa) and okapi (Okapi johnstoni): comparison with other ruminants. Placenta 36:77–87
Wu G (2009) Amino acids: metabolism, functions, and nutrition. Amino Acids 37:1–17
Wu G (2018) Principles of animal nutrition. CRC Press, Boca Raton, FL
Wu G (2021) Amino Acids: biochemistry and nutrition, 2nd edition. CRC Press, Boca Raton, Florida
Wu G (2022) Nutrition and metabolism: foundations for animal growth, development, reproduction, and health. Adv Exp Med Biol 1354:1–24
Wu G, Bazer FW, Wallace JM, Spencer TE (2006) BOARD-INVITED REVIEW: Intrauterine growth retardation: Implications for the animal sciences. J Anim Sci 84:2316–2337
Wu G, Bazer FW, Satterfield MC, Li X, Wang X, Johnson GA, Burghardt RC, Dai Z, Wang J, Wu Z (2013) Impacts of arginine nutrition on embryonic and fetal development in mammals. Amino Acids 45:241–245
Wu G, Bazer FW, Johnson GA, Herring C, Seo H, Dai ZL, Wang JJ, Wu ZL, Wang XL (2017) Functional amino acids in the development of the pig placenta. Mol Reprod Dev 84:879–882
Wu G, Bazer FW, Johnson GA, Knabe DA, Burghardt RC, Spencer TE, Li XL, Wang JJ (2011) TRIENNIAL GROWTH SYMPOSIUM: Important roles for L-glutamine in swine nutrition and production. J Anim Sci 89:2017–2030
Wu G, Bazer FW, Johnson GA, Hou YQ (2018) Arginine nutrition and metabolism in growing, gestating and lactating swine. J Anim Sci 96:5035–5051
Wu G, Bazer FW, Satterfield MC, Gilbreath KR, Posey EA, Sun YX (2022) L-Arginine nutrition and metabolism in ruminants. Adv Exp Med Biol 1354:177–206
Wu G, Meininger CJ, McNeal CJ, Bazer FW, Rhoads JM (2021) Role of L-arginine in nitric oxide synthesis and health in humans. Adv Exp Med Biol 1332:167–188
Xiong Y, Tan YJ, Xiong YM, Huang YT, Hu XL, Lu YC, Ye YH, Wang TT, Zhang D, Jin F, Huang HF, Sheng JZ (2013) Expression of aquaporins in human embryos and potential role of AQP3 and AQP7 in preimplantation mouse embryo development. Cell Physiol Biochem 31:649–658
Yamamoto T, Sasaki S, Fushimi K, Ishibashi K, Yaoita E, Kawasaki K, Fujinaka H, Marumo F, Kihara I (1997) Expression of AQP family in rat kidneys during development and maturation. Am J Physiol 272:F198–F204
Yang B, Song Y, Zhao D, Verkman AS (2005) Phenotype analysis of aquaporin-8 null mice. Am J Physiol 288: C1161–1170
Yasui M, Hazama A, Kwon TH, Nielsen S, Guggino WB, Agre P (1999) Rapid gating and anion permeability of an intracellular aquaporin. Nature 402:184–187
Zelenina M, Tritto S, Bondar AA, Zelenin S, Aperia A (2004) Copper inhibits the water and glycerol permeability of aquaporin-3. J Biol Chem 279:51939–51943
Zhang JM, He WL, Yi D, Zhao D, Song Z, Hou YQ, Wu G (2019) Regulation of protein synthesis in porcine mammary epithelial cells by L-valine. Amino Acids 51:717–726
Zhang Q, Hou YQ, Bazer FW, He WL, Posey EA, Wu G (2021) Amino acids in swine nutrition and production. Adv Exp Med Biol 1285:81–107
Zhao Y, Lin L, Lai A (2018) Expression and significance of aquaporin-2 and serum hormones in placenta of patients with preeclampsia. J Obstet Gynaecol 38:42–48
Zheng Z, Liu H, Beall M, Ma T, Hao R, Ross MG (2014) Role of aquaporin 1 in fetal fluid homeostasis. J Matern Fetal Neonatal Med 27:505–510
Zhu XQ, Jiang SS, Zhu XJ, Zou SW, Wang YH, Hu YC (2009) Expression of aquaporin 1 and aquaporin 3 in fetal membranes and placenta in human term pregnancies with oligohydramnios. Placenta 30:670–676
Zhu X, Jiang S, Hu Y, Zheng X, Zou S, Wang Y, Zhu XJ (2010) The expression of aquaporin 8 and aquaporin 9 in fetal membranes and placenta in term pregnancies complicated by idiopathic polyhydramnios. Early Hum Dev 86:657–663
Zhu C, Jiang Z, Bazer FW, Johnson GA, Burghardt RC, Wu G (2015) Aquaporins in the female reproductive system of mammals. Front Biosci 20:838–871
Zhu C, Ye J, Bazer F, Johnson G, Bai Y, Yang J, Chen Z, Jiang Z (2018a) L-Arginine upregulates aquaporin-3 expression and water transport by porcine trophectoderm cells through NO- and cAMP-dependent mechanisms. J Anim Sci 96(Suppl 3):303–304
Zhu C, Jiang Z, Bazer F, Johnson G, Burghardt R, Wu G (2018b) Expression of AQP 1, 3, 5, and 9 in the porcine placenta and uterine endometrium during the estrous cycle and gestation. J Anim Sci 96(Suppl 3):482–483
Zhu C, Li XL, Bazer FW, Johnson GA, Burghardt RC, Jiang ZY, Wu G (2016) Dietary L-arginine supplementation during Days 14 and 25 of gestation enhances placental aquaporin-1 expression and water transport by the pig placenta. The Society of the Study of Reproduction (SSR) Annual Meeting, July 16–20, 2016, San Diego, CA
Zhu C, Li XL, Bazer FW, Johnson GA, Burghardt RC, Jiang ZY, Wu G (2021) Dietary L-arginine supplementation during days 14-25 of gestation enhances aquaporin expression in the placentae and endometria of gestating gilts. Amino Acids 53:1287–1295

Amino Acids in Microbial Metabolism and Function

7

Zhaolai Dai, Zhenlong Wu, Weiyun Zhu, and Guoyao Wu

Abstract

Amino acids (AAs) not only serve as building blocks for protein synthesis in microorganisms but also play important roles in their metabolism, survival, inter-species crosstalk, and virulence. Different AAs have their distinct functions in microbes of the digestive tract and this in turn has important impacts on host nutrition and physiology. Deconjugation and re-conjugation of glycine- or taurine-conjugated bile acids in the process of their enterohepatic recycling is a good example of the bacterial adaptation to harsh gut niches, inter-kingdom cross-talk with AA metabolism, and cell signaling as the critical control point. It is also a big challenge for scientists to modulate the homeostasis of the pools of AAs and their metabolites in the digestive tract with the aim to improve nutrition and regulate AA metabolism related to anti-virulence reactions. Diversity of the metabolic pathways of AAs and their multi-functions in modulating bacterial growth and survival in the digestive tract should be taken into consideration in recommending nutrient requirements for animals. Thus, the concept of functional amino acids can guide not only microbiological studies but also nutritional and physiological investigations. Cutting edge discoveries in this research area will help to better understand the mechanisms responsible for host-microbe interactions and develop new strategies for improving the nutrition, health, and well-being of both animals and humans.

Keywords

Amino acids · Bacteria · Intestine · Metabolism · Function

Abbreviations

AAs	Amino acids
ADI	Arginine deiminase
Agr1	Accessory gene regulator 1
AI-2	Autoinducer-2
AHR	Aryl hydrocarbon receptor
BA	Bile acids
BCAA	Branched-chain amino acids
BCFA	Branched-chain fatty acids
CBA	Conjugated bile acids
EC	Enterochromaffin

Z. Dai (✉) · Z. Wu
State Key Laboratory of Animal Nutrition, College of Animal Science and Technology, China Agricultural University, Beijing 100193, China
e-mail: daizhaolai@cau.edu.cn

W. Zhu
National Center for International Research On Animal Gut Nutrition, College of Animal Science and Technology, Nanjing Agricultural University, Nanjing 210095, Jiangsu, China

G. Wu
Department of Animal Science, Texas A&M University, College Station 77843, TX, USA

G. Wu (ed.), *Recent Advances in Animal Nutrition and Metabolism*, Advances in Experimental Medicine and Biology 1354,
https://doi.org/10.1007/978-3-030-85686-1_7

GABA	Gamma-aminobutyric acid
GLP-1	Glucagon-like pepetide-1
5-HT	Serotonin
Hyp	Trans-4-hydroxyl-L-proline
IAA	Indole-3-acetic acid
IPA	Indole-3-propionic acid
LuxS	S-ribosylhomocysteine lyase
MPN	Most probable number counts
PXR	Pregnane X receptor
SAM	S-adenosylmethionine
SCFA	Short-chain fatty acids
TLR4	Toll-like receptor 4
TnaA	Tryptophanase

7.1 Introduction

Amino acids (AAs) are important nitrogen sources for microorganisms (Wu 2018). These nutrients are essential for the synthesis of microbial proteins and also participate in the microbial adaptation to the surrounding niches through cascades cell signaling and metabolic changes. One of the important niches for microorganisms is the digestive tract. In the digestive tract of monogastric animals and humans, there exist a large amount of bacterial species that have the ability to hydrolyze and metabolize peptides and AAs (Smith and Macfarlane 1997; Dai et al. 2010, 2012; Richardson et al. 2013; Wu 2009). The majority of these microbes belong to *Clostridium*, *Bacteroides*, *Streptococcus,* and enterobacteria (Macfarlane and Macfarlane 2012; Dai et al. 2011, 2015; Louis and Flint 2017). AA metabolism in gut microbes supports their growth and adaption to their harsh environment, while modulating the nutrition, metabolism and physiology of the host animal.

Early studies with the difference in metabolism between germ-free and conventional mice showed important contributions of the gut microbiota to whole-body nitrogen metabolism (Claus et al. 2008). Many of the identified metabolites of dietary AAs were the newly synthesized AAs and their derivatives, suggesting the critical role of the gut microbiota in host AA metabolism (Claus et al. 2008). Metabolomic analysis of the plasma from germ-free mice and conventional mice showed that the presence of the gut microbiota affected the concentrations of AA metabolites in the blood of mice (Wikoff et al. 2009). These findings underscore the important role of microbial AA metabolites in host metabolism as well as their nutrition, health, and well-being (Wu 2022).

The interactions between the gut microbiota and the host is important for host health (Ren et al. 2020). However, the important role of AA metabolism and cell signaling involved in host-microbe interactions is not well understood. Stressful conditions, such as high acid loads in the stomach and the large intestine, excess bile acids in the small intestine, bile-induced acid and oxidative stress, competition for substrates, the imbalanced and inadequate provision of nutrients, and colonization resistance present many challenges to the gut microbiota. AAs and their metabolites serve as important metabolic intermediates and signaling molecules that are crucial for the survival of bacteria in the digestive tract. However, one should keep in mind that this kind of adaptation can bring in either the health-promoting effects of the commensal microbiota or the virulence of intestinal pathogens. This review aims to summarize the current progress in the functional roles of amino acids in bacterial metabolism as well as the nutrition and health of the host.

7.2 Contribution of AA Metabolism in Intestinal Bacteria to Host Nutrition

Both plant- and animal-sourced feedstuffs provide AAs (Li et al. 2021). Studies of protein and AA requirements in humans and animals over the last decades showed a significant contribution of the de novo lysine synthesis in the ileal microbiota to host AA nutrition (Metges 2000; Fuller 2012). It was estimated that the absorption of the microbe-derived lysine from the intestine can be up to 0.9–1.3 g/day in growing pigs (Torrallardona et al. 2003). However, unlike ruminants, dietary AAs must be provided to nonruminant animals, including swine (Zhang et al. 2021) and poultry (He et al. 2021). Further evaluations of the nutritional significance of the microbial lysine synthesis are warranted in future experiments.

Except for arginine and branched-chain AAs (BCAAs), enterocytes of pigs have no ability to degrade AAs that are not synthesized de novo (Chen et al. 2007, 2009). A series of studies with the pig small-intestinal bacteria showed that when using a single AA as the nitrogen source, a large amount of arginine, lysine, threonine and glutamate were utilized by the mixed bacteria during in vitro incubation and subculture (Dai et al. 2010, 2011). Work with neonatal pigs receiving the administration of the stable-isotopic labeled threonine revealed that the treatment with antibiotics decreased AA utilization and catabolism in the gut microbiota and the plasma concentration of urea, while increasing the plasma concentrations of glycine, threonine, tyrosine and cystine (Puiman et al. 2013). Although it is technically challenging to quantify the microbial metabolism and transformation of dietary AAs in vivo, the above findings open a new area for investigating the critical role of the gut microbiota in AA nutrition and health under various conditions, such as weaning, pregnancy, and obesity. The long-term effects of antibiotics treatment or probiotics supplementation on body nitrogen metabolism need further investigation.

7.3 Bacterial AA Metabolism and Survival in the Digestive Tract

Studies of the characteristics and functional aspects of AA utilization and metabolism in the gut microbiota showed that AAs served as not only the building blocks for protein synthesis in the gut bacteria but also substrates for the production of many AA-derived compounds (Dai et al. 2011, 2015). These functional AAs and metabolites [including H_2S, agmatine, polyamines (putrescine, spermidine, and spermine), and indoles] play important roles in the intestinal and whole-body growth, as well as the niche adaptation of the intestine to physiological and nutritional alterations. The latter include the imbalanced and inadequate intakes of AAs, low pH in the lumen of the stomach and other parts of the digestive tract, high concentrations of short-chain fatty acids, low carbohydrate intake, elevated levels of bile acids, impaired crosstalk among intestinal microbes, and the improper colonization of microbes in the gut (Table 7.1). Those factors are highlighted in the following paragraphs.

7.3.1 AA Metabolism and Acid Resistance

Acid resistance has an adverse impact on Gram-positive bacteria in the gut (Cotter and Hill 2003). Low pH in the digestive tract can be induced either by hydrochloride produced by the parietal cell in the stomach of non-ruminants or by short-chain fatty acids (SCFA) produced by microbial fermentation. Studies over the last decades showed that glutamate metabolism in the bacteria of nonruminant animals not only serves as a strategy to a response to low pH but also plays an important role in maintaining gut health through glutamate-derived metabolites. Multiple metabolites are produced from the metabolism of glutamate in the digestive tract, which included gamma-aminobutyric acid (GABA), butyrate, and propionate (Louis and Flint 2017). Functional study of the decarboxylation of glutamate and production of GABA showed that this metabolic route was used by the bacteria to maintain intracellular pH in the stomach and in the hindgut (Cotter and Hill 2003; Barrett et al. 2012; Feehily and Karatzas 2013; Tsai et al. 2013). Moreover, when carbohydrate intake is limited, GABA can be further degraded to propionate and butyrate for ATP production (Louis and Flint 2017). Interestingly, a recent study suggested that the acid-resistance role of glutaminase (the key enzyme that catalyze the bacterial conversion of glutamine into glutamate) in host-adapted lactobacilli species did not depend

Table 7.1 Examples of the metabolism of amino acids and derivatives in bacteria and related function

Amino acids / derivatives	Products	Key enzymes	Genes	Bacteria	Function	References
L-Arginine	Ornithine	Arginine deiminase	*arcA*	*Streptococcus gordonii*, *S. parasanguis*, *S. pyogenes*, *Lactobacillus sakei*, *L. sanfranciscensis*, etc	Acid resistance, biofilm formation and antibiotic resistance	Cotter and Hill (2003); Freiberg et al. (2020)
	Agmatine	Arginine decarboxylase	*adiA*, *speA*	*Escherichia coli*, *Pseudomonas aeruginosa*, *Salmonella*, *Shigella*, etc	Acid resistance, antibiotic resistance	Tsai and Miller (2013); McCurtain et al. (2019)
L-Aspartate	Fumarate	Aspartase	*aspA*	*Campylobacter jejuni*	Support growth under microaerobic and oxygen-limited conditions	Guccione et al. (2008)
L-Glutamate	2-Oxoglutarate	Glutamate dehydrogenase	*gudB*	*Staphylococcus aureus*	Support growth under carbohydrate-limited condition	Halsey et al. (2017)
	GABA	Glutamate decarboxylase	*gad*, varied among different bacteria	*Bifidobacterium dentium*, *Escherichia coli*, *Lactobacillus brevis*, *Lactococcus lactis*, *Listeria monocytogenes*, etc	Colonization and adaptation to the gut environment especially survival at low pH of the digestive tract and macrophage phagosome	Barrett et al. (2012); Cotter and Hill (2003); Feehily and Karatzas (2013); Louis and Flint (2017); Tsai et al. (2013)
	Butyrate	3-Methylaspartate pathway; 4-aminobutyrate pathway; 2-hydroxyglutarate pathway	Varied among different bacteria	*Acidaminococcus fermentans*, *Clostridium limosum*, *C. sporosphaeroides*, *C. symbiosum*, *Fusobacterium* spp., *Peptostreptococcus asaccharolyticus* etc	Energy harvesting, growth and survival in the digestive tract	Louis and Flint (2017)
Trans-4-hydroxy-L-proline	P5C, proline	4-Hydroxyproline dehydratase, P5C reductase	*hypD*, *proC*	*Clostridioides difficile* (formerly *Clostridium difficile*), *Clostridium* spp., etc	Proline act as an electron acceptor in the Stickland fermentation for energy harvesting and may also use by the host for protein synthesis	Levin et al. (2017); Huang et al. (2018)

(continued)

Table 7.1 (continued)

Amino acids / derivatives	Products	Key enzymes	Genes	Bacteria	Function	References
S-adenosyl-methionine (SAM)	Autoinducer-2 (AI-2)	S-ribosylhomocysteine lyase (LuxS)	*luxS*	*Bifidobacterium* spp., *Campylobacter jejuni*, *Lactobacillus* spp., *Salmonella* spp., *Vibrio* spp., etc	Regulate quorum-sensing, exopolysaccharide production, biofilm formation, response to oxidative stress and adherence to intestinal epithelial cells	Plummer (2012); Wilson et al. (2012); Christiaen et al. (2014)
Taurine	H_2S	Taurine: pyruvate aminotransferase, isethionate sulfite-lyase, dissimilatory sulfite reductase	*Tpa*, *IslA*, *dsrA*	*Bacteroides* spp., *Bilophila wadsworthia*, *Clostridium* spp., *Desulfovibrio* spp.	Regulate 7α-dehydroxylation of bile salt and improve growth of the *Bacteroides* spp.; Sulfite respiration in *B. wadsworthia* and *Desulfovibrio* spp.	Van Eldere et al. (1996); Ridlon et al. (2016); Peck et al. (2019)
Tryptophan	Indole	Tryptophanase	*tnaA*	*Bacteroides avatus*, *Clostridium bifermentans*, *Escherichia coli*, *Enterococcus faecalis*, *Klebsiella* spp., *Vibrio cholerae*, etc	Response to oxidative stress; increase antibiotic tolerance and drug resistance; increase spore formation, decrease cell motility, adhesion to epithelium and biofilm formation	Lee and Lee (2010); Lee et al. (2015)

GABA, gamma-aminobutyric acid; P5C, Δ^1-pyrroline-5-carboxylate on glutamine metabolism (Li et al. 2020b). Nonetheless, dietary L-glutamine supplementation modulates microbial community and activates innate immunity in the mouse intestine (Ren et al. 2014a, b).

Arginine metabolism regulates acid resistance and biofilm formation- related antibiotic resistance in bacteria (Cotter and Hill 2003; Freiberg et al. 2020; Tsai and Miller 2013). Early studies with oral bacteria showed that the metabolism of arginine by arginine deiminase (ADI) and the production of ornithine through the ADI system is crucial for the resistance of the host to acid produced by bacteria and for maintaining the homeostasis of the plaque microbiota (Cotter and Hill 2003). Interestingly, in the *Streptococcus*, the ADI pathway contributes to pH resistance, bacterial growth, and resistance to penicillin in the biofilm growth phase (Freiberg et al. 2020). Meanwhile, the agmatine pathway of arginine metabolism also contributed to acid resistance and antibiotic tolerance in many bacteria, such as *Escherichia coli*, *Pseudomonas aeruginosa*, and *Salmonella* (Tsai and Miller 2013; McCurtain et al. 2019). Further studies are warranted to uncover the metabolic regulation of arginine metabolism in gut bacteria in acid resistance, as well as its link to the intestinal colonization of microbes and antibiotic resistance. Nonetheless, adequate intakes of dietary arginine are essential for the optimum growth and health of the small intestine and the whole body (Wu et al. 2018, 2021).

7.3.2 AA Metabolism in Bacterial Adaptation to Conditions in the Large Intestine

In the large intestine, the extent of peptide and AA fermentation by gut bacteria increases in response to a limited intake of dietary carbohydrates. Studies on the metabolism of nitrogenous compounds by the human fecal bacteria showed that the production of ammonia (a potentially toxic metabolite) from peptides (pancreatic casein hydrolysates) was greater than that from casein and AAs (Richardson et al. 2013). Interestingly, although some harmful bacteria were identified from the enrichment study, the authors failed to isolate the hyper-ammonia-producing bacteria in the human fecal samples (Richardson et al. 2013). More importantly, the AA-utilizing bacteria were found to be more abundant than the peptide-utilizers (Richardson et al. 2013). These observations suggest that gut bacteria can adapt to the carbohydrates-limiting condition in the large intestine and generate ATP from peptides much faster than from proteins and AAs. The notion of a high activity but a low abundance vs a low activity but high abundance in the utilization and metabolism of peptides and AAs by gut bacteria should be take into consideration in nutritional studies, as originally suggested by Wallace (1996).

Interestingly, the gut bacteria have their preference for AA utilization and metabolism. It was shown that bacteria of the human colon origin could metabolize a large amount of aspartate, serine, arginine, lysine and glutamate (Smith and Macfarlane 1997; Richardson et al. 2013). Further studies are warranted to uncover the function of the above-mentioned AAs in the growth and adaption of bacteria (including pathogenic and probiotic bacteria) in the digestive tract (Guccione et al. 2008; Halsey et al. 2017). When the probiotic bacteria *Propionibacterium freudenreichii* grow in the pig colon, genes related to the transport and degradation of aspartate, glycine and alanine were down-regulated and genes related to the transport of BCAAs and cell division were up-regulated (Saraoui et al. 2013).

In the gut microbiota, the metabolism of tryptophan is complex. Deamination and decarboxylation can both take place in different gut bacteria and sometimes cross-feeding is required. An early study with the human colonic bacteria showed that the indoleacetate-forming bacteria were the major tryptophan-utilizing organisms (Smith and Macfarlane 1996). Using the most probable number (MPN) counts technique and the incubation of different species of the human colonic bacteria, the authors discovered that indole-3-propionic acid (IPA) and indole-3-acetic acid (IAA) could be produced by the dominant human colonic bacteria and that many *Bifidobacterium* spp. produced indolelactate and IPA but many *Clostridium* spp. generated IAA (Smith and Macfarlane 1996). Meanwhile, in *Escherichia coli* and many other bacteria, indole can be produced from tryptophan by tryptophanase (TnaA) in association with the production of pyruvate and ammonia (Lee and Lee 2010). This process was promoted by high cell density, low carbon source and high pH (Lee and Lee 2010), which simulated the conditions in the large intestine when individuals would consume a high-protein and low- carbohydrate diet. Indole can be used as a signaling molecule to stimulate spore formation and inhibit energy-consuming processes, such as cell motility, adhesion to epithelium and biofilm formation in the bacteria (Lee and Lee 2010; Lee et al. 2015).

7.3.3 AA Metabolism and Bile Resistance

The deconjugation of bile acids in bacteria plays an important role in the resistance to bile acid and the survival of bacteria in the intestine (Ruiz et al. 2013; Dawson and Karpen 2015). The deconjugation of bile acids is carried out by bile salt hydrolase, which de-conjugates glycine/taurine from the conjugated bile acids in bacteria (Ruiz et al. 2013). The released glycine and taurine can be further utilized and metabolized by the bacteria for protein synthesis and ATP production as part of the bile resistance strategy (Ruiz et al. 2013; Dawson and Karpen 2015). Studies of bile acid metabolism in *Clostridium* and *Bacteroides* of the human origin showed that the 7α-dehydroxylation of the taurocholic acid depended on the reduction of taurine to H_2S (Van Eldere et al. 1996). This, in turn, regulates not only the enterohepatic cycling of bile acids but also the availability of taurine for the use of both the gut microbiota and the host (Ridlon et al. 2016). In anaerobic bacteria, glycine (an electron receptor) can be reduced to acetate and ammonia through the Stickland reaction that also involves alanine as an electron donor (Barker 1981; Wu 2021). Taurine was utilized as another electron acceptor to improve the growth of *Clostridium* (Begley et al. 2006). Thus, an adequate intake of dietary taurine may be crucial for intestinal health. However, it should be borne in mind that the extra amount of H_2S produced from the fermentation of taurine in the large intestine by gut bacteria such as *Bilophila wadsworthia*, *Desulfovibrio* spp. may be toxic and contribute to the occurrence of gut and systemic inflammation (Peck et al. 2019). Besides, bile acid challenges may induce oxidative and acid stress in bacteria due to the production of reactive oxygen/nitrogen species and protons (Ruiz et al. 2013). This may trigger acid-induced AA metabolism and downstream signaling in bacteria as described above.

Bile acids regulate AA metabolism in many bacteria. Transcriptome analysis of ox-bile challenged *Akkermansia muciniphila* showed that genes related to threonine transformation (threonine dehydrogenase) and glutamate metabolism (glutamate 5-kinase) were downregulated (Hagi et al. 2020). Studies with *Lactobacillus johnsonii* showed that bile stress increased the production of enzymes related to D-alanine metabolism and peptidoglycan biosynthesis (Lee et al. 2013). In vitro and in vivo experiments with the mice cecal microbiota also revealed that the bacterial metabolism of AAs such as BCAAs, alanine, methionine, tyrosine, phenylalanine, lysine, glutamate, glycine and taurine were altered by bile acids in a bacterial species-dependent manner (Tian et al. 2020). However, the mechanisms responsible for the up- and down-regulation of AA metabolism in bacteria by different species of bile acids in the intestine and the link to host nutrition and health need further investigation.

7.3.4 AA Metabolism in Inter-Species Crosstalk

Studies over the last decades have shown that autoinducer-2 (AI-2) can be produced from a methionine derivative, *S*-adenosylmethionine (SAM), by the key enzyme *S*-ribosylhomocysteine lyase (LuxS), to modulate the quorum-sensing and pathogenesis of many bacteria

(Plummer 2012). However, studies with the probiotic bacteria (e.g., *Lactobacillus* or *Bifidobacterium*) indicated that AI-2 and the key enzyme LuxS regulate exopolysaccharide production, biofilm formation, responses to oxidative stress, and adherence to intestinal epithelial cells in the bacteria (Wilson et al. 2012; Christiaen et al. 2014). The above-mentioned metabolic routes of AAs in gut bacteria, together with their metabolites, not only serve as strategies for the regulation of their growth and survival under the harsh conditions of the digestive tract but also play important roles in the regulation of microbial metabolism as well as gut homeostasis and function under conditions of symbiosis and dysbiosis which will be discussed below.

7.4 Regulation of Microbial Metabolism by AAs

The composition of the dominant AAs in bacterial proteins varied in different bacterial species (summarized Dai et al. 2011). Besides, some bacteria prefer to utilize peptides rather than AAs for growth (Arakawa et al. 2015). To date, little is known about the driving force for variations in the AA abundance of bacterial proteins as well as nitrogen utilization preference during evolution. However, the findings suggest that the abundant AAs either in bacterial proteins or present in the free form in bacteria may play an important role in the regulation of bacterial metabolism, survival, and growth.

Our in vitro studies with the pig small-intestinal bacteria showed that glutamine or arginine regulated the AA utilization profiles in different bacterial species (Dai et al. 2012, 2013). For example, the net utilization of lysine, threonine, isoleucine, leucine, glycine and alanine decreased in jejunal or ileal mixed bacteria with increased concentration of arginine in the media (Dai et al. 2012). Of particular note, glutamine reduced the net utilization of lysine, leucine, valine, ornithine and serine in jejunal or ileal mixed bacteria (Dai et al. 2013). The above regulatory role of arginine and glutamine may beneficially contribute to acid resistance in enterobacteria (Tsai and Miller 2013; Tsai et al. 2013). These findings suggest that the nutritional and functional properties of arginine and glutamine to the animals (Wu et al. 2011, 2018) are partially mediated by their regulation of AA metabolism and cell signaling in gut bacteria (Forchhammer 2007; Wu 2013).

Animal studies showed that dietary supplementation with AAs regulated the metabolism of gut microbiota. For example, supplementation with monosodium glutamate to the diet of suckling pigs at the dose of 0.5 g/kg body weight per day increased the concentrations of acetate, butyrate, isobutyrate in the cecum and colon (Tan et al. 2019). Intragastric administration of tryptophan (0.2 g/kg body weight) for 7 days decreased trimethylamine in the feces of rats (Ruan et al. 2014). Furthermore, supplementation of 0.2% tryptophan to the diet of weaning piglets for 30 days increased the concentrations of propionate, isobutyrate, isovalerate, indole-3-acetate, and indole in the colonic content (Liang et al. 2018a). Likewise, dietary supplementation with tryptophan beneficially influenced the number and species of the gut microbes, such as *Lactobacillus* and *Prevotella* in these study (Liang et al. 2018a,b). Further investigations are warranted to uncover the regulatory role of AAs in the metabolism and cell signaling of specific bacteria species in the microbial community.

7.5 AA-Derived or Induced Microbial Metabolites in Health

7.5.1 Aromatic AA Metabolites and Signaling in Health

Mounting evidence has shown that AA metabolites produced by the gut microbiota regulate gut function and whole-body metabolism through different pathways (Table 7.2). Let's use tryptophan metabolites as examples. Studies with germ-free and conventional mice revealed that indole-3-propionic acid (IPA) was produced from tryptophan only by the gut microbiota (Wikoff et al. 2009). This is consistent with our

Table 7.2 Examples of microbial amino acid metabolites and the link to host health

Amino acids	Metabolites	Related bacteria	Function	References
L-Tryptophan	IPA	*Bacteroides fragilis*, *Bifidobacterium* spp., *Clostridium sporogenes*, *Clostridium cadaveris*, *Peptostreptococcus anaerobius*	Antioxidant; regulate gut barrier through PXR and gut immune function through TLR4	Smith and Macfarlane (1996); Wikoff et al. (2009); Venkatesh et al. (2014); Dodd et al. (2017)
	IAA	*Bacteroides fragilis*, *Bacteroides thetaiotaomicron*, *Bifidobacterium pseudolongum*, *Clostridioides difficile*, *Clostridium* spp.	Interact with AHR in the intestine and increase IL-22 production	Smith and Macfarlane (1996); Roager and Licht (2018); Dong et al. (2020)
	Indole	*Bacteroides avatus*, *Clostridium bifermentans*, *Escherichia coli*, *Enterococcus faecalis*, *Klebsiella* spp., *Proteus* spp., *Vibrio cholerae*, etc	Interact with AHR and PXR in the intestine, regulate serotonin production and transport, modulate gut barrier and immune function	Smith and Macfarlane (1996); Bansal et al. (2010); Lee et al. (2015); Roager and Licht (2018); Dong et al. (2020)
	Tryptamine	*Lactobacillus bulgaricus*, *Clostridium sporogenes*, *Ruminococcus gnavus*	Regulate tryptophan metabolism in gut lumen; modulate brain function and gut motility through TAARs and EC cells; activate epithelial 5-HT_4 receptor	Williams et al. (2014); Bhattarai et al. (2018)
L-Phenylalanine	PPA	*Peptostreptococcus anaerobius*, *Clostridium sporogenes*, *Clostridium cadaveris*	Function as chemical messenger interacting with host	Dodd et al. (2017); Ku et al. (2020)
	4-OH-PPA	Same as PPA	Same as PPA	Dodd et al. (2017); Ku et al. (2020)
L-Proline	5-Aminovalerate	*Clostridioides difficile* and *Clostridium* spp.	Analog of GABA and act as a weak antagonist of $GABA_B$ receptor	Muhyaddin et al. (1982); Huang et al. (2018)
	Valerate	*Clostridioides difficile* and *Clostridium* spp.	Lower arterial blood pressure in rats; protect against radiation injuries	Muhyaddin et al. (1982); Huang et al. (2018); Onyszkiewicz et al. (2020); Li et al. (2020b)

(continued)

Table 7.2 (continued)

Amino acids	Metabolites	Related bacteria	Function	References
L-Glutamate	GABA	*Lactobacillus brevis*, *Bifidobacterium dentium*, *Listeria monocytogenes*, *Bacteroides thetaiotaomicron*, etc	Regulate gut motility and inflammation through the interaction with GABAergic system; blood pressure and heart rate; sensation of pain and anxiety, corelated with body amino acids metabolism in obesity	Barrett et al. (2012); Louis and Flint (2017); Mazzoli and Pessione (2016); Chen et al. (2019); Liu et al. (2017)
Taurine	H_2S	*Clostridium* spp., *Bacteroides* spp., *Bilophila wadsworthia*, *Desulfovibrio* spp.	Regulate bile metabolism and enterohepatic cycling; involve in gut and systemic inflammation	Van Eldere et al. (1996); Ridlon et al. (2016); Peck et al. (2019)
Tyrosine	Tyramine	*Clostridium* spp., etc	Regulate tryptophan hydroxylase 1 activity and serotonin production in RIN14B cell	Yano et al. (2015)

AHR, aryl hydrocarbon receptor; EC, enterochromaffin; GABA, gamma-aminobutyric acid; IAA, indole-3-acetic acid; IPA, indole-3-propionic acid; 4-OH-PPA, 4-hydroxyphenylpropionic acid; PPA, phenylpropionic acid; PXR, pregnane X receptor; TAARs, trace amine-associated receptors; TLR4, toll-like receptor 4

findings that enterocytes do not oxidize this AA (Chen et al. 2007, 2009). IPA produced by the gut bacteria *Clostridium sporogenes* could modulate intestinal barrier function and immunity through the regulation of the xenobiotic sensor, pregnane X receptor (PXR) and Toll-like receptor 4 (TLR4) in mice with a defined gut microbiota (Venkatesh et al. 2014; Dodd et al. 2017). Similar mechanisms were reported for indole in the regulation of intestinal tight junction and mucosal homeostasis (Bansal et al. 2010; Roager and Licht 2018). Moreover, indole could regulate the secretion of glucagon-like pepetide-1 (GLP-1) from immortalized and primary mouse colonic L cells. The increase in GLP-1 release occurred during a short period of exposure to indole, but an opposite effect was observed during a long period of exposure to indole (Chimerel et al. 2014). Different from IPA and indole, IAA can serve as a ligand for the aryl hydrocarbon receptor (AHR) and stimulate the production of IL-22 by immune cells in the intestine (Roager and Licht 2018).

Besides, as the tryptophan decarboxylation product, tryptamine can be produced by gut bacteria such as *Clostridium sporogenes*, *Ruminococcus gnavus* and *Lactobacillus bulgaricus* (Williams et al. 2014). Tryptamine is a neurotransmitter and can activate the gut epithelial 5-HT_4 receptor and increase colonic secretion and transition in germ-free mice colonized with either the human stool microbiota or genetically engineered tryptamine-producing *Bacteroides thetaiotaomicron* (Bhattarai et al. 2018). However, the concentration of tryptamine used in the study was quite high (3 mM) and the conclusion needs further validation.

Many microbial metabolites can interact with receptors located on the cell membrane of the gut enterochromaffin cells and immune cells to regulate serotonin (5-HT) production and gut immune function (Yano et al. 2015; Bellono et al. 2017). Studies with germ-free mice colonized with spore-forming *Clostridium* spp. of the human and mice origin showed that the spore-forming bacteria increased colonic serotonin production and systemic serotonin availability in mice (Yano et al. 2015). Cell culture experiments revealed that tyrosine metabolites (e.g., tyramine) upregulated tryptophan hydroxylase-1 activity and serotonin production in RIN14B cells (Yano et al. 2015).

Recent studies have demonstrated that indole, 2-oxindole, indole-3-acetic acid, and kynurenic acid are the dominant AHR activators in the digestive tract (Dong et al. 2020). Interestingly, although many tryptophan metabolites have been detected in the cecum content of mice and stool of humans, only some of the metabolites (e.g., indole, 2-oxindole, indole-3-acetic acid, 3-methylindole, kynurenic acid) can upregulate CYP1A1 gene expression in Caco2 cells at concentrations observed in human feces (Dong et al. 2020). Notably, dietary interventions by changing either the types of diet or tryptophan supplementation can affect the profiles of tryptophan metabolites and substantially regulate AHR activation and downstream cell signaling to maintain gut barrier function (Liang et al. 2018a; Dong et al. 2020).

7.5.2 Glutamate Metabolites

Glutamate is crucial for intestinal metabolism and function (Hou and Wu 2018; Li et al. 2020a). Interestingly, comparative studies with the microbiome of the lean and obese individuals showed that the relative abundance of the glutamate-utilizer *Bacteroides thetaiotaomicron* was decreased in obese individuals and this was correlated with deceased gene expression of glutamate decarboxylase (the GABA-producing enzyme) and increased serum glutamate levels (Liu et al. 2017). These findings suggested that the abundance and AA metabolism of the commensal bacteria are crucial for the regulation of gut function and body metabolism partially through their AA metabolites. Furthermore, the interactions between above mentioned AA metabolites and other microbial metabolites in the crosstalk between gut microbiota and intestinal epithelial cells in health and disease need further investigation (Chimerel et al. 2014; Yano et al. 2015; Bellono et al. 2017; Agus et al. 2018; Dong et al. 2020). There is evidence that dietary supplementation with glutamate improves the growth, morphology, and antioxidative responses in the small intestine of weanling pigs, while reducing the incidence of diarrhea during the first two weeks post weaning (Rezaei et al. 2013). Thus, adequate intakes of glutamate, which is abundant in both animal and plant proteins (Hou et al. 2019; Li and Wu 2020), are crucial for the intestinal health and function of the gut.

7.6 AA Metabolism in the Virulence of Pathogens

7.6.1 Glutamate and Proline Metabolism in Virulence

The human gut microbiota contains the glycyl radical enzyme (*trans*-4-hydroxyl-L-proline dehydratase), which can dehydrate *trans*-4-hydroxyl-L-proline (Hyp, an AA in collagen) to form Δ^1-pyrroline-5-carboxylate (Levin et al. 2017). This microbial reaction may be of nutritional important for animals consuming a diet with animal-sourced foods (Li and Wu 2018). In *Clostridioides difficile* and *Clostridium* spp., one of the major fermentation products of proline (product of the Hyp metabolism by 4-hydroxyproline dehydratase and Δ^1-pyrroline-5-carboxylate reductase in the bacteria) is 5-aminovalerate (Huang et al. 2018). 5-aminovalerate is the analog of GABA and act as a weak antagonist of the $GABA_B$ receptor that plays an important role in gut motility and inflammation (Muhyaddin et al. 1982; Hyland and Cryan 2010; Auteri et al. 2015). There is evidence that dietary L-proline supplementation confered immuno-stimulatory effects on mice immunized with the inactivated *Pasteurella multocida* vaccine (Ren et al. 2013b).

The glutamate-derived metabolite GABA can be produced by gut bacteria such as *Lactobacillus brevis* and *Bifidobacterium dentium* through the decarboxylation of glutamate (Barrett et al. 2012). GABA plays an important role in maintaining gut homeostasis through the regulation of gut motility and immune responses by complex interactions within the gut GABAergic system (Mazzoli and Pessione 2016). This is of utmost importance in intestinal inflammation in association with the dysbiosis of the gut microbiota under specific physiological conditions such as weaning (Chen et al. 2019).

7.6.2 Arginine Metabolism and Cell Signaling in Virulence

Studies with pathogenic bacteria showed that the metabolism of arginine and the production of ornithine or agmatine from arginine contribute to biofilm formation and antibiotic resistance. An extra amount of arginine added to culture medium supports the biofilm growth of *Streptococcus pyogenes* especially in a low-pH environment (Freiberg et al. 2020). Mutation of the arginine deiminase gene in the bacteria reduced tolerance to antibiotics (penicillin, ampicillin, cefoperazone, rifampin, erythromycin, clindamycin) and reduced the rate of infection in penicillin-treated mice (Freiberg et al. 2020). In *Pseudomonas aeruginosa*, the overproduction of agmatine from arginine was shown to increase tolerance to cationic antibiotics such as tobramycin, gentamicin, and colistin (McCurtain et al. 2019). Studies with mice showed that increased arginine concentration in the colon promoted virulence gene expression in *Citrobacter rodentium* and that the arginine sensor ArgR regulated the virulence of enterohemorrhagic *E. coli* and *C. rodentium* (Menezes-Garcia et al. 2020). In addition, dietary arginine supplementation promoted immune responses to inactivated *Pasteurella multocida* vaccination in mice (Ren et al. 2013a). Furthermore, dietary supplementation with arginine beneficially induced a shift in the ratio of *Firmicutes* to *Bacteroidetes* to favor Bacteroidetes in the jejunum (Ren et al. 2014a, b). Thus, it is imperative to understand the regulation of microbial arginine metabolism and its underlying mechanisms, as well as the role of arginine in the antibiotic resistance and virulence of pathogenic bacteria in the gut (Tsai and Miller 2013).

7.6.3 Tryptophan Metabolism in Virulence

The production of indole from the degradation of tryptophan in *Escherichia coli* and some indole-producing bacteria not only modulates the growth and virulence of the bacteria but also contributes to interspecies signaling that affects gut health and disease (Lee and Lee 2010; Lee et al. 2015). For example, early studies with *Pseudomonas aeruginosa* showed that indole and its bacterial oxidized compound 7-hydroxyindole decreased the production of pyocyanin, rhamnolipid, 2-heptyl-3-hydroxy-4(1H)-quinolone, and pyoverdine, while enhancing antibiotic resistance in the bacteria (Lee et al. 2009). Similarly, both exogenous indole or indole produced by *E. coli* in mixed-microbial communities enhanced antibiotic tolerance in *Salmonella typhimurium* (Vega et al. 2013). Interestingly, the protective range of indole for *S. typhimurium* in response to different antibiotics (e.g., carbenicillin or ciprofloxacin) was different (Vega et al. 2013). Besides, a recent study showed that fecal indole concentration was increased in patients with *Clostridium* (*Clostridioides*) *difficile* infection possibly because this bacterium stimulated the production of indole by the indole-producing gut microbiota (including *E. coli*) through the release of accessory gene regulator 1 (Agr1) quorum signaling peptide. (Darkoh et al. 2019). In addition, an increased in colonic indole concentration inhibited the growth of the indole-sensitive gut bacteria (Darkoh et al. 2019). Furthermore, the Agr1 quorum signaling system was crucial for the regulation of the production of toxins A and B by *C. difficile* (Darkoh et al. 2019). These novel findings suggest that pathogenic bacteria can use a series of strategies to maintain their growth and virulence in the digestive tract partially through the modulation of tryptophan metabolism in the gut lumen.

7.7 Conclusions and Perspectives

AAs not only serve as building blocks for protein synthesis in microorganisms but also play important roles in the regulation of microbial metabolism and function. The adaption of intestinal bacteria to the surrounding environment through AA metabolism and the production of AA metabolites have big impacts on the niches for their maintenance. This is of utmost importance for the microecology of the digestive system, which regulates nutrition and gut physiology (Fig. 7.1). Interspecies chemo-sensing among microbes through the production of small molecules (e.g., autoinducer-2, polyamines, agmatine, and indole) is crucial for the regulation of gut microbiota composition and gut physiology (Thompson et al. 2016). Microbial metabolism of AAs (e.g., arginine, glutamine, glutamate, and proline) comntributes to the bacterial acid resistance and virulence of pathogens. The release of glycine and taurine from the conjugated bile salts by bacteria serves as a strategy for bacterial bile acid resistance and this, in turn, regulates bacterial bile acid metabolism and responses to acid and oxidative stresses (Begley et al. 2006; Ruiz et al. 2013). Such an interaction is important for the enterohepatic cycle for the intestinal reabsorption of bile acids and AAs as well as the homeostasis of the pool of bile acids, glycine and taurine to prevent bile acid-related gut diseases (Ridlon et al. 2016; Peck et al. 2019; Wu 2020). Tryptophan metabolism per se and tryptophan-regulated microbial metabolism are complex and unique in the intestine. Bacterial metabolites, such as isovalerate and indole compounds, interact with the sensory receptor of the enterochromaffin cells and gut serotonin receptors that regulate gut serotonin homeostasis in health and disease (Bellono et al. 2017; Wang et al. 2020). Thus, the concept of functional AAs in nutrition (Wu 2018) is important for the study of microbial AA metabolism as well as its roles in nutrition and health. Towards this goal, multi-disciplinary knowledge and a holistic view of AA metabolism in gut microbiota are required. New

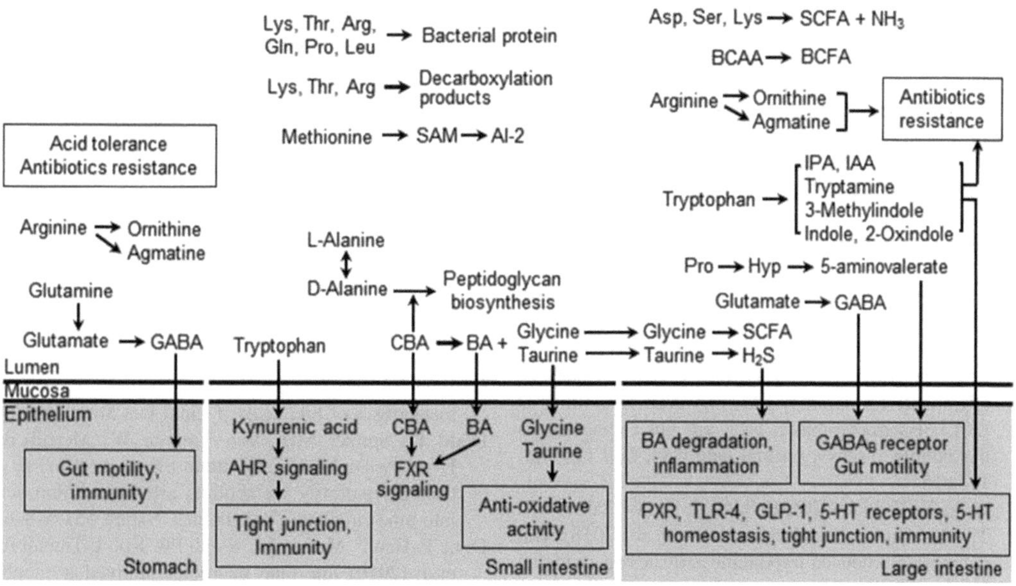

Fig. 7.1 Examples of amino acid metabolism in gut bacteria and the link to gut nutrition and health in animals. AI-2, autoinducer-2; AHR, aryl hydrocarbon receptor; BA, bile acids; BCAA, branched-chain amino acids; BCFA, branched-chain fatty acids; CBA, conjugated bile acids; GABA, gamma-aminobutyric acid; GLP-1, glucagon-like pepetide-1; 5-HT, serotonin; Hyp, *trans*-4-hydroxyl-L-proline; IAA, indole-3-acetic acid; IPA, indole-3-propionic acid; PXR, pregnane X receptor; SAM, S-adenosylmethionine; SCFA, short chain fatty acids; TLR4, Toll-like receptor 4. Standard abbreviations of amino acids are used in the figure

discoveries and concepts in this area will help to better understand the mechanisms responsible for the host-microbe interactions and to develop new strategies for improving the nutrition and health in humans and other animals.

Acknowledgements ZD, ZW and WZ were supported by the National Key R&D Program of China (2017YFD0500501), National Natural Science Foundation of China (32072689, 31301979), National Key Basic Research Program of China (2013CB127303) and the open research fund of the National Center for International Research on Animal Gut Nutrition. GW was supported by Texas A&M AgriLife Research (H-8200).

References

Agus A, Planchais J, Sokol H (2018) Gut microbiota regulation of tryptophan metabolism in health and disease. Cell Host Microbe 23:716–724

Arakawa K, Matsunaga K, Takihiro S, Moritoki A, Ryuto S, Kawai Y, Masuda T, Miyamoto T (2015) *Lactobacillus gasseri* requires peptides, not proteins or free amino acids, for growth in milk. J Dairy Sci 98:1593–1603

Auteri M, Zizzo MG, Serio R (2015) GABA and GABA receptors in the gastrointestinal tract: from motility to inflammation. Pharmacol Res 93:11–21

Bansal T, Alaniz RC, Wood TK, Jayaraman A (2010) The bacterial signal indole increases epithelial-cell tight-junction resistance and attenuates indicators of inflammation. Proc Natl Acad Sci USA 107:228–233

Barker HA (1981) Amino acid degradation by anaerobic bacteria. Annu Rev Biochem 50:23–40

Barrett E, Ross RP, O'Toole PW, Fitzgerald GF, Stanton C (2012) γ-Aminobutyric acid production by culturable bacteria from the human intestine. J Appl Microbiol 113:411–417

Begley M, Hill C, Gahan CG (2006) Bile salt hydrolase activity in probiotics. Appl Environ Microbiol 72:1729–1738

Bellono NW, Bayrer JR, Leitch DB, Castro J, Zhang C, O'Donnell TA, Brierley SM, Ingraham HA, Julius D (2017) Enterochromaffin cells are gut chemosensors that couple to sensory neural pathways. Cell 170:185-198.e16

Bhattarai Y, Williams BB, Battaglioli EJ, Whitaker WR, Till L, Grover M, Linden DR et al (2018) Gut microbiota-produced tryptamine activates an epithelial G-protein-coupled receptor to increase colonic secretion. Cell Host Microbe 23:775–85.e5

Chen LX, Yin YL, Jobgen WS, Jobgen SC, Knabe DA, Hu WX, Wu G (2007) In vitro oxidation of essential amino acids by intestinal mucosal cells of growing pigs. Livest Sci 109:19–23

Chen LX, Li P, Wang JJ, Li XL, Gao HJ, Yin YL, Hou YQ, Wu G (2009) Catabolism of nutritionally essential amino acids in developing porcine enterocytes. Amino Acids 37:143–152

Chen S, Tan B, Xia Y, Liao S, Wang M, Yin J et al (2019) Effects of dietary gamma-aminobutyric acid supplementation on the intestinal functions in weaning piglets. Food Funct 10:366–378

Chimerel C, Emery E, Summers DK, Keyser U, Gribble FM, Reimann F (2014) Bacterial metabolite indole modulates incretin secretion from intestinal enteroendocrine L cells. Cell Rep 9:1202–1208

Claus SP, Tsang TM, Wang Y, Cloarec O, Skordi E, Martin FP et al (2008) Systemic multicompartmental effects of the gut microbiome on mouse metabolic phenotypes. Mol Syst Biol 4:219

Cotter PD, Hill C (2003) Surviving the acid test: responses of gram-positive bacteria to low pH. Microbiol Mol Biol Rev 67:429–453

Christiaen SE, O'Connell Motherway M, Bottacini F, Lanigan N, Casey PG, Huys G et al. (2014) Autoinducer-2 plays a crucial role in gut colonization and probiotic functionality of *Bifidobacterium breve* UCC2003. PLoS One 9:e98111

Dai ZL, Zhang J, Wu G, Zhu WY (2010) Utilization of amino acids by bacteria from the pig small intestine. Amino Acids 39:1201–1215

Dai ZL, Wu G, Zhu WY (2011) Amino acid metabolism in intestinal bacteria: links between gut ecology and host health. Front Biosci (Landmark Ed) 16:1768–1786

Dai ZL, Li XL, Xi PB, Zhang J, Wu G, Zhu WY (2012) Metabolism of select amino acids in bacteria from the pig small intestine. Amino Acids 42:1597–1608

Dai ZL, Li XL, Xi PB, Zhang J, Wu G, Zhu WY (2013) L-Glutamine regulates amino acid utilization by intestinal bacteria. Amino Acids 45:501–512

Dai Z, Wu Z, Hang S, Zhu W, Wu G (2015) Amino acid metabolism in intestinal bacteria and its potential implications for mammalian reproduction. Mol Hum Reprod 21:389–409

Darkoh C, Plants-Paris K, Bishoff D, DuPont HL (2019) *Clostridium difficile* modulates the gut microbiota by inducing the production of indole, an interkingdom signaling and antimicrobial molecule. mSystems 4: e00346–18

Dawson PA, Karpen SJ (2015) Intestinal transport and metabolism of bile acids. J Lipid Res 56:1085–1099

Dodd D, Spitzer MH, Van Treuren W, Merrill BD, Hryckowian AJ, Higginbottom SK et al (2017) A gut bacterial pathway metabolizes aromatic amino acids into nine circulating metabolites. Nature 551:648–652

Dong F, Hao F, Murray IA, Smith PB, Koo I, Tindall AM et al (2020) Intestinal microbiota-derived tryptophan metabolites are predictive of Ah receptor activity. Gut Microbes 12:1–24

Feehily C, Karatzas KA (2013) Role of glutamate metabolism in bacterial responses towards acid and other stresses. J Appl Microbiol 114:11–24

Freiberg JA, Le Breton Y, Harro JM, Allison DL, McIver KS, Shirtliff ME (2020) The arginine deiminase pathway impacts antibiotic tolerance during biofilm-mediated *Streptococcus pyogenes* infections. mBio 11:e00919–20

Forchhammer K (2007) Glutamine signalling in bacteria. Front Biosci 12:358–370

Fuller M (2012) Determination of protein and amino acid digestibility in foods including implications of gut microbial amino acid synthesis. Br J Nutr 108(Suppl 2):S238-246

Guccione E, Leon-Kempis Mdel R, Pearson BM et al (2008) Amino acid-dependent growth of *Campylobacter jejuni*: key roles for aspartase (AspA) under microaerobic and oxygen-limited conditions and identification of AspB (Cj0762), essential for growth on glutamate. Mol Microbiol 69:77–93

Hagi T, Geerlings SY, Nijsse B, Belzer C (2020) The effect of bile acids on the growth and global gene expression profiles in *Akkermansia muciniphila*. Appl Microbiol Biotechnol 104:10641–10653

Halsey CR, Lei S, Wax JK, Lehman MK, Nuxoll AS, Steinke L et al. (2017) Amino acid catabolism in *Staphylococcus aureus* and the function of carbon catabolite repression. mBio 8:e01434–16

He WL, Li P, Wu G (2021) Amino acid nutrition and metabolism in chickens. Adv Exp Med Biol 1285:109–131

Hou YQ, Wu G (2018) L-Glutamate nutrition and metabolism in swine. Amino Acids 50:1497–1510

Hou YQ, He WL, Hu SD, Wu G (2019) Composition of polyamines and amino acids in plant-source foods for human consumption. Amino Acids 51:1153–1165

Huang YY, Martínez-Del Campo A, Balskus EP (2018) Anaerobic 4-hydroxyproline utilization: Discovery of a new glycyl radical enzyme in the human gut microbiome uncovers a widespread microbial metabolic activity. Gut Microbes 9:437–451

Hyland NP, Cryan JF (2010) A gut feeling about GABA: Focus on $GABA_B$ receptors. Front Pharmacol 1:124

Ku K, Park I, Kim D, Kim J, Jang S, Choi M et al (2020) Gut Microbial metabolites induce changes in circadian oscillation of clock gene expression in the mouse embryonic fibroblasts. Mol Cells 43:276–285

Lee J, Attila C, Cirillo SL, Cirillo JD, Wood TK (2009) Indole and 7-hydroxyindole diminish *Pseudomonas aeruginosa* virulence. Microb Biotechnol 2:75–90

Lee JH, Lee J (2010) Indole as an intercellular signal in microbial communities. FEMS Microbiol Rev 34:426–444

Lee JH, Wood TK, Lee J (2015) Roles of indole as an interspecies and interkingdom signaling molecule. Trends Microbiol 23:707–718

Lee JY, Pajarillo EA, Kim MJ, Chae JP, Kang DK (2013) Proteomic and transcriptional analysis of *Lactobacillus johnsonii* PF01 during bile salt exposure by iTRAQ shotgun proteomics and quantitative RT-PCR. J Proteome Res 12:432–443

Levin BJ, Huang YY, Peck SC, Wei Y, Martínez-Del Campo A, Marks JA et al. (2017) A prominent glycyl radical enzyme in human gut microbiomes metabolizes trans-4-hydroxy-l-proline. Science 355:eaai8386

Li P, Wu G (2018) Roles of dietary glycine, proline and hydroxyproline in collagen synthesis and animal growth. Amino Acids 50:29–38

Li P, Wu G (2020) Composition of amino acids and related nitrogenous nutrients in feedstuffs for animal diets. Amino Acids 52:523–542

Li Q, Tao Q, Teixeira JS, Shu-Wei Su M, Gänzle MG (2020a) Contribution of glutaminases to glutamine metabolism and acid resistance in *Lactobacillus reuteri* and other vertebrate host adapted lactobacilli. Food Microbiol 86:103343

Li X, Zheng S, Wu G (2020b) Nutrition and metabolism of glutamate and glutamine in fish. Amino Acids 52:671–691

Li P, He WL, Wu G (2021) Composition of amino acids in foodstuffs for humans and animals. Adv Exp Med Biol 1332:189–209

Liang H, Dai Z, Liu N, Ji Y, Chen J, Zhang Y et al (2018a) Dietary L-tryptophan modulates the structural and functional composition of the intestinal microbiome in weaned piglets. Front Microbiol 9:1736

Liang H, Dai Z, Kou J, Sun K, Chen J, Yang Y et al (2018b) Dietary L-tryptophan supplementation enhances the intestinal mucosal barrier function in weaned piglets: implication of tryptophan-metabolizing microbiota. Int J Mol Sci 20:20

Liu R, Hong J, Xu X, Feng Q, Zhang D, Gu Y et al (2017) Gut microbiome and serum metabolome alterations in obesity and after weight-loss intervention. Nat Med 23:859–868

Louis P, Flint HJ (2017) Formation of propionate and butyrate by the human colonic microbiota. Environ Microbiol 19:29–41

Macfarlane GT, Macfarlane S (2012) Bacteria, colonic fermentation, and gastrointestinal health. J AOAC Int 95:50–60

Mazzoli R, Pessione E (2016) The neuro-endocrinological role of microbial glutamate and GABA signaling. Front Microbiol 7:1934

McCurtain JL, Gilbertsen AJ, Evert C, Williams BJ, Hunter RC (2019) Agmatine accumulation by *Pseudomonas aeruginosa* clinical isolates confers antibiotic tolerance and dampens host inflammation. J Med Microbiol 68:446–455

Menezes-Garcia Z, Kumar A, Zhu W, Winter SE, Sperandio V (2020) L-Arginine sensing regulates virulence gene expression and disease progression in enteric pathogens. Proc Natl Acad Sci USA 117:12387–12393

Metges CC (2000) Contribution of microbial amino acids to amino acid homeostasis of the host. J Nutr 130:1857S-1864S

Muhyaddin M, Roberts PJ, Woodruff GN (1982) Presynaptic gamma-aminobutyric acid receptors in the rat anococcygeus muscle and their antagonism by 5-aminovaleric acid. Br J Pharmacol 77:163–168

Onyszkiewicz M, Gawrys-Kopczynska M, Sałagaj M, Aleksandrowicz M, Sawicka A, Koźniewska E et al.

(2020) Valeric acid lowers arterial blood pressure in rats. Eur J Pharmacol 877:173086

Peck SC, Denger K, Burrichter A, Irwin SM, Balskus EP, Schleheck D (2019) A glycyl radical enzyme enables hydrogen sulfide production by the human intestinal bacterium *Bilophila wadsworthia*. Proc Natl Acad Sci USA 116:3171–3176

Plummer PJ (2012) LuxS and quorum-sensing in *Campylobacter*. Front Cell Infect Microbiol 2:22

Puiman P, Stoll B, Mølbak L, de Bruijn A, Schierbeek H, Boye M et al (2013) Modulation of the gut microbiota with antibiotic treatment suppresses whole body urea production in neonatal pigs. Am J Physiol 304:G300-310

Ren WK, Zou LX, Li NZ, Wang Y, Peng YY, Ding JN et al (2013a) Dietary arginine supplementation promotes immune responses to inactivated *Pasteurella multocida* vaccination in mice. Br J Nutr 109:867–872

Ren WK, Zou LX, Ruan Z, Li NZ, Wang Y, Peng Y et al (2013b) Dietary L-proline supplementation confers immuno-stimulatory effects on inactivated *Pasteurella multocida* vaccine immunized mice. Amino Acids 45:555–561

Ren WK, Chen S, Yin J, Duan JL, Li TJ, Liu G et al (2014a) Dietary arginine supplementation of mice alters the microbial population and activates intestinal innate immunity. J Nutr 144:988–995

Ren WK, Duan JL, Yin J, Liu G, Cao Z, Xion X et al (2014b) Dietary L-glutamine supplementation modulates microbial community and activates innate immunity in the mouse intestine. Amino Acids 46:2403–2413

Ren WK, Bin P, Yin YL, Wu G (2020) Impacts of amino acids on the intestinal defensive system. Adv Exp Med Biol 1265:133–151

Rezaei R, Knabe DA, Tekwe CD, Dahanayaka S, Ficken MD, Fielder SE et al (2013) Dietary supplementation with monosodium glutamate is safe and improves growth performance in postweaning pigs. Amino Acids 44:911–923

Richardson AJ, McKain N, Wallace RJ (2013) Ammonia production by human faecal bacteria, and the enumeration, isolation and characterization of bacteria capable of growth on peptides and amino acids. BMC Microbiol 13:6

Ridlon JM, Wolf PG, Gaskins HR (2016) Taurocholic acid metabolism by gut microbes and colon cancer. Gut Microbes 7:201–215

Roager HM, Licht TR (2018) Microbial tryptophan catabolites in health and disease. Nat Commun 9:3294

Ruan Z, Yang Y, Wen Y, Zhou Y, Fu X, Ding S et al (2014) Metabolomic analysis of amino acid and fat metabolism in rats with L-tryptophan supplementation. Amino Acids 46:2681–2691

Ruiz L, Margolles A, Sánchez B (2013) Bile resistance mechanisms in *Lactobacillus* and *Bifidobacterium*. Front Microbiol 4:396

Saraoui T, Parayre S, Guernec G, Loux V, Montfort J, Le Cam A et al (2013) A unique *in vivo* experimental approach reveals metabolic adaptation of the probiotic *Propionibacterium freudenreichii* to the colon environment. BMC Genomics 14:911

Smith EA, Macfarlane GT (1997) Dissimilatory amino Acid metabolism in human colonic bacteria. Anaerobe 3:327–337

Smith EA, Macfarlane GT (1996) Enumeration of human colonic bacteria producing phenolic and indolic compounds: effects of pH, carbohydrate availability and retention time on dissimilatory aromatic amino acid metabolism. J Appl Bacteriol 81:288–302

Tan X, Zhang J, Yang H, Li J, Li Y, Ding X et al (2019) Glutamate effects on sucking piglet intestinal morphology and luminal metabolites. J Anim Physiol Anim Nutr (berl) 103:612–617

Thompson JA, Oliveira RA, Xavier KB (2016) Chemical conversations in the gut microbiota. Gut Microbes 7:163–170

Tian Y, Gui W, Koo I, Smith PB, Allman EL, Nichols RG et al (2020) The microbiome modulating activity of bile acids. Gut Microbes 11:979–996

Torrallardona D, Harris CI, Fuller MF (2003) Pigs' gastrointestinal microflora provide them with essential amino acids. J Nutr 133:1127–1131

Tsai MF, Miller C (2013) Substrate selectivity in arginine-dependent acid resistance in enteric bacteria. Proc Natl Acad Sci USA 110:5893–5897

Tsai MF, McCarthy P, Miller C (2013) Substrate selectivity in glutamate-dependent acid resistance in enteric bacteria. Proc Natl Acad Sci USA 110:5898–5902

Van Eldere J, Celis P, De Pauw G, Lesaffre E, Eyssen H (1996) Tauroconjugation of cholic acid stimulates 7 alpha-dehydroxylation by fecal bacteria. Appl Environ Microbiol 62:656–661

Vega NM, Allison KR, Samuels AN, Klempner MS, Collins JJ (2013) *Salmonella typhimurium* intercepts *Escherichia coli* signaling to enhance antibiotic tolerance. Proc Natl Acad Sci USA 110:14420–14425

Venkatesh M, Mukherjee S, Wang H, Li H, Sun K, Benechet AP et al (2014) Symbiotic bacterial metabolites regulate gastrointestinal barrier function via the xenobiotic sensor PXR and Toll-like receptor 4. Immunity 41:296–310

Wallace RJ (1996) Ruminal microbial metabolism of peptides and amino acids. J Nutr 126(Suppl 4):1326S-1334S

Wang B, Sun S, Liu M, Chen H, Liu N, Wu Z, Wu G, Dai Z (2020) Dietary L-tryptophan regulates colonic serotonin homeostasis in mice with dextran sodium sulfate-induced colitis. J Nutr 150:1966–1976

Wikoff WR, Anfora AT, Liu J, Schultz PG, Lesley SA, Peters EC, Siuzdak G (2009) Metabolomics analysis reveals large effects of gut microflora on mammalian blood metabolites. Proc Natl Acad Sci USA 106:3698–3703

Williams BB, Van Benschoten AH, Cimermancic P, Donia MS, Zimmermann M, Taketani M et al (2014) Discovery and characterization of gut microbiota decarboxylases that can produce the neurotransmitter tryptamine. Cell Host Microbe 16:495–503

Wilson CM, Aggio RB, O'Toole PW, Villas-Boas S, Tannock GW (2012) Transcriptional and metabolomic consequences of *luxS* inactivation reveal a metabolic rather than quorum-sensing role for LuxS in *Lactobacillus reuteri* 100–23. J Bacteriol 194:1743–1746
Wu G (2009) Amino acids: metabolism, functions, and nutrition. Amino Acids 37:1–17
Wu G (2013) Functional amino acids in nutrition and health. Amino Acids 45:407–411
Wu G (2018) Principles of Animal Nutrition. CRC Press, Boca Raton, Florida
Wu G (2020) Important roles of dietary taurine, creatine, carnosine, anserine and 4-hydroxyproline in human nutrition and health. Amino Acids 52:329–360
Wu G (2021) Amino Acids: Biochemistry and Nutrition. CRC Press, Boca Raton, Florida
Wu G (2022) Nutrition and metabolism: Foundations for animal growth, development, reproduction, and health. Adv Exp Med Biol 1354:1–24
Wu G, Bazer FW, Johnson GA, Knabe DA, Burghardt RC, Spencer TE et al (2011) Important roles for L-glutamine in swine nutrition and production. J Anim Sci 89:2017–2030
Wu G, Bazer FW, Johnson GA, Hou YQ (2018) Arginine nutrition and metabolism in growing, gestating and lactating swine. J Anim Sci 96:5035–5051
Wu G, Meininger CJ, McNeal CJ, Bazer FW, Rhoads JM (2021) Role of L-arginine in nitric oxide synthesis and health in humans. Adv Exp Med Biol 1332:167–188
Yano JM, Yu K, Donaldson GP, Shastri GG, Ann P, Ma L et al (2015) Indigenous bacteria from the gut microbiota regulate host serotonin biosynthesis. Cell 161:264–276
Zhang Q, Hou YQ, Bazer FW, He WL, Posey EA, Wu G (2021) Amino acids in swine nutrition and production. Adv Exp Med Biol 1285:81–107

8 Potential Replacements for Antibiotic Growth Promoters in Poultry: Interactions at the Gut Level and Their Impact on Host Immunity

Christina L. Swaggerty, Cristiano Bortoluzzi, Annah Lee, Cinthia Eyng, Gabriela Dal Pont, and Michael H. Kogut

Abstract

The chicken gastrointestinal tract (GIT) has a complex, biodiverse microbial community of ~ 9 million bacterial genes plus archaea and fungi that links the host diet to its health. This microbial population contributes to host physiology through metabolite signaling while also providing local and systemic nutrients to multiple organ systems. In a homeostatic state, the host-microbial interaction is symbiotic; however, physiological issues are associated with dysregulated microbiota. Manipulating the microbiota is a therapeutic option, and the concept of adding beneficial bacteria to the intestine has led to probiotic and prebiotic development. The gut microbiome is readily changeable by diet, antibiotics, pathogenic infections, and host- and environmental-dependent events. The intestine performs key roles of nutrient absorption, tolerance of beneficial microbiota, yet responding to undesirable microbes or microbial products and preventing translocation to sterile body compartments. During homeostasis, the immune system is actively preventing or modulating the response to known or innocuous antigens. Manipulating the microbiota through nutrition, modulating host immunity, preventing pathogen colonization, or improving intestinal barrier function has led to novel methods to prevent disease, but also resulted in improved body weight, feed conversion, and carcass yield in poultry. This review highlights the importance of adding different feed additives to the diets of poultry in order to manipulate and enhance health and productivity of flocks.

Keywords

Butyrate · Enzymes · Prebiotics · Probiotics · Intestinal health · Phytobiotics · Poultry

8.1 Introduction

In the United States (US), traditional poultry management depends on husbandry practices, biosecurity, vaccination, and when medically necessary, application of broad-spectrum antibiotics. Historically, continual feeding of low-dose antibiotic growth

Christina L. Swaggerty and Cristiano Bortoluzzi—These authors contributed equally for this publication.

C. L. Swaggerty (✉) · M. H. Kogut (✉)
Southern Plains Agricultural Research Center, USDA-ARS, College Station, TX, USA
e-mail: christi.swaggerty@usda.gov

M. H. Kogut
e-mail: Mike.kogut@usda.gov

C. Bortoluzzi · A. Lee · GabrielaD. Pont
Department of Poultry Science, Texas A&M University, College Station, TX, USA

C. Eyng
Department of Animal Science, Western Parana University, Marechal C. Rondon, PR, Brazil

G. Wu (ed.), *Recent Advances in Animal Nutrition and Metabolism*, Advances in Experimental Medicine and Biology 1354,
https://doi.org/10.1007/978-3-030-85686-1_8

promoters (AGP) was a standard practice in the poultry industry (Dibner and Richards 2005). However, public opinion is now driving the industry to pursue natural alternatives to support animal health. This trend resulted in the European Union (EU) withdrawing the use of growth-promoting, low-dose AGP in 2006 (Castanon 2007). In the US only non-medically important antibiotics are allowed in the feed as AGP, including bacitracin, flavomycin, ionophores, and avilamycin (Smith 2019). In Brazil, colistin was restricted as an AGP in 2016 (Cardoso 2019), and tylosin, lincomycin, virginiamycin, bacitracin, and tiamulin were restricted in 2018 (Brasil 2018); whereas avilamycin, enramycin, flavomycin, and halquinol are still permitted as AGP (Cardoso 2019). However, the increased awareness of antibiotic resistance genes in bacteria, particularly zoonotic bacterial strains, has further driven some consumers to switch from traditionally raised poultry to those raised without any antibiotics and other drugs.

Because of this and other factors, the poultry industry needs to find suitable alternative control and preventative measures. Antibiotic alternatives can refer to "any substance that can be substituted for therapeutic drugs" in their absence because of mandated withdrawal or reduced efficacy (Kogut 2017). Therefore, new approaches to achieve flock health and performance is essential if breeding companies are to meet global consumption, fulfill consumer demands, and comply with increased regulations all the while improving robustness, livability, production efficiency, and animal welfare.

8.2 What is Intestinal Health

Optimal gut health is a critical facet for an animal to achieve its genetic potential for performance and production and is equated with animal health. There seems to be a direct relationship between performance and a "healthy" gastrointestinal tract (GIT). However, there is not a simple definition or parameter for a "healthy gut" that considers all of the physiological functions including nutrient digestion and absorption, host metabolism and energy generation, a stable microbiome, mucus layer development, barrier function, and mucosal immune responses. In broad terms, the gut is responsible for regulating physiological homeostasis providing the host with the ability to withstand infectious and non-infectious stressors (Garriga et al. 2006; Maslowski and Mackay 2011; Quinteiro-Filho et al. 2012).

8.3 Intestinal Immune Response

The immune system is divided into innate and acquired responses with each having distinct functions yet working in conjunction to offer protection to the host. Innate immune pattern recognition receptors (PRR) on host cells recognize pathogen- and danger-associated molecular patterns (PAMPs and DAMPs, respectively) including polysaccharides, glycolipids, lipoproteins, nucleotides, and nucleic acids and triggers the immediate defense against the threat. In chickens, innate immune cells mainly include macrophages, heterophils, and B1-type lymphocytes that produce natural antibodies while cells associated with acquired immunity are typically B2 and T lymphocytes (Klasing 2007; Genovese et al. 2013).

The intestinal immune system has two distinct functions: the ability to respond to potentially pathogenic microbes, invasive pathogens, and microbial products, yet all the while, maintaining a balanced state of tolerance to the diverse and beneficial commensal microbes that reside within the intestine (Broom and Kogut 2018). As seen with the systemic immune response, sensing and detecting also uses PRRs on epithelial cells and professional immune cells in the lamina propria (dendritic cells and macrophages), triggers immune pathways that result in microbial killing and activation of acquired immune effector T cells (Th1, Th2, Th17, T regulatory [Treg]) while keeping the resident microbiota in check without generating an overt inflammatory response.

The intestinal innate defenses are comprised of a system of four barriers (i.e., a "mucosal firewall") that separates the luminal side of the intestine from the subepithelial tissues and are essential for the interactions between the immune system and intestinal contents (Macpherson et al. 2009; Belkaid and Hand 2014). One component of the mucosal firewall is the microbiological barrier where the microbiota colonizes and protects the upper mucus layer. Colonization by commensal bacteria functions by blocking pathogenic bacteria from being able to colonize the mucus layer. Further, these commensals produce metabolites and other components that modulate immune signaling and, in general, promote immune homeostasis (Belkaid and Hand 2014; Belkaid and Harrison 2017). Another firewall is the chemical barrier consisting of the mucus overlaying the gut epithelium which regulates contact between the commensal bacteria and the epithelial cells. This separation is facilitated when mucus is produced by goblet cells in the epithelium, release of antimicrobial peptides by epithelial cells, and mucosal IgA production by intestinal dendritic cells (Garrett et al. 2010). Another two components of this barrier consist of specialized epithelial cells that function in conjunction with lymphoid, myeloid, and stromal cells to secrete mucus, antimicrobial peptides, IgA, and chemokines; thereby limiting direct contact between the epithelium and infectious agents and activation of innate defense mechanisms (Medzhitov and Janeway 2002; Abreu et al. 2005; Akira et al. 2006; Kawai and Akira 2010).

8.4 Components of a Healthy Intestinal System

The intestine has the greatest surface area separating the environmentally exposed lumen and the internal subepithelial tissue and is therefore continually exposed to infectious and non-infectious triggers (Ren et al. 2020). As a result of this constant exposure, the gut is a highly active immune organ with more resident immune cells than any other organ in the host. As previously stated, the gut mucosal immune system is a highly regulated network of innate and acquired components allowing it to thrive and protect the host under such dynamic conditions and exposures resulting in the maintenance of a symbiotic microbiota population (Thaiss et al. 2014; Honda and Littman 2016). Collectively, the gut microbiota is responsible for training, stimulating, and functionally adjusting the host immune system (Hooper and Macpherson 2010; Hooper et al. 2012).

Proper nutrition plays a vital role in maintaining a healthy GIT, a stable gut microbiota, enables the animal to perform to its genetic potential, and improves animal health all the while regulating the immune system and providing metabolites for host nutrition (Sergeant et al. 2014; Roberts et al. 2015; Wu 2022). When discussing gut health, we are really talking about a number of physiological, microbiological, and physical functions that work together to maintain intestinal homeostasis and it is through evolution with the host that the gut microbiota directly influences each of these components in the host (Sergeant et al. 2014; Roto et al. 2015; Levy et al. 2017).

The primary function of the gut is the digestion and absorption of dietary nutrients to sustain the host (Dibner and Richards 2005; Kairie et al. 2013) while still providing the proper barrier in the form of the epithelial lining and reducing host exposure to environmental toxins and pathogenic microbes (Turner 2009). Glutamate and aspartate, which are abundant in plant and animal proteins (Hou et al. 2019; Li and Wu 2020; Li et al. 2021), are major energy substrates for the chicken small intestine (He et al. 2021a, b). In addition, taurine, which is a functional amino acid abundant in animal product but absent from plants, has an immune-modulatory effect in the intestine (Wu 2020). The breakdown of indigestible feed and feed compounds by the gut microbiota provides essential amino acids and vitamins to the host. Further, the microbiota, using a number of biochemical pathways, metabolize diet- and host-derived metabolites that can have a direct impact on the intestinal immune system (Wu 2018). Additionally, bacterial metabolites including short-chain fatty acids (SCFA) serve as an energy source to the epithelial cells that line the intestine, but these SCFA may also be antimicrobial and limit virulence factor expression on pathogenic bacteria (Sergeant et al. 2014; Roto et al. 2015). Other examples include the degradation of dietary tryptophan to promote epithelial cell barrier function and the breakdown of dietary arginine which inhibits pro-inflammatory cytokine production (Postler and Ghosh 2017). The host has also developed immune signaling pathways (inflammasomes) expressed in various intestinal cell subsets capable of recognizing microbial-mediated metabolic activity which can then stimulate antimicrobial activity that promotes stable colonization of the intestine (Levy et al. 2015a, b; Birchenough et al. 2016). Collectively, these studies demonstrate the choreographed crosstalk between the host and the microbiota that is directly influenced by metabolite secretion and immune signaling and the impact on animal health and disease.

The gut is more than a large complex immune organ, and it is also thought to be the largest neuroendocrine organ in the body because of the large numbers of neurons, gut hormones, and secondary messengers involved in regulating an array of physiological functions in the host (Neuman et al. 2015; Cani and Knauf 2016). Therefore, a favorable gut microbiota is important for the optimal growth and performance of chickens, while an unfavorable microbiota may promote enteric infections that lead to decreased growth rates and increased mortality. The future of animal production will be dependent on a better understanding of the interactions of the gut microbiome and the host physiology processes that aid in maintaining homeostasis as well as our ability to manipulate it for the advantage of the animal.

As we have stated, the chicken GIT is a complex society that is rich in microbial biodiversity, playing home to about nine million bacterial genes that are pivotal linkages between diet and health (Huang et al. 2018). Over 90% of all bacteria in the chicken cecum are not culturable in the laboratory and are identifiable through molecular-biology techniques (Wei et al. 2013; Sergeant et al. 2014). These communities serve the host in a positive way; however, when this balance is upset (dysbiosis), pathogens may have the opportunity to colonize the GIT or multiply to numbers sufficient to cause disease. For this reason, the chicken gut microbiome can also be a source of human infections including *Salmonella* and *Campylobacter* or antibiotic resistant strains of bacteria. Manipulation of the flora to enhance the beneficial components represents a promising therapeutic strategy for the future.

8.5 Dysregulation of Gut Functionality

As we have discussed, gut function is controlled by at least three distinct factors including environmental, host, and microbial-mediated factors. The microbiota in the GIT is further impacted, and potentially altered, when exposed to antibiotics, infection of the host by pathogens, diet, and a number of other host- and environmental-dependent influences. In a healthy gut, the bacteria work together to produce beneficial metabolites that are used by the host. Dysbiosis occurs when there is a perturbation in the microbiota composition or function that leads to an imbalance between beneficial and harmful bacteria resulting in an unwanted immune response against commensal bacteria. Dysbiosis leads to a decrease in bacterial diversity and essential functions directly affecting metabolic activity which can then deprive the host of valuable end products leading to poor gut health and bird performance (Ducatelle et al. 2015). Intestinal inflammation is likely involved in dysbiosis in poultry (Ducatelle et al. 2018; Kogut et al. 2018), but it is not clear if the inflammation is the cause or the effect of the disruption in gut homeostasis (Singh et al. 2016; Sommer et al. 2017; Zeng et al. 2017). Targeted use of prebiotics, probiotics, symbiotics, postbiotics, butyrate, enzymes, phytobiotics, and other compounds may constitute an effective therapeutic strategy to re-balance gut dysbiosis and mitigate the negative impact on the performance of poultry flocks. The crosstalk between microorganisms, immune-system and nutritional interventions has been discussed elsewhere (Ma et al. 2018) and will not be the described in the present review.

8.6 Feed Additives to Improve Gut Health

8.6.1 Probiotics

The internationally recognized definition of a probiotic is live microorganisms that, when administered in adequate amounts, confer a health benefit on the host (Sanders 2008). Oftentimes in agriculture, the term probiotic and direct-fed microbial (DFM) are used interchangeably to describe live or dead microbes or their microbial components fed to animals for a health benefit (FDA 1995). Some of the most common probiotics and DFM include *Bacillus, Lactobacillus, Enterococcus, Bifidobacterium, Escherichia,* yeast (*Saccharomyces cerevisiae*), and mold (*Aspergillus oryzae* and *A. niger*) (Roberts et al. 2015; Joerger and Ganguly 2017). According to the Microbial Compendium, there are about 130 commercial DFM products sold in the US (http://www.microbialcompendium.com/publication/). As might be expected, probiotics and DFM may work in multiple mechanisms collectively called colonization resistance and include: prevention of colonization of pathogenic bacteria via competitive exclusion, enhancement of immune responses, improvement of the gut barrier function, production of bacteriocins, improvement of microbiota homeostasis, and generation of toxic microbial fermentation products including volatile fatty acids, ethanol, and organic acids (Wolin 1969; Russell 1992; Jack et al. 1995; Ricke 2003; Rodriguez-Palacios et al. 2009). In combination with the host intestinal microbiota, probiotics may also promote the intestinal epithelium to release bioactive compounds such as mucins, defensins, bacteriocins, and cytokines and chemokines which, in turn, promote the adaptive immune response and production of secretory IgA and regulatory T cells (Corthesy et al. 2007; Neish 2009; Tellez et al. 2015). Collectively, these mechanisms demonstrate the complex host–pathogen interaction involved in disrupting the cycle of transmission and colonization of pathogenic bacteria and emphasize implementing diverse intervention strategies.

8.6.2 Prebiotics

Prebiotics are "a nonviable food component that confers a health benefit on the host associated with modulation of the microbiota" (Piniero et al. 2008). Specifically, prebiotics can be dietary supplements not digested by the host that selectively stimulate favorable growth or metabolic activity of indigenous beneficial gut bacteria (Ricke 2018). A number of products are available for poultry and include fructooligosaccharides (FOS), inulin, mannanoligosaccharides, xylooligosaccharides, and yeast including the cell wall and/or metabolites (Ricke 2018). According to the Microbial Compendium (http://www.microbialcompendium.com/publication/), there are over 50 products sold as prebiotics in the US. Many of these commercial products lack a defined mode of action, yet some do improve performance, intestinal integrity, and host immune responses, while increasing beneficial bacteria (Huff et al. 2010; Roto et al. 2015; Adhikari et al. 2018; Bortoluzzi et al. 2018). It is possible that prebiotics provide a competitive advantage to select host gut microbiota that are able to exclude colonization by pathogenic bacteria via direct competition for nutrients and/or binding sites (Zopf and Roth 1996). There are a number of studies in the literature demonstrating the efficacy of prebiotics. Some examples include FOS, a sugar that is not degraded by intestinal enzymes, which allows it to reach the cecum and colon and become "colonic food" for host bacteria and have beneficial effect against *Salmonella* Enteritidis (Adhikari et al. 2018), and supplementation with beta-glucan from yeast cell walls results in enhanced heterophil function and increased resistance against *Salmonella* Enteritidis organ invasion in young chicks (Lowry et al. 2005). In another study, a prebiotic mixture altered gut microflora and boosted innate immune responsiveness in *Salmonella* Typhimurium challenged chickens (Faber et al. 2012). Collectively, these studies indicate that prebiotic feed additives may be used to successfully expedite and boost the development of early defense mechanisms to protect young chicks potentially serving as an alternative to antibiotics.

8.6.3 Symbiotics and Postbiotics

Symbiotics are nutritional supplements that combine a prebiotic with a probiotic to produce a more beneficial result than either product alone. Since prebiotics serve as a food source for probiotic organisms, it is thought that when used in combination the probiotic strain has a better chance for survival because of the food source provided by the prebiotic. The United Nations Food and Agriculture Organization (FAO) recommends that the term symbiotic is used only when the resulting health benefit is synergistic (Cencic and Chingwaru 2010). Symbiotics are designed to present beneficial microbial populations and to promote the proliferation of specific endogenous microbiota in the gut, thereby functioning as modulators of the gut microbiota composition (Scavuzzi et al. 2014; Huynh et al. 2017).

Postbiotics are substances produced by the metabolic processes of beneficial bacteria and have positive effects on the host health (Johnson et al. 2019). The several metabolic products produced by bacteria and classified as postbiotic include SCFA, bacteriocins, peptides and proteins (Klemashevich et al. 2014). One strategy to improve the microbiota balance is to first identify the molecules that are depleted in a particular disease and then supplement the diet with the depleted molecule to restore the deficiency bringing the environment back to a homeostatic state (Klemashevich et al. 2014). For example, a postbiotic product containing organic acids produced from *Pediococcus acidilactici*, *L. reuteri*, *E. faecium*, and *L. acidophilus* modulated the activation of the innate immune response and inhibited *Clostridium perfringens* induced necrotic enteritis (Johnson et al. 2019).

8.6.4 Butyrate

The SCFA are a group of molecules that contain from one to seven carbons produced within the intestinal lumen by bacterial fermentation of plant materials such as cellulose, fibers, starches, and sugars for which animals lack the necessary enzymes to break these compounds down (Guilloteau et al. 2010). The SCFA with higher abundance in the GIT are acetate, propionate, and butyrate (Bedford and Gong 2018). Butyrate is essential to maintain the proper interaction between the host and its intestinal microbiota (Byndloss et al. 2019). It has been hypothesized that the butyrate endogenously produced by the microbiota is metabolized by the intestinal epithelial cells through beta-oxidation, which reduces the availability of oxygen on the epithelial surface. Dysbiosis may decrease the number of butyrate-producing bacteria, change the epithelial metabolism, lead to accumulation of oxygen, and drive further expansion of pathogens such as *Salmonella* (Gillis et al. 2018).

Butyrate can be supplied as a feed supplement as Na, K, Mg, or Ca salts which are odorless and easier to be incorporated into the feed (Guilloteau et al. 2010). Sodium butyrate (SB) is the sodium salt of butyric acid which contains a sodium atom in place of the hydrogen atom in the -OH group (Ahsan et al. 2016). Butyric acid has received particular attention as a feed supplement in the diet of broiler chickens, especially because its production in the small intestine is limited (Levy et al. 2015a, b), suggesting that its dietary inclusion would be beneficial. However, it has been described that uncoated butyrate could be absorbed or metabolized before reaching the distal portions of the small intestine (van der Wielen et al. 2002). Therefore, dietary supplementation with an encapsulated source of butyrate may delay the release of the substance along the GIT, thereby having plausible functional effects on the lower GIT. Besides the location of the GIT where butyrate is released, the dose of butyrate is also a factor that should be considered when using it as a feed supplement (Liu et al. 2019; Tugnoli et al. 2020).

The efficacy in which SB will be absorbed and used depends on the pH of the GIT location. With a pH lower than 4.82 (pKa of butyric acid), most of the molecules of butyric acid remain undissociated which is the desirable form for higher antimicrobial activity (Ahsan et al. 2016). Sodium butyrate is converted into butyric acid after ingestion due to the acidic pH found in the proximal regions of the GIT. Thereafter, when butyric acid reaches the small intestine (alkaline pH), it dissociates into butyrate and hydrogen ions, and butyrate is absorbed as a source of energy (Ahsan et al. 2016), as well as exerting many other functions on the host (Song et al. 2017). Sodium butyrate supplementation partially counteracted the impairment in performance of broilers fed a diet formulated with reduced energy and amino acid concentrations, modulated the immune system and the diversity, composition, and predicted function of the cecal microbiota (Bortoluzzi et al. 2017).

Butyrate can have several effects on the host, including the ability to regulate the production of cytokines, antimicrobial peptides, mucin, and tight junction proteins (Guilloteau et al. 2010; Song et al. 2017). Butyrate seems to have an anti-inflammatory effect mediated by signaling pathways (Meijer et al. 2010), such as modulating pro-inflammatory cytokines via impairment in NF-kB activation (Guilloteau et al. 2010). Butyrate leads to epigenetic adaptations, which can result in hyperacetylation of histones and can change the expression of a large number of genes (Marks et al. 2000). This epigenetic effect caused by butyrate may be responsible for the changes in the expression of genes involved in the inflammatory process, reducing pro-inflammatory cytokine expression and upregulating anti-inflammatory cytokines (Meijer et al. 2010; Fung et al. 2012).

Zhou et al. (2017) observed that coated SB had no significant effect on the cecal microbiota of healthy chickens but balanced the shifts of microbial composition caused by *Eimeria tenella* infection and decreased cecal colonization of *Salmonella* after experimental infection (Van Immerseel et al. 2005). In addition to the direct effects that butyrate has on the function and metabolism of the intestine itself, we hypothesize that microencapsulated sources of butyrate released throughout the GIT can diminish the impact of these pathogens in the lower gut.

8.6.5 Exogenous Feed Enzymes

The feed used by the poultry industry usually contains anti-nutritional factors that may trigger an immune response, leading to energy waste and reduced animal performance (Dal Pont et al. 2020). However, ingredients of lower quality and consequently higher concentration of undesirable components are used due to their lower cost and availability. In this scenario, the supplementation of exogenous enzymes (carbohydrases, proteases, and phytases) may be beneficial to break down these components, improve the digestibility of the diets, modulate the intestinal microbiota, and mitigate the activation of the immune system.

8.6.5.1 Carbohydrases

Non-starch polysaccharides (NSP) can be recognized by PRR and lead to feed-induced inflammation (Kogut et al. 2018). For instance, the major NSP in soybean meal, β-galactomannan, is recognized by the mannose receptor. It has been demonstrated that β-mannanase, that degrades β-galactomannans, beneficially modulated the immune and metabolic phenotype of the intestine (Arsenault et al. 2017). The dietary inclusion of the xylanase and β-glucanase to high NSP diets (wheat, barley, and sunflower meal) can also reduce the viscosity of the intestinal digesta (Wu et al. 2004), improve nutrient digestibility (Saleh et al. 2018), and modulate the microbiota (Aftab and Bedford 2018). Xylanase may break down the arabinoxylan backbone producing short-chain xylans and xylo-oligosaccharides (Morgan et al. 2019). These smaller particles may act as prebiotics, increasing *Lactobacillus* and *Bifidobacterium* species and reducing pathogenic bacteria such as *C. perfringens* (Sun et al. 2015). Similar effects have been observed with other NSP enzymes (Craig et al. 2020). Therefore, the use of carbohydrases not only releases more energy from the diet, but may also improve the heatlh of the inestine by modulating the immune system and the microbiota.

8.6.5.2 Proteases

The use of proteases in the feed has been shown to reduce the attachment of enterotoxigenic *Escherichia coli* to the intestinal mucosa of rabbits (Mynott et al. 1991) and pigs (Mynott et al. 1996). In broilers, the addition of proteases to corn-soyben meal based diets mitigated the effects of coccidiosis, increased the adherent mucus layer, improved weight gain (Peek et al. 2009), and improved the epithelial integrity by up-regulating Claudin-1 (Cowieson et al. 2017a, b). Several authors reported that the benefits of proteases is due to their "extra-proteinaceous" effects, such as better protein digestion in the proximal gut (Liu et al. 2013), reduction of protein fermentation, and formation of putrefactive molecules in the lower gut (Windey et al. 2012), mucin synthesis (Cowieson and Roos 2014), and improved enterocytes metabolism and intestinal integrity (Cowieson et al. 2017a, b). However, more detailed studies should be conducted on the function of the intestinal microbiota, and immune and metabolic changes of the gut driven by the use of dietary proteases.

8.6.5.3 Phytases

Phytases are enzymes that break the bond between phytate and phosphorus (P), releasing P and the other nutrients complexed in the molecule. The effects of phytase in improving gut health have been associated with its extra-phosphoric effects that includes increased expression of Mucin-2 (Ajuwon et al. 2020) and reduced expression of IL-1β (Jiang et al. 2018). Also, the complete dephosphorylation of the phytate, due to high phytase dosages in the diet (Cowieson et al. 2017a, b), releases the myo-inositol ring in the center of the phytate molecule. Myo-inositol concentration in the GIT of animals has been correlated with upregulation of nutrient transporters (Ajuwon et al. 2020). Cowieson et al. (2016) reported there is a relationship between the presence of low esters of inositol and free myo-inositol and several biochemicals pathways, including those responsible for muscle deposition. Therefore, in addition to the beneficial effects of phytase on the digestive and metabolic processes of animals, we hypothesize that phytase may exert anti-inflammatory effects in the intestine by degrading phytate molecules that could be recognized by the immune system and trigger an undesirable immune-response; however, additional studies are required to confirm this hypothesis.

8.6.6 Phytobiotics

The term "phytobiotic" includes a range of plant-derived products, such as herbs, essential oils, and oleoresins (Lillehoj et al. 2018). There are considerable variations with their chemical composition, depending mostly on the weather, season of harvest, location or storage conditions which are responsible for the differences in efficacy of these compounds (Applegate et al. 2010). There has been increased interest in utilizing plant-based compounds or phytobiotics as AGP alternatives (Diaz Carrasco et al. 2018; Ren et al. 2019). The current knowledge, as summarized by Lillehoj et al. (2018), is that the active ingredients in phytobiotics alters the host microbiota, providing antimicrobial activities against pathogens, and reduces oxidative stress to improve overall health of the animal. Extensive studies have been performed in other food production species, particularly in swine, which showed phytobiotics as suitable alternatives to AGP to improve growth performance and health (Michiels et al. 2010; Liu et al. 2014). Although the data vary between each type of phytobiotic classification, the ideal antibiotic alternative would alter the host microbiota to guide protein and lipid metabolism, promote effective nutrient utilization, and prevent harmful infections to the host (Allen and Stanton 2014).

Remmal et al. (2011) explained the need to look for substances with a reduced potential to develop antimicrobial resistance, as well as leaving no residues in the final product. For example, these authors investigated the anticoccidial effect, in vitro, of 10 different phytobiotic molecules and verified that all of them have action against *Eimeria* oocysts. However, essential oils from *Artemisia absinthium*, Tea tree, *Thymus vulgaris* (thyme), and *Syzygium aromaticum* (clove oil) were more effective than salinomycin in reducing the number of viable oocysts. On the other hand, in vivo studies have demonstrated beneficial effects of the plant compounds against coccidiosis, such as *Artemisia annua* (Almeida et al. 2012) and carvacrol (Giannenas et al. 2003). *Capsicum* and *Curcuma longa* oleoresins modulated the gut microbiota of broilers induced to necrotic enteritis (NE) by increasing *Candidatus* Arthromitus and *Lactobacillus* and reduced the losses on body weight gain (Kim et al. 2013). Even though additional studies concerning the use of phytobiotics on gut microbiota and host–pathogen interactions are necessary, the use of essential oils is becoming a common practice primarily due to the improvements of gut functions, which include the stabilization of the microbiota and better nutrient utilization and absorption (Diaz-Sanchez et al. 2015).

The antimicrobial activity of a phytobiotic varies greatly and will depend on the bioactive compound and may include disruption of the cellular membrane of pathogens, modification of cells affecting the virulence capacity of the microorganism, and protection against the pathogen binding to the intestinal mucosa (Diaz-Sanchez et al. 2015). These plant metabolites have also been shown to exert immunomodulatory effects on the host, including induction of heat shock proteins, induction of Toll-like receptors, and proliferation and maturation of T-Helper cells to maintain a balance between cellular and humoral immune response. Many studies focus on the effects of thymol, capsaicin, and carvacrol and show a reduction of *E. coli*, *Salmonella*, and *C. perfringens* with the use of these substances in the diets of chickens (Diaz-Sanchez et al. 2015). Bortoluzzi et al. (2014) observed that dietary supplementation of 30 mg/kg of beta-acids isolated from hops led to the same feed conversion ratio as zinc bacitracin in broilers fed diets containing poultry by-product meal and wheat bran, modulated the microbiota (Bortoluzzi et al. 2015), and altered cytokines production (Bortoluzzi et al. 2016).

Another promising phytobiotic extensively studied in the literature is tannins, a readily found plant-based compound with antimicrobial activities and growth performance promotion effects (Lillehoj et al. 2018), and has been the subject of studies in our laboratory. Plant-based tannins can be categorized into two major groups: condensed or hydrolysable tannins (Huang et al. 2018) and are found in many plant species, mostly in the inedible portions such as the bark or wood (Brus et al. 2018; Molino et al. 2020). The previous knowledge was that tannins possess anti-nutritional effects in livestock species but with new evidence, the benefits show promising results across livestock species depending on the dosages and quality of tannins in feed (Schiavone et al. 2008; Diaz Carrasco et al. 2018). The bioactive compounds, polyphenols, allow tannins to have immunomodulatory effects (Molino et al. 2020). These phenol compounds added to poultry diets have also stimulated biochemical pathways to enhance growth performance, such as decreased lipid oxidation (Cejas et al. 2011) and increased beneficial fatty acids (Starčević et al. 2015). Tannins have also been shown to improve feed efficiency, growth performance, and intestinal health in poultry (Gai et al. 2010; Redondo et al. 2014).

Previous studies evaluated the functionality of different tannins against pathogenic infections across livestock species, showing anti-microbial activity in concentrations ranging between 0.5–1 kg/ton (Costabile et al. 2011; Liu et al. 2014; Lillehoj et al. 2018). The addition of tannins in the diet significantly reduces the incidence of problematic poultry pathogens such as coccidia, *Salmonella* Typhimurium, *Campylobacter jejuni,* and *C. perfringens* (McDougald et al. 2008; Cejas et al. 2011; Diaz Carrasco et al. 2016). Diaz Carrasco et al. (2018) observed that chestnut tannins affected *Bifidobacterium* in the ceca of mammals and chickens, which have been shown to alter carbohydrate metabolism and other metabolic processes in the host by modifying enzymes and sugar transport pathways (Pokusaeva et al. 2011). In the same study by Diaz Carrasco et al. (2018), older birds treated with tannins consistently had increased populations of order Clostridiales and family Ruminococcaceae, which have been of interest in the poultry industry as potential probiotic options. Molino et al. (2018) have shown preliminary data on the importance of tannins in gut microbial fermentation and the nutritional importance in human application as well, showing the importance of studying tannins further across different species to improve overall health. When sufficient concentrations of a quality tannin are used in production, they have shown the potential to be an effective AGP (Redondo et al. 2014).

8.7 Importance of Understanding the Mode-of-Action

From an industry perspective, a product's impact on performance and production traits will be the driving factors as to whether a product is used or not. Most studies found in the literature lack comprehensive data showing the mechanism(s) that produced the enhanced performance, or reduced bacterial load, or increased resistance. In order to clarify the mode-of-action of these additives, in addition to functional assays, newer technologies, such as a kinome array, are emerging as powerful research tools to dissect mechanisms. Kinome analysis using peptide arrays provide site-specific information, display similar biochemical properties to the full-length protein, and provide a means for defining phosphorylation-mediated events (Ouyang et al. 2003). Phosphorylation is the predominant mechanism of post-translational modification regulating protein function, has a central role in virtually every cellular event, is essential for all cell signaling networks, and regulates fundamental biological processes (Manning et al. 2002; Wang 2014). Global analysis of the kinome provides information on the abundance, activity, substrate specificity, phosphorylation pattern, and mutational status of a given peptide (Wang 2014), and chicken-specific kinome arrays are available (Arsenault and Kogut 2012). Studies show the usefulness of kinome arrays to dissect key immuno-metabolic pathways associated with different feed additives, such as β-mannanase and postbiotic in broiler diets (Arsenault et al. 2017; Johnson et al. 2019). Kinome studies can also provide insight while examining the complex interplay between the host and dietary supplementation under control and challenged conditions. From a purely scientific standpoint, understanding mechanism(s) is important, but even more so, this understanding will be vital for the poultry industry moving forward so they can make decisions based on sound science.

Additionally, there are other methodologies to study mode-of-action in nutritional and feed additive studies including molecular, genomic, microbiological, and immunological technologies such as RNA sequencing, proteomics, metagenomics, transcriptomics, metabolomics, and epigenomics. For example, utilizing transcriptomics allows for the identification of the level of mRNA transcripts in response to different stimuli, while proteomics facilitates the study of proteins and post-translational modifications during under specific conditions, and metabolomics to study either endogenous or exogenous metabolites (Zampiga et al. 2018). Therefore, these technologies will drive a deeper understanding of the mechanism of action of feed additives and nutrients which will be of paramount importance to develop strategies and formulate diets without the use of AGP.

8.8 Future Trends and Conclusions

The maintenance of a healthy gut status is complex and relies on a delicate balance between the immune system and the normal endogenous microbiota. The normal microbiota confers many benefits to the intestinal physiology of the host. However, when this balance is upset in the form of dysbiosis, pathogens that arrive or that have already been present but in numbers too small to cause disease will take the opportunity to multiply. Future studies building on the gene and organism catalogues established thus far will need to include increasingly detailed investigations of meta-transcriptomes and meta-proteomes and allow us to more fully understand the links between the chicken microbiome, health, and disease.

Acknowledgements The USDA is an equal opportunity provider and employer. Mention of trade names or commercial products in this publication is solely for the purpose of providing specific information and does not imply recommendation or endorsement by the USDA.

References

Abreu MT, Fukata M, Arditi M (2005) TLR signaling in the gut in health and disease. J Immunol 174:4453–4460

Adhikari PA, Cosby DE, Cox NA, Franca MSS, Williams M, Gogal RM Jr, Ritz CW, Kim WW (2018) Effect of dietary fructooligosaccharides supplementation on internal organs *Salmonella* colonization, immune response, ileal morphology, and ileal immunohistochemistry in laying hens challenged with *Salmonella* Enteritidis. Poult Sci 97:2525–2533

Aftab U, Bedford MR (2018) The use of NSP enzymes in poultry nutrition: myths and realities. World's Poult Sci J 74:277–286

Ahsan U, Cengiz O, Raza I, Kuter E, Chacher MFA, Iqbal Z, Umar S, Cakir S (2016) Sodium butyrate in chicken's nutrition: the dynamics of performance, gut microbiota, gut morphology, and immunity. Worlds Poult Sci J 72:265–275

Ajuwon KM, Sommerfeld V, Paul V, Däuber M, Schollenberger M, Kühn I, Adeola O, Rodehutscord M (2020) Phytase dosing affects phytate degradation and Muc2 transporter gene expression in broiler starters. Poult Sci 99:981–991

Akira S, Uematsu S, Takeuchi O (2006) Pathogen recognition and innate immunity. Cell 124:783–801

Allen HK, Stanton TB (2014) Altered egos: antibiotic effects on food animal microbiomes. Annu Rev Microbiol 68:297–315

Almeida GF, Horsted K, Thamsborg SM, Kyvsgaard NC, Ferreira JFS, Hermansen JE (2012) Use of *Artemisia annua* as a natural coccidiostat in free-range broilers and its effects on infection dynamics and performance. Vet Parasitol 186:178–187

Applegate TJ, Klose V, Steiner T, Ganner A, Schatzmayr G (2010) Probiotics and phytogenics for poultry: Myth or reality? J Appl Poult Res 19:194–210

Arsenault RJ, Kogut MH (2012) Chicken-specific peptide arrays for kinome analysis: flight for the flightless. Curr Top Biotechnol 7:79–89

Arsenault RJ, Lee JT, Latham R, Carter B, Kogut MH (2017) Changes in immune and metabolic gut response in broilers fed beta-mannanase in beta-mannan-containing diets. Poult Sci 96:4307–4316

Bedford A, Gong J (2018) Implications of butyrate and its derivatives for gut health and animal production. Anim Nutr 4:151–159

Belkaid Y, Hand TW (2014) Role of the microbiota in immunity and inflammation. Cell 157:121–141

Belkaid Y, Harrison OJ (2017) Homeostatic immunity and the microbiota. Immunity 46:562–576

Birchenough GM, Nystrom EE, Johansson ME, Hansson GC (2016) A sentinel goblet cell guards the colonic crypt by triggering Nlrp6-dependent Muc2 secretion. Science 352:1535–1542

Bortoluzzi C, Barbosa JGM, Pereira R, Fagundes NS, Rafael JM, Menten JFM (2018) Autolyzed yeast (*Saccharomyces cerevisiae*) supplementation improves performance while modulating the intestinal immune-system and microbiology of broiler chickens. Front Sustain Food Syst 2:85

Bortoluzzi C, Menten JFM, Pereira P, Fagundes NS, Napty GS, Pedroso AA, Bigaton AD, Andreote FD (2015) Hops β-acids and zinc bacitracin affect the performance and intestinal microbiota of broilers challenged with *Eimeria acervulina* and *Eimeria tenella*. Anim Feed Sci Technol 207:181–189

Bortoluzzi C, Menten JFM, Romano GG, Pereira R, Napty GS (2014) Effect of hops beta-acids (*Humulus lupulus*) on the performance and intestinal health of broiler chickens. J Appl Poult Res 23:437–443

Bortoluzzi C, Menten JFM, Silveira H, Melo ADB, Rostagno MH (2016) Hops beta-acids (*Humulus Lupulus*) decrease intestinal gene expression of proinflammatory cytokines in an *ex-vivo* model. J Appl Poult Res 25:191–196

Bortoluzzi C, Pedroso AA, Mallo JJ, Puyalto M, Kim WK, Applegate TJ (2017) Sodium butyrate improved performance while modulating the cecal microbiota and regulating the expression of intestinal-immune related genes of broiler chickens. Poult Sci 96:3981–3993

BRASIL (2018) Ordinance number 171 of December 13th, 2018. Ministry of Agriculture, Livestock, and Supply – Secretary of Agricultural Defense. Diário Oficial da União, December 19th, 2018

Broom LJ, Kogut MH (2018) Inflammation: friend or foe for animal production? Poult Sci 97:510–514

Brus M, Gradisnik L, Trapecar M, Skorjanc D, Frangez R (2018) Beneficial effects of water-soluble chestnut (Castanea sativa Mill.) tannin extract on chicken small intestinal epithelial cell culture. Poult Sci 97:1271–1282

Byndloss MX, Litvak Y, Baumler AJ (2019) Microbiota-nourishing immunity and its relevance for ulcerative colitis. Inflamm Bowell Dis 25:811–815

Cani PD, Knauf C (2016) How gut microbes talk to organs: The role of endocrine and nervous routes. Mol Metab 5:743–752

Cardoso M (2019) Antimicrobial use, resistance and economic benefits and costs to livestock producers in Brazil. OECD Food, Agriculture and Fisheries Papers (2019) 135. OECD Publishing, Paris

Castanon JIR (2007) History of the use of antibiotic as growth promoters in European poultry feeds. Poult Sci 86:2466–2471

Cejas E, Pinto S, Prosdocimo F, Batalle M, Barrios H, Tellez G, Franceschi M (2011) Evaluation of quebracho red wood (*Schinopsis lorentzii*) polyphenols vegetable extract for the reduction of coccidiosis in broiler chicks. Int J Poult Sci 10:344–349

Cencic A, Chingwaru W (2010) The role of functional foods, nutraceuticals, and food supplements in intestinal health. Nutrients 2:611–625

Corthesy B, Gaskins HR, Mercenier A (2007) Cross-talk between probiotic bacteria and the host immune system. J Nutr 137:781–790

Costabile A, Sanghi S, Martin-Pelaez S, Mueller-Harvey I, Gibson GR, Rastall RA, Klinder A (2011) Inhibition of *Salmonella* Typhimurium by tannins in vitro. J Food Agric Environ 9:119–124

Cowieson AJ, Roos FF (2014) Bioefficacy of a mono-component protease in the diets of pigs and poultry: a meta-analysis of effect on ileal amino acid digestibility. J Appl Anim Nutr 2:13–21

Cowieson AJ, Ruckebusch JP, Knap I, Guggenbuhl P, Fru-Nji F (2016) Phytate-free nutrition: a new paradigm in monogastric animal production. Anim Feed Sci Technol 222:180–189

Cowieson AJ, Lu H, Ajuwon K, Knap I, Adeola O (2017a) Interactive effects of dietary protein source and exogenous protease on growth performance, immune competence and jejunal health of broiler chickens. Anim Prod Sci 57:252–261

Cowieson AJ, Ruckebusch JP, Sorbara JOB, Wilson JW, Guggenbuhl P, Roos FF (2017b) A systematic view on the effect of phytase on ileal amino acid digestibility in broilers. Anim Feed Sci Tech 225:182–194

Craig AD, Khattak F, Hastie P, Bedford MR, Olukosi OA (2020) The similarity of the effect of carbohydrase or prebiotic supplementation in broilers aged 21 days, fed mixed cereal diets and challenged with coccidiosis infection. PLoS ONE 15:e0229281

Dal Pont GC, Farnell M, Farnell Y, Kogut MH (2020) Dietary factors as triggers of low-grade chronic intestinal inflammation in poultry. Microorganisms 8:139

Diaz Carrasco JM, Redondo EA, Pin Viso ND, Redondo LM, Farber MD, Fernandez Miyakawa ME (2018) Tannins and bacitracin differentially modulate gut microbiota of broiler chickens. BioMed Res Int 2018:1879168

Diaz Carrasco JM, Redondo LM, Redondo EA, Dominguez JE, Chacana AP, Fernandez Miyakawa ME (2016) Use of plant extracts as an effective manner to control Clostridium perfringens induced Necrotic Enteritis in poultry. BioMed Res Int 2016:3278359

Diaz-Sanchez S, Souza DD, Biswas D, Hanning I (2015) Botanical alternatives to antibiotics for use in organic poultry production. Poult Sci 94:1419–1430

Dibner JJ, Richards JD (2005) Antibiotic growth promoters in agriculture: history and mode of action. Poult Sci 84:634–643

Ducatelle R, Eeckhaut V, Haesebrouk F, Van Immerseel F (2015) A review on prebiotics and probiotics for the control of dysbiosis: present status and future perspectives. Animal 9:43–48

Ducatelle R, Goossens E, De Meyer F, Eeckhaut V, Antonissen G, Haesebrouck F, Van Immerseel F (2018) Biomarkers for monitoring intestinal health in poultry: present status and future perspectives. Vet Res 49:43

Faber TA, Dilger RN, Iakiviak M, Hopkins AC, Price NP, Fahey GC Jr (2012) Ingestion of a novel galactoglucomannan oligosaccharide-arabinoxylan (GGMO-AX) complex affected growth performance and fermentative and immunological characteristics of broiler chicks challenged with *Salmonella* Typhimurium. Poult Sci 91:2241–2254

FDA (1995) Direct-fed microbial products. In: SERVICES, U. S. D. O. H. A. H., Washington DC

Fung KY, Cosgrove L, Lockett T, Head R, Topping DL (2012) A review of the potential mechanisms for the lowering of colorectal oncogenesis by butyrate. Br J Nutr 108:820–831

Gai F, Gasco L, Schiavone A, Zoccarato I (2010) Nutritional effects of chestnut tannins in poultry and rabbit. Types, foods containing, and nutrition. Nova Science Publishers, Hauppauge, NY, pp 297–306

Garrett WS, Gordon JI, Glimcher LH (2010) Homeostasis and inflammation in the intestine. Cell 140:859–870

Garriga C, Hunter RR, Amat C, Planas JM, Mitchell MA, Moreto M (2006) Heat stress increases apical glucose transport in the chicken jejunum. Am J Physiol 290:195–201

Genovese KJ, He H, Swaggerty CL, Kogut MH (2013) The avian heterophil. Dev Comp Immunol 41:334–340

Giannenas I, Florou-Paneri P, Papazahariadou M, Christaki F, Botsoglou NA, Spais AB (2003) Effect of dietary supplementation with oregano essential oil on performance of broilers after experimental infection with *Eimeria tenella*. Arch Anim Nutr 57:99–106

Gillis CC, Hughes ER, Spiga L, Winter MG, Zhu W, Carvalho TF, Chanin R, Behrendt CL, Hooper LV, Santos RL, Winter SE (2018) Dysbiosis-associated change in host metabolism generates lactate to support *Salmonella* growth. Cell Host Microbe 23:54–64

Guilloteau P, Martin L, Eeckhaut V, Ducatelle R, Zabielski R, Van Immerseel F (2010) From the gut to the peripheral tissues: the multiple effects of butyrate. Nutr Res Rev 23:366–384

He WL, Li P, Wu G (2021a) Amino acid nutrition and metabolism in chickens. Adv Exp Med Biol 1285:109–131

He WL, Furukawa K, Toyomizu M, Nochi T, Bailey CA, Wu G (2021b) Interorgan metabolism, nutritional impacts, and safety of dietary L-glutamate and L-glutamine in poultry. Adv Exp Med Biol 1332:107–128

Honda K, Littman DR (2016) The microbiota in adaptive immune homeostasis and disease. Nature 535:75–84
Hooper LV, Littman DR, Macpherson AJ (2012) Interactions between the microbiota and the immune system. Science 336:1268–1273
Hooper LV, Macpherson AJ (2010) Immune adaptations that maintain homeostasis with the intestinal microbiota. Nat Rev Immunol 10:159–169
Hou YQ, He WL, Hu SD, Wu G (2019) Composition of polyamines and amino acids in plant-source foods for human consumption. Amino Acids 51:1153–1165
Huang Q, Liu G, Hu T, Wang Y (2018) Potential and challenges of tannins as an alternative to in-feed antibiotics for farm animal production. Anim Nutr 4:137–150
Huff GR, Huff WE, Farnell MB, Rath NC, De Los S, Santos F, Donoghue AM (2010) Bacterial clearance, heterophil function, and hematological parameters of transport-stressed turkey poults supplemented with dietary yeast extract. Poult Sci 89:447–456
Huynh TG, Shiu YL, Nguyen TP, Truong QP, Chen JC, Liu CH (2017) Current applications, selection, and possible mechanisms of actions of synbiotics in improving the growth and health status in aquaculture: a review. Fish Shellfish Immunol 64:367–382
Jack RW, Tagg JR, Ray B (1995) Bacteriocins of gram-positive bacteria. Microbiol Rev 59:171–200
Jiang SQ, Lamont SJ, Persia ME (2018) Differential growth performance and intestinal immune gene expression in diverse genetic lines of growing chickens fed a high concentration of supplemental phytase. J Agric Sci 156:258–264
Joerger RD, Ganguly A (2017) Current status of the preharvest application of pro- and prebiotics to farm animals to enhance the microbial safety of animal products. Microbiol Spectr 2017:5
Johnson CN, Kogut MH, Genovese K, He H, Kazemi S, Arsenault RJ (2019) Administration of a postbiotic causes immunomodulatory responses in broiler gut and reduces disease pathogenesis following challenge. Microorganisms 7:268
Kairie E, Romero LF, Nyachoti CM (2013) The role of feed enzymes in promoting gut health in swine and poultry. Nutr Res Rev 26:71–88
Kawai T, Akira S (2010) The role of pattern-recognition receptors in innate immunity: update on Toll-like receptors. Nat Immunol 11:373–384
Kim DK, Lillehoj HS, Lee SH, Jang SI, Lillehoj EP, Bravo D (2013) Dietary *Curcuma longa* enhances resistance against *Eimeria maxima* and *Eimeria tenella* infections in chickens. Poult Sci 92:2635–2643
Klasing KC (2007) Nutrition and the immune system. Br Poult Sci 48:525–537
Klemashevich C, Wu C, Howsmon D, Alaniz RC, Lee K, Jayaraman A (2014) Rational identification of diet-derived postbiotics for improving intestinal microbiota function. Curr Opin Biotech 26:85–90
Kogut MH, Genovese KJ, Swaggerty CL, He H, Broom L (2018) Inflammatory phenotypes in the intestine of poultry: not all inflammation is created equal. Poult Sci 97:2339–2346
Kogut MH (2017) Issues and consequences of using nutrition to modulate the avian immune response. J Appl Poult Res 26:605–612
Levy AW, Kessler JW, Fuller L, Williams S, Mathis GF, Lumpkins B, Valdez F (2015a) Effect of feeding an encapsulated source of butyric acid (ButiPEARL) on the performance of male Cobb broilers reared to 42 d of age. Poult Sci 94:1864–1870
Levy M, Thaiss CA, Katz MN, Suez J, Elinav E (2015b) Inflammasomes and the microbiota–partners in the preservation of mucosal homeostasis. Semin Immunopathol 37:39–46
Levy M, Blacher E, Elinav E (2017) Microbiome, metabolites and host immunity. Curr Opin Microbiol 35:8–15
Li P, Wu G (2020) Composition of amino acids and related nitrogenous nutrients in feedstuffs for animal diets. Amino Acids 52:523–542
Li P, He WL, Wu G (2021) Composition of amino acids in foodstuffs for humans and animals. Adv Exp Med Biol 1332:189–209
Lillehoj H, Liu Y, Calsamiglia S, Fernandez-Miyakawa ME, Chi F, Cravens RL, Oh S, Gay CG (2018) Phytochemicals as antibiotic alternatives to promote growth and enhance host health. Vet Res 49:76
Liu JD, Lumpkins B, Mathis G, Williams SM, Fowler J (2019) Evaluation of encapsulated sodium butyrate with varying releasing times on growth performance and necrotic enteritis mitigation in broilers. Poult Sci 98:3240–3245
Liu SY, Selle PH, Court SG, Cowieson AJ (2013) Protease supplementation of sorghum-based broiler diets enhances amino acid digestibility coefficients in four small intestinal sites and accelerates their rates of digestion. Anim Feed Sci Technol 183:175–183
Liu Y, Song M, Che TM, Lee JJ, Bravo D, Maddox CW, Pettigrew JE (2014) Dietary plant extracts modulate gene expression profiles in ileal mucosa of weaned pigs after an infection. J Anim Sci 92:2050–2062
Lowry VK, Farnell MB, Ferro PJ, Swaggerty CL, Bahl A, Kogut MH (2005) Purified beta-glucan as an abiotic feed additive up-regulates the innate immune response in immature chickens against *Salmonella enterica* serovar enteritidis. Int J Food Microbiol 98:309–318
Ma N, Guo P, Zhang J, He T, Kim SW, Zhang G, Ma X (2018) Nutrients mediate intestinal bacteria-mucosa immune crosstalk. Front Immunol 9:5
Macpherson AJ, Slack E, Geuking MB, Mccoy KD (2009) The mucosal firewalls against commensal intestinal microbes. Semin Immunopathol 31:145–149
Manning G, Whyte DB, Martinez R, Hunter T, Sudarsanam S (2002) The protein kinase complement of the human genome. Science 298:1912–1934
Marks PA, Richon VM, Rifkind RA (2000) Histone deacetylase inhibitors: inducers of differentiation or apoptosis of transformed cells. J Natl Cancer Int 92:1210–1216

Maslowski KM, Mackay CR (2011) Diet, gut microbiota and immune responses. Nat Immunol 12:5–9
McDougald LR, Hofacre EC, Mathis G, Fuller L, Hargrove JL, Greenspan P, Hartle DK (2008) Enhancement of resistance to coccidiosis and necrotic enteritis in broiler chickens by dietary muscadine pomace. Avian Dis 52:646–651
Medzhitov R, Janeway CA (2002) Decoding the patterns of self and nonself by the innate immune response. Science 296:298–300
Meijer K, de Vos P, Priebe MG (2010) Butyrate and other short-chain fatty acids as modulators of immunity: what relevance for health? Curr Opin Clin Nutr Metab Care 13:715–721
Michiels J, Missotten J, Van Hoorick A, Ovyn A, Fremaut D, De Smet S, Dierick N (2010) Effects of dose and formulation of carvacrol and thymol on bacteria and some functional traits of the gut in piglets after weaning. Arch Anim Nutr 64:136–154
Molino S, Casanova NA, Henares JAR, Fernandez Miyakawa ME (2020) Natural tannin wood extracts as a potential food ingredient in the food industry. J Agric Food Chem 68:2836–2848
Molino S, Fernandez-Miyakawa M, Giovando S, Rufian-Henares JA (2018) Study of antioxidant capacity and metabolization of quebracho and chestnut tannins through in vitro gastrointestinal digestion-fermentation. J Funct Foods 49:188–195
Morgan NK, Keerqin C, Wallace A, Wu SB, Choct M (2019) Effect of arabinoxylan-oligosaccharides and arabinoxylans on net energy and nutrient utilization in broilers. Anim Nutr 5:56–62
Mynott TL, Chandler DS, Luke RKJ (1991) Efficacy of enteric-Coated protease in preventing attachment of enterotoxigenic *Escherichia coli* and diarrheal disease In the RITARD model. Infect Immun 59:3708–3714
Mynott TL, Luke RK, Chandler DS (1996) Oral administration of protease inhibits enterotoxigenic *Escherichia coli* receptor activity in piglet small intestine. Gut 38:28–32
Neish AS (2009) Microbes in gastrointestinal health and disease. Gastroenterology 136:65–80
Neuman H, Debelius JW, Knight R, Koren O (2015) Microbial endocrinology: the interplay between the microbiota and the endocrine system. FEMS Microbiol Rev 39:509–521
Ouyang Z, Takats Z, Blake TA, Gologan B, Guymon AJ, Wiseman JM, Oliver JC, Davisson VJ, Cooks RG (2003) Preparing protein microarrays by soft-landing of mass-selected ions. Science 301:1351–1354
Peek HW, Van Der Klis JD, Vermeulen B, Landman WJM (2009) Dietary protease can alleviate negative effects of a coccidiosis infection on production performance in broiler chickens. Anim Feed Sci Technol 150:151–159
Piniero M, Asp N-G, Reid G (2008) FAO technical meeting on prebiotics. J Clin Gastroenterol 42:S156–S162
Pokusaeva K, Fitzgerald GF, van Sinderen D (2011) Carbohydrate metabolism in Bifiobacteria. Genes Nutr 6:285–306
Postler TS, Ghosh S (2017) Understanding the holobiont: How microbial metabolites affect human health and shape the immune system. Cell Metab 26:110–130
Quinteiro-Filho WM, Rodrigues MV, Ribeiro A, Ferraz-De-Paula V, Pinheiro ML, Sa LR, Ferreira AJ, Palermo-Neto J (2012) Acute heat stress impairs performance parameters and induces mild intestinal enteritis in broiler chickens: role of acute hypothalamic-pituitary-adrenal axis activation. J Anim Sci 90:1986–1994
Redondo LM, Chacana PA, Dominguez JE, Fernandez Miyakawa ME (2014) Perspectives in the use of tannins as alternative to antimicrobial growth promoter factors in poultry. Front Microbiol 5:118
Remmal A, Achahbar S, Bouddine L, Chami N, Chami F (2011) *In vitro* destruction of *Eimeria* oocysts by essential oils. Vet Parasitol 182:121–126
Ren H, Vahjen W, Dadi T, Saliu EM, Boroojeni FG, Zenrek J (2019) Synergistic effects of probiotics and phytobiotics on the intestinal microbiota in young broiler chicken. Microorganisms 7:684
Ren WK, Bin P, Yin YL, Wu G (2020) Impacts of amino acids on the intestinal defensive system. Adv Exp Med Biol 1265:133–151
Ricke SC (2018) Impact of prebiotics on poultry production and food safety. Yale J Biol Med 91:151–159
Ricke SC (2003) Perspectives on the use of organic acids and short chain fatty acids as antimicrobials. Poult Sci 82:632–639
Roberts T, Wilson J, Guthrie A, Cookson K, Vancraeynest D, Schaeffer J, Moody R, Clark S (2015) New issues and science in broiler chicken intestinal health: emerging technology and alternative interventions. J Appl Poult Res 24:257–266
Rodriguez-Palacios A, Staempfli HR, Duffield T, Weese JS (2009) Isolation of bovine intestinal Lactobacillus plantarum and *Pediococcus acidilactici* with inhibitory activity against *Escherichia coli* O157 and F5. J Appl Microbiol 106:393–401
Roto SM, Rubinelli PM, Ricke SC (2015) An introduction to the avian gut microbiota and the effects of yeast-based prebiotic-type compounds as potential feed additives. Front Vet Sci 2:28
Russell JB (1992) Another explanation for the toxicity of fermentation acids at low pH: anion accumulation versus uncoupling. J Appl Bacteriol 73:363–370
Saleh AA, Ali H, Abdel-Latif MA, Emam MA, Ghanem R, El- Hamid HSA (2018) Exogenous dietary enzyme formulations improve growth performance of broiler chickens fed a low-energy diet targeting the intestinal nutrient transporter genes. PLoS ONE 13: e0198085
Sanders ME (2008) Probiotics: definition, sources, selection, and uses. Clin Infect Dis 46:58–61

Scavuzzi BM, Henrique FC, Miglioranza LHS, Simao ANC, Dichi I (2014) Impact of prebiotics, probiotics, and synbiotics on components of the metabolic syndrome. Ann Nutr Disord Ther 2014:1009

Schiavone A, Guo K, Tassone S, Gasco L, Hernandez E, Denti R, Zoccarato I (2008) Effects of a natural extract of chestnut wood on digestibility, performance traits, and nitrogen balance of broiler chicks. Poult Sci 87:521–552

Sergeant MJ, Constantinidou C, Cogan TA, Bedford MR, Penn CW, Pallen MJ (2014) Extensive microbial and functional diversity within the chicken cecal microbiome. PLoS ONE 9:e91941

Singh VP, Proctor SD, Willing BP (2016) Koch's postulates, microbial dysbiosis and inflammatory bowel disease. Clin Microbiol Infect 22:594–599

Smith JA (2019) Broiler production without antibiotic: United States field perspectives. Anim Feed Sci Technol 250:93–98

Sommer F, Anderson JM, Bharti R, Raes J, Rosenstiel, (2017) The resilience of the intestinal microbiota influences health and disease. Nat Rev Microbiol 15:630–638

Song B, Li H, Wu Y, Zhen W, Wang Z, Xia Z, Guo Y (2017) Effect of microencapsulated sodium butyrate dietary supplementation on growth performance and intestinal barrier function of broiler chickens infected with necrotic enteritis. Anim Feed Sci Technol 232:6–15

Starčević K, Krstulović L, Brozić D, Maurić M, Stojević Z, Mikulec Ž, Bajić M, Mašek T (2015) Production performance, meat composition and oxidative susceptibility in broiler chicken fed with different phenolic compounds. J Sci Food Agric 95:1172–1178

Sun Q, Liu D, Guo S, Chen Y, Guo Y (2015) Effects of dietary essential oil and enzyme supplementation on growth performance and gut health of broilers challenged by *Clostridium perfringens*. Anim Feed Sci Technol 207:234e244

Tellez G, Laukova A, Latorre JD, Hernandez-Velasco X, Hargis BM, Callaway TR (2015) Food-producing animals and their health in relation to human health. Microb Ecol Health Dis 26:25876

Thaiss CA, Levy M, Suez J, Elinav E (2014) The interplay between the innate immune system and the microbiota. Curr Opin Immunol 26:41–48

Tugnoli B, Piva A, Sarli G, Grilli E (2020) Tributyrin differentially regulates inflammatory markers and modulates goblet cells number along the intestinal tract segments of weaning pigs. Livest Sci 234:103996

Turner JR (2009) Intestinal mucosal barrier function in health and disease. Nat Rev Immunol 9:799–809

van der Wielen PW, van Knapen F, Biesterveld S (2002) Effect of administration of *Lactobacillus crispatus*, *Clostridium lactatifermentans* and dietary lactose on the development of the normal microflora and volatile fatty acids in the caeca of broiler chicks. Br Poult Sci 43:545–550

Van Immerseel F, Boyen F, Gantois I, Timbermont L, Bohez L, Pasmans F, Haesebrouck F, Ducatelle R (2005) Supplementation of coated butyric acid in the feed reduces colonization and shedding of *Salmonella* in poultry. Poult Sci 84:1851–1856

Wang X-H (2014) Progress in chemical proteomics-based kinome study. J Int Pharmaceut Res 41:259–267

Wei S, Morrison M, Yu Z (2013) Bacterial census of poultry intestinal microbiome. Poult Sci 92:671–683

Windey K, De Preter V, Verbeke K (2012) Relevance of protein fermentation to gut health. Mol Nutr Food Res 56:184–196

Wolin MJ (1969) Volatile fatty acids and the inhibition of *Escherichia coli* growth by rumen fluid. Appl Microbiol 17:83–87

Wu G (2018) Principles of animal nutrition. CRC Press, Boca Raton, Florida

Wu G (2020) Important roles of dietary taurine, creatine, carnosine, anserine and hydroxyproline in human nutrition and health. Amino Acids 52:329–360

Wu G (2022) Nutrition and metabolism: Foundations for animal growth, development, reproduction, and health. Adv Exp MedBiol 1354:1–24

Wu Y, Ravindran V, Hendriks W (2004) Influence of exogenous enzyme supplementation on energy utilisation and nutrient digestibility of cereals for broilers. J Sci Food Agric 84:1817–1822

Zampiga M, Flees J, Meluzzi A, Dridi S, Sirri F (2018) Application of omics technologies for a deeper insight into quali-quantitative production traits in broiler chickens: A review. J Anim Sci Technol 9:61

Zeng MY, Inohara N, Nunez G (2017) Mechanisms of inflammation-driven bacterial dysbiosis in the gut. Mucosal Immunol 10:18–26

Zhou Z, Nie K, Huang Q, Li K, Sun Y, Zhou R, Wang Z, Hu S (2017) Changes of cecal microflora in chickens following *Eimeria tenella* challenge and regulating effect of coated sodium butyrate. Exp Parasitol 177:73–81

Zopf D, Roth S (1996) Oligosaccharide anti-infective agents. Lancet 347:1017–1021

Microbiomes in the Intestine of Developing Pigs: Implications for Nutrition and Health

9

Chunlong Mu, Yu Pi, Chuanjian Zhang, and Weiyun Zhu

Abstract

The past decade has seen an expansion of studies on the role of gut microbiome in piglet nutrition and health. With the help of culture-independent sequencing techniques, the colonization of gut microbiota and their implication in physiology are being investigated in depth. Immediately after birth, the microbes begin to colonize following an age-dependent trajectory, which can be modified by maternal environment, diet, antibiotics, and fecal microbiota transplantation. The early-life gut microbiome is relatively simple but enriched with huge metabolic potential to utilize milk oligosaccharides and affect the epithelial function. After weaning, the gut microbiome develops towards a gradual adaptation to the introduction of solid food, with an enhanced ability to metabolize amino acids, fibers, and bile acids. Here we summarize the compositional and functional difference of the gut microbiome in the keystone developing phases, with a specific focus on the use of different nutritional approaches based on the phase-specific gut microbiome.

Keywords

Gut · Microbes · Amino acids · Fiber · Nutrition · Health

C. Mu · Y. Pi · C. Zhang · W. Zhu (✉)
Laboratory of Gastrointestinal Microbiology, Jiangsu Key Laboratory of Gastrointestinal Nutrition and Animal Health, College of Animal Science and Technology, Nanjing Agricultural University, Nanjing 210095, China
e-mail: zhuweiyun@njau.edu.cn

C. Mu · Y. Pi · C. Zhang · W. Zhu
National Center for International Research On Animal Gut Nutrition, Nanjing Agricultural University, Nanjing 210095, China

9.1 Introduction

Gut microbiome is a collection of microorganisms (e.g., bacteria, viruses, fungi, and protozoa) and their collective genetic components in the gastrointestinal tract. An increasing number of studies have shown that the gut microbiome serves as an inner organ regulating a diverse of physiological processes, including nutrient metabolism (Wu 2009; Zmora et al. 2019), microbe–host immune interaction (Mu et al. 2015), and gut–brain dialogue (Mu et al. 2016a). Recent advances in gut microbiology have greatly expanded our insights into the composition and function of the gut microbiome in pigs. The gut microbiome develops gradually from a simple and vulnerable status to a complex and stable assembly. Correspondingly, the microbial functionality also changes with dietary and physiological shifts, especially at the early

G. Wu (ed.), *Recent Advances in Animal Nutrition and Metabolism*, Advances in Experimental Medicine and Biology 1354,
https://doi.org/10.1007/978-3-030-85686-1_9

suckling and post-weaning periods (Guevarra et al. 2019). The suckling-to-weaning transition introduces extra stress for piglets. Considering the fundamental role of microbiome in gut health, adequate nutritional interventions that optimize the microbiome composition and function may protect the piglets from stressful events and maintain systemic health (Wu 2022). In this review, we provide an overview of microbial colonization at early life and the phase-specific metabolic properties in the gut microbiome of the developing pigs.

9.2 Microbiome Structure and Succession

9.2.1 Longitudinal and Temporal Distribution of Piglet's Gut Microbiota

The microbial colonization in the intestine of the piglet begins immediately following birth, which depends on the sow and environmental exposures. Initial colonizers including facultative anaerobes *Escherichia* and *Streptococcus* spp. create an anaerobic environment for subsequent colonization by strict anaerobes *Bacteroides*, *Bifidobacterium*, *Clostridium*, and *Lactobacillus* (Petri et al. 2010; Bian et al. 2016). As the piglets grow, the bacterial community becomes more complex as reflected by the increase in richness and evenness (Bian et al. 2016). A previous study showed that the abundances of *Bacteroides*, *Butyricimonas*, genera from the *Clostridiales* (*Oscillibacter*, *Clostridium* sensu stricto, *Clostridium* IV, *Clostridium* XIVa) and *Escherichia/Shigella*, exhibited significant decline with age in the feces of piglets (Mach et al. 2015). The enrichment of *Bacteroides* and *Oscillibacter* in the feces of suckling pigs suggests that these genera are adapted to use a wide range of both milk oligosaccharides and host-derived glycans (e.g., sulfomucin) as a carbon source (Poroyko et al. 2010; Marcobal et al. 2011). However, at weaning, the introduction of cereal-based diets modifies the substrate availability and the physiological conditions in the gastrointestinal tract (Boudry et al. 2002; Kim et al. 2011), probably leading to the increased abundances of *Prevotella*, *Acetivibrio*, *Oribacterium*, *Paraprevotella*, *Roseburia,* and *Succinivibrio* in the feces (Mach et al. 2015). A higher relative abundance of *Prevotella* observed at weaning might be due to the capacity of this genus to degrade the polysaccharides in the cereal cell wall (Ivarsson et al. 2012). At 10 weeks of age, *Prevotella* was the most dominant genera in the feces of pigs, contributing up to 30% of all the classifiable bacteria. However, at 22 weeks of age, the relative abundance of *Prevotella* was only 3.5–4.0%. As the abundance of *Prevotella* decreased, there was a marked increase in *Anaerobacter* (Kim et al. 2011). Overall, microbial colonization changes with age and physiological condition.

Inside the gut, the microbiome shows a compartment-specific distribution. At the longitudinal locations, there are differences in the number and composition of the pig intestinal microbiota from the proximal end of the intestinal tract to the distal end (Looft et al. 2014; Mu et al. 2017a). Using denatured gradient gel electrophoresis, we have shown that the bacterial community in the stomach and small intestine (jejunum and ileum) is markedly different before weaning, but being highly similar after weaning, characterized by an increase in *Streptococcus suis* in the stomach and small intestine (Su et al. 2008). The microbiota diversity of gastric and duodenal lumen samples from the piglets is higher than that of ileal digesta samples (Mu et al. 2017a). Compared to the microbial numbers in the lumen of the large intestine (cecum and colon) in growing pigs (Zhang et al. 2016a, b, 2017), jejunum and ileum have relatively low numbers of bacteria, with approximately $10^{9.5-11}$ copies per gram of dry matter (Zhang et al. 2020). The lumen of the small intestine is dominated by *Streptococcus* and *Lactobacillus* belonging to Firmicutes, while those of the colon is dominated by *Subdoligranulum*. Even in the small intestine, the relative abundances of *Escherichia*, *Pseudomonas*, and *Haemophilus* are higher in the lumen of the stomach and duodenum than that in the lumen of the jejunum and ileum (Mu et al. 2017a). In growing pigs, *Anaerobacter* and *Turicibacter* are the most dominant genera in the ileal lumen, and *Prevotella, Oscillibacter,* and *Succinivibrio* in the colonic lumen (Looft et al. 2014). In addition to the microbiota variation in longitudinal locations, in the radial locations, the microbial communities of the gut lumen and mucosa are also markedly different. At the genus level, *Escherichia* dominates in the mucosa of the small intestine in piglets, whereas its abundance decreased in the lumen (Mu et al. 2017a). In growing pigs, the ileum harbors bacteria both on the mucosa and in the lumen (Looft et al. 2014). Since the intestinal sites have different physiological functions in vivo (Wu 2018), the compartment-specific communities may be further involved in the metabolism of different nutrients such as protein and carbohydrate.

9.2.2 Hydrogen-Utilizing Microbes

Besides the dominant bacteria, the piglet intestine also contains minor hydrogenotrophic populations that utilize hydrogen and reduce hydrogen pressure in the gut. Succession of hydrogen-utilizing bacteria has also been found in piglets. Methanogen, namely, methanogenic archaea and sulfate-reducing bacteria are major hydrogen-utilizing members. In Meishan and Yorkshire piglets, from postnatal day 1–14, the diversity of methanogens decrease but the amount increase, with *Methanobrevibacter smithii*-related operational taxonomic units (OTUs) increased significantly and the abundances of *M. thaueri*- and *M. millerae*-related OTUs decreased with age (Su et al. 2014). Using the same experimental design, we further use a targeted sequencing of dissimilatory sulfite reductase subunit A (*dsrA*) gene to profile sulfate-reducing bacteria. We identify *dsrA*-containing bacteria within *Proteobacteria, Actinobacteria,* and *Firmicutes* at the phylum level, and *Proteobacteria* as the predominant taxa in the cecum of Meishan and Yorkshire piglets (Ran et al. 2019). *Desulfovibrio intestinalis* within *Desulfovibrio* is the dominant species from postnatal day 14–49, followed by *Desulfovibrio piger* and *Bilophila wadsworthia* (Ran et al. 2019). Age rather than breed mainly affects the colonization of sulfate-reducing bacteria, for example, bacteria belonging to *Faecalibacterium* increase with age from day 14 to day 49, while breed has no effects on the colonization of sulfate-reducing bacteria (Ran et al. 2019). Obviously, the change in the microbial colonization at early life is accompanied by changes in hydrogen-utilizing microbes.

Metabolism by the methanogens may further affect the microbial communities and host metabolism. Inhibition of the methanogenesis by bromochloromethane reduces the abundance of methanogen populations but increases sulfate-reducing bacteria in the colonic digesta of rats, together with a decrease in the abundance of *Actinobacteria* and *Proteobacteria* and the concentration of carbohydrate metabolites (Yang et al. 2016). These evidences indicate a potential role of hydrogen-utilizing microbes in regulating host physiology.

9.2.3 Non-Bacterial Components-Phages and Virome

It is well known that phages potentially help determine microbial colonization and function (Labrie et al. 2010). Representative phages of the *Myoviridae, Siphoviridae,* and *Podoviridae* families have been found by electron microscopy in fecal samples of young pigs (Allen et al. 2011). Metagenomes from the ileums of pigs contain phage-related contigs over 10 Kb. Many of the 12-Kb contig with 16 putative open-reading frames have full-length homologs in Gram-positive gut bacteria (such as *Clostridium* and *Lactobacillus*). The other phage-like contig (nearly 21 Kb) contains a mere seven open-reading frames, only one of which has a full-length homolog (34% amino-acid identity) to a metallophosphoesterase in *Bacillus* phage (Allen et al. 2011). There is a greater proportion of phages in the ileal metagenomes compared with those of the large intestine, which may be connected to different gut physiology in the small and large intestine. The ileal microbiota may undergo cyclic feast or famine conditions due to the role of phages in nutrient release (Abedon 2009), leading to an appropriate cost of phage resistance for ileal bacteria.

Viruses present in the intestine are called the intestinal virome. An average of 4.2 different mammalian viruses have been detected in the feces of healthy piglets and 5.4 in the feces of diarrheic piglets (Shan et al. 2011). Ninety-nine percent of the viral sequences are assigned to the RNA virus families *Picornaviridae*, *Astroviridae*, and *Coronaviridae*, while the other 1% belongs to DNA virus families *Circoviridae* and *Parvoviridae* (Shan et al. 2011). Furthermore, eight mammalian virus families (*Adenoviridae*, *Anelloviridae*, *Astroviridae*, *Caliciviridae*, *Circoviridae*, *Parvoviridae*, *Picornaviridae*, and *Reoviridae*) have been detected in the distal jejunum of healthy pigs compared to four in diarrhoeic pathogens (*Anelloviridae*, *Circoviridae*, *Picornaviridae*, and *Reoviridae*) (Karlsson et al. 2016). However, the role of these viromes is still unknown. Future studies involving analyses of the viromes in the intestine of pigs are necessary to discover new viruses that might be important in control of clinical enteric diseases and growth retardation.

9.3 Suckling Period as a Key Window for Microbial Colonization and Manipulation

9.3.1 Milk Glycans and Microbial Utilization

Milk is the priority nutrient for newborn piglets. It provides a set of bioactive substrates, such as oligosaccharides, glycoproteins (e.g., lactoferrin), and immunoglobulins. Free oligosaccharides and *N*-glycans are major sources of milk oligosaccharides in pigs. Investigations on microbial physiology discover that some microbes are capable of degrading the oligosaccharide substrate. For example, *Bacteroides vulgatus* has been found to exert *N*-linked deglycosylation activities in a medium containing human transferrin (Cao et al. 2014), and can grow in a minimal medium containing human milk oligosaccharides as the sole carbon source (Marcobal et al. 2011). *B. thetaiotaomicron* may consume highly mannosylated *N*-glycan GlcNAc2-Man9 using enzymes encoded by polysaccharide utilization loci (Cuskin et al. 2015). By analyzing the *N*-glycome profiles in sow milk and offspring microbiota succession, we identify the structures of 22 *N*-glycans in sow milk (Mu et al. 2019a). Fucosylated (8 out of 22, 36%) and sialylated (9 out of 22, 41%) *N*-glycans are the major forms followed by high mannosylated (3 out of 22, 14%). *N*-glycans such as fucosylated GlcNAc4-Man3-Fuc and sialylated GlcNAc4-Man3-Gal2-NeuAc increase with age. Many statistical correlations are identified, such as the positive correlation between mannosylated GlcNAc2-Man9 and a *Lactobacillus amylovorus*-related species (Mu et al. 2019a). The capacity to consume certain *N*-glycoproteins may be responsible for the different colonization trajectories in piglets. Although direct evidence of *N*-glycan utilization by piglet gut microbes is limited, the presence of species with oligosaccharide-degrading ability in the gut of newborn piglets implicates their potential role in oligosaccharide metabolism.

In analyzing the *N*-glycan composition, we note that breed also affects the sow milk *N*-glycan compositions (Mu et al. 2019b). To study if breed and maternal milk affect the gut environment, we have employed an interbreed piglet model through fostering neonatal Yorkshire and Meishan piglets to the same or another breed of sows. We find that piglets nursed by Meishan sows have a lower abundance of *Streptococcus suis* and a higher abundance of *Cloacibacillus* in the colonic digesta, and higher abundances of interleukin 10 and Foxp3-positive cells in the colonic mucosa than Yorkshire sow-nursed piglets before weaning. After weaning, the maternal effects decline and the effects of breed persist (Mu et al. 2019b). Therefore, the environment provided by the nursing mother is a key factor that affects preweaning colonic microbiota and immune status.

9.3.2 Milk-Related Substrates as Dietary Additives for Suckling Piglets

Considering the benefit of milk substrates in regulating gut health, different milk components and related substances have been supplemented to piglets, such as lactoferrin, galactooligosaccharides (GOS), and fructooligosaccharide (FOS). For example, lactoferrin supplementation to newborn piglets is efficient to reduce *Escherichia-Shigella*, increase butyrate concentrations in the colonic digesta, and upregulate the epithelial barrier function (Hu et al. 2020), thus leading to an improved intestinal function. Piglets given orally GOS daily during the first week after birth could have significantly higher abundances of *Lactobacillus* and unclassified *Lactobacillaceae*, and a lower abundance of *Clostridium* sensu stricto in the ileum on day 8 and 21. In addition, the oral administration of GOS to the suckling piglets increases the concentrations of propionate and butyrate in the ileal digesta on day 8 and of butyrate on day 21 (Tian et al. 2019). Early-life GOS supplementation also increases *Prevotella*, *Barnesiella*, and *Parabacteroides* and the concentrations of short-chain fatty acids in the colon digesta of suckling piglets (Wang et al. 2019a). FOS supplementation from postnatal days 2–14 has bifidogenic effects by increasing the abundances of *Lactobacillus* and *Bifidobacterium* in the colonic digesta at postnatal day 14 but less effects on the colonic gene expression at day 25 (Schokker et al. 2018). Interestingly, the effects on the epithelium are more pronounced in the jejunum by increasing the villi height and crypt depth and downregulating the gene expressions involved in immune-related processes (Schokker et al. 2018). All of these interventions tend to foster a healthy gut microbiome in suckling piglets, which provides references for feeding practice.

9.3.3 Fiber Inclusion for Suckling Piglets

It is generally considered that the gut of suckling piglets has a limited ability to degrade fibers. In a trial to investigate fiber inclusion in suckling piglets, alfalfa supplementation (1.3%) resulted in favorable alterations in the gut microbiota composition compared to pure cellulose and wheat bran, as reflected by the lowest abundance of the potential pathogen *Streptococcus suis* in the cecum and distal colon (Zhang et al. 2016a, b). Further studies of microbial functionality demonstrate that dietary alfalfa supplementation could increase the abundance and activity of butyrate-producing bacteria (*Clostridium* cluster XIVa) in the proximal colon and enhance the gene expression of butyryl-CoA: acetate CoA-transferase and butyrate production compared with piglets supplemented with wheat bran (Mu et al. 2017b). Since butyrate is known to be anti-inflammatory and protective in the gut, the alterations induced by a moderate supplementation of alfalfa may provide gut benefits via increased delivery of butyrate to the mucosa. A recent study also supports the usage of fiber inclusion in suckling piglets. Supplementation of a largely non-fermentable purified cellulose from day 2 of age could increase the concentration of total volatile fatty acids in the cecum and mid-colon, and decrease the abundance of *Escherichia-Shigella* in the mid-colon compared to the supplementation of a fermentable long-chain arabinoxylan and a low-fiber control diet (Van Hees et al. 2019). With the increasing evidence on the beneficial effects of fiber, it is promising to further use adequate inclusion to foster gut health in suckling piglets.

9.3.4 Microbiota Manipulation Affects Host Health During Suckling Period

Early exposure to antibiotics could disturb the normal microbial colonization and gut health, as reviewed by Mu and Zhu (2019). An antibiotic mixture administration in suckling piglets decreases the microbial diversity and the abundance of *Lactobacillus* and increases the abundance of *Streptococcus*, unclassified *Enterococcaceae* in the ileum of piglets at weaning (Yu et al. 2018). In the cecum, metabolic byproducts from microbial carbohydrate fermentation decrease while those from protein fermentation increase (Yu et al. 2018), indicating an imbalanced utilization of nutrients in the gut. Microbiome intervention at newborn period has long-lasting effects on fecal microbiome (Yu et al. 2017a), epithelial amino acid transporter expression (Yu et al. 2020), and immune response (Fouhse et al. 2019). Newborn piglets receiving amoxicillin from postnatal days 0–14 show a transient change in gut microbiota by increasing *Enterobacteriaceae* species, but persistent alterations in circulating immune response, as reflected by the higher percentage of $CD3^{+}CD4^{+}$ T cells and a pronounced inflammatory response to pathogen challenges (Fouhse et al. 2019). Given these effects observed, early usage of antibiotics should be avoided to introduce unnecessary insults to the newborns.

Fecal microbiota transplantation (FMT) is an alternative approach to restore microbiome balance in suckling piglets. Oral administration of sow fecal suspension to newborn piglets from day 1–3 of age could alter the gut microbiota and metabolic phenotype such as reducing the incidence of diarrhea and endotoxin levels, increasing quantities of *Lactobacillus* and *Faecalibacterium prausnitzii*, and elevating plasma immunoglobulin G and fecal sIgA on day 21 of age (Cheng et al. 2019). FMT also protects newborn piglets from lipopolysaccharide-induced damage of epithelial integrity (Geng et al. 2018), dextran sulphate sodium-induced colitis (Xiao et al. 2017), and the incidence of necrotizing enterocolitis (Brunse et al. 2019), suggesting a great potential of FMT in treating gut disorders at early life.

9.4 Post-weaning Microbiome: A Functionally Diverse Community

Relative to the microbiome in suckling piglet, the post-weaning microbiome is more complex and gradually develops into an adult-like microbial consortium. The microbial functionalities also expand correspondingly, such as amino acid metabolism, oligosaccharide/fiber degradation, and bile acid metabolism, as shown in Fig. 9.1.

9.5 Amino Acid Metabolism

9.5.1 *Microbiome Affects Amino Acid Utilization* in Vitro *and* in Vivo

Both plant- and animal-sourced feedstuffs in diets provide amino acids and small peptides for intestinal microbes (Hou et al. 2019; Li and Wu 2020; Li et al. 2021; Wu 2020). Extensive studies have proved that the gut microbiota is actively involved in amino acid metabolism. Nearly 30–60% of dietary essential amino acids were disappeared in the first-pass intestinal metabolism by gut epithelial cells in pigs (Wu et al. 2014). However, intestinal epithelial cells have limited ability to metabolize amino acids. Therefore, the substantial utilization of essential amino acids in the first-pass intestinal metabolism might be mainly due to the gut bacteria. Employing a well-established approach of anaerobic cultures, we have found several rationales underlying the amino acid metabolism by gut microbes: (1) luminal bacteria from pig small intestine have the ability to metabolize essential amino acids, but the ability varies depending on gut location and bacteria species (Dai et al. 2010, 2012); (2) luminal bacteria and gut wall-associated bacteria differ in the amino acid utilization ability (Yang et al. 2014); (3) bacteria in the small and large intestine differ in the amino acid metabolism (Ma et al. 2016). For example, serine and cysteine are highly incorporated into the bacteria in the small intestine than the colon (Dai et al. 2011). Enterocytes and gut bacteria have different tasks in amino acid metabolism. Enterocytes can degrade branched-chain amino acids to a high degree by encoding branched-chain α-ketoacid dehydrogenase, while the oxidation of lysine and aromatic amino acids is limited (Wu et al. 2014). Interestingly, the metabolism by gut microbiota may complement the metabolism. For example, the microbes from the small intestine can degrade lysine to a high

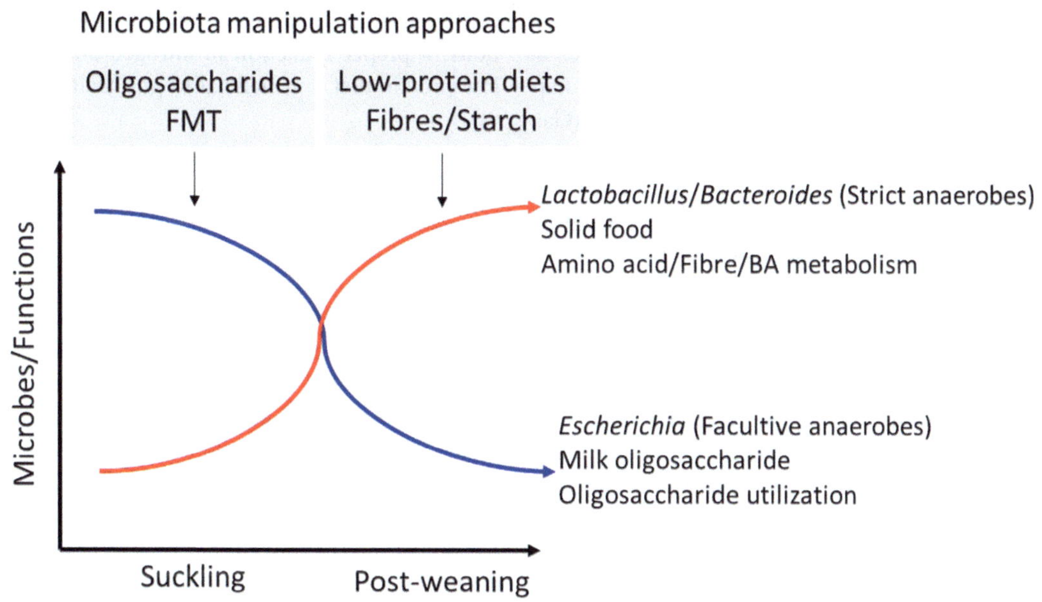

Fig. 9.1 Overview of the compositional and functional difference of gut microbiota at suckling and post-weaning periods. Several approaches aiming to manipulate the gut microbiota have been developed, depending on age and health condition. BA, bile acid; FMT, fecal microbiome transplantation

degree. During the subculture of gut microbes, the disappearance rate of lysine in both luminal and mucosal microbiota as inocula is around 90% in 24 h, mostly due to the catabolism of gut microbes (Yang et al. 2014). Meanwhile, the small intestinal microbes have a limited ability to degrade branched-chain amino acids, with the disappearance rate less than 10% or negligible in piglets (Yang et al. 2014). Given such considerations, the enterocytes and gut bacteria seem to have evolved a mutual relationship in utilizing amino acids without interfering with one another.

Direct evidence for microbial effects on amino acid metabolism has been observed in vivo. In growing piglets receiving an antibiotic cocktail, the amount of bacteria is decreased in the small intestine, characterized by the decrease of *Clostridium*, *Bacillus*, and *Sharpea* in the digesta of the stomach, duodenum, and jejunum (Mu et al. 2017a). Interestingly, the concentrations of most amino acids decrease in the digesta of jejunum and ileum, while the concentration of amino acids, including lysine, phenylalanine, valine, and aspartate, increase in the serum (Mu et al. 2017c), which is tightly related to a high expression of amino acid transporters in the epithelium of jejunum and ileum (Yu et al. 2017b). Antibiotic treatment for 2 weeks also increases the terminal ileum apparent digestibility of crude protein, phenylalanine, valine, alanine, and tyrosine while decreases *Bifidobacterium* and *Lactobacillus* quantities in the ileum digesta and feces, consequently leading to an increased total nitrogen excretion in piglets (Pi et al. 2019). These evidences provide reference to the involvement of the gut microbiome in regulating amino acid partition.

9.5.2 Gut Microbiome and Amino Acid Nutrition

Excess dietary proteins can introduce detrimental effects on gut health. In adult rats fed with high-protein diet at 45% protein level, the opportunistic bacteria *Escherichia*, microbial metabolites (sulfide, amines), and epithelial expressions of pro-inflammatory cytokines increase in the colon (Mu et al. 2016a, b), while the carbohydrate-fermenting gut microbiota is significantly declined (Mu et al. 2017d), which exposes the colon to a high risk of disease. In piglets, high-protein diet further increases diarrhea ratios and disturbs the gut microbiome (Gao et al. 2020a, b). Considering the increasing cost of protein source and the demand for sustainable industry development, many studies have investigated the use of the low-protein diet. The rationale is to reduce the partition of protein to microbial fermentation in the large intestine and retain a healthy gut environment.

Low-protein diets have diverse benefits via reducing post-weaning diarrhea and improving intestinal morphology, microbiota, and immune responses (Wang et al. 2018). A 6% decrease in crude protein level from 20 to 14% impairs the growth performance and increases the feed-to-gain ratio, together with the reduction of *Firmicutes* and *Clostridium* cluster IV species in the cecum of growing piglets (Luo et al. 2015), suggesting the over-reduction is inadequate for the growing piglets. By a stepwise reduction in dietary crude protein level for growing pigs, we find that reducing from 20 to 15.3% could retain the growth performance and increase the concentrations of short-chain fatty acids and decrease those from microbial protein fermentation; however, when reducing to 13.9% crude protein level, the growth performance is compromised and the amount of *Escherichia coli* increases in the colon (Peng et al. 2017). This study provides an important reference for the threshold of the dietary protein level that can be reduced in the growing pigs.

Why does an adequate decrease in dietary nitrogen supply confers favorable effects on the gut? Intestinal microbes mainly use dietary nitrogen and host-secreted nitrogen as major sources. An ecological evidence proves that in the large intestine, the host epithelium has a high ability to absorb dietary nitrogen, leading to a low-nitrogen microenvironment. Specific microbes such as *Bacteroidales* can readily consume host-secreted nitrogen and affect nitrogen metabolism (Reese et al. 2018). The nitrogen limitation status spans across the gut of 30 mammal species, such as mice, sheep, and elephants (Reese et al. 2018). Probably due to the evolutionary adaptation to low-nitrogen conditions, a proper reduction in nitrogen supply by low-protein diet is beneficial to gut health.

9.5.3 Microbial Metabolism of Amino Acids: Effects Beyond Gut

In addition to the effects of microbial amino acid metabolism in the gut, we further uncover a mechanism that how the microbiota in the large intestine regulates amino acids and neurotransmitters in the central nervous system, connected by the term “microbiota-gut-brain axis” (Mu et al. 2016b; Gao et al. 2020b). In a piglet model, ileum antibiotic infusion increases the relative abundance of *Streptococcus*, *Lactobacillus* and decreases those of *Ruminococcus*, *Clostridium*, *Christensenella*, *Methanobrevibacter,* and *Prevotella* in the feces. Meanwhile, the concentrations of tryptophan decrease in feces, blood, and hypothalamus, in parallel with the decrease in the neurotransmitters serotonin and dopamine in the hypothalamus (Gao et al. 2018), suggesting the linkage between gut microbiota and brain through amino acids. To further study the mechanism behind, we employ a cecal-cannulated piglet model and infuse starch to change the hindgut microbiota, considering the fact that starch supplementation reduces microbial protein fermentation (He et al. 2017) and affects the microbial metabolism of amino acids (Sun et al. 2016). Interestingly, the starch infusion decreases the relative abundances of *Lactobacillus* and *Streptococcus* and increases the concentrations of aromatic amino acids, including tryptophan, tyrosine, and phenylalanine in the serum and hypothalamus. Correspondingly, the concentrations of serotonin, dopamine, and brain-derived neurotrophic factor also increase in the hypothalamus (Gao et al. 2019). Employing mice models and neural cell cultures, we further found that tryptophan and tyrosine stimulated serotonin and dopamine production, as well as the generation of brain-derived neurotrophic factors via the 5-serotonin 1A receptor/D1 dopamine receptor-cyclic adenosine monophosphate response element-binding protein signaling (Gao et al.

2019). The role of the gut microbiome in gut–brain interaction further suggests the application of using amino acids to regulate brain function.

9.6 Utilization of Oligosaccharides

An increase in oligosaccharide metabolism capability is a keystone shift in microbial functions. During suckling-to-weaning transition, the diets shift from liquid milk to solid food that contains plant oligosaccharides. Correspondingly, the function of gut microbiota also shifts with this change. Compared with the microbial functionalities in the suckling period, the weaned piglets have an increased capacity to metabolize plant-derived oligosaccharides and simple sugars, such as fructooligosaccharides, xylose, mannose, L-rhamnose, and maltodextrin, but a decrease in the lactose and galactose uptake by the fecal microbiota (Guevarra et al. 2019). Together with the change of microbial functions, the gut can gradually adapt to a solid diet.

Oligosaccharides are compounds containing three to nine monomeric sugar residues, and many of them are prebiotics. Commercially available prebiotics, GOS and FOS, are abundant in some foods and are mainly consumed by species of *Lactobacillus* and *Bifidobacterium* (Rastall and Gibson 2015). Selective stimulation of bifidobacteria and lactobacilli by these nondigestible oligosaccharides has been well documented both in vitro and in vivo (Moro et al. 2002; Davis et al. 2011).

GOS can stimulate the growth of both *Lactobacillus amylovorus* and *Bifidobacterium animalis* in fecal material from adult female pigs in vitro (Martinez et al. 2013). FOS can be fermented by both *Bifidobacterium* and *Lactobacillus*, which can be reflected by the growth of these bacteria in the gut of growing pigs after dietary FOS supplementation (Xu et al. 2002). FOS utilization appears to occur via one of the two catabolic pathways: (a) The substrate is transported and hydrolyzed by a cytoplasmic Glycoside Hydrolase Family 32 (GH32) β-FFase, or (b) extracellular hydrolysis of the substrate is catalyzed by a cell surface–associated GH32 β-FFase, followed by the uptake of the hydrolytic products (i.e., fructose, sucrose, and glucose) via transporters. The majority of FOS-utilizing *Lactobacillus* and *Bifidobacterium* species possesses transporters and intracellular β-FFase for the catabolism of mainly FOS substrates. The ability of bifidobacteria to ferment FOS, specifically shorter-chain oligofructose, is a universal metabolic feature (Rossi et al. 2005). The general assumption is that bifidobacteria degrade long-chain fructans such as inulin because of their diverse sugar metabolic gene repertoire and specialized niche in the colon. Unexpectedly, most *Bifidobacterium* species grow poorly on inulin as a carbon source, and extracellular enzymes with specificities for long-chain fructans (DP $> \sim 8$) are rarely detected among bifidobacteria (Rossi et al. 2005), suggesting their preference for short-chain FOS substrates. Overall, the genetic mechanisms and regulation of FOS utilization in the genera are less well defined, particularly in terms of the transport systems responsible for the uptake of these oligomers.

Xylo-oligosaccharides (XOS) has also been widely reported to promote *Bifidobacterium* proliferation and improve host immunity in pigs (Yin et al. 2019). Dietary XOS supplementation during the growing and fattening periods (GFP) (30–100 kg BW) of pigs significantly reduced the relative abundances of *Proteobacteria* and *Citrobacter*, and enhanced the relative abundances of *Lactobacillus* (Pan et al. 2019). Meanwhile, the XOS supplementation during the GFP increased acetic acid, straight-chain fatty acids, and total SCFA concentrations in the intestinal digesta (Pan et al. 2019). Administration of XOS also decreased the abundance of the fecal *Escherichia coli*, while increasing the abundance of lactobacilli on day 14 of weanling pigs (Liu et al. 2018). In summary, the oligosaccharides GOS, FOS, and XOS can provide beneficial effects on gut microbiota and be used as growth-promoting additives.

9.7 Fibers

9.7.1 Fiber Fermentation by Gut Microbiome

Dietary fiber (DF) is a broad term, and the impact of fiber consumption on the gastrointestinal microbiota varies based on the type of fiber consumed. Broadly, DF includes plant cell wall compounds such as cellulose, hemicelluloses, mixed linked β-glucan, pectins, and mucilages. Lignin, a complex phenolic compound, is also included in DF because it is a constituent of the plant cell walls that can greatly affect the digestibility of plant-derived foods (Jha and Berrocoso 2015). From a physiological point of view, non-starch polysaccharides (NSP) and nondigestible oligosaccharides are grouped in the soluble DF fraction because they are not hydrolyzed by endogenous enzymes, and consequently, become available as substrates for microbial fermentation in the large intestine (Cummings and Stephen 2007). DF escapes enzymatic digestion in the small intestine and becomes available for fermentation by bacteria in the colon. DF fermentation in the hindgut results in the production of SCFAs (including acetate, propionate, and butyrate), along with some gases (such as hydrogen, carbon dioxide, and methane), all of which regulate the intestinal environment (Koh et al. 2016). The susceptibility of DF to microbial fermentation varies depending on the accessibility of DF to the microbial population in the hindgut. In pigs, the large intestine is the most important site of fermentation (Williams et al. 2001). Fermentation of soluble DF occurs mainly at the proximal colon, whereas fermentation of insoluble DF is sustained at the distal colon. However, fermentation of soluble DF has also been observed in the pig's small intestine (Jha et al. 2010; Jha and Leterme 2012), although the contribution to the overall fiber fermentation is limited.

The microbial conversions of dietary fiber to monosaccharides in the gut involve a number of principal events mediated by the enzymatic repertoire of specific members of the gut microbiota. Fermentation of DF is more variable than the digestion of the macronutrients such as starch, fat, and CP (generally above 80.0%). The variation in fermentability is mainly due to changes in physico-chemical properties of DF such as bulk, viscosity, solubility, and fermentation degree. For example, the consumption of apple pectin with high viscosity alters microbiota composition by increasing the relative abundance of *Megasphaera elsdenii* and *Anaerovibrio* in the pig colon, thereby exerting beneficial impacts on gut health (Xu et al. 2019). Thus, the property of fibers should be considered when including in diet.

9.7.2 Gut Microbiome Relates to Fiber Digestibility

The role of different microbes in degrading fibers has been studied in vivo. Diet is one of the most important factors in shaping the gut microbiota relative to age and gender in pigs (Wang et al. 2019b). Among the dietary components, fiber ranks first in explaining the microbiome variation (Wang et al. 2019b). It is now clear that the apparent digestibility of crude fibers increases with age in pigs (Niu et al. 2015). What is more interesting is that the relative abundances of some genera (including Anaeroplasma and Campylobacter) showed a positive correlation with the apparent digestibility of crude fibers (Niu et al. 2015). Further studies are needed to understand how these microbes affect the apparent digestibility of dietary fiber, which can provide more insights into the underlying mechanisms responsible for its health effects.

9.8 Bile Acid Metabolism

9.8.1 Bile Acid Pool

Bile acids (BAs) are saturated and hydroxylated C24 cyclopentanepheznanthrene sterols, and metabolized mainly in the liver, linking the gut-liver axis. Primary BAs are synthesized from cholesterol in the liver and conjugated with either taurine or glycine via an amide linkage at the C24 carboxyl (Wu 2018). They are then secreted to the biliary system through the canaliculi (Hou et al. 2020). More than 95% of the BAs secreted in bile are reabsorbed in the distal ileum and return to the liver (Zwicker and Agellon 2013). This process is known as enterohepatic circulation and four to twelve cycles occur per day. The BAs that escape the enterohepatic circulation enter the colon where they are used for bacterial metabolism. Gut microbiota also plays an important role in regulating BAs homeostasis (Mu and Zhu 2019).

The main bile salt conversions in the gut include deconjugation, oxidation, epimerization, esterification, and desulfation, resulting in the formation of over 20 different secondary BAs in the gut (Gerard 2013). In humans, cholic acid (CA) and chenodeoxycholic acids (CDCA) are the two primary BAs, whereas in pigs, CA, CDCA, and hyocholic acid (HCA) are the main primary BAs (Eggink et al. 2018). In the feces of piglets, the proportion of primary BAs is about 60%, whereas the proportion of secondary BAs is about 40% (Fig. 2a, unpublished data). Among the primary BAs, HCA, CDCA, and CA are the main BAs, accounting for about 84%, 11%, and 4%, respectively (Fig. 2b). Among the secondary BAs, hyodeoxycholic acid (HDCA), ursodeoxycholic acid (UDCA), 3-dehydrocholic acid (3-DHCA); dehydro-lithocholic acid (Dehydro-LCA), ursocholic acid (UCA), 3β-ursodeoxycholic acid (β-UDCA), β-muricholic acid (β-MCA) are the main BAs, accounting for about 26%, 21%, 14%, 10%, 7%, 5%, and 4%, respectively (Fig. 2c).

9.8.2 Bile Salt Hydrolases in Gut Microbiome

The hydrolysis of the C24 *N*-acyl amide bond of conjugated BAs is catalyzed by bile salt hydrolases (BSHs). Most BSHs hydrolyze both glycine- and taurine-conjugated BAs, whereas a few display strong specificity. BSH genes have been detected in the predominant bacterial genera of the gut microbiota (Jones et al. 2008) and the enzyme has been purified from *Bacteroides fragilis*, *B. vulgatus*, and several species of *Lactobacillus* and *Bifidobacterium*. In addition, *Lactobacillus reuteri* isolated from feces of pigs exerts BSH activities (Rodriguez et al. 2003). A living *Bifidobacterium animalis* DN-173 010 also has BSH activities in the gut of pigs, probably in the small intestine (Lepercq et al. 2004), which may contribute to its probiotic property.

9.8.3 Bile Acid Dihydroxylation by Gut Microbiome and Health Relevance

7α-dehydroxylation is the process of metabolizing primary BAs (CA and CDCA) into DCA and LCA, which is the most quantitatively important and the physiologically significant conversion of BAs (Hamilton et al. 2007). DCA accounts for up to 25% of the total BA pool. The known bacterial species possessing 7α-dehydroxylation activity are taxa within the *Firmicutes* phylum (such as *Clostridium*, *Eubacterium*). The BA-inducible (bai) enzyme system which dehydroxylates 7α-hydroxy BAs has been extensively studied in the human intestinal isolate *Clostridium scindens* and *C. hylemonae* (Ridlon et al. 2006). Recently, dehydroxylation metabolism of bile acid by gut microbiota is found to mediate the diet-induced change in the intestinal epithelial barrier function in pigs. By infusing corn starch or casein hydrolysate to the cecum of piglets, we find that corn starch increases carbohydrate/nitrogenous compound ratio in the colonic digesta and decreases the abundance of bacteria capable of bile acid 7α-dehydroxylation (baiJ), baiJ expression, and secondary bile acids including deoxycholic acid and lithocholic acid, all of which show the opposite direction of changes after casein hydrolysate infusion (Pi et al. 2020). Further studies use Caco-2 cell cultures demonstrate that deoxycholic acid and lithocholic acid serve as a major driver of the compromised barrier function by reducing the

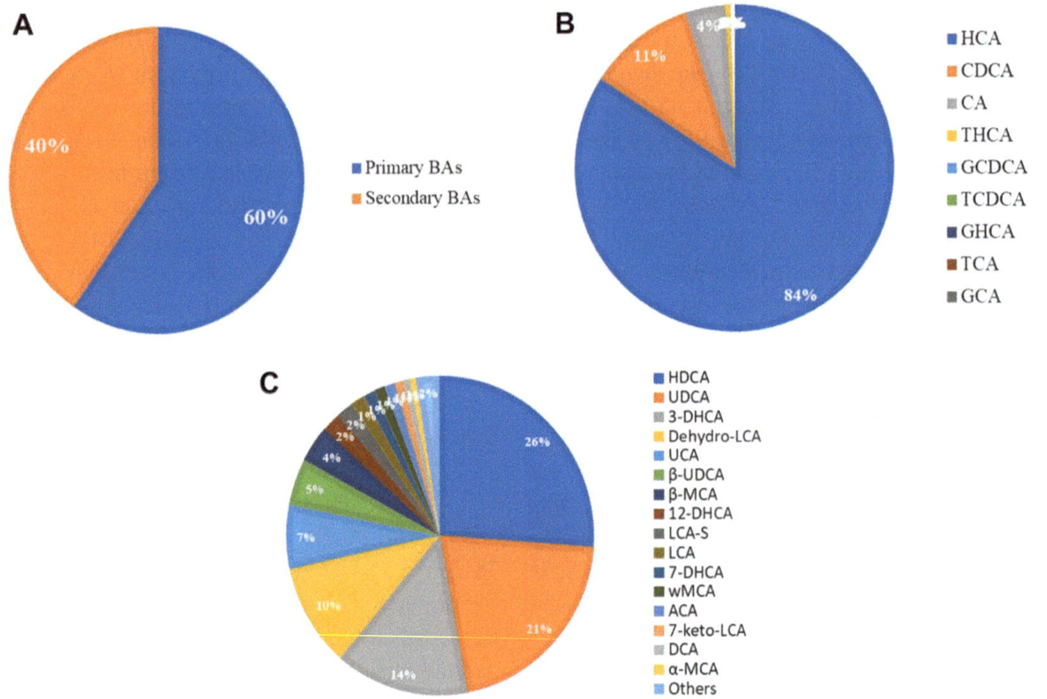

Fig. 9.2 Bile acid composition in the feces of piglets. **a** The proportion of primary and secondary bile acids in total bile acids. **b** The proportion of individual bile acids in primary bile acids. **c** The proportion of individual bile acid in secondary bile acids

expression of tight junction proteins via epidermal growth factor receptor signaling (Pi et al. 2020). These evidences indicate that secondary bile acid metabolism by gut microbiota probably mediates the interplay between diet and gut barrier function. Considering the important physiological relevance of microbial bile acid metabolism, more investigations are needed to define the role of the gut microbiome in mediating these processes.

9.9 Conclusion

As discussed above, the gut microbiome develops different functionalities in an age-dependent manner that is tightly related to diet, environment, breed, and other factors. Although the newborn piglets tend to have a simple microbiome, there is a great potential to manipulate the early colonizers towards a healthy-promoting direction, such as using prebiotics and probiotic bacteria to reshape the microbiome. After weaning, the gut microbiome develops into a functionally abundant consortium that contributes to amino acid partition, fiber degradation, and bile acid metabolism. Based on these rationales, a diverse of nutritional approaches have been developed to foster a favorable microbiome. Since the gut microbiome affects intestinal homeostasis and host health in pigs, it is promising to use microbiome manipulation to promote animal wellness and health in the future.

Acknowledgements This work was funded by the Natural Science Foundation of China (32030104, 31902166, 31430082) and National Key Basic Research Program of China (2013CB127300).

References

Abedon ST (2009) Phage evolution and ecology. Adv Appl Microbiol 67:1–45

Allen HK, Looft T, Bayles DO, Humphrey S, Levine UY, Alt D, Stanton TB (2011) Antibiotics in feed induce prophages in swine fecal microbiomes. mBio 2: e00260–11

Bian G, Ma S, Zhu Z et al (2016) Age, introduction of solid feed and weaning are more important determinants of gut bacterial succession in piglets than breed and nursing mother as revealed by a reciprocal cross-fostering model. Environ Microbiol 18:1566–1577

Boudry G, Lallès JP, Malbert CH, Bobillier E, Sève B (2002) Diet-related adaptation of the small intestine at weaning in pigs is functional rather than structural. J Pediatr Gastroenterol Nutr 34:180–187

Brunse A, Martin L, Rasmussen TS et al (2019) Effect of fecal microbiota transplantation route of administration on gut colonization and host response in preterm pigs. ISME J 13:720–733

Cao YL, Rocha ER, Smith CJ (2014) Efficient utilization of complex *N*-linked glycans is a selective advantage for Bacteroides fragilis in extraintestinal infections. Proc Natl Acad Sci USA 111:12901–12906

Cheng CS, Wei HK, Wang P, Yu HC, Zhang XM, Jiang SW, Peng J (2019) Early intervention with faecal microbiota transplantation: an effective means to improve growth performance and the intestinal development of suckling piglets. Animal 13:533–541

Cummings JH, Stephen AM (2007) Carbohydrate terminology and classification. Eur J Clin Nutr 61:S5–S18

Cuskin F, Lowe EC, Temple MJ et al (2015) Human gut *Bacteroidetes* can utilize yeast mannan through a selfish mechanism. Nature 517:165-U186

Dai Z, Wu G, Zhu W (2011) Amino acid metabolism in intestinal bacteria: links between gut ecology and host health. Front Biosci 16:1768–1786

Dai Z, Zhang J, Wu G, Zhu W (2010) Utilization of amino acids by bacteria from the pig small intestine. Amino Acids 39:1201–1215

Dai Z, Li XL, Xi PB, Zhang J, Wu G, Zhu W (2012) Metabolism of select amino acids in bacteria from the pig small intestine. Amino Acids 42:1597–1608

Davis LMG, Martinez I, Walter J, Goin C, Hutkins RW (2011) Barcoded pyrosequencing reveals that consumption of galactooligosaccharides results in a highly specific bifidogenic response in humans. PloS one 6:e25200

Eggink HM, van Nierop ES, Schooneman MG et al (2018) Transhepatic bile acid kinetics in pigs and humans. Clin Nutr 37:1406–1414

Fouhse JM, Yang K, More-Bayona J, Gao Y, Goruk S, Plastow G, Field CJ, Barreda DR, Willing BP (2019) Neonatal exposure to amoxicillin alters long-term immune response despite transient effects on gut-microbiota in piglets. Front Immunol 10:2059

Gao J, Yin J, Xu K, Han H, Liu Z, Wang C, Li T, Yin Y (2020a) Protein level and infantile diarrhea in a postweaning piglet model. Mediators Inflamm 2020:1937387

Gao K, Mu C, Farzi A, Zhu W (2020b) Tryptophan metabolism: a link between the gut microbiota and brain. Adv Nutr 11:709–723

Gao K, Pi Y, Mu C, Peng Y, Huang Z, Zhu W (2018) Antibiotics-induced modulation of large intestinal microbiota altered aromatic amino acid profile and expression of neurotransmitters in the hypothalamus of piglets. J Neurochem 146:219–234

Gao K, Pi Y, Mu C, Farzi A, Liu Z, Zhu W (2019) Increasing carbohydrate availability in the hindgut promotes hypothalamic neurotransmitter synthesis: aromatic amino acids linking the microbiota-brain axis. J Neurochem 149:641–659

Geng S, Cheng S, Li Y, Wen Z, Ma X, Jiang X, Wang Y, Han X (2018) Faecal microbiota transplantation reduces susceptibility to epithelial injury and modulates tryptophan metabolism of the microbial community in a piglet model. J Crohns Colitis 12:1359–1374

Gerard P (2013) Metabolism of cholesterol and bile acids by the gut microbiota. Pathogens 3:14–24

Guevarra RB, Lee JH, Lee SH, Seok MJ, Kim DW, Kang BN, Johnson TJ, Isaacson RE, Kim HB (2019) Piglet gut microbial shifts early in life: causes and effects. J Anim Sci Biotechnol 10:1

Hamilton JP, Xie GF, Raufman JP, Hogan S, Griffin TL, Packard CA, Chatfield DA, Hagey LR, Steinbach JH, Hofmann AF (2007) Human cecal bile acids: concentration and spectrum. Am J Physiol 293:G256–G263

He X, Sun W, Ge T, Mu C, Zhu W (2017) An increase in corn resistant starch decreases protein fermentation and modulates gut microbiota during in vitro cultivation of pig large intestinal inocula. Anim Nutr 3:219–224

Hou YQ, He WL, Hu SD, Wu G (2019) Composition of polyamines and amino acids in plant-source foods for human consumption. Amino Acids 51:1153–1165

Hou YQ, Hu SD, Li XY, He WL, Wu G (2020) Amino acid metabolism in the liver: nutritional and physiological significance. Adv Exp Med Biol 1265:21–37

Hu P, Zhao F, Wang J, Zhu W (2020) Early-life lactoferrin intervention modulates the colonic microbiota, colonic microbial metabolites and intestinal function in suckling piglets. Appl Microbiol Biotechnol 104:6185–6197

Ivarsson E, Liu HY, Dicksved J, Roos S, Lindberg JE (2012) Impact of chicory inclusion in a cereal-based diet on digestibility, organ size and faecal microbiota in growing pigs. Animal 6:1077–1085

Jha R, Leterme P (2012) Feed ingredients differing in fermentable fibre and indigestible protein content affect fermentation metabolites and faecal nitrogen excretion in growing pigs. Animal 6:603–611

Jha R, Berrocoso JD (2015) Dietary fiber utilization and its effects on physiological functions and gut health of swine. Animal 9:1441–1452

Jha R, Rossnagel B, Pieper R, Van Kessel A, Leterme P (2010) Barley and oat cultivars with diverse carbohydrate composition alter ileal and total tract nutrient

digestibility and fermentation metabolites in weaned piglets. Animal 4:724–731
Jones BV, Begley M, Hill C, Gahan CGM, Marchesi JR (2008) Functional and comparative metagenomic analysis of bile salt hydrolase activity in the human gut microbiome. Proc Natl Acad Sci USA 105:13580–13585
Karlsson OE, Larsson J, Hayer J, Berg M, Jacobson M (2016) The intestinal eukaryotic virome in healthy and diarrhoeic neonatal piglets. PloS One 11:e0151481
Kim HB, Borewicz K, White BA, Singer RS, Sreevatsan S, Tu ZJ, Isaacson RE (2011) Longitudinal investigation of the age-related bacterial diversity in the feces of commercial pigs. Vet Microbiol 153:124–133
Koh A, De Vadder F, Kovatcheva-Datchary P, Backhed F (2016) From dietary fiber to host physiology: short-chain fatty acids as key bacterial metabolites. Cell 165:1332–1345
Labrie SJ, Samson JE, Moineau S (2010) Bacteriophage resistance mechanisms. Nat Rev Microbiol 8:317–327
Lepercq P, Relano P, Cayuela C, Juste C (2004) Bifidobacterium animalis strain DN-173 010 hydrolyses bile salts in the gastrointestinal tract of pigs. Scand J Gastroenterol 39:1266–1271
Li P, Wu G (2020) Composition of amino acids and related nitrogenous nutrients in feedstuffs for animal diets. Amino Acids 52:523–542
Li P, He WL, Wu G (2021) Composition of amino acids in foodstuffs for humans and animals. Adv Exp Med Biol 1332:189–209
Liu JB, Cao SC, Liu J, Xie YN, Zhang HF (2018) Effect of probiotics and xylo-oligosaccharide supplementation on nutrient digestibility, intestinal health and noxious gas emission in weanling pigs. Asian-Australas J Anim Sci 31:1660–1669
Looft T, Allen HK, Cantarel BL, Levine UY, Bayles DO, Alt DP, Henrissat B, Stanton TB (2014) Bacteria, phages and pigs: the effects of in-feed antibiotics on the microbiome at different gut locations. ISME J 8:1566–1576
Luo Z, Li C, Cheng Y, Hang S, Zhu W (2015) Effects of low dietary protein on the metabolites and microbial communities in the caecal digesta of piglets. Arch Anim Nutr 69:212–226
Ma M, He X, Zhu W (2016) Metabolic pattern of pig hindgut bacteria on aromatic amino acids by an in vitro fermentation method. Acta Microbiol Sin 56:1786–1793
Mach N, Berri M, Estelle J et al (2015) Early-life establishment of the swine gut microbiome and impact on host phenotypes. Environ Microbiol Rep 7:554–569
Marcobal A, Barboza M, Sonnenburg ED et al (2011) Bacteroides in the infant gut consume milk oligosaccharides via mucus-utilization pathways. Cell Host Microbe 10:507–514
Martinez RC, Cardarelli HR, Borst W et al (2013) Effect of galactooligosaccharides and *Bifidobacterium animalis* Bb-12 on growth of *Lactobacillus amylovorus* DSM 16698, microbial community structure, and metabolite production in an in vitro colonic model set up with human or pig microbiota. FEMS Microbiol Ecol 84:110–123
Moro G, Minoli I, Mosca M, Fanaro S, Jelinek J, Stahl B, Boehm G (2002) Dosage-related bifidogenic effects of galacto- and fructooligosaccharides in formula-fed term infants. J Pediatr Gastroenterol Nutr 34:291–295
Mu C, Zhu W (2019) Antibiotic effects on gut microbiota, metabolism, and beyond. Appl Microbiol Biotechnol 103:9277–9285
Mu C, Yang Y, Zhu W (2015) Crosstalk between the immune receptors and gut microbiota. Curr Protein Pept Sc 16:622–631
Mu C, Yang Y, Zhu W (2016a) Gut microbiota: the brain peacekeeper. Front Microbiol 7:345
Mu C, Yang Y, Luo Z, Guan L, Zhu W (2016b) The colonic microbiome and epithelial transcriptome are altered in rats fed a high-protein diet compared with a normal-protein diet. J Nutr 146:474–483
Mu C, Bian G, Su Y, Zhu W (2019a) Differential effects of breed and nursing on early-life colonic microbiota and immune status as revealed in a cross-fostering piglet model. Appl Environ Microbiol 85:e02510-e2518
Mu C, Cai Z, Bian G, Du Y, Ma S, Su Y, Liu L, Voglmeir J, Huang R, Zhu W (2019b) New insights into porcine milk *N*-glycome and the potential relation with offspring gut microbiome. J Proteome Res 18:1114–1124
Mu C, Yang Y, Su Y, Zoetendal EG, Zhu W (2017a) Differences in microbiota membership along the gastrointestinal tract of piglets and their differential alterations following an early-life antibiotic intervention. Front Microbiol 8:797
Mu C, Zhang L, He X, Smidt H, Zhu W (2017b) Dietary fibres modulate the composition and activity of butyrate-producing bacteria in the large intestine of suckling piglets. Antonie Van Leeuwenhoek 110:687–696
Mu C, Yang Y, Yu K, Yu M, Zhang C, Su Y, Zhu W (2017c) Alteration of metabolomic markers of amino-acid metabolism in piglets with in-feed antibiotics. Amino Acids 49:771–781
Mu C, Yang Y, Luo Z, Zhu W (2017d) Temporal microbiota changes of high-protein diet intake in a rat model. Anaerobe 47:218–225
Niu Q, Li PH, Hao SS et al (2015) Dynamic distribution of the gut microbiota and the relationship with apparent crude fiber digestibility and growth stages in pigs. Sci Rep 5:9938
Pan J, Yin J, Zhang K, Xie PF, Ding H, Huang XG, Blachier F, Kong XF (2019) Dietary xylo-oligosaccharide supplementation alters gut microbial composition and activity in pigs according to age and dose. AMB Express 9:134
Peng Y, Yu K, Mu C, Hang S, Che L, Zhu W (2017) Progressive response of large intestinal bacterial community and fermentation to the stepwise decrease

of dietary crude protein level in growing pigs. Appl Microbiol Biotechnol 101:5415–5426

Petri D, Hill J, Van Kessel A (2010) Microbial succession in the gastrointestinal tract (GIT) of the preweaned pig. Livest Sci 133:107–109

Pi Y, Gao K, Peng Y, Mu C, Zhu W (2019) Antibiotic-induced alterations of the gut microbiota and microbial fermentation in protein parallel the changes in host nitrogen metabolism of growing pigs. Animal 13:262–272

Pi Y, Mu C, Gao K, Liu Z, Peng Y, Zhu W (2020) Increasing the hindgut carbohydrate/protein ratio by cecal infusion of corn starch or casein hydrolysate drives gut microbiota-related bile acid metabolism to stimulate colonic barrier function. mSystems 5: e00176–20

Poroyko V, White JR, Wang M, Donovan S, Alverdy J, Liu DC, Morowitz MJ (2010) Gut microbial gene expression in mother-fed and formula-fed piglets. PloS One 5:e12459

Ran S, Mu C, Zhu W (2019) Diversity and community pattern of sulfate-reducing bacteria in piglet gut. J Anim Sci Biotechnol 10:40

Rastall RA, Gibson GR (2015) Recent developments in prebiotics to selectively impact beneficial microbes and promote intestinal health. Curr Opin Biotech 32:42–46

Reese AT, Pereira FC, Schintlmeister A et al (2018) Microbial nitrogen limitation in the mammalian large intestine. Nat Microbiol 3:1441–1450

Ridlon JM, Kang DJ, Hylemon PB (2006) Bile salt biotransformations by human intestinal bacteria. J Lipid Res 47:241–259

Rodriguez E, Arques JL, Rodriguez R, Nunez M, Medina M (2003) Reuterin production by lactobacilli isolated from pig faeces and evaluation of probiotic traits. Lett Appl Microbiol 37:259–263

Rossi M, Corradini C, Amaretti A, Nicolini M, Pompei A, Zanoni S, Matteuzzi D (2005) Fermentation of fructooligosaccharides and inulin by bifidobacteria: a comparative study of pure and fecal cultures. Appl Environ Microbiol 71:6150–6158

Schokker D, Fledderus J, Jansen R, Vastenhouw SA, de Bree FM, Smits MA, Jansman A (2018) Supplementation of fructooligosaccharides to suckling piglets affects intestinal microbiota colonization and immune development. J Anim Sci 96:2139–2153

Shan T, Li L, Simmonds P, Wang C, Moeser A, Delwart E (2011) The fecal virome of pigs on a high-density farm. J Virol 85:11697–11708

Su Y, Yao W, Perez-Gutierrez ON, Smidt H, Zhu W (2008) Changes in abundance of *Lactobacillus* spp. and *Streptococcus suis* in the stomach, jejunum and ileum of piglets after weaning. FEMS Microbiol Ecol 66:546–555

Su Y, Bian G, Zhu Z, Smidt H, Zhu W (2014) Early methanogenic colonisation in the faeces of Meishan and Yorkshire piglets as determined by pyrosequencing analysis. Archaea 2014:547908

Sun Y, Su Y, Zhu W (2016) Microbiome-metabolome responses in the cecum and colon of pig to a high resistant starch diet. Front Microbiol 7:779

Tian S, Wang J, Yu H, Wang J, Zhu W (2019) Changes in ileal microbial composition and microbial metabolism by an early-life galacto-oligosaccharides intervention in a neonatal porcine model. Nutrients 11:1153

Van Hees HMJ, Davids M, Maes D, Millet S, Possemiers S, den Hartog LA, van Kempen T, Janssens GPJ (2019) Dietary fibre enrichment of supplemental feed modulates the development of the intestinal tract in suckling piglets. J Anim Sci Biotechnol 10:83

Wang J, Tian S, Yu H, Wang J, Zhu W (2019a) Response of colonic mucosa-associated microbiota composition, mucosal immune homeostasis, and barrier function to early life galactooligosaccharides intervention in suckling piglets. J Agric Food Chem 67:578–588

Wang P, Zhong HJ, Song YM et al (2019b) Targeted metabolomics analysis of maternal-placental-fetal metabolism in pregnant swine reveals links in fetal bile acid homeostasis and sulfation capacity. Am J Physiol 317:G8–G16

Wang X, Tsai T, Deng F et al (2019c) Longitudinal investigation of the swine gut microbiome from birth to market reveals stage and growth performance associated bacteria. Microbiome 7:109

Wang Y, Zhou J, Wang G, Cai S, Zeng X, Qiao S (2018) Advances in low-protein diets for swine. J Anim Sci Biotechnol 9:60

Williams BA, Verstegen MWA, Tamminga S (2001) Fermentation in the large intestine of single-stomached animals and its relationship to animal health. Nutr Res Rev 14:207–227

Wu G (2009) Amino acids: metabolism, functions, and nutrition. Amino Acids 37:1–17

Wu G (2018) Principles of animal nutrition. CRC Press, Boca Raton, Florida

Wu G (2020) Important roles of dietary taurine, creatine, carnosine, anserine and hydroxyproline in human nutrition and health. Amino Acids 52:329–360

Wu G (2022) Nutritionand metabolism: Foundations for animal growth, development, reproduction, and health.Adv Exp Med Biol 1354:1–24

Wu G, Bazer FW, Dai Z, Li D, Wang J, Wu Z (2014) Amino acid nutrition in animals: protein synthesis and beyond. Annu Rev Anim Biosci 2:387–417

Xiao Y, Yan H, Diao H, Yu B, He J, Yu J, Zheng P, Mao X, Luo Y, Chen D (2017) Early gut microbiota intervention suppresses DSS-induced inflammatory responses by deactivating TLR/NLR signalling in pigs. Sci Rep 7:3224

Xu RY, Lu Y, Wang J, Liu JJ, Su Y, Zhu W (2019) Effects of the different dietary fibers on luminal microbiota composition and mucosal gene expression in pig colons. J Funct Foods 59:71–79

Xu ZR, Zou XT, Hu CH, Xia MS, Zhan XA, Wang MQ (2002) Effects of dietary fructooligosaccharide on digestive enzyme activities, intestinal microflora and morphology of growing pigs. Asian-Australas J Anim Sci 15:1784–1789

Yang Y, Dai Z, Zhu W (2014) Important impacts of intestinal bacteria on utilization of dietary amino acids in pigs. Amino Acids 46:2489–2501

Yang Y, Mu C, Luo Z, Zhu W (2016) Bromochloromethane, a methane analogue, affects the microbiota and metabolic profiles of the rat gastrointestinal tract. Appl Environ Microbiol 82:778–787

Yin J, Li FN, Kong XF et al (2019) Dietary xylo-oligosaccharide improves intestinal functions in weaned piglets. Food Funct 10:2701–2709

Yu M, Mu C, Zhang C, Yang Y, Su Y, Zhu W (2018) Marked response in microbial community and metabolism in the ileum and cecum of suckling piglets after early antibiotics exposure. Front Microbiol 9:1166

Yu M, Mu C, Zhang C, Yang Y, Su Y, Zhu W (2020) Long-term effect of early antibiotic exposure on amino acid profiles and gene expression of transporters and receptors in the small intestinal mucosa of growing pigs with different dietary protein levels. J Sci Food Agric 100:235–244

Yu M, Zhang C, Yang Y, Mu C, Su Y, Yu K, Zhu W (2017a) Long-term effects of early antibiotic intervention on blood parameters, apparent nutrient digestibility, and fecal microbial fermentation profile in pigs with different dietary protein levels. J Anim Sci Biotechnol 8:60

Yu M, Mu C, Yang Y, Zhang C, Su Y, Huang Z, Yu K, Zhu W (2017b) Increases in circulating amino acids with in-feed antibiotics correlated with gene expression of intestinal amino acid transporters in piglets. Amino Acids 49:1587–1599

Zhang C, Yu M, Yang Y, Mu C, Su Y, Zhu W (2016a) Effect of early antibiotic administration on cecal bacterial communities and their metabolic profiles in pigs fed diets with different protein levels. Anaerobe 42:188–196

Zhang L, Mu C, He X, Su Y, Mao S, Zhang J, Smidt H, Zhu W (2016b) Effects of dietary fibre source on microbiota composition in the large intestine of suckling piglets. FEMS Microbiol Lett 363: fnw138.

Zhang C, Yu M, Yang Y, Mu C, Su Y, Zhu W (2017) Differential effect of early antibiotic intervention on bacterial fermentation patterns and mucosal gene expression in the colon of pigs under diets with different protein levels. Appl Microbiol Biotechnol 101:2493–2505

Zhang C, Yu M, Yang Y, Mu C, Su Y, Zhu W (2020) Effect of early antibiotic intervention on specific bacterial communities and immune parameters in the small intestine of growing pigs fed different protein level diets. Animal 1–12

Zmora N, Suez J, Elinav E (2019) You are what you eat: diet, health and the gut microbiota. Nat Rev Gastroenterol Hepatol 16:35–56

Zwicker BL, Agellon LB (2013) Transport and biological activities of bile acids. Int J Biochem Cell B 45:1389–1398

L-Arginine Nutrition and Metabolism in Ruminants

10

Guoyao Wu, Fuller W. Bazer,
M. Carey Satterfield, Kyler R. Gilbreath,
Erin A. Posey, and Yuxiang Sun

Abstract

L-Arginine (Arg) plays a central role in the nitrogen metabolism (e.g., syntheses of protein, nitric oxide, polyamines, and creatine), blood flow, nutrient utilization, and health of ruminants. This amino acid is produced by ruminal bacteria and is also synthesized from L-glutamine, L-glutamate, and L-proline via the formation of L-citrulline (Cit) in the enterocytes of young and adult ruminants. In pre-weaning ruminants, most of the Cit formed de novo by the enterocytes is used locally for Arg production. In post-weaning ruminants, the small intestine-derived Cit is converted into Arg primarily in the kidneys and, to a lesser extent, in endothelial cells, macrophages, and other cell types. Under normal feeding conditions, Arg synthesis contributes 65% and 68% of total Arg requirements for nonpregnant and late pregnany ewes fed a diet with $\sim$12% crude protein, respectively, whereas creatine production requires 40% and 36% of Arg utilized by nonpregnant and late pregnant ewes, respectively. Arg has not traditionally been considered a limiting nutrient in diets for post-weaning, gestating, or lactating ruminants because it has been assumed that these animals can synthesize sufficient Arg to meet their nutritional and physiological needs. This lack of a full understanding of Arg nutrition and metabolism has contributed to suboptimal efficiencies for milk production, reproductive performance, and growth in ruminants. There is now considerable evidence that dietary supplementation with rumen-protected Arg (e.g., 0.25–0.5% of dietary dry matter) can improve all these production indices without adverse effects on metabolism or health. Because extracellular Cit is not degraded by microbes in the rumen due to the lack of uptake, Cit can be used without any encapsulation as an effective dietary source for the synthesis of Arg in ruminants, including dairy and beef cows, as well as sheep and goats. Thus, an adequate amount of supplemental rumen-protected Arg or unencapsulated Cit is necessary to support maximum survival, growth, lactation, reproductive performance, and feed efficiency, as well as optimum health and well-being in all ruminants.

Keywords

Amino acids · Function · Growth · Lactation · Nutrition · Pregnancy

G. Wu (✉) · F. W. Bazer ·
M.C. Satterfield · K. R. Gilbreath · E. A. Posey ·
Y. Sun
Departments of Animal Science and Nutrition, Texas A&M University, College Station, TX 77843, USA
e-mail: g-wu@tamu.edu

G. Wu (ed.), *Recent Advances in Animal Nutrition and Metabolism*,
Advances in Experimental Medicine and Biology 1354,
https://doi.org/10.1007/978-3-030-85686-1_10

Abbreviations

AA	Amino acid
Arg	L-Arginine
BW	Body weight
Cit	L-Citrulline
IUGR	Intrauterine growth restriction
MTOR	Mechanistic target of rapamycin
NAG	N-Acetylglutamate
NCG	N-Carbamoylglutamate
NO	Nitric oxide
NRC	National Research Council
RPA	Rumen-protected arginine product

10.1 Introduction

L-Arginine (Arg) is a basic amino acid (AA) in the physiological fluids of all animals. As shown in Fig. 10.1, this nutrient activates the mechanistic target of rapamycin (MTOR) cell signaling to increase protein synthesis, inhibit protein degradation, and promote the development of brown adipose tissue in the conceptuses of ruminant mammals, including sheep (Bazer et al. 2012; Kim et al. 2011a, b; Ma et al. 2017; McKnight et al. 2020; Sales et al. 2016; Satterfield et al. 2012, 2013; Wang et al. 2014a). Arg also enhances the expression of genes for the synthesis of polyamines (putrescine, spermidine, and spermine), nitric oxide (NO), and interferon tau that are essential for the proliferation and migration of ovine trophectoderm cells involved in placental formation (Bazer et al. 2011; Kim et al. 2011c; Wang 2015a, b, 2016). Through the production of NO (Jobgen et al. 2006), Arg can modulate the post-translational modifications of histone proteins (including H3; Fig. 10.1) that are crucial for chromosome condensation to enable gene transcription (Palczewski et al. 2019). Furthermore, Arg plays a crucial role in the detoxification of ammonia that is particularly toxic to the embryos of mammals, including those of cattle, sheep and goats (Herring et al. 2018). Finally, Arg is essential for vasodilation, blood flow, angiogenesis (the growth of blood vessels from the existing vasculature), spermatogenesis, and embryonic survival in all animals, including ruminants (Gao 2020; Peine et al. 2020; Reynolds et al. 2006; Wu et al. 2009, 2021). Thus, through multiple mechanisms, Arg is vital to the growth, development, fertility, lactation, and health of all animals.

Based on research on AA biochemistry, nutrition and physiology in nonruminants, there has been in recent years a growing interest in the role of Arg and its immediate precursor L-citrulline (Cit) in the metabolism and adaptations to physiological conditions, such as pregnancy and lactation, in ruminants, including cattle, sheep, and goats (Bazer et al. 2018; Cao et al. 2021; Gilbreath et al. 2021; McKnight et al. 2020; Meyer et al. 2018). Ruminants have a large rumen that contains many different species of bacteria to extensively degrade Arg and other AAs (Bergen 2021; Lewis and Emery 1962; Recabarren et al. 1996). Thus, in post-weaning ruminants (> 7 months of age in beef cattle and > 3 months in lambs), nearly all dietary unprotected Arg is degraded in the rumen and, therefore, does not reach the small intestine, making this AA unavailable for absorption into the portal vein (Wu 2018). In contrast, in preruminant beef calves (i.e., prior to the presence of full microbial population; < 8 months of age) and lambs (< 4 months of age), a significant proportion of dietary Arg escapes the rumen (Pelaez et al. 1978; Williams and Hewitt 1979), and oral administration of Arg increases its concentration in blood (Fligger et al. 1997; Hüsier and Blum 2002). Given the important role of dairy products (e.g., milk, yogurt, and cheese), beef, and other ruminant-derived foods in improving human nutrition and health (Smith et al. 2020; Wu 2020), it is imperative to gain new knowledge about the role of Arg in nutrient metabolism, affecting the reproduction, lactation, growth, and survival of ruminants. This new knowledge will then be translated into strategies for improving their productivity and health, minimizing their potential impacts on the environment, and sustaining animal agriculture worldwide (Wu et al. 2020a).

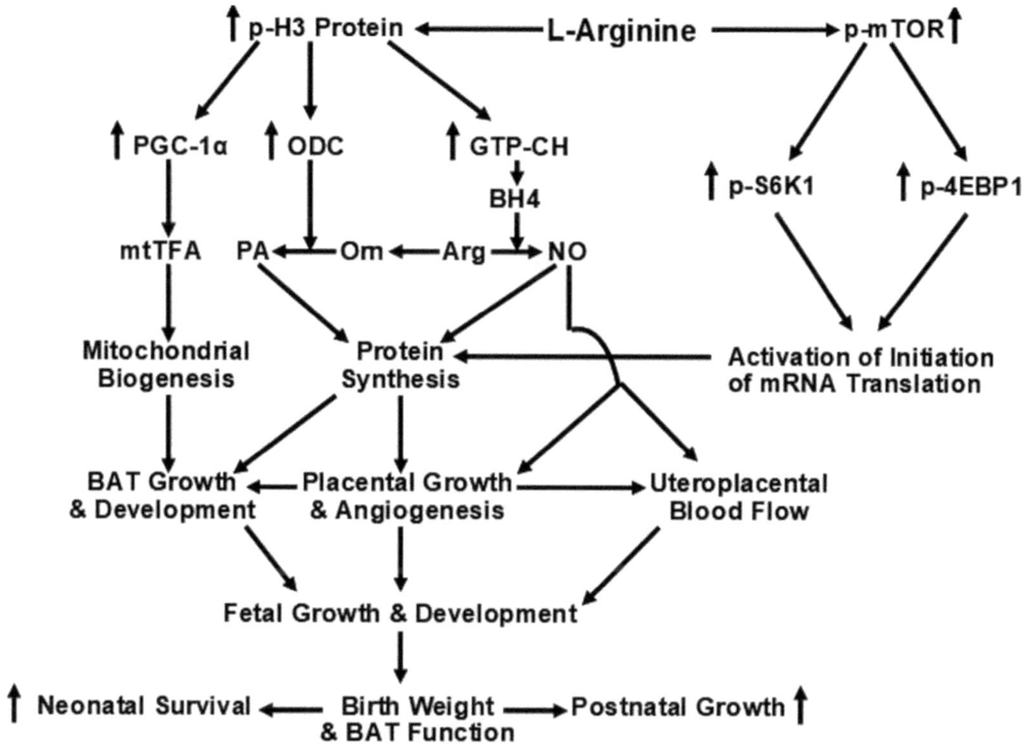

Fig. 10.1 Mechanisms whereby L-arginine enhances fetal growth and development of brown adipose tissue in mammals. Through the production of nitric oxide (NO), arginine can modulate post-translational modifications (including phosphorylation, acetylation, and methylation) of histone proteins (including H3) required for the condensation of chromosomes that enables gene transcription. In addition, arginine activates the mechanistic target of rapamycin (MTOR) cell signaling pathway to increase mRNA translation and protein synthesis, while inhibiting protein degradation in the conceptus. Furthermore, arginine enhances uterine blood flow, placental angiogenesis, and placental growth to promote the transfer of nutrients from mother to fetus required for fetal survival and growth

10.2 Arginine Metabolism in Ruminants

As for most nonruminant mammals (including humans, pigs, rats, and mice), enterocytes (the columnar absorptive epithelial cells of the small intestine) can synthesize Cit from glutamine, glutamate, and proline (Tables 10.1 and 10.2). All the necessary reactions occur in the same enterocytes, as extracellular ornithine is a poor substrate for citrulline formation in these cells (Wu and Morris 1998; Wu et al. 2021). At present, few studies have been conducted to assess quantitative aspects of Arg synthesis and catabolism in the whole body or tissues of young or adult ruminants (e.g., cattle, goats, and sheep). As for nonruminants, ruminants express various cell-specific transporters to transport neutral (e.g., Cit, glutamine, glycine, proline, and serine), acidic (e.g., glutamate and aspartate), and basic (e.g., Arg and ornithine) AAs that participate in interorgan and intracellular metabolism of Arg (Crouse et al. 2017, 2021; Gao et al. 2009a, b, c, d; Liao et al. 2008). With respect to ruminants, the following sections describe the role of the small intestine in the synthesis and release of Cit and Arg, Arg catabolism in mammary tissue, Arg metabolites in conceptuses, and whole-body creatine synthesis.

Table 10.1 Production of CO_2, ornithine, citrulline, and arginine from glutamine by enterocytes of cattle, sheep, swine, rats, and mice

Animals	Number of animals	Production of metabolites from glutamine (nmol/mg DNA/30 min)			
		CO_2	Ornithine	Citrulline	Arginine
2-day-old calves[e]	5	3903 ± 228^a	15.8 ± 1.4^a	149 ± 13^a	185 ± 20^a
7-day-old calves[e]	5	3725 ± 301^a	15.0 ± 1.2^a	132 ± 16^a	169 ± 22^a
6-month-day-old calves[e]	5	906 ± 54^b	5.13 ± 0.09^b	44.7 ± 2.5^b	9.22 ± 0.61^b
24-month-old NP beef cattle[e]	5	410 ± 291^c	3.76 ± 0.07^c	30.2 ± 1.1^c	2.30 ± 0.14^c
0-day-old lambs	6	4582 ± 170^a	21.4 ± 0.79^a	205 ± 9.0^a	390 ± 13^a
3-month-old lambs	6	1359 ± 51^b	10.6 ± 0.35^b	86.8 ± 3.8^b	40.2 ± 1.2^b
24-month-old NP ewes[f]	6	406 ± 16^d	6.12 ± 0.19^d	39.0 ± 1.3^d	2.84 ± 0.15^d
24-month-old pregnant ewes[f]	6	493 ± 19^c	7.56 ± 0.22^c	48.7 ± 1.6^c	3.72 ± 0.20^c
2-day-old pigs[g]	8	18694 ± 1772^a	36.4 ± 3.2^a	231 ± 10^a	505 ± 31^a
6-month-old pigs[h]	8	512 ± 30^b	4.08 ± 0.24^b	45.2 ± 2.3^b	5.38 ± 0.14^b
12-month-old NP gilts[i]	8	236 ± 9.8^d	1.92 ± 0.07^d	18.6 ± 0.68^d	1.47 ± 0.06^d
12-month-old pregnant gilts[i]	8	302 ± 12^c	2.65 ± 0.09^c	22.8 ± 0.77^c	1.83 ± 0.07^c
4-week-old rats[j]	8	3396 ± 107^a	351 ± 17^a	793 ± 29^a	128 ± 7.6^a
3-month-old rats[j]	8	1824 ± 84^b	163 ± 9.1^b	362 ± 25^b	28.4 ± 1.2^b
8-month-old rats[j]	8	856 ± 50^c	81.2 ± 3.8^c	178 ± 10^c	13.2 ± 0.74^c
4-month old mice[k]	5	3228 ± 204^a	245 ± 15^a	318 ± 21^a	25.1 ± 1.8^a
8-month old mice[k]	5	1488 ± 131^b	114 ± 5.6^b	153 ± 8.7^b	11.9 ± 0.74^b
20-month old mice[k]	5	634 ± 56^c	80.7 ± 3.3^c	16.7 ± 1.0^c	ND

Values are means ± SEM. All animals were used for metabolic studies in the fed state as described by Wu (1997). Enterocytes (3×10^6) freshly isolated from the jejunum of the indicated animals were incubated at 37 °C for 30 min in 1 ml of oxygenated (95% O_2/5% CO_2) Krebs bicarbonate buffer (pH 7.4) containing 5 mM D-glucose and 0 or 2 mM L-glutamine plus L-[U-^{14}C]glutamine (150 dpm/nmol). The incubation was terminated by the addition of 0.2 ml of 1.5 M $HClO_4$, followed by the collection of $^{14}CO_2$ for measurement by a liquid scintillation counter. The neutralized medium was analyzed for amino acids by high-performance liquid chromatography. The rates of production of amino acids from glutamine were calculated on the basis of the differences in their concentrations in cell extracts between 0 and 2 mM L-glutamine. Data were analyzed by one-way analysis of variance and the Student–Newman–Keuls multiple comparison (Assaad et al. 2014). In all age groups of an animal species, there were no differences ($P > 0.05$) in the concentrations of phenylalanine (an amino acid that is neither synthesized nor degraded by pig enterocytes) in cell extracts between 0 and 2 mM L-glutamine

[a–d]: Within a column for each animal species, means not sharing the same superscript letters differ ($P < 0.05$), as analyzed by one-way-analysis of variance and the Student–Newman–Keuls multiple comparison

[e] The jejunum of various ages of Brahman cattle (*Bos indicus*) was obtained from the Rosenthal Meat Science Center, Department of Animal Science, Texas A&M University, College Station, TX, USA

[f] Suffolk sheep were used in the study. Adult ewes were fed a normal soybean hulls-, wheat middlings-, and corn-based diet (Satterfield et al. 2013). Ewes at 125 days of gestation and age-matched nonpregnant ewes were used for the study

[g] Female pigs (F1 crosses of Yorkshire × Landrace sows and Duroc × Hampshire boars) were fed an 18%-crude protein diet as described by Hu et al. (2015)

[h] Female pigs (F1 crosses of Yorkshire × Landrace sows and Duroc × Hampshire boars) were fed a 14%-crude protein diet as described by Li et al. (2014)

[i] Female pigs (F1 crosses of Yorkshire × Landrace sows and Duroc × Hampshire boars) were fed a 12%-crude protein diet as described by Li et al. (2010). Gilts (12 months of age) at 90 days of gestation and age-matched nonpregnant gilts were used for the study

[j] Male Sprague–Dawley rats were housed and fed a 19%-casein diet as described by Jobgen et al. (2009)

[k] Wild-type male C57BL/6 J mice were housed and fed as described by Wu et al. (2020b)

ND, not detected; NP, nonpregnant

Table 10.2 Production of CO_2, ornithine, citrulline, and arginine from proline by enterocytes of cattle, sheep, swine, rats, and mice

Animals	Number of animals	Production of metabolites from proline (nmol/mg DNA/30 min)			
		CO_2	Ornithine	Citrulline	Arginine
2-day-old calves[e]	5	12.8 ± 0.67[a]	107 ± 5.4[a]	165 ± 8.8[a]	221 ± 13[a]
7-day-old calves[e]	5	12.4 ± 0.55[a]	101 ± 4.1[a]	154 ± 7.6[a]	209 ± 10[a]
6-month-day-old calves[e]	5	4.31 ± 0.18[b]	32.8 ± 1.7[b]	49.0 ± 2.5[b]	9.52 ± 0.68[b]
24-month-old NP beef cattle[e]	5	2.76 ± 0.10[c]	14.5 ± 0.78[c]	26.8 ± 2.1[c]	1.96 ± 0.10[c]
0-day-old lambs	6	15.4 ± 0.67[a]	143 ± 5.8[a]	227 ± 11[a]	349 ± 16[a]
3-month-old lambs	6	11.2 ± 0.48[b]	70.6 ± 3.9[b]	99.3 ± 4.7[b]	31.8 ± 1.4[b]
24-month-old NP ewes[f]	6	5.09 ± 0.18[c]	24.1 ± 0.68[d]	29.0 ± 1.0[d]	2.16 ± 0.08[d]
24-month-old pregnant ewes[f]	6	5.27 ± 0.21[c]	30.8 ± 0.95[c]	36.4 ± 1.2[c]	2.77 ± 0.09[c]
2-day-old pigs[g]	8	1.72 ± 0.15[c]	75.0 ± 4.6[a]	71.8± 6.5[a]	240 ± 11[a]
6-month-old pigs[h]	8	9.96 ± 0.46[a]	27.4 ± 1.2[b]	36.1 ± 1.7[b]	4.22 ± 0.19[b]
12-month-old NP gilts[i]	8	6.52 ± 0.31[b]	14.2 ± 0.67[d]	20.3 ± 0.72[d]	1.58 ± 0.06[d]
12-month-old pregnant gilts[i]	8	7.08 ± 0.35[b]	18.6 ± 0.79[c]	25.8 ± 0.77[c]	1.92 ± 0.08[c]
4-week-old rats[j]	8	214 ± 11[a]	286 ± 17[a]	475 ± 2.6[a]	74.0 ± 4.9[a]
3-month-old rats[j]	8	107 ± 7.0[b]	119 ± 7.4[b]	206 ± 1.2[b]	15.5 ± 0.71[b]
8-month-old rats[j]	8	58.9 ± 4.2[c]	55.2 ± 3.6[c]	97.4 ± 5.0[c]	7.06 ± 0.40[c]
4-month old mice[k]	5	202 ± 14[a]	20.4 ± 1.3[a]	218 ± 13[a]	17.0 ± 1.0[a]
8-month old mice[k]	5	115 ± 6.3[b]	163 ± 5.8[b]	106 ± 7.4[b]	8.06 ± 0.51[b]
20-month old mice[k]	5	40.6 ± 2.5[c]	58.9 ± 3.4[c]	11.2 ± 0.86[c]	ND

Values are means ± SEM, n = 8. All animals were used for metabolic studies in the fed state as described by Wu (1997). Enterocytes (3 × 10^6) freshly isolated from the jejunum of the indicated animals were incubated at 37 °C for 30 min in 1 ml of oxygenated (95% O_2/5% CO_2) Krebs bicarbonate buffer (pH 7.4) containing 5 mM D-glucose, 2 mM L-glutamine, and 0 or 2 mM proline plus L-[U-^{14}C]proline (150 dpm/nmol). The incubation was terminated by the addition of 0.2 ml of 1.5 M $HClO_4$, followed by the collection of $^{14}CO_2$ for measurement by a liquid scintillation counter. The neutralized medium was analyzed for amino acids by high-performance liquid chromatography. The rates of production of amino acids from L-[U-^{14}C] proline were calculated on the formation of ^{14}C-labeled ornithine, citrulline, and arginine. Data were analyzed by one-way analysis of variance and the Student–Newman–Keuls multiple comparison (Assaad et al. 2014). In all age groups of an animal species, there were no differences ($P > 0.05$) in the concentrations of phenylalanine (an amino acid that is neither synthesized nor degraded by pig enterocytes) in cell extracts between 0 and 2 mM L-proline

[a–d]: Within a column for each animal species, means not sharing the same superscript letters differ ($P < 0.05$), as analyzed by one-way-analysis of variance and the Student–Newman–Keuls multiple comparison

[e] The jejunum of various ages of Brahman cattle (*Bos indicus*) was obtained from the Rosenthal Meat Science Center, Department of Animal Science, Texas A&M University, College Station, TX, USA

[f] Suffolk sheep were used in the study. Adult ewes were fed a normal soybean hulls-, wheat middlings-, and corn-based diet (Satterfield et al. 2013). Ewes at 125 days of gestation and age-matched nonpregnant ewes were used for the study

[g] Female pigs (F1 crosses of Yorkshire × Landrace sows and Duroc × Hampshire boars) were fed an 18%-crude protein diet as described by Hu et al. (2015)

[h] Female pigs (F1 crosses of Yorkshire × Landrace sows and Duroc × Hampshire boars) were fed a 14%-crude protein diet as described by Li et al. (2014)

[i] Female pigs (F1 crosses of Yorkshire × Landrace sows and Duroc × Hampshire boars) were fed a 12%-crude protein diet as described by Li et al. (2010). Gilts (12 months of age) at 90 days of gestation and age-matched nonpregnant gilts were used for the study

[j] Male Sprague–Dawley rats were housed and fed a 19%-casein diet as described by Jobgen et al. (2009)

[k] Wild-type male C57BL/6 J mice were housed and fed as described by Wu et al. (2020b)

ND, not detected; NP, nonpregnant; PN, pregnant

10.2.1 Arginine Synthesis in Ruminants

10.2.1.1 Arginine Synthesis in Cattle

Based on a low concentration of Arg in cow's milk and the much lower Arg:lysine ratio in the milk (0.40:1.00; g/g) than that in the body protein of calves (1.09:1.00; g/g), Davis et al. (1994) and Li et al. (2021c) suggested that cow's milk is remarkably deficient in Arg. On the basis of Arg supplied from cow's milk and Arg deposition in the whole body, Williams and Hewitt (1979) estimated that the bovine milk provides at most 50% of the total daily Arg requirement for calves prior to the formation of a functional rumen. This means that the endogenous synthesis of Arg must provide at least 50% of the daily Arg required by the calf. This value of endogenous Arg synthesis is likely underestimated because the catabolism of Arg by various tissues of the animals was not previously considered (Williams and Hewitt 1979). Thus, young calves must be able to synthesize a large amount of Arg to meet requirements for their growth and development. In support of this view, enterocytes of 2- and 7-day-old calves synthesize Cit and Arg from glutamine and proline (Wu et al. 2005a), as is the case for most mammals including pigs, rats, and mice (Table 10.1). The underlying biochemical pathways involve both the mitochondria and cytosol of the same enterocytes, with Δ^1-pyrroline-5-carboxylate synthase (an NADPH-dependent bifunctional protein) being a rate-controlling enzyme (Fig. 10.2). This indicates an important role of the small intestine in AA metabolism for the endogenous synthesis of Cit and Arg in most mammals, including ruminants (Wu et al. 1994; 2005a). It is important to note that (1) there is no net synthesis of Arg by the liver via the urea cycle under physiological conditions and (2) extracellular ornithine is a poor substrate for the synthesis of Cit and Arg by their enterocytes. Therefore, dietary supplementation or intravenous administration of ornithine is ineffective in increasing Arg in blood in vivo (Wu and Morris 1998). Interestingly, like swine, the enterocytes of adult cattle also synthesize Cit, and beef cattle have an increased rate of intestinal synthesis of Cit plus Arg from glutamine and proline during pregnancy (Table 10.1). It is likely that gestating dairy cows are capable of synthesizing more Cit de novo than nonpregnant cows, but experimental data are lacking. As for swine, the rates of the endogenous synthesis of Cit and Arg are less for adult cattle compared with young calves (Table 10.1).

There is evidence that the endogenous synthesis of Arg alone is inadequate for (a) maximal growth in preruminant calves, as adding Arg to a milk replacer diet is required for optimal weight gain, protein deposition, and feed efficiency (Williams and Hewitt 1979); and (b) maximum embryonic survival in beef cows, as increasing Arg provision in their diets during pregnancy enhances the number of live-born calves (Gilbreath et al. 2021). Chacher et al. (2013) reported that dietary supplementation with N-carbamoyl glutamate [NCG; a metabolically stable analogue of N-acetyl glutamate (NAG) which is an allosteric activator of carbamoylphosphate synthase I for intestinal synthesis of Cit from both glutamine and proline (Dillon and Wu 2021; Wu et al. 2004)] increased the concentration of Arg in the plasma of lactating cows. These authors suggested that Arg synthesis via the intestinal-renal axis plays an important role in milk production by cattle.

10.2.1.2 Arginine Synthesis in Sheep

Sheep also synthesize Cit and Arg in the small intestine (Fig. 10.2). Bergman and Heitmann (1978) found that the small intestine of sheep takes up glutamine from arterial blood and releases Cit. In fed sheep, the small intestine also releases Cit into the portal vein (Tagari and Bergman 1978). Consistent with the intestinal synthesis of Cit, oral administration of NCG to sheep [2.5 g/day; 40 kg body weight (BW)] increased the concentration of Cit and Arg in the plasma of underfed ewes by 51% and 35%, respectively (Zhang et al. 2016). Similar results were reported by Sun et al. (2018). The cell responsible for the intestinal synthesis of Cit is the enterocyte, and its substrates are glutamine, glutamate, and proline (Tables 10.1 and 10.2). As for swine, rats, and cattle, glucose metabolism via the pentose phosphate pathway (pentose

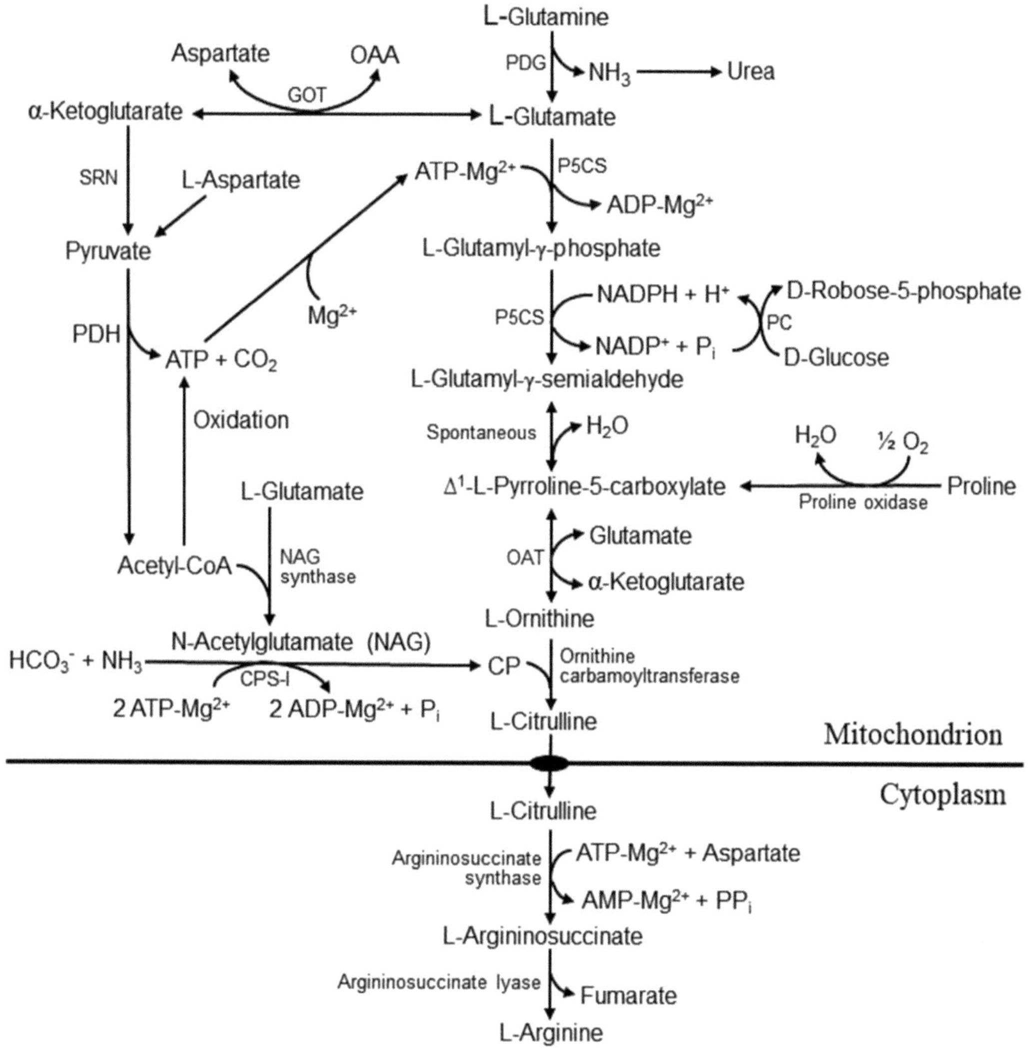

Fig. 10.2 Arginine synthesis in ruminants. The conversion of glutamine, glutamate, and proline into citrulline occurs exclusively in the mitochondria of enterocytes. All the reactions occur in the same enterocytes, as extracellular ornithine is a poor substrate for citrulline formation in these cells. Arginine is formed from citrulline in the cytoplasm of almost all cell types. Oxidation of amino acids (primarily glutamate, glutamine, and aspartate) provides acetyl-CoA and ATP for citrulline and arginine synthesis. Glucose metabolism via the pentose cycle generates NADPH that is a cofactor of pyrroline-5-carboxylate synthase for the conversion of glutamate into L-glutamyl-γ-semialdehyde. In the neonatal small intestine, the near absence of arginase maximizes the output of arginine into the portal circulation. In adults, most of the intestine-derived citrulline is released into the portal circulation, bypasses the liver, and is extracted primarily by the kidneys for arginine synthesis. Adapted from Wu and Morris (1998) and Wu (2021). CP, carbamoyl phosphate; CPS-I, carbamoylphosphate synthetase I; GOT, glutamate-OAA transaminase; OAA, oxaloacetate; OAT, ornithine aminotransferase; PC, pentose cycle; PDG, phosphate-activated glutaminase; P5CS, Δ^1-pyrroline-5-carboxylate synthase (a bifunctional enzyme); and SRN, a series of enzyme-catalyzed reactions (Wu 2021)

cycle; the major source of cellular NADPH) is necessary for the conversion of glutamine and glutamate into citrulline in the enterocytes of sheep because no citrulline or Arg is produced from glutamine or glutamate by these cells in the absence of extracellular glucose (Wu 1998; Wu et al. 2005a). We found that smooth muscle cells, immunocytes (lymphocytes and macrophages), and connective tissue in the small intestine of sheep, cattle, swine, and rats do not convert glutamine, glutamate, or proline into Cit or Arg (Wu 1998; Wu et al. 2005a). There is no detectable uptake of Cit within the portal vein by the liver of adult sheep, and the rate of hepatic uptake of Arg is low (0.86 mmol/h) in comparison with that of glycine (3.88 mmol/h) and alanine (3.19 mmol/h) (Bergman et al. 1974). Because the milk of sheep and goats also contains much less Arg than needed for the growth of their neonates (Davis et al. 1994), it is likely that these two species are capable of de novo synthesis of Arg in extra-intestinal tissues. In support of this view, Bergman et al. (1974) reported that the kidneys of fed and fasted pregnant sheep take up 1.41 and 0.44 mmol/h Cit from the arterial blood, respectively, and release 1.46 and 0.77 mmol/h Arg, respectively. At present, little is known about quantitative aspects of Arg synthesis in other ruminants (e.g., goats and deer). We estimated that Arg synthesis contributes 65% and 68% of total Arg requirements by nonpregnant and late pregnant ewes fed a diet with ~ 12% crude protein, respectively (Table 10.3). Assuming that the rate of endogenous Arg synthesis does not differ between sheep fed a 12%-crude protein diet and a 14%-crude protein diet, Arg synthesis accounts for 61% and 64% of total Arg requirements by nonpregnant and late pregnant ewes fed a diet with 14% crude protein, respectively. Based on the recent discovery that extracellular Cit is not degraded by ruminal microbes in adult cattle and sheep (Gilbreath et al. 2021), this AA can be supplemented without rumen protection to ruminants to enhance Arg synthesis and availability in the blood and other tissues of ruminants (Gilbreath et al. 2019, 2020a, b; Gootwine et al. 2020; Greene et al. 2017, 2020; Peine et al. 2020).

Table 10.3 Synthesis of creatine in adult pregnant and nonpregnant sheep

Variable	Adult nonpregant sheep [60 kg body weight (BW)]	Adult pregnant sheep (60 kg BW at the start of gestation; Day 130 of gestation)	
		Mother	Fetus (13.2 kg BW)
1. *Estimated according to the urinary excretion of Cr plus creatinine*			
Urinary excretion of Cr + creatinine, mmol/kg BW/day	0.327[a]	0.326[b]	0.548[c]
Arginine required for creatine synthesis, mg/kg BW/day	57.0	56.8	95.5
Arginine required for creatine synthesis, g/animal/day	3.42	3.41	0.305
Arginine required for creatine synthesis, mother plus fetus, g/whole body/day	–	3.72	–
Feed intake (11.7% crude protein; as-fed basis)[d], kg/day	1.20	1.33	–
Crude protein intake, g/day	140	156	–
Microbial and feed proteins entering the duodenum[e], g/day	95.2	106	–
Nitrogen leaving the rumen as ammonia and reentering the rumen via N recycling for microbial protein synthesis[f], g/day	18.2	20.3	–
Total digestible microbial and feed proteins in the SI, g/day	113.4	126.3	
Digestible microbial and feed proteins in the SI[g], g/day	96.4	107.4	–
Arginine (Arg) content in microbial and feed protein[h], %	5.12	5.12	–
Arg released from microbial and feed protein in the SI, g/day	4.94	5.50	–
Arg entering the portal vein (40% extraction by the SI), g/day	2.96	3.30	–

(continued)

Table 10.3 (continued)

Variable	Adult nonpregant sheep [60 kg body weight (BW)]	Adult pregnant sheep (60 kg BW at the start of gestation; Day 130 of gestation)	
		Mother	Fetus (13.2 kg BW)
Endogenous synthesis of Arg, g/day	5.54[i]	6.93[j]	–
Total amount of Arg utilized by the whole body, g/day	8.50	10.2	–
Contribution of Arg synthesis to Arg requirements, %	65.2	67.7	–
Percent of utilized Arg for creatine synthesis, %	40.2	36.4	–
2. *Estimated according to the concentrations of Cr plus Cr-P in skeletal muscle*			
Skeletal muscle (wet mass), kg	27.0	27.0	1.28
Skeletal muscle (dry mass), kg	8.1	8.1	0.384
Content of Cr + Cr-P in skeletal muscle, mmol/kg dry mass	134.6 [k]	135.0 [k]	96.2 [k]
Cr + Cr-P in total skeletal muscles, mmol	1090	1094	36.94
Cr + Cr-P in the whole body, mmol	147	1152	38.9
Loss of Cr + Cr-P (1.7%/day) as creatinine, mmol/kg BW/day	0.325	0.326	0.206
Skeletal muscle (wet mass) gain (BW gain of 90 mg/day), g/day	–	–	36
Skeletal muscle (dry mass) gain (BW gain of 90 g/day), g/day	–	–	10.8
Cr + Cr-P accretion in skeletal muscle, mmol/kg BW/day	–	–	0.325
Cr + Cr-P accretion in the whole body, mmol/kg BW/day	–	–	0.342

[a] Xue et al. (1988)
[b] Our calculated value. Skeletal muscle represents 40% and 45% of the whole body weight in the fetus and moth, respectively
[c] Calculated as loss of Cr + Cr-P as creatinine/day plus Cr + Cr-P accretion in the whole body/day (0.206 + 0.342)
[d] Lassala et al. (2010). The dietary content of arginine was 0.6% (as-fed basis)
[e] 68% of dietary crude protein intake (Wu 2018)
[f] 13% of dietary nitrogen intake (Wu 2018)
[g] The true digestibility of proteins in the small intestine is 85% (Wu 2018)
[h] Gilbreath et al. (2021). It is assumed that microbial protein and feed protein account for 90% and 10% of the total proteins, respectively
[i] 80% of the rate for pregnant sheep (Table 10.1)
[j] Calculated from the renal extraction rate of 1.41 mmol citrulline/h in the adult sheep (Bergman et al. 1974) that represents 85% of the total citrulline synthesized de novo by the small intestine
[k] Determined by high-performance liquid chromatography (Li and Wu 2020).
Cr, creatine; Cr-P, creatine phosphate; SI, small intestine

10.2.1.3 Concentrations of Arg and Related AAs in the Plasma and Conceptuses of Ruminants

The concentrations of Cit in the plasma of ruminants such as cattle and sheep (e.g., 0.1 to 0.3 mM; Crouse et al. 2019; Kwon et al. 2003a, 2004a, b) are unusually much greater than those in the plasma of nonruminant mammals such as pigs and rats (e.g., about 0.05–0.1 mM, depending on age; Jobgen et al. 2009; Wu 2018; Zhang et al. 2021). In addition, the concentrations of Cit and Arg in the plasma of fed pregnant ewes are 64% and 48% greater than for age-matched nonpregnant ewes, respectively (Bergman and Heitmann 1978). These findings may be explained, in part, by the higher rates of the syntheses of Cit and Arg from both glutamine and proline in the enterocytes of adult ruminants compared with adult swine, as well as in the enterocytes of pregnant ewes compared with age-matched nonpregnant ewes (Tables 10.1 and 10.2). It is possible that ruminal microbes produce more Cit and Arg in pregnant than nonpregnant ruminants, but experimental data are needed to support this possibility.

Concentrations of Cit in ovine allantoic fluid are particularly high during gestation (e.g., approximately 10 mM on Day 60 of gestation) compared with that of Arg (about 0.8 mM; Kwon et al. 2003a), reflecting the placental transport of large amounts of Cit from mother to fetus. The maternal transfer of Cit and its storage in fetal fluids of the ovine conceptus is of physiological significance (Fig. 10.1) as this strategy is used by sheep to limit Arg catabolism in their placentomes that possess high arginase activity (Kwon et al. 2003b). Ovine fetuses effectively synthesize Arg from Cit via argininosuccinate synthase and lyase (Lassala et al. 2009). In contrast to sheep, concentrations of Cit in the allantoic fluid of cattle have been reported to be very low on Days 34 and 50 of gestation (0.06 and 0.05 mM, respectively) at only 5.5 and 11% of those for Arg, respectively (Crouse et al. 2019). It is unknown whether this is due to species differences in the intestinal synthesis of Cit and its placental transport between sheep and cattle. Nonetheless, results of our studies clearly indicate that adult sheep synthesize greater amounts of Cit from both glutamine and proline than adult cattle (Tables 10.1 and 10.2).

10.2.2 Arginine Catabolism in Ruminants

10.2.2.1 Catabolism of Arg in the Mammary Tissue of Ruminants

The mammary glands of lactating cows extract a large amount of Arg, but the output of Arg in milk is much less than its uptake (Clark et al. 1978; Mepham 1982). Similar results were obtained for lactating goats (Alumot et al. 1983) and sheep (Davis et al. 1978). These results suggest the extensive catabolism of Arg by the lactating mammary glands of ruminants and are supported by the work of Clark et al. (1975) who found that the mammary tissue of lactating cows had high arginase activity and actively degraded Arg to form ornithine, glutamate, and proline. In addition, the mammary glands of lactating sheep and goats actively convert Cit into Arg, ornithine, and proline (Roets et al. 1974). Taken together, the findings help to explain why Arg is remarkably deficient, but proline and glutamate are highly abundant, in the milk of cows (Davis et al. 1994), sheep (Davis et al. 1978), goats (Sawaya et al. 1984), and sows (Wu and Knabe 1994).

10.2.2.2 Arginine Metabolites in the Ruminant Conceptus

Baetz et al. (1975) reported that the concentrations of Cit and Arg in bovine allantoic fluid did not change between Days 25 and 90 of gestation (term = 283 days) and ranged from 0.40 to 0.44 mM. However, in this study, the timing of sample collection from the local abattoir relative to the slaughter of the cattle was not provided by the authors. In contrast, in a well-controlled experiment, Kwon et al. (2003a) discovered an unusually high abundance of the Arg-family AAs in ovine allantoic fluid. Specifically, the concentrations of Cit and glutamine in ovine allantoic fluid increased 34- and 18-fold, respectively, between Days 30 and 60 of gestation. At Day 60 of gestation, concentrations of Cit and glutamine in ovine allantoic fluid were approximately 10 and 24 mM, respectively (Fig. 10.4). Such an abundance of Cit has not been previously reported for body fluids of any animals. The concentration of Arg was also relatively high in ovine allantoic fluid (0.5–2 mM), compared with its maternal plasma level (0.1–0.3 mM) during gestation, and increased fourfold between Days 30 and 100 of gestation (Kwon et al. 2003a, b). Cit plus Arg plus glutamine accounts for more than 50% of the total α-AAs in allantoic fluid during early gestation (Kwon et al. 2003a). Cit is not a building block for tissue protein synthesis and is virtually absent from plant-based diets (Gilbreath et al. 2020a; Hou et al. 2019; Li and Wu 2020). Thus, the high abundance of Cit in fetal fluids likely reflects its endogenous synthesis from glutamine and proline, as well as the progesterone-induced expression of Cit and Arg transporters in the endometrium and placenta (Satterfield et al. 2010). Furthermore, the unusual abundance of the Arg-family AAs in allantoic fluid supports many Arg-dependent pathways necessary for fetal survival, growth and development.

It should be borne in mind that Arg is oxidized to NO and Cit by NO synthase in ruminant conceptuses (Kwon et al. 2004a). This reaction does not contribute to de novo synthesis of Arg in animals (Wu and Morris 1998). In nearly all mammalian cell types (e.g., endothelial cells and macrophages), Cit is recycled into Arg to sustain NO production. Despite the essential physiological functions of NO, the synthesis of NO accounts for less than 1% of the Arg catabolized in the whole body (Wu et al. 2016). Consistent with alterations in uterine blood flow and placental angiogenesis, there are dynamic changes in the placental NO synthase and GTP cyclohydrolase-I activities as well as the concentrations of tetrahydrobiopterin and NO synthesis in ovine placentomes during pregnancy (Fig. 10.4). Likewise, a small amount of Arg is utilized via Arg decarboxylase to form agmatine, which is a substrate for the production of putrescine, spermidine, and spermine in ruminant conceptuses (Halloran et al. 2021; Hoskins et al. 2021; Wang et al. 2014b, c). Such an alternative pathway to the formation of putrescine from Arg-derived ornithine via arginase and ornithine decarboxylase (Kwon et al. 2004b) for polyamine synthesis is critical for the survival and growth of ruminant conceptuses (Lenis et al. 2018; Wang et al. 2014b, c). In contrast, porcine placentae, endometria, and embryos do not have Arg decarboxylase activity; therefore, porcine uterine fluid lacks agmatine. This illustrates another species difference in Arg metabolism between swine and ruminants.

Unlike the porcine placenta that lacks arginase (Wu et al. 2005b), the ovine placenta expresses arginase to actively degrade Arg to ornithine, polyamines, and proline (Kwon et al. 2003b). Thus, the placental transport of Cit into fetal blood helps to conserve Arg in the ovine conceptus where Cit serves as a reservoir of an effective precursor for the synthesis of Arg in the developing ovine fetus (Figs. 10.3 and 10.4), as noted previously. In addition, the Arg-derived ornithine is converted into proline, which is incorporated into collagen (a major protein in all animals), with some proline residues undergoing post-translational hydroxylation to form 4-hydroxyproline and 3-hydroxyproline (Li and Wu 2018; Wu et al. 2011). Degradation of collagen releases 4-hydroxyproline and 3-hydroxyproline. 4-Hydroxyproline is metabolized to glycine in multiple tissues (e.g., the liver, kidney, intestine, placentomes, and skeletal muscle) via the 4-hydroxyproline oxidase pathway (Wu et al. 2019), whereas 3-hydroxyproline is converted into proline in multiple tissues and gut microbes via the 3-hydroxyproline dehydratase pathway (Visser et al. 2012). Glycine is then used for the syntheses of serine, porphyrins, heme, creatine, glutathione, and nucleic acids (Bazer et al. 2021; Seo et al. 2021; Wu et al. 2004, 2018), whereas proline can be metabolized to Arg in enterocytes (Wu 1997, 1998). The recycling of 4-hydroxyproline and 3-hydroxyproline into glycine and proline, respectively, helps to converse both Arg and proline in animals, including ruminants (Hu et al. 2021).

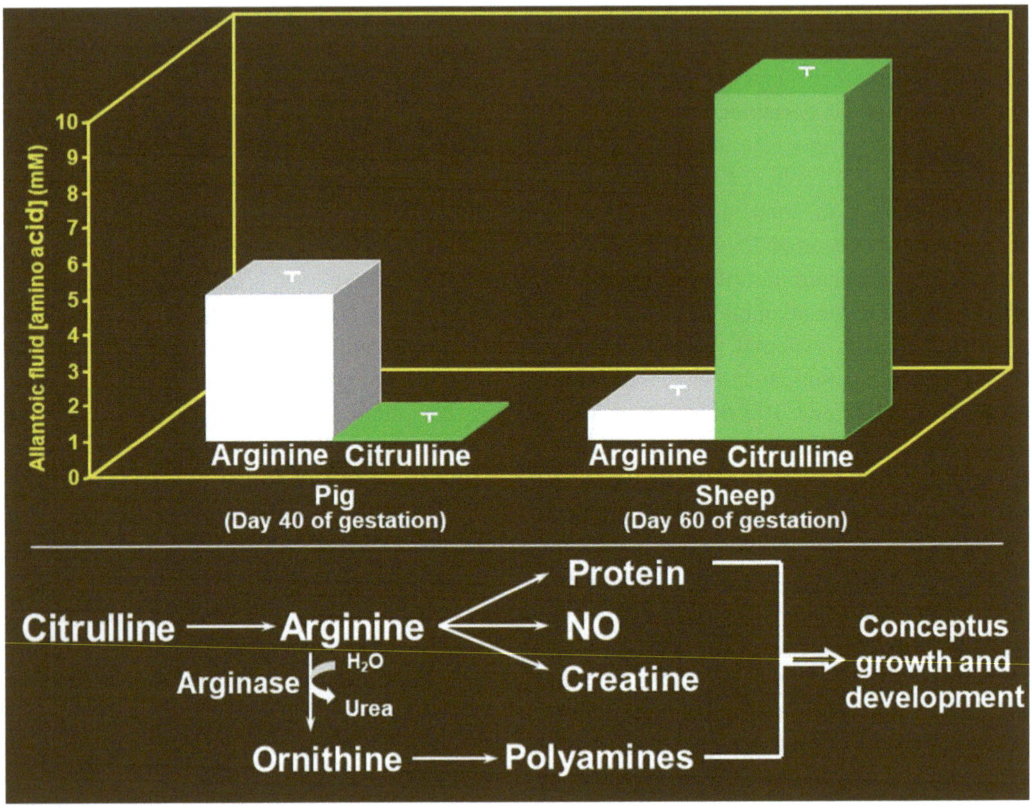

Fig. 10.3 Abundances of arginine and citrulline in the allantoic fluids of gestating sheep and pigs. Pregnant gilts and sows have high concentrations of L-arginine (4–6 mM) in porcine allanotic fluid during early gestation, but arginine is not a major amino acid in the allantoic fluid of ewes during early and mid-gestation. The opposite is true for citrulline. This may be explained by the expression of arginase in ruminant placentae. First, arginase activity is present in ovine allantoic fluid and placentomes on Days 30 and 60 of gestation but is not detectable in porcine allantoic fluid and placentae on Days 30 to 60 of gestation. The presence of arginase in ovine allantoic fluid and placentomes would hydrolyse arginine and reduce its availability to the fetus. Because citrulline is an effective precursor for arginine synthesis in mammals via argininosuccinate synthase and lyase, high concentrations of citrulline in ovine allantoic fluid serve as an efficient reservoir of precursor for the synthesis of arginine in the fetus. In contrast, the absence of arginase in porcine allantoic fluid and placentae limits arginine catabolism and maximizes the concentration of arginine in porcine conceptuses. Second, citrulline is a neutral amino acid. Thus, unlike arginine (a basic amino acid), citrulline does not disturb the acid–base balance in ovine allantoic fluid even at high concentrations. Third, citrulline is an efficient antioxidant, protecting DNA, lipids and proteins from hydroxyl radical-induced oxidative damage. This effect of citrulline may contribute to a protective environment for fetal growth and development in sheep. Thus, different strategies are used by different animal species to conserve arginine, the most abundant nitrogen carrier in tissue proteins. Adapted from Wu et al. (1996) and Kwon et al. (2003a)

10.2.2.3 Catabolism of Arg and Cit by Ruminal Microbes

We recently reported results of a series of experiments to determine AA metabolism by bovine and ovine ruminal microbes (Gilbreath et al. 2019, 2020a, b). In the first study, whole ruminal fluid (3 mL, containing microorganisms) from rumen-fistulated adult steers (~500 kg) was incubated at 37 °C with 5 mM Arg or Cit for 0, 0.5, 1, and 2 h to determine time-dependent changes in the metabolism of these AAs. Additional ruminal fluid was incubated with 0, 0.5, 2 or 5 mM Arg or Cit for 2 h to determine dose-dependent changes in their metabolism. An aliquot (50 μL) of the incubation solution was collected at the predetermined time points for AA

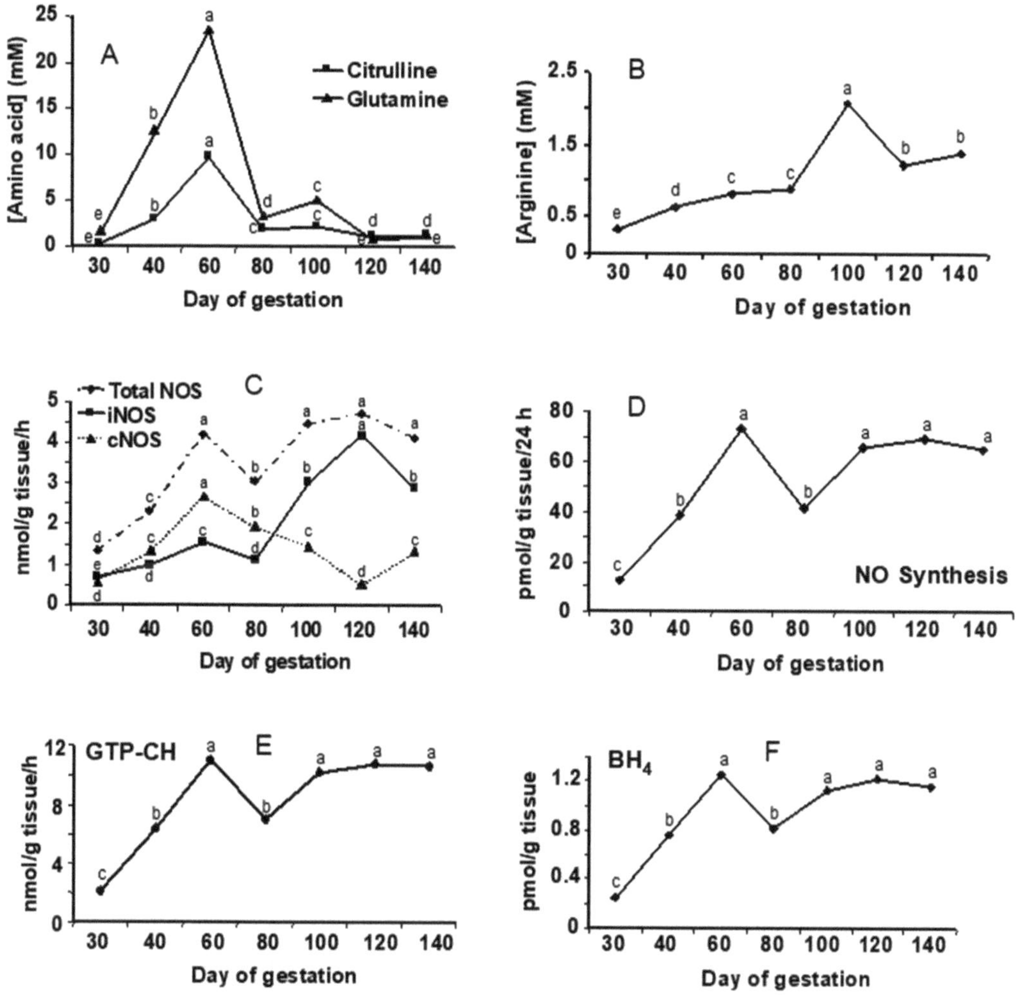

Fig. 10.4 Concentrations of amino acids in ovine allantoic fluid (panels A and B), NOS activity and NO synthesis in ovine placentomes (panels C and D), and GTP-CH activity and BH4 levels in ovine placentomes (panels E and F). Data are the mean for 4 ewes per day of gestation (Kwon et al. 2003a, b, 2004a). NOS activity and NO synthesis in ovine placentomes. Data are the mean for 4 ewes per gestational age (Kwon et al. 2004a). For each variable, means with different letters (a-e) are different ($P < 0.01$)

analyses. There was extensive hydrolysis of Arg into ornithine, proline, and ammonia, but no degradation of extracellular Cit by ruminal microbes due to no detectable uptake of ^{14}C-labeled Cit by the microbes (Gilbreath et al. 2019). The second study involved both i*n vitro* and in vivo experiments during which whole ruminal fluid (3 mL, containing microorganisms) from steers was incubated at 37 °C with 5 mM Cit in a rumen-protected or unprotected form for 0, 0.5, 2, or 4 h, after which time points 50 μL samples were collected for AA and ammonia analyses. In the in vivo experiment, at 0.5 h before and 0, 0.5, 1, 2, 4, and 6 h after rumen-fistulated adult steers consumed 0.56 kg dried-distillers' grain mixed with 70 g Cit plus 70 g glutamine (in a rumen-protected or unprotected form), samples of ruminal fluid and jugular venous blood were obtained for AA analyses. Results from both in vitro and in vivo experiments revealed that there was little degradation of rumen-protected and unprotected Cit by ruminal microbes. Concentrations of Cit and Arg in the plasma of steers consuming rumen-protected or unprotected Cit increased at 1 and 2 h after the meal, respectively, when compared with values at 0 h, indicating that ruminal microbes of adult steers do not degrade extracellular Cit in a rumen-protected or unprotected form (Gilbreath et al. 2020a). In the third study, whole rumen fluid (3 mL) from six adult Suffolk sheep was incubated at 37 °C with 5 mM Arg or Cit for 0, 0.5, 1, and 2 h, or with 0, 0.5, 2, or 5 mM Arg or Cit for 2 h. An aliquot (50 μL) of the incubation solution was collected at the predetermined time points for AA analyses. The results indicated the extensive hydrolysis of Arg into ornithine, proline, and ammonia, but no degradation of extracellular Cit by ovine ruminal microbes (Gilbreath et al. 2020b). To confirm the in vitro findings, six adult sheep were individually fed separate supplements (8 g Cit or urea) along with regular feed (800 g/animal). Blood (2 mL) was sampled from the jugular vein prior to feeding (time 0) and at 0.5, 1, 2, and 4 h after the sheep consumed the supplement. Plasma was analyzed for AAs, glucose, ammonia, and urea. The concentrations of Cit in the plasma of sheep consuming this AA increased progressively by 117% at 4 h, and those of Arg increased gradually by 23% at 4 h, compared with baseline values. Consistent with our observations, Greene et al. (2020) demonstrated that oral administration of Cit to nonpregnant Suffolk ewes (81 mg/kg/d) increased the concentrations of Cit in plasma within 2 h and values remained elevated for 6 h. Feeding urea for the short period did not affect the concentrations of Cit or Arg in plasma during the sampling period. Collectively, these results indicate, for the first time, that ruminal microbes of adult steers do not degrade extracellular Cit. Thus, although it has long been believed ruminal microbes have the capability of extensively degrading all dietary AAs (Wu 2018), this dogma is not correct in the case of Cit.

Because extracellular Cit is not degraded by ruminal microbes due to the absence of its uptake by those organisms (Gilbreath et al. 2019, 2020a, b), dietary Cit in either a rumen-protected or unprotected form escaped the rumen, entered the portal circulation, and served as the immediate precursor for Arg synthesis in extrahepatic tissues of beef and dairy cows, as well as other ruminants. Of note, the mammalian liver does not take up extracellular Cit, and thus the gut-derived Cit is effectively available for Arg production by extrahepatic tissues in all terrestrial animals studied, including pigs, sheep, cattle, and chickens (Cao et al. 2021; Gilbreath et al. 2021; Wu and Morris 1998). The inability of ruminal microbes to utilize extracellular Cit is consistent with a previous report that few bacteria can utilize extracellular Cit as a nitrogen source for their growth (Stalon and Merceniner 1984). Cit, without encapsulation (that is highly costly in manufacturing) or protection from ruminal microbes, may be effectively supplemented to the diets of ruminants to increase the concentrations of Cit and Arg in plasma. This is also important because the binder [long-chain saturated fatty acids, e.g., hydrogenated soy oil] used for manufacturing rumen-protected Arg or Cit may have an epigenetic effect on the postnatal growth of offspring. For example, Gootwine et al. (2020) reported that lambs born to ewes fed either rumen-protected Arg or unprotected Cit plus the binder (e.g., soybean hydrogenated oil that contained long-chain saturated fatty acids) weighed about 3.7 kg less at 5 months of age than contemporary lambs born to control ewes without receiving the supplement. In support of this view, supplementing 7% hydrogenated vegetable lipids to rats during gestation adversely impaired gene expression in the white adipose tissue of adult offspring, and the responsible dietary substance may be *trans* fatty acids (Pisani et al. 2008). Similarly, the maternal consumption of hydrogenated vegetable lipids by rats during pregnancy reduced the abundances of insulin receptor and insulin receptor substrate-1 in the hypothalamus of 3-month-old male progeny by 26% and 50%, respectively (Albuquerque et al. 2006). Thus, we suggest that dietary supplementation with Cit (a stable and neutral AA) without any encapsulation is a safe and effective means to increase the availability of Arg for utilization by ruminants to enhance their productivity.

10.2.2.4 Creatine Synthesis

Creatine was discovered in 1832 as a water-soluble, abundant component of skeletal muscle in cattle (Wu 2020). Through the reversible reaction catalyzed by creatine kinase, creatine plays an important role in storing energy as creatine phosphate in excitable tissues (particularly skeletal muscle and brain), as well as the reproductive tract (Wyss and Kaddurah-Daouk 2000). In mammals (including ruminants), creatine synthesis requires the interorgan metabolism of Arg, glycine, and methionine (Wu 2021). This metabolic pathway is initiated by arginine:glycine amidinotransferase, which transfers the guanidino group of Arg to glycine to form guanidinoacetate and ornithine. In terrestrial mammals (including cattle, sheep, and pigs) and birds, arginine:glycine amidinotransferase is expressed primarily in the renal tubules, pancreas, and to a much lesser extent in the liver and other organs (Brosnan and Brosnan 2007). The guanidinoacetate released by the site of its synthesis is methylated by guanidinoacetate *N*-methyltransferase located predominantly in the liver and pancreas to produce creatine. Stoichiometrically, the synthesis of 1 mol creatine requires 1 mol each of Arg, glycine, and methionine. There are reports on the urinary excretion of creatine and creatinine, the concentrations of creatine and creatine phosphate in skeletal muscle, and the accumulation of creatine plus phosphocreatine in porcine (Wu et al. 2018) and ovine (Baharom et al. 2017; Xue et al. 1988) conceptuses. In addition, much is known about protein synthesis by ruminal microbes in vivo, their true digestibility in the terminal ileum of ruminants (e.g., sheep and cattle; Wu 2018), and Arg content in microbial proteins (Gilbreath et al. 2021). Based on these data, it is estimated that creatine synthesis represents 40% and 36% of Arg utilization in nonpregnant and late pregnant ewes fed a diet with ~ 12% crude protein, respectively (Table 10.3). For comparison, creatine synthesis represents 52% and 50% of Arg utilization in nonpregnant and late pregnant gilts fed a diet with ~ 12%-rude protein, respectively (Wu et al. 2018). Creatine in the arterial blood is actively taken up through creatine transporters by many tissues, including the skeletal muscle (which account for 40% and 45% of BW in fetuses and adults, respectively; Sandoval et al. 2020), heart, and brain. Approximately 95% of creatine in the body is present in skeletal muscle. A small amount of creatine and phosphocreatine (1.7%/day) is spontaneously converted into creatinine (Brosnan and Brosnan 2007).

10.3 Arginine Nutrition in Ruminants

At present, the National Research Council (NRC) has not established dietary requirements of preruminants or ruminants (e.g., calves, beef cattle, dairy cows, lambs, and ewes) for Arg (NRC 2000, 2001, 2007). However, available evidence indicates that milk-fed pre-ruminants, gestating sheep, and lactating cows require dietary Arg (rumen-protected Arg in post-weaning ruminants) for maximal growth and production performance (Gilbreath et al. 2021; Table 10.4). The relevant data are summarized in the following sections. Improving Arg nutrition plays an important role in sustaining ruminant production worldwide (Satterfield et al. 2021; Wu 2022).

10.3.1 Arginine Supplementation to Preruminants

Because Arg can increase the secretion and plasma concentrations of growth hormone and insulin in cattle and sheep (Hertelendy et al. 1970), some studies have determined the effect of Arg supplementation on the growth of calves. For example, supplementing Arg to a liquid milk replacer for calves at 0.5 g/kg BW/day tended ($P < 0.10$) to increase daily weight gain during weeks 1, 3, and 5 of life (Fligger et al. 1997), but the difference was not statistically significant due to the small number of animals used in the study ($n = 8$/group). These authors further reported that Arg supplementation reduced the number of leukocytes in blood, likely as a result of reduced inflammation (Fligger et al. 1997). In another study, Hill et al. (2011) noted that adding 0.4% Arg to a whey-based milk replacer did not affect growth performance in young calves. However, whether the Arg supplementation treatment could effectively increase the concentration of Arg in the blood of the calves is not known because the authors did not analyze AAs in any tissue, plasma or serum samples. It is also possible that one or more AAs in the diet may limit the response of the calves to dietary Arg supplementation. Nonetheless, Arg supplementation may be beneficial for the growth and health of calves fed a low-protein diet (Williams and Hewitt 1979). At present, little is known about Arg nutrition in neonatal sheep or goats. This line of research is warranted to better define the requirements of preruminant animals for dietary AAs.

10.3.2 Arg Nutrition in Growing/Finishing Ruminants

Choi et al. (2014) reported that intra-abomasal administration of Arg (50 g/day) to growing Angus steers (405–457 kg BW) for 14 days increased the expression of stearoyl-CoA desaturase in white adipose tissue. This enzyme converts stearic acid (a saturated fatty acid) into oleic acid (a monosaturated fatty acid). A high level of oleic acid in beef can improve meat quality for human consumption because dietary intake of oleic acid can beneficially increase the concentrations of high-density lipoprotein in the blood of adult humans (Gilmore et al. 2011). Within the short 2-week period of the study, Arg administration did not affect the BW gain of the steers (Choi et al. 2014). However, the number of post-weaning beef cattle used in the experiment was small (n = 11 or 13 per group) due to their high costs. More animals are necessary to draw a definite conclusion on their growth performance. In a recent study, Teixeira et al. (2019) reported that dietary supplementation with rumen-protected Arg to beef cattle increased the proportion of choice-grade carcasses and meat quality.

Post-weaning lambs appear to respond more sensitively to dietary Arg provision than cattle in terms of growth. For example, supplementing rumen-protected Arg to growing lambs (7 g/head/day) for 3 months improved their growth performance and feed efficiency (Hassan et al. 2011), as well as the water-holding capacity and quality of their meat (Al-Rubeii et al. 2015). Likewise, dietary supplementation with rumen-protected Arg enhanced the mammary gland development of ewes and the growth rate of newborn lambs (Ashour et al. 2018). This line of research indicates that growing lambs cannot synthesize sufficient Arg to meet their metabolic needs for optimal development or maximal growth performance and feed efficiency.

In a recent feeding trial, Johnson (2018) found that supplementing 24% rumen-bypass Arg supplement (feather meal) to a corn chop-, soybean hull-, and cottonseed hull-based diet for Limousin heifers (277 kg initial BW) for 98 days did not affect their growth performance during summer, as compared with a corn chop-, soybean hull-, cottonseed hull-, cottonseed hull-, and corn distillers grains-based diet (Control). However, the Control and Arg-supplemented diets contained very different nitrogen content, i.e., 17.0% and 34.7% crude protein (on the dry matter basis), respectively. Thus, caution should be exercised in interpreting the results and drawing a definite conclusion from this study. As with other nutrients, more does not necessarily mean better in the nutrition and growth of ruminants and the increased production of ammonia in animals fed excessive protein or AAs can be toxic (Wu 2018). Future experiments should involve graded levels of feather meal [an abundant source of Arg (Li et al. 2021c)] in the diets of heifers and other ruminants.

10.3.3 Studies with Nonpregnant Cattle and Sheep

Greene et al. (2017) reported effects of dietary supplementation with rumen-protected Arg on blood flow parameters and the concentration of luteinizing hormone in the blood of cyclic nonlactating beef cows (539 kg BW) consuming toxic endophyte-infected tall fescue seed. This work is of practical significance in beef production because Tall fescue [*Lolium arundinaceum (Schreb.)* Darbysh] is a common cool-season perennial forage in the southeastern United

States and is commonly infected with an endophytic fungus (*Epichloe coenophiala*). Although the fungus and Tall fescue have a symbiotic relationship to provide the plant with a desirable hardiness property, the fungus synthesizes ergot alkaloids that result in fescue toxicosis, including vasoconstriction, placental insufficiency, intrauterine and postnatal growth restriction, and infertility, in ruminants. Of note, supplementing rumen-protected Arg (180 mg/kg of BW) to cyclic beef cows consuming toxic endophyte-infected fescue seed increased the systemic generation of NO, peripheral blood flow, and the circulating level of luteinizing hormone (Greene et al. 2017).

Dietary supplementation with rumen-protected Arg increases the concentration of Arg in plasma and alters hemodynamics in non-pregnant ruminants (Reynolds et al. 2019). In further support of this notion, Peine et al. (2020) conducted a study with nonpregnant, nulliparous Rambouillet ewes (51 kg initial BW) that were housed individually and randomly assigned to one of four treatments: a control group without supplemental Arg (Control; 50 g of finely ground corn, only), or Arg-supplemented groups that received 90, 180, or 360 mg rumen-protected ARG/kg BW/day mixed in 50 g of finely ground corn. Supplements were administered orally once daily for 14 days immediately before consuming a pelleted diet. Results of Doppler ultrasound assessments indicated that Arg supplementation improved peripheral tissue blood perfusion, with the dose of 180 mg/kg BW/day being the optimal dose for the nonpregnant ewes (Peine et al. 2020). However, it is unknown whether this is an ideal dose of dietary Arg supplementation for maximum embryonic or fetal survival and growth in sheep and other ruminants.

10.3.4 Arg Nutrition in Gestating Ruminants

10.3.4.1 Studies with Gestating Sheep

Intravenous administration of Arg improves pregnancy outcomes in sheep

Intravenous administration of Arg can improve embryonic survival in pregnant sheep (de Chávez et al. 2015; Luther et al. 2009) and their ovarian blood flow (Saevre et al. 2011), while reducing ovarian resistance index and increasing the systemic concentrations of progesterone (Luther et al. 2008). It is important to note that the intravenous administration of arginine-HCl prevents intrauterine growth restriction (IUGR) in underfed ewes (Lassala et al. 2010). Specifically, beginning on Day 28 of gestation, Suffolk ewes were fed a diet providing 100% or 50% (underfed) of NRC nutrient requirements and, between Day 60 of gestation and parturition, underfed ewes received intravenous infusions of either saline or 155 μmol Arg per kg BW three times daily, whereas control ewes received saline only. The birth weights of lambs from saline-infused underfed ewes were 23% lower than those of lambs from the control-fed dams (Lassala et al. 2010). The administration of Arg to underfed ewes between Days 100 and 125 of gestation increased the concentrations of Arg in the maternal plasma (69%), fetal brown adipose tissue mass (48%), and birth weights of lambs by 21%, compared to the saline-infused underfed ewes (Satterfield et al. 2013). Similar results were observed for diet-induced obese ewes (Satterfield et al. 2012). Furthermore, intravenous administration of Arg-HCl (345 μmol per kg BW three times daily) between Days 100 and 121 of gestation reduced the percentage of lambs born dead by 23%, increased the percentage of lambs born alive by 59%, and enhanced the birth weights of quadruplets by 23%, without affecting maternal BW (Lassala et al. 2011). Furthermore, intravenous administration of Arg-HCl to underfed, overweight, or prolific ewes enhanced fetal growth and fetal brown fat, as well as thermogenesis in neonatal lambs (Lassala et al. 2010, 2011; McKnight et al. 2020; Satterfield et al. 2012, 2013). Of note, lambs from Arg-supplemented ewes had a higher rate of neonatal survival than lambs from control ewes (Lassala et al. 2011). Beneficial effects of Arg on pregnancy outcomes in sheep have also been reported by McCoard et al. (2013, 2014, 2016), Sales et al. (2016), and Reynolds et al. (2019). For example, intravenous administration of Arg-HCl to twin-bearing ewes during late gestation enhanced placental growth and development (van der Linden et al. 2015), reduced mammary gland infections during early lactation (Sciascia et al. 2019), enhanced heat production by newborn lambs (McCoard et al. 2014), and promoted the postnatal growth of lambs (Sales et al. 2016). Furthermore, Zeitoun et al. (2016) reported that daily intravenous administration of Arg to sheep (75 mg/kg BW/day) during the first 56 days of pregnancy resulted in a 35% increase in lamb birth weight and a 37% increase in lamb weaning weight, as well as improved maternal health. Collectively, these results support an important role for the use of dietary Arg to improve the reproductive efficiency of ruminants.

Dietary supplementation with rumen-protected Arg improves pregnancy outcomes in sheep

While the intravenous administration of Arg to gestating sheep provides the proof-of-principle that increasing exogenous Arg provision is beneficial for improving pregnancy outcomes, this approach is not practical for production settings on farms. Dietary supplementation with rumen-protected Arg (e.g., equivalent to 0.25 g Arg/100 g dietary dry matter) to gestating sheep increased the concentration of Arg in plasma, while reducing the concentrations of both ammonia and urea in plasma (Gootwine et al. 2020). Those results are consistent with the facts that (1) Arg is key to the urea cycle in mammals for the conversion of ammonia into urea and, therefore, to prevent the occurrence of potentially lethal hyperammonemia (Herring et al. 2018); (2) an adequate provision of Arg promotes tissue protein synthesis in animals, thereby reducing the oxidation of AAs to ammonia (Wu et al. 2009).

Some feeding trials have documented the beneficial effects of Arg on reproductive performance in ewes. For example, in order to minimize the catabolism of Arg in the ruminal bacteria, de Chávez et al. (2015) developed rumen-protected Arg for its supplementation to the regular maternal diet for gestating ewes. They found that daily supplementation with 7.8 g Arg (as arginine-HCl) to sheep (45 kg BW) between the onset of estrus and Day 25 after breeding enhanced embryonic and fetal survival during early pregnancy. Likewise, Zhang et al. (2016, 2018) reported that dietary supplementation with rumen-protected Arg (10 g/day) to underfed ewes (40 kg BW; 50% of NRC-recommended nutrient requirements) between Days 35 and 110 of gestation enhanced the development of fetal immune and fetal BW by 18%. Similar results were noted in response to dietary supplementation with NCG (2.5 g/day) to underfed ewes (40-kg BW; Zhang et al. 2016). Consistent with these findings, dietary supplementation with rumen-protected Arg (180 mg/kg BW/day) to underfed ewes between Day 54 of gestation and parturition (term = 147 days) increased the expression of the proopiomelanocortin protein in the hypothalamus, as well as fetal and neonatal growth, in comparison with underfed ewes without Arg supplementation (Peine et al. 2018; Prezotto et al. 2018). Finally, dietary supplementation with 0.1% NCG or 1% rumen-protected Arg to IUGR neonatal lambs improved intestinal energy status and ameliorated intestinal injury (Zhang et al. 2019). Collectively, results of these studies indicate that pregnant sheep do not synthesize sufficient Arg for optimum reproductive performance and show the promise of Arg in improving fertility and fetal growth in sheep under production conditions.

10.3.4.2 Studies with Gestating Cattle

Intravenous supplementation with Arg enhances circulating levels of prolactin, growth hormone, and insulin in the blood of gestating cattle

Intravenous administration of Arg (0.1 g/kg BW/day) into dairy cows during the last 7 days of gestation resulted in dramatic but transient increases in the concentrations of prolactin, growth hormone, and insulin in blood (Chew et al. 1984). These hormonal changes tended to enhance ($P < 0.10$) the development and function of mammary glands, as indicated by a 10% increase in the subsequent 22-week milk yield (Chew et al. 1984). At present, little is known about the effects of Arg supplementation on the reproductive performance of dairy or beef cattle.

Dietary supplementation with rumen-protected Arg improves pregnancy outcomes in cows

Before our research on AA metabolism in ruminal microbes was conducted (Gilbreath et al. 2019, 2020a, b), we believed that a nutritionally significant amount of dietary unprotected Arg and Cit would not pass the rumen intact into the small intestine (Wu 2018; Wu et al. 2006, 2009). Thus, based on our studies with swine (e.g., Li et al. 2010; 2014; Mateo et al. 2007), we determined an effect of dietary supplementation with rumen-protected AA (RPAA; Cit + L-glutamine) or rumen-unprotected AA (RUAA; Cit + L-glutamine) on embryonic survival in lactating beef cows that were fed a diet meeting NRC (2000) nutrient requirements (Gilbreath et al. 2018). Cit was used because it is a neutral AA and an effective precursor of Arg in ruminants (Lassala et al. 2009). During the entire experimental period, multiparous Brangus cows grazed green pasture and had free access to drinking water and mineral blocks. At the onset of lactation, cows received dried distillers grain (DDG) only, DDG top-dressed with the RUAA product, or DDG top-dressed with the RPAA product. After two months of lactation, all cows (BW = approximately 470 kg; body condition score = 4.5) were synchronized to estrus and artificially inseminated once following the detection of estrus. From day 1 to day 60 after artificial insemination, cows were fed daily either 0.64 kg DDG, 0.56 kg DDG + 0.28 kg RUAA (2% of estimated daily intake of 14 kg dry matter from pasture; 0.07 kg Cit + 0.07 kg L-glutamine), or 0.56 kg DDG + 0.28 kg RPAA (2% of estimated daily intake of 14 kg dry matter from pasture; 0.07 kg Cit + 0.07 kg L-glutamine). Once on each day of the supplementation period, cows were moved to pens to receive their respective supplement and then returned to their original pasture. Dietary supplementation with RUAA or RPAA enhanced the birth rate of live-born calves from 22% in cows fed DGG alone to 34% and 36%, respectively, for the two treatment groups (Gilbreath et al. 2018). The beneficial effects of the AA supplement were associated with increases in the concentrations of insulin in serum (67–96%) and Cit (18–19%), Arg (19 –21%), ornithine (20 – 60%), and proline (16–18%) in plasma, but decreases in the concentrations of ammonia (14 –15%) in plasma (Gilbreath 2018). The concentrations of glucose and urea in plasma were not different among the three groups of cattle. These findings further support the notion that ruminal microbe do not degrade extracellular Cit and have important implications for improving both lactation and fertility in both beef and dairy cows. We are excited about the possibility that Cit can be used extensively to improve the productivity of all ruminants in the future.

Although Cit was used along with L-glutamine in our previous studies with beef cows, dietary supplementation with 0.5% Cit alone was sufficient to increase the concentrations of both Arg and Cit in the plasma of adult sheep and steers (Table 10.5). Dietary supplementation with Cit [a neutral AA and an effective precursor of Arg (Lassala et al. 2009)] is expected to increase the reproductive efficiency of beef cows and their profitability. A successful pregnancy in beef or dairy cows is currently estimated to be worth $750 (Dr. Jason Cleere, Texas A&M AgriLife Extension, personal communication). Based on the cost of Cit + L-glutamine ($10/kg) and the daily use of 0.14 kg/day for 60 days, the total expense for feeding one cow would be $84. For an operation with 1,000 beef cows, the net incomes would be $68,970, $103,720, and $138,470, respectively, assuming a value of either $750, $1000, or $1250 per calf (Table 10.6). Considering the large numbers of beef and dairy (milk) cows in Texas (5.15×10^6), the United States (40.651×10^6), and the world (649×10^6), the net incomes from the dietary supplementation with Cit to their diets between Days 1 and 60 of gestation would be 534×10^6, 4.22×10^9, and 67.3×10^9, respectively, at the price of $1,000/calf (Table 10.6). Additional benefits that are not included in the margin of profit calculation include reductions in management and labor costs, improvements in herd health, an increase in cow numbers, and the prospect of improved fertility in the next breeding period. The results of this research indicate that unencapsulated Cit is able to bypass the rumen

Table 10.4 Effects of dietary supplementation with rumen-protected arginine on milk production by dairy cows[a]

Treatment group	Number of cows	Milk yield (kg/cow/day)		Changes in milk yield (kg/cow/7 days)	Coefficient of variation in milk
		Day 0	Day 7	Between Days 0 and 7	Yield change (%)
Non-handling control	14	27.7 ± 2.3	31.5 ± 3.4	3.77 ± 3.22 [NS]	321
Isonitrogenous control (Ala)	13	26.3 ± 3.5	30.9 ± 3.0	4.58 ± 3.36 [NS]	264
Unprotected Arg product	14	26.0 ± 3.3	31.3 ± 2.3	5.36 ± 3.59 [NS]	259
Rumen-protected Arg product	13	26.2 ± 2.6	33.6 ± 2.1	7.35 ± 2.70 *	132

[a] Taken from Keith et al. (2018). Data are means ± SEM. This study was conducted on HAW Farms (Belen, NM). Before parturition, 54 healthy Holstein cows (parities 1 to 4), weighing 550–600 kg, were assigned randomly to receive either no dietary supplementation (non-handling control) or dietary supplementation of 500 g rumen-protected L-arginine (Arg) product (RPA), 500 g unprotected arginine product or an isonitrogenous amount of L-alanine (Ala) per day beginning on days 1 to 4 after parturition (the initial day of supplementation = day 0 of the trial). Each cow was fed twice daily a typical silage- and alfalfa hay-based lactation diet containing 17% crude protein (25 kg DM/day) to meet the National Research Council (NRC)-recommended requirements of nutrients and had free access to drinking water. A supplement was administered to cows twice daily (equally divided doses at 8 am and 6 pm) by gavaging immediately after consumption of their regular meals. On Days 0 and 7 of the trial, milk yields of cows were determined using an auto-milking system. Changes in milk yields between Days 7 and 0 were analyzed by the paired T-test. Feed intake did not differ ($P > 0.05$) among the four groups of cows, as analyzed by one-way analysis of variance (Assaad et al. 2014)
* Changes in milk yields of cows between Days 7 and 0 differ ($P = 0.016$)

and, therefore, will be more affordable for use by producers of ruminants. Thus, the price for feed-grade Cit without encapsulation will be substantially reduced (e.g., $5/kg), similar to that for feed-grade Arg (Wu et al. 2018). Thus, a nutrition-based management system to increase embryonic survival will have an enormous impact on the global beef industry, as well as sheep and goat enterprises. These findings also have important implications for enhancing both milk production and fertility in lactating dairy cows because they also have very low pregnancy rates (e.g., 16% in Texas and Florida of the United States in the summer; Stewart et al. 2011). Large-scale experiments are warranted to optimize the supplemental doses and estimate economic returns from dietary supplementation of diets for beef and dairy cows with Cit.

10.3.5 Arg Nutrition in Lactating Ruminants

An acute increase in the circulating concentrations of Arg through intravenous or intragastric administration of Arg to lactating cows (520–557 kg BW; 0.2 g arginine-HCl/kg BW per day) did not affect milk composition or milk production (Vicini et al. 1988). By contrast, supplementing rumen-protected Arg (130 g/day) to the diets of lactating cows for 8 weeks increased milk production by 0.9 kg/day without affecting the composition of protein and lactose, compared with the control group (Kirchgessner et al. 1993). Recently, Haque et al. (2013) reported that duodenal infusions of Arg (21.4 g/day) plus isoleucine (23.6 g/day) plus valine (271 g/day) to cows (at week 22 ± 6 of lactation; 634 kg BW) for 7 days did not affect milk composition or yield. Whether a possible inadequacy of other synthesizable AAs (e.g., glycine, glutamate, glutamine, proline, and serine) may limit the response of lactating cows to Arg plus isoleucine plus valine remains to be determined.

We conducted a proof-of-concept study to test the hypothesis that Arg can enhance milk production by lactating dairy cows (Keith et al. 2018). A rumen-protected Arg product (RPA, which contained a by-pass matrix) and an unprotected Arg product that had the same composition as RPA were manufactured by

Table 10.5 Concentrations of amino acids in the plasma of adult sheep and steers after the consumption of L-citrulline without any encapsulation

Amino acid	Amount of L-citrulline	Time after the consumption of L-citrulline (h)					
	(g)	0	0.5	1	2	4	6
Adult sheep[e]							
Citrulline	0	143 ± 11	140 ± 8.7	142 ± 9.1	142 ± 9.2	149 ± 9.9	–
	8	140 ± 11^{d}	147 ± 12^{cd}	$162 \pm 12^{c*}$	$213 \pm 16^{b*}$	$304 \pm 29^{a*}$	–
Arginine	0	188 ± 23	189 ± 23	189 ± 21	184 ± 25	193 ± 23	–
	8	190 ± 25^{d}	195 ± 27^{cd}	$202 \pm 28^{c*}$	$217 \pm 31^{b*}$	$233 \pm 32^{a*}$	–
Adult cattle[f]							
Citrulline	0	90 ± 2.5^{a}	89 ± 2.3^{a}	88 ± 2.7^{a}	83 ± 2.1^{b}	78 ± 1.9^{c}	72 ± 1.6^{d}
	70	89 ± 2.0^{c}	94 ± 2.6^{c}	$103 \pm 3.1^{b*}$	$105 \pm 3.8^{b*}$	$113 \pm 3.5^{a*}$	$95 \pm 3.2^{c*}$
Arginine	0	124 ± 4.5^{a}	122 ± 5.0^{a}	121 ± 5.7^{a}	114 ± 4.9^{b}	106 ± 4.0^{c}	99 ± 4.3^{d}
	70	126 ± 4.9^{c}	132 ± 5.3^{c}	$140 \pm 6.2^{b*}$	$144 \pm 5.1^{ab*}$	$147 \pm 5.8^{a*}$	$130 \pm 6.6^{c*}$

[e] Adapted from Gilbreath et al. (2020b). Values, expressed as nmol/mL, are means ± SEM, n = 6. Six adult Suffolk female sheep (60 to 65 kg) were individually fed a supplement consisting of 8 g urea (0 g L-citrulline) plus 800 g of a soybean hulls-, wheat middlings-, and corn-based diet (Satterfield et al. 2013) on Day 0. Blood (2 mL) was sampled from the jugular vein of the sheep prior to feeding (time 0) and at 0.5, 1, 2, and 4 h after consuming the supplement for the analysis of amino acids in plasma. On Day 2, the same animals were individually fed a supplement consisting of 8 g L-citrulline plus 800 g of the soybean hulls-, wheat middlings-, and corn-based diet (Satterfield et al. 2013); blood sampling and the analysis of amino acids were performed as described for Day 0

[f] Values, expressed as nmol/mL, are means ± SEM, n = 8. Sixteen Angus × Hereford steers (~540 kg body weight were individually fed a supplement consisting of 70 g urea (0 g L-citrulline) plus 0.56 kg of dried distillers' grains with solubles (DDGS; Gilbreath et al. 2020a) on Day 0. Blood (2 mL) was sampled from the jugular vein of the cattle prior to feeding (time 0) and at 0.5, 1, 2, 4, and 6 h after consuming the supplement for the analysis of amino acids in plasma. On Day 2, the same animals were individually fed a supplement consisting of 70 g L-citrulline plus 0.56 kg of DDGS (Gilbreath et al. 2020a); blood sampling and the analysis of amino acids in plasma were performed as described for Day 0.

a-d: Within a row, means not sharing the same superscript letter differ ($P < 0.05$), as analyzed by one-way ANOVA for repeated measure data, followed by the Student–Newman–Keuls multiple comparison test (Lee et al. 2019)

* $P < 0.05$ vs the corresponding 0 g L-citrulline group, as analyzed by the paired-test

Table 10.6 Estimated economic returns from dietary supplementation with L-citrulline to lactating cows between Days 1 and 60 of gestation[a]

Dietary supplementation with L-citrulline	1,000 Cows				All cows (5.15×10^6) in Texas, USA[d], $	All cows (40.651×10^6) in USA[d], $	All cows (649×10^6) in the world[e], $
	Live-born calves[b]	Gross income, $	Supplement cost[c], $	Net income gain, $			
$750/calf							
Control	222	166,500	0	166,500	–	–	–
L-Citrulline	361	270,750	35,280	235,470	–	–	–
Difference	139	104,250	35,280	68,970	355×10^6	2.80×10^9	44.8×10^9
$1,000/calf							
Control	222	222,000	0	222,000	–	–	–
L-Citrulline	361	361,000	35,280	325,720	–	–	–
Difference	139	139,000	35,280	103,720	534×10^6	4.22×10^9	67.3×10^9
$1,250/calf							
Control	222	277,500	0	277,500	–	–	–
L-Citrulline	361	451,250	35,280	415,970	–	–	–
Difference	139	173,750	35,280	138,470	713×10^6	5.63×10^9	89.9×10^9

[a] A total amount of 4.2 kg of L-citrulline (70 g/day) is supplemented to one cow for 60 days (70 g/day × 60 days = 4,200 g; Gilbreath et al. 2021)
[b] Taken from Gilbreath et al. (2018). These values assume an increase in the number of live-born calves by 14 per 100 beef cows due to the dietary supplementation with L-citrulline plus L-glutamine
[c] Based on the cost of L-citrulline ($8.4/kg) available from Promois International Lt. (http://www.promoisinternational.com). Thus, the cost of L-citrulline supplemented to beef cows for 60 days is $35.28 per cow ($8.4/kg × 4.2 kg = $35.28) or $35,280 per 1000 cows
[d] United Department of Agriculture (USDA, 2020) for beef plus dairy (milk) cows. The numbers of beef and dairy cows that calved in 2020 were 4.57 and 0.58 million head in Texas, respectively, and were 31.317 and 9.335 million head in the United States, respectively. https://www.nass.usda.gov/Statistics
[e] Food and Agriculture Organization of the United Nations (FAO, 2020). http://www.fao.org/faostat/en/#data/QA The total number of cattle in the world is 1.51×10^9 in 2019. Assuming that 43% of them are beef cows plus dairy cows (USDA 2020), the total number of cows in the world is 649×10^6 in 2019

Biotechnology Services & Consulting Inc. (Coppell, TX). Before parturition, 54 healthy Holstein cows (parities 1 to 4), weighing 550–600 kg, were assigned randomly to receive either no dietary supplementation (non-handling control) or dietary supplementation of 500 g RPA, 500 g unprotected Arg product or an isonitrogenous amount of L-alanine (Ala) per day beginning on days 1 to 4 after parturition (the initial day of supplementation = Day 0 of the trial). Each cow was fed twice daily a typical silage- and alfalfa hay-based lactation diet containing 17% crude protein (25 kg dry matter/day) to meet the National Research Council (NRC)-recommended requirements of nutrients and had free access to drinking water. A supplement was administered to cows twice daily (equally divided doses at 8 am and 6 pm) by gavaging immediately after the consumption of their regular meals. On Days 0 and 7 of the trial, milk yields of cows were determined using an automilking system. On either day, the composition of nutrients did not differ among the 4 groups of cows. Changes in milk yields between Days 0 and 7 were significant for the RPA group but nonsignificant for the other groups (Table 10.4). Interestingly, the variation in changes of milk yield was much less for RPA-supplemented cows, compared with the other groups of cows. These results indicate that cows fed RPA had more consistent lactation performances and produced more milk.

10.4 Safety of Arg Supplementation in Ruminants

10.4.1 Safety of Arg Supplementation in Sheep

As noted previously, dietary Arg is extensively catabolized by bacteria in the rumen, and some studies have determined the safety of intravenously administered Arg in gestating sheep (Lassala et al. 2009, 2010, 2011; Satterfield et al. 2012, 2013; McCoard et al. 2013). Between Day 60 of gestation and parturition, intravenous infusions of 81 mg Arg (as Arg-HCl)/kg BW/day) into underfed ewes, three times daily (0800, 1500 and 2200 h), for 82 or 87 days did not affect: (1) feed intake; (2) concentrations of lysine, histidine, insulin, or growth hormone in the maternal serum; or (3) maternal and fetal health (Lassala et al. 2010). Rather, the Arg treatment beneficially reduced concentrations of ammonia, free fatty acids, and triglycerides and enhanced birth weights of lambs, compared with saline-infused underfed ewes (Lassala et al. 2010). Similar results were obtained for intravenous infusions of (a) 180 mg Arg (as arginine-HCl)/kg BW/day) into Booroola Rambouillet ewes carrying 2 to 4 fetuses between Days 100 and 121 of gestation (Lassala et al. 2011); (b) 180 mg Arg (as arginine-HCl)/kg BW/day) into ewes with twin fetuses between Days 100 and 140 of gestation (McCoard et al. 2013; Sales et al. 2016); or (c) 81 mg Arg (as arginine-HCl)/kg BW/day) into underfed or obese ewes between Days 100 and 125 of gestation (Satterfield et al. 2013). Likewise, ewes can tolerate dietary supplementation with 81 mg Cit/kg BW/day between Day 86 of gestation and parturition (Greene et al. 2020). These results indicate that a large amount of supplemental Arg provided via intravenous administration is well tolerated by pregnant sheep and can improve fetal growth and development without adverse effects on the mother or the fetus. Likewise, ewes can well tolerate 0.25 g/kg BW supplemental Arg (as rumen-protected Arg) per day between Days 35 and 110 of gestation (Zhang et al. 2016). Furthermore, dietary supplementation with rumen-protected Arg to post-weaning sheep (7 g/animal per day) is safe for 3 months (3.5 to 6.5 months of age) (Hassan et al. 2011). As reported for non-ruminants such as pigs (Hu et al. 2015; Wu et al. 2018), chickens (He et al. 2021), and aquatic animals (Li et al. 2021a, b), growing lambs can tolerate dietary supplementation with at least 1% Arg or Cit (on the basis of dietary dry matter; Gilbreath et al. 2020b).

10.4.2 Safety of Arg Supplementation in Cattle

Intravenous administration of Arg-HCl (0.5 g Arg/kg BW over a 30-min period) had no adverse effect on cattle (Hertelendy et al. 1970). Likewise, constant intravenous administration of Arg-HCl (0.1 g Arg/kg BW per day) for 7 days did not affect feed intake, physiological variables in blood, or health of dairy cows (Chew et al. 1984). Furthermore, oral administration of rumen-protected Arg to dairy cows (~0.3 g/kg BW/day) for 8 weeks did not affect their feed intake or health (Kirchgessner et al. 1993). Similar results were obtained for the intra-gastric administration of Arg to beef cattle (Davenport et al. 1990a, b). Likewise, based on the content of Arg in feather meal (5.83% on the as-fed basis; Li and Wu 2020) and the inclusion of 34.7% feather meal in the diet (Johnson 2018), growing cattle can tolerate dietary supplementation with at least 2% Arg (on the as-fed basis; dry matter content = 89.1%). Finally, dietary supplementation with 70 g Cit/day to beef cattle between Days 1 and 60 of gestation did not have adverse effects on the mother or the fetus (Gilbreath et al. 2018).

10.5 Summary and Perspectives

Arg is synthesized de novo and utilized via multiple pathways in ruminants. Arg and its immediate precursor Cit are abundant in fetal fluids of ruminants during pregnancy. The small intestine of adult ruminants appears to have a greater ability to synthesize Cit plus Arg from glutamine, glutamate, and proline than adult nonruminants (e.g., pigs, rats, and mice). This may explain why the concentrations of Cit plus Arg in the plasma of ruminants are much greater than those in nonruminants when fed diets containing similar content of protein. Arg synthesis contributes to 65% and 68% of total Arg requirements for nonpregnant and late pregnant ewes fed a diet with ~ 12% crude protein, respectively. About 40% of the dietary Arg is catabolized by the small intestine during the first pass, and endogenous synthesis via inter-organ metabolism of AAs is crucial for maintaining Arg homeostasis in the whole body. At the cellular level, Arg is physiologically essential for the synthesis of proteins and other nitrogenous substances (including protein, creatine, NO, agmatine, polyamines, and homoarginine) with key metabolic functions in the body. There is extensive evidence that dietary supplementation with rumen-protected Arg (e.g., 0.25 to 0.5% of dietary dry matter) can improve all these indices of production traits without adverse effects. Because extracellular Cit is not degraded by microbes in the rumen, it can be used without encapsulation as an effective means to increase Arg availability in the plasma and tissues (including those of the reproductive and mammary systems) of ruminants. This can eliminate the high cost of manufacturing rumen-protected Arg and Cit. In addition, feeding unencapsulated Cit to ruminants can abolish the potential negative epigenetic effect of the binder (e.g., hydrogenated vegetable lipids) used to encapsulate an AA on prenatal and postnatal growth, metabolism, and health of offspring. An adequate amount of dietary Arg in the rumen-protected form or unencapsulated Cit is necessary to maximize the growth, lactation performance, reproductive performance, health, and environmental adaptations of cattle, sheep, goats, and other ruminants. Dietary supplementation with Cit holds great promise for improving the productivity of all ruminants worldwide.

Acknowledgements This work was supported by the Texas A&M AgriLife Research Beef Program and the Department of Animal Science Mini-Grant Program. The authors thank our research assistants (Dr. Gayan I. Nawaratna, W. David Long, Daniel Long, Neil D. Wu, and Dr. Shengdi Hu) for their technical assistance, as well as Dr. Jason J. Cleere and Dr. Tryon A. Wickersham for helpful discussion.

References

Albuquerque KT, Sardinha FLC, Telles MM, Watanabe RLH, Nascimento CMO, Tavares do Carmo MG, Ribeiro EB, (2006) Intake of *trans* fatty acid–rich hydrogenated fat during pregnancy and lactation inhibits the hypophagic effect of central insulin in the adult offspring. Nutrition 22:820–829

Al-Rubeii AMS, Al-Badri AAHD, Taha SA (2015) Effect of protected arginine supplementation to ration of Awassi lamb on the chemical and physical analysis of carcasses meat. Anim Sci LVIII:86–93

Alumot E, Bruckental I, Tadmor A, Kennit C, Holstein P (1983) Effect of proline on arginine uptake and nitrogen metabolism of lactating goats. J Dairy Sci 66:1243–1247

Assaad H, Zhou L, Carroll RJ, Wu G (2014) Rapid publication-ready MS-Word tables for one-way ANOVA. Springerplus 3:474

Ashour AM, Hussein HA, Fahmy S, Ali EM (2018) Significance of rumen protected L-arginine –supplementation on certain blood parameters, mammary gland functions and growth rate of newly born lambs. Arch Agric Sci J 1:59–67

Baetz AL, Hubbert WT, Graham CK (1975) Developmental changes of free amino acids in bovine fetal fluids with gestational age and the interrelationships between the amino acid concentrations in the fluid compartments. J Reprod Fertil 44:437–444

Baharom S, De Matteo R, Ellery S, Gatta PD, Bruce CR, Kowalski GM, Hale N, Dickinson H, Harding R, Walker D, Snow RJ (2017) Does maternal-fetal transfer of creatine occur in pregnant sheep? Am J Physiol 313:E75-83

Bazer FW, Wu G, Johnson GA, Kim JY, Song GW (2011) Uterine histotroph and conceptus development: Select nutrients and secreted phosphoprotein 1 affect MTOR cell signaling in ewes. Biol Reprod 85:1094–1107

Bazer FW, Song GH, Kim JY, Erikson DW, Johnson GA, Burghardt RC, Gao HJ, Satterfield MC, Spencer TE, Wu G (2012) Mechanistic mammalian target of rapamycin (MTOR) cell signaling: Effects of select nutrients and secreted phosphoprotein 1 on development of mammalian conceptuses. Mol Cell Endocrinol 354:22–33

Bazer FW, Burghardt RC, Johnson GA, Spencer TE, Wu G (2018) Mechanisms for the establishment and maintenance of pregnancy: Synergies from scientific collaborations. Biol Reprod 99:225–241

Bazer FW, Seo H, Johnson GA, Wu G (2021) One-carbon metabolism and development of the conceptus during pregnancy: Lessons from studies with sheep and pigs. Adv Exp Med Biol 1285:1–15

Bergen WG (2021) Amino acids in beef cattle nutrition and production. Adv Exp Med Biol 1285:29–42

Bergman EN, Heitmann RN (1978) Metabolism of amino acids by the gut, liver, kidneys, and peripheral tissues. Fed Proc 37:1228–1232

Bergman EN, Kaufman CF, Wolff JE, Williams HH (1974) Renal metabolism of amino acids and ammonia in fed and fasted pregnant sheep. Am J Physiol 226:833–837

Brosnan JT, Brosnan ME (2007) Creatine: endogenous metabolite, dietary, and therapeutic supplement. Annu Rev Nutr 27:241–261

Cao Y, Yao J, Sun X, Liu S, Martin GB (2021) Amino acids in the nutrition and production of sheep and goats. Adv Exp Med Biol 1285:63–79

Chacher B, Liu H, Wang D, Liu J (2013) Potential role of N-carbamoyl glutamate in biosynthesis of arginine and its significance in production of ruminant animals. J Anim Sci Biotechnol 4:16

Chew BP, Eisenman JR, Tanaka TS (1984) Arginine infusion stimulates prolactin, growth hormone, insulin, and subsequent lactation in pregnant dairy cows. J Dairy Sci 67:2507–2518

Choi SH, Wickersham TA, Wu G, Gilmore LA, Edwards HD, Park SK, Kim KH, Smith SB (2014) Abomasal infusion of arginine stimulates *SCD* and *C/EBPβ* gene expression, and decreases *CPT1β* gene expression in bovine adipose tissue independent of conjugated linoleic acid. Amino Acids 46:353–366

Clark JH, Derrig RG, Davis CL, Spires HR (1975) Metabolism of arginine and ornithine in the cow and rabbit mammary tissue. J Dairy Sci 58:1808–1813

Clark JH, Spires HR, Davis CL (1978) Uptake and metabolism of nitrogenous components by the lactating mammary gland. Fed Proc 37:1233–1238

Crouse MS, McLean KJ, Greseth NP, Crosswhite MR, Pereira NN, Ward AK, Reynolds LP, Dahlen CR, Neville BW, Borowicz PP, Caton JS (2017) Maternal nutrition and stage of early pregnancy in beef heifers: Impacts on expression of glucose, fructose, and cationic amino acid transporters in utero-placental tissues. J Anim Sci 95:5563–5572

Crouse MS, Greseth NP, McLean KJ, Crosswhite MR, Pereira NN, Ward AK, Reynolds LP, Dahlen CR, Neville BW, Borowicz PP, Caton JS (2019) Maternal nutrition and stage of early pregnancy in beef heifers: Impacts on hexose and AA concentrations in maternal and fetal fluids. J Anim Sci 97:1296–1316

Crouse MS, McLean KJ, Dwamena J, Neville TL, Menezes ACB, Ward AK, Reynolds LP, Dahlen CR, Neville BW, Borowicz PP, Caton JS (2021) The effects of maternal nutrition during the first 50 days of gestation on the location and abundance of hexose and cationic amino acid transporters in beef heifer utero-placental tissues. J Anim Sci 99:skaa386

Davenport GM, Boling JA, Schillo KK (1990a) Nitrogen metabolism and somatotropin secretion in beef heifers receiving abomasal arginine infusions. J Anim Sci 68:1683–1692

Davenport GM, Boling JA, Schillo KK, Aaron DK (1990b) Nitrogen metabolism and somatotropin secretion in lambs receiving arginine and ornithine via abomasal infusion. J Anim Sci 68:222–232

Davis SR, Bickerstaffe R, Hart DS (1978) Amino acid uptake by the mammary gland of the lactating ewe. Aust J Biol Sci 31:123–132

Davis TA, Nguyen HV, Garciaa-Bravo R, Fiorotto ML, Jackson EM, Lewis DS, Lee DR, Reeds PJ (1994) Amino acid composition of human milk is not unique. J Nutr 124:1126–1132

de Chávez JAR, Guzmán A, Zamora-Gutiérrez D, Mendoza GD, Melgoza LM, Montes S, Rosales-Torres AM (2015) Supplementation with rumen-protected L-arginine-HCl increased fertility in sheep with synchronized estrus. Trop Anim Health Prod 47:1067–1073

Dillon EL, Wu G (2021) Cortisol enhances ctrulline synthesis from proline in enterocytes of suckling piglets. Amino Acids. https://doi.org/10.1007/s00726-021-03039-y

Fligger JM, Gibson CA, Sordillo LM, Baumrucker CR (1997) Arginine supplementation increases weight gain, depresses antibody production, and alters circulating leukocyte profiles in preruminant calves without affecting plasma growth hormone concentrations. J Anim Sci 75:3019–3025

Gao H (2020) Amino acids in reproductive nutrition and health. Adv Exp Med Biol 1265:111–131

Gao HJ, Wu G, Spencer TE, Johnson GA, Li XL, Bazer FW (2009a) Select nutrients in the ovine uterine lumen: I. Amino acids, glucose and ions in uterine lumenal flushings of cyclic and pregnant ewes. Biol Reprod 80:86–93

Gao HJ, Wu G, Spencer TE, Johnson GA, Bazer FW (2009b) Select nutrients in the ovine uterine lumen: III. Expression of cationic amino acid transporters in ovine uterus and peri-implantation conceptuses. Biol Reprod 80:602–609

Gao HJ, Wu G, Spencer TE, Johnson GA, Bazer FW (2009c) Select nutrients in the ovine uterine lumen: V. Nitric oxide synthase, GTP cyclohydrolase and ornithine decarboxylase in ovine uteri and peri-implantation conceptuses. Biol Reprod 81:67–76

Gao HJ, Wu G, Spencer TE, Johnson GA, Bazer FW (2009d) Select nutrients in the ovine uterine lumen: VI. Expression of FK506-binding protein 12-rapamycin complex-associated protein 1 (FRAP1) and associated regulators and effectors of mTORC1 and mTORC2 complexes in ovine uteri and peri-implantation conceptuses. Biol Reprod 81:87–100

Gilbreath KR (2018) Role of a rumen-protected arginine product in improving fertility in beef cows. M.S. Thesis, Texas A&M University. College Station, TX, USA

Gilbreath KR, Bazer FW, Satterfield MC, Cleere JJ, Wu G (2018) Dietary supplementation with an arginine product between Days 1 and 60 of gestation enhances embryonic survival in lactating beef cows. J Anim Sci 96(Suppl 3):373

Gilbreath KR, Nawaratna GI, Wickersham TA, Satterfield MC, Bazer FW, Wu G (2019) Ruminal microbes of adult steers do not degrade extracellular L-citrulline and have a limited ability to metabolize extra-cellular L-glutamate. J Anim Sci 97:3611–3616

Gilbreath KR, Nawaratna GI, Wickersham TA, Satterfield MC, Bazer FW, Wu G (2020a) Metabolic studies reveal that ruminal microbes of adult steers do not degrade rumen-protected or unprotected L-citrulline. J Anim Sci 98:skz370

Gilbreath KR, Bazer FW, Satterfield MC, Cleere JJ, Wu G (2020b) Ruminal microbes of adult sheep do not degrade extracellular L-citrulline. J Anim Sci 98: skaa164

Gilbreath KR, Bazer FW, Satterfield MC, Wu G (2021) Amino acid nutrition and reproductive performance in ruminants. Adv Exp Med Biol 1285:43–61

Gilmore LA, Walzem RL, Crouse SF, Smith DR, Adams TH, Vaidyanathan V, Cao X, Smith SB (2011) Consumption of high-oleic acid ground beef increases HDL-cholesterol concentration but both high- and low-oleic acid ground beef decrease HDL particle diameter in normocholesterolemic men. J Nutr 141:1188–1194

Gootwine E, Rosov A, Alon T, Stenhouse C, Halloran KM, Wu G, Bazer FW (2020) Effect of supplementation of un-protected or protected arginine to prolific ewes on maternal amino acids profile, lamb survival at birth and pre- and post-weaning lamb growth. J Anim Sci 98:skaa284

Greene MA, Whitlock BK, Edwards JL, Scholljegerdes EJ, Mulliniks JT (2017) Rumen-protected arginine alters blood flow parameters and luteinizing hormone concentration in cyclic beef cows consuming toxic endophyte-infected tall fescue seed. J Anim Sci 95:1537–1544

Greene MA, Klotz JL, Goodman JP, May JB, Harlow BE, Baldwin WS, Strickland JR, Britt JL, Schrick FN (2020) Duckett SK (2020) Evaluation of oral citrulline administration as a mitigation strategy for fescue toxicosis in sheep. Transl Anim Sci 4:1–16

Halloran KM, Hoskins EC, Stenhouse C, Moses RM, Dunlap KA, Satterfield MC, Seo H, Johnson GA, Wu G, Bazer FW (2021) Pre-implantation exogenous progesterone and pregnancy in sheep. II. Effects on fetal-placental development and nutrient transporters in late pregnancy. J Anim Sci Biotechnol 12:4

Haque MN, Rulquin H, Lemosquet S (2013) Milk protein responses in dairy cows to changes in postruminal supplies of arginine, isoleucine, and valine. J Dairy Sci 96:420–430

Hassan SA, Ahmed ARA, Al-Samaraae WH, Al-Badri AA (2011) Effects of rumen protected arginine supplementation on growth rate, rumen fermentation and blood biochemicals of Awassi lambs. KSU J Nat Sci 14:38–45

He WL, Li P, Wu G (2021) Amino acid nutrition and metabolism in chickens. Adv Exp Med Biol 1285:109–131

Herring CM, Bazer FW, Johnson GA, Wu G (2018) Impacts of maternal dietary protein intake on fetal survival, growth and development. Exp Biol Med 243:525–533

Hertelendy F, Takahashi K, Machlin LJ, Kipnis DM (1970) Growth hormone and insulin secretory responses to arginine in the sheep, pig, and cow. Gen Comp Endocrinol 14:72–77

Hill TM, Bateman HG II, Aldrich Pas JM, Schlotterbeck RL (2011) Effects of adding arginine and histidine to dairy calf milk replacers. Prof Anim Scientists 27:565–570

Hoskins EC, Halloran KM, Stenhouse C, Moses RM, Dunlap KA, Satterfield MC, Seo H, Johnson GA, Wu G, Bazer FW (2021) Pre-implantation exogenous progesterone and pregnancy in sheep: I. polyamines, nutrient transport, and progestamedins J Anim Sci Biotechnol 12:39

Hou YQ, He WL, Hu SD, Wu G (2019) Composition of polyamines and amino acids in plant-source foods for human consumption. Amino Acids 51:1153–1165

Hu SD, Li XL, Rezaei R, Meininger CJ, McNeal CJ, Wu G (2015) Safety of long-term dietary supplementation with L-arginine in pigs. Amino Acids 47:925–936

Hu SD, He WL, Wu G (2021) Hydroxyproline in animal metabolism, nutrition, and cell signaling. Amino Acids. https://doi.org/10.1007/s00726-021-03056-x

Hüsier BR, Blum JW (2002) Metabolic and endocrine changes in response to endotoxin administration with or without oral arginine supplementation. J Dairy Sci 85:1927–1935

Jobgen WS, Fried SK, Fu WJ, Meininger CJ, Wu G (2006) Regulatory role for the arginine-nitric oxide pathway in metabolism of energy substrates. J Nutr Biochem 17:571–588

Jobgen WJ, Meininger CJ, Jobgen SC, Li P, Lee M-J, Smith SB, Spencer TE, Fried SK, Wu G (2009) Dietary L-arginine supplementation reduces white-fat gain and enhances skeletal muscle and brown fat masses in diet-induced obese rats. J Nutr 139:230–237

Johnson KA (2018) Alternative Feeding Strategies for Growing Cattle Grazing Endophyte-Infected Tall Fescue During the Summer. Missouri State University Graduate Theses. 3258

Keith AB, Satterfield MC, Bazer FW, Wu G (2018) Dietary supplementation with a rumen-protected L-arginine product enhances milk production by dairy cows. J Dairy Sci 101(Suppl 2):408

Kim JY, Burghardt RC, Wu G, Johnson GA, Spencer TE, Bazer FW (2011a) Select nutrients in the ovine uterine lumen: VII. Effects of arginine, leucine, glutamine, and glucose on trophectodem cell signaling, proliferation, and migration. Biol Reprod 84:62–69

Kim JY, Burghardt RC, Wu G, Johnson GA, Spencer TE, Bazer FW (2011b) Select nutrients in the ovine uterine lumen: VIII. Arginine stimulates proliferation of ovine trophectoderm cells through mTOR-RPS6K-RPS6 signaling cascade and synthesis of nitric oxide and polyamines. Biol Reprod 84:70–78

Kim JY, Burghardt RC, Wu G, Johnson GA, Spencer TE, Bazer FW (2011c) Select nutrients in the ovine uterine lumen: IX. Differential effects of arginine, leucine, glutamine and glucose on interferon tau, orinithine decarboxylase and nitric oxide synthase in the ovine conceptus. Biol Reprod 84:1139–1147

Kirchgessner M, Maierhofer R, Schwarz FJ, Eidelsburger U, (1993) Effect of feeding protected arginine on food intake, milk yield and growth hormone and amino acid levels in blood plasma of cows during the summer feeding period with grass. Arch Tierernahr 45:57–69

Kwon H, Spencer TE, Bazer FW, Wu G (2003a) Developmental changes of amino acids in ovine fetal fluids. Biol Reprod 68:1813–1820

Kwon H, Wu G, Bazer FW, Spencer TE (2003b) Developmental changes in polyamine levels and synthesis in the ovine conceptus. Biol Reprod 69:1626–1634

Kwon H, Wu G, Meininger CJ, Bazer FW, Spencer TE (2004a) Developmental changes in nitric oxide synthesis in the ovine conceptus. Biol Reprod 70:679–686

Kwon H, Ford SP, Bazer FW, Spencer TE, Nathanielsz PW, Nijland MJ, Hess BW, Wu G (2004b) Maternal undernutrition reduces concentrations of amino acids and polyamines in ovine maternal and fetal plasma and fetal fluids. Biol Reprod 71:901–908

Lassala A, Bazer FW, Cudd TA, Li P, Li XL, Satterfield MC, Spencer TE, Wu G (2009) Intravenous administration of L-citrulline to pregnant ewes is more effective than L-arginine for increasing arginine availability in the fetus. J Nutr 139:660–665

Lassala A, Bazer FW, Cudd TA, Datta S, Keisler DH, Satterfield MC, Spencer TE, Wu G (2010) Parenteral administration of L-arginine prevents fetal growth restriction in undernourished ewes. J Nutr 140:1242–1248

Lassala A, Bazer FW, Cudd TA, Datta S, Keisler DH, Satterfield MC, Spencer TE, Wu G (2011) Parenteral administration of L-arginine enhances fetal survival and growth in sheep carrying multiple pregnancies. J Nutr 141:849–855

Lee U, Garcia TP, Carroll RJ, Gilbreath KR, Wu G (2019) Analysis of repeated measures data in nutrition research. Front Biosci 24:1378–1390

Lenis YY, Elmetwally MA, Tang WJ, Satterfield C, Dunlap K, Wu G, Bazer FW (2018) Functional roles of agmatinase during the peri-implantation period of pregnancy in sheep. Amino Acids 50:293–308

Lewis TR, Emery RS (1962) Metabolism of amino acids in the bovine rumen. J Dairy Sci 45:1487–1492

Li P, Wu G (2018) Roles of dietary glycine, proline and hydroxyproline in collagen synthesis and animal growth. Amino Acids 50:29–38

Li P, Wu G (2020) Composition of amino acids and related nitrogenous nutrients in feedstuffs for animal diets. Amino Acids 52:523–542

Li XL, Bazer FW, Johnson GA, Burghardt RC, Erikson DW, Frank JW, Spencer TE, Shinzato I, Wu G (2010) Dietary supplementation with 0.8% L-arginine between days 0 and 25 of gestation reduces litter size in gilts. J Nutr 140:1111–1116

Li XL, Bazer FW, Johnson GA, Burghardt RC, Frank JW, Dai ZL, Wang JJ, Wu ZL, Shinzato I, Wu G (2014) Dietary supplementation with L-arginine between days 14 and 25 of gestation enhances embryonic development and survival in gilts. Amino Acids 46:375–384

Li XY, Zheng SX, Wu G (2021a) Nutrition and functions of amino acids in fish. Adv Exp Med Biol 1285:133–168

Li XY, Han T, Zheng SX, Wu G (2021b) Nutrition and functions of amino acids in aquatic crustaceans. Adv Exp Med Biol 1285:169–198

Li P, He WL, Wu G (2021c) Composition of amino acids in foodstuffs for humans and animals. Adv Exp Med Biol 1332:189–210

Liao SF, Vanzant ES, Boling JA, Matthews JC (2008) Identification and expression pattern of cationic amino acid transporter-1 mRNA in small intestinal epithelia of Angus steers at four production stages. J Anim Sci 86:620–631

Luther JS, Windorski EJ, Schauer CS, Kirsch JD, Vonnahme KA, Reynolds LP, Caton JS, Wu G (2008) Impacts of L-arginine on ovarian function and reproductive performance in ewes. J Anim Sci 86(E-Suppl 2)/ J Dairy Sci 9(E-Suppl 1): ii (Abstr LB5)

Luther JS, Windorski EJ, Caton JS, Wu G, Kirsch JD, Vonnahme KA, Reynolds LP, Schauer CS (2009) Effects of arginine supplementation on reproductive performance in Rambouillet ewes. Sheep Res Report 50:11–13

Ma X, Han M, Li D, Hu S, Gilbreath KR, Bazer FW, Wu G (2017) L-Arginine promotes protein synthesis and cell growth in brown adipocyte precursor cells via the mTOR signal pathway. Amino Acids 49:957–964

Mateo RD, Wu G, Bazer FW, Park JC, Shinzato I, Kim SW (2007) Dietary L-arginine supplementation enhances the reproductive performance of gilts. J Nutr 137:652–656

McCoard S, Sales F, Wards N, Sciascia Q, Oliver M, Koolaard J, van der Linden D (2013) Parenteral administration of twin-bearing ewes with L-arginine enhances the birth weight and brown fat stores in sheep. Springerplus 2:684

McCoardA S, Wards N, Koolaard J, Salerno MS (2014) The effect of maternal arginine supplementation on the development of the thermogenic programin the ovine fetus. Anim Prod Sci 54:1843–1847

McCoard SA, Sales FZ, Sciascia QL (2016) Amino acids in sheep production. Front Biosci E8:264–288
McKnight SM, Simmons RM, Wu G, Satterfield MC (2020) Maternal arginine supplementation enhances thermogenesis in the newborn lamb. J Anim Sci 98: skaa118
Mepham TB (1982) Amino acid utilization by lactating mammary gland. J Dairy Sci 65:287–298
Meyer AM, Klein SI, Kapphahn M, Dhuyvetter DV, Musser RE, Caton JS (2018) Effects of rumen-protected arginine supplementation and arginine-HCl injection on site and extent of digestion and small intestinal amino acid disappearance in forage-fed steers. Transl Anim Sci 2:205–215
NRC (National Research Council (2000) Nutrient Requirements of Beef Cattle. National Academy of Sciences National Research Council, Washington DC
NRC (National Research Council (2001) Nutrient Requirements of Dairy Cattle, 7th, Revised. National Academy Press, Washington DC
NRC (National Research Council (2007) Nutrient Requirements of Small Ruminants: Sheep, Goats, Cervids, and New World Camelids. National Academy Press, Washington DC
Palczewski MB, Petraitis H, Thomas DD (2019) Nitric oxide is an epigenetic regulator of histone post-translational modifications in cancer. Curr Opion Physiol 9:94–99
Peine JL, Jia GQ, Van Emon ML, Neville TL, Kirsch JD, Hammer CJ, O'Rourke ST, Reynolds LP, Caton JS (2018) Effects of maternal nutrition and rumen-protected arginine supplementation on ewe performance and postnatal lamb growth and internal organ mass. J Anim Sci 96:3471–3481
Peine JL, Neville TL, Klinkner EE, Egeland KE, Borowicz PP, Meyer AM, Reynolds LP, Caton JS (2020) Rumen-protected arginine in ewe lambs: effects on circulating serum amino acids and carotid artery hemodynamics. J Anim Sci 98:skaa196
Pelaez R, Phillips DD, Walker DM (1978) Amino acid supplementation of isolated soybean protein in milk replacers for preruminant lambs. Adv Exp Med Biol 105:443–452
Pisani LP, Oller do Nascimento CM, Bueno AA, Biz C, Albuquerque KT, Ribeiro EB, Lila M Oyama LM, (2008) Hydrogenated fat diet intake during pregnancy and lactation modifies the PAI-1 gene expression in white adipose tissue of offspring in adult life. Lipids Health Dis 7:13
Prezotto LD, Thorson JF, Borowicz PP, Peine JL, Bedenbaugh M, Hileman SM, Lents CA, Caton JS, Swanson KC (2018) Influences of maternal nutrient restriction and arginine supplementation on visceral metabolism and hypothalamic circuitry of offspring. Dom Anim Endocrinol 65:71–79
Recabarren SE, Jofré A, Lobos A, Orellana P, Parilo J (1996) Effect of arginine and ornithine infusions on luteinizing hormone secretion in prepubertal ewes. J Anim Sci 74:162–166
Reynolds LP, Caton JS, Redmer DA, Grazul-Bilska AT, Vonnahme KA, Borowicz PP, Luther JS, Wallace JM, Wu G, Spencer TE (2006) Evidence for altered placental blood flow and vascularity in compromised pregnancies. J Physiol 572:51–58
Reynolds LP, Borowicz PP, Caton JS, Crouse MS, Dahlen CR, Ward AK (2019) Developmental programming of fetal growth and development. Vet Clin North Am Food Anim Pract 35:229–247
Roets E, Verbeke R, Massart-Leën A-M, Peeters G (1974) Metabolism of [^{14}C]citrulline in the perfused sheep and goat udder. Biochem J 144:435–446
Saevre CB, Caton JS, Luther JS, Meyer AM, Dhuyvetter DV, Musser RE, Kirsch JD, Kapphahn M, Redmer DA, Schauer CS (2011) Effects of rumen-protected arginine supplementation on ewe serum-amino-acid concentration, circulating progesterone, and ovarian blood flow. Sheep Goats Res J 26:8–12
Sales F, Sciascia Q, van der Linden DS, Wards NJ, Oliver MH, McCoard SA (2016) Intravenous maternal arginine administration to twin-bearing ewes, during late pregnancy, is associated with increased fetal muscle mTOR abundance and postnatal growth in twin female lambs. J Anim Sci 94:2519–2531
Sandoval C, Wu G, Smith SB, Dunlap KA, Satterfield MC (2020) Maternal nutrient restriction and skeletal muscle development: consequences for postnatal health. Adv Exp Med Biol 1265:153–165
Satterfield MC, Gao HJ, Li XL, Wu G, Johnson GA, Spencer TE, Bazer FW (2010) Select nutrients and their associated transporters are increased in the ovine uterus following early progesterone administration. Biol Reprod 82:224–231
Satterfield MC, Dunlap KA, Keisler DH, Bazer FW, Wu G (2012) Arginine nutrition and fetal brown adipose tissue development in diet-induced obese sheep. Amino Acids 43:1593–1603
Satterfield MC, Dunlap KA, Keisler DH, Bazer FW, Wu G (2013) Arginine nutrition and fetal brown adipose tissue development in nutrient-restricted sheep. Amino Acids 45:489–499
Satterfield MC, Edwards AK, Bazer FW, Dunlap KA, Steinhauser CB, Wu G (2021) Placental adaptation to maternal malnutrition. Reproduction 162:R73–R83
Sawaya WN, Safi WJ, Al-Shalhat AF, Al-Mohammad MM (1984) Chemical composition and nutritive value of goat milk. J Dairy Sci 67:1655–1659
Sciascia QL, van der Linden DS, Sales FA, Wards NJ, Blair HT, Pacheco D, Oliver MH, McCoard SA (2019) Parenteral administration of L-arginine to twin-bearing Romney ewes during late pregnancy is associated with reduced milk somatic cell count during early lactation. J Dairy Sci 102:3071–3081
Seo H, Johnson GA, Bazer FW, Wu G, McLendon BA, Kramer AC (2021) Cell-specific expression of enzymes for serine biosynthesis and glutaminolysis in farm animals. Adv Exp Med Biol 1285:17–28
Smith SB, Lunt DK, Smith DR, Walzem RL (2020) Producing high-oleic acid beef and the impact of

ground beef consumption on risk factors for cardiovascular disease: A review. Meat Sci 163:108076

Stalon V, Merceniner A (1984) L-Arginine utilization by Pseudomonas species. J Gen Microbiol 130:69–76

Stewart BM, Block J, Morelli P, Navarette AE, Amstalden M, Bonilla L, Hansen PJ, Bilby TR (2011) Efficacy of embryo transfer in lactating dairy cows during summer using fresh or vitrified embryos produced in vitro with sex-sorted semen. J Dairy Sci 94:3437–3445

Sun L, Zhang H, Wang Z, Fan Y, Guo Y, Wang F (2018) Dietary rumen-protected arginine and N-carbamylglutamate supplementation enhances fetal growth in underfed ewes. Reprod Fertil Dev 30:1116–1127

Tagari H, Bergman EN (1978) Intestinal disappearance and portal blood appearance of amino acids in sheep. J Nutr 108:790–803

Teixeira PD, Tekippe JA, Rodrigues LM, Ladeira MM, Pukrop JR, Kim YHB, Schoonmaker JP (2019) Effect of ruminally protected arginine and lysine supplementation on serum amino acids, performance and carcass traits of feedlot steers. J Anim Sci 97:3511–3522

United States Department of Agriculture (USDA, 2020) Cattle Inventory. USDA National Agricultural Statistics Service. Washington, DC.

van der Linden DS, Sciascia Q, Sales F, Wards NJ, Oliver MH, McCoard SA (2015) Intravenous maternal arginine addition to twin-bearing ewes during late pregnancy enhances placental growth and development. J Anim Sci 93:4917–4925

Vicini JL, Clark JH, Hurley WL, Bahr JM (1988) Effects of abomasal or intravenous administration of arginine on milk production, milk composition, and concentrations of somatotropin and insulin in plasma of dairy cows. J Dairy Sci 71:658–665

Visser WF, Verhoeven-Duif NM, de Koning TJ (2012) Identification of a human *trans*-3-hydroxy-L-proline dehydratase, the first characterized member of a novel family of proline racemase-like enzymes. J Biol Chem 287:21654–21662

Wang XQ, Frank JW, Little DR, Dunlap KA, Satterfield MC, Burghardt RC, Hansen TR, Wu G, Bazer FW (2014a) Functional role of arginine during the peri-implantation period of pregnancy. I. Consequences of loss of function of arginine transporter *SLC7A1* mRNA in ovine conceptus trophectoderm. FASEB J 28:2852–2863

Wang XQ, Frank JW, Xu J, Dunlap KA, Satterfield MC, Burghardt RC, Romero JJ, Hansen TR, Wu G, Bazer FW (2014b) Functional role of arginine during the peri-implantation period of pregnancy. II. Consequences of loss of function of nitric oxide synthase *NOS3* mRNA in ovine conceptus trophectoderm. Biol Reprod 91:59

Wang XQ, Ying W, Dunlap KA, Lin G, Satterfield MC, Burghardt RC, Wu G, Bazer FW (2014c) Arginine decarboxylase and agmatinase: An alternative pathway for de novo biosynthesis of polyamines for development of mammalian conceptuses. Biol Reprod 90:84

Wang XQ, Johnson GA, Burghardt RC, Wu G, Bazer FW (2015a) Uterine histotroph and conceptus development. I. Cooperative effects of arginine and secreted phosphoprotein 1 on proliferation of ovine trophectoderm cells via activation of the PDK1-Akt/PKB-TSC2-MTORC1 signaling cascade. Biol Reprod 92:51

Wang XQ, Burghardt RC, Romero JJ, Hansen TR, Wu G, Bazer FW (2015b) Functional roles of arginine during the peri-implantation period of pregnancy. III. Arginine stimulates proliferation and interferon tau production by ovine trophectoderm cells via nitric oxide and polyamine-TSC2-MTOR signaling pathways. Biol Reprod 92:75

Wang XQ, Johnson GA, Burghardt RC, Wu G, Bazer FW (2016) Uterine histotroph and conceptus development. II. Arginine and secreted phosphoprotein 1 cooperatively stimulate migration and adhesion of ovine trophectoderm cells via focal adhesion-MTORC2 mediated cytoskeleton reorganization. Biol Reprod 95:71

Williams AP, Hewitt D (1979) The amino acid requirements of the preruminant calf. Br J Nutr 41:311–319

Wu G (1997) Synthesis of citrulline and arginine from proline in enterocytes of postnatal pigs. Am J Physiol 272:G1382–G1390

Wu G (1998) Amino acid metabolism in the small intestine. Trends Comp Biochem Physiol 4:39–74

Wu G (2018) Principles of Animal Nutrition. CRC Press, Boca Raton, Florida

Wu G (2020) Important roles of dietary taurine, creatine, carnosine, anserine and 4-hydroxyproline in human nutrition and health. Amino Acids 52:329–360

Wu G (2021) Amino acids: biochemistry and nutrition, 2nd edn. CRC Press, Boca Raton, Florida

Wu G (2022) Nutrition and metabolism: Foundations for animal growth, development, reproduction, and health. Adv Exp Med Biol 1354:1-24

Wu G, Knabe DA (1994) Free and protein-bound amino acids in sow's colostrum and milk. J Nutr 124:415–424

Wu G, Morris SM Jr (1998) Arginine metabolism: nitric oxide and beyond. Biochem J 336:1–17

Wu G, Knabe DA, Flynn NE (1994) Synthesis of citrulline from glutamine in pig enterocytes. Biochem J 299:115–121

Wu G, Bazer FW, Tuo W, Flynn SP (1996) Unusual abundance of arginine and ornithine in porcine allantoic fluid. Biol Reprod 54:1261–1265

Wu G, Bazer FW, Cudd TA, Meininger CJ, Spencer TE (2004) Maternal nutrition and fetal development. J Nutr 134:2169–2172

Wu G, Knabe DA, Flynn NE (2005a) Amino acid metabolism in the small intestine: biochemical bases and nutritional significance. In: Biology of Metabolism of Growing Animals (Burrin DG and Mersmann HJ eds), Elsevier, New York. pp 107–126

Wu G, Bazer FW, Hu J, Johnson GA, Spencer TE (2005b) Polyamine synthesis from proline in the developing porcine placenta. Biol Reprod 72:842–850

Wu G., Bazer FW, Wallace JM, Spencer TE (2006) Intrauterine growth retardation: implications for the animal sciences. J Anim Sci 84:2316–2337

Wu G, Bazer FW, Davis TA, Kim SW, Li P, Rhoads JM, Satterfield MC, Smith SB, Spencer TE, Yin YL (2009) Arginine metabolism and nutrition in growth, health and disease. Amino Acids 37:153–168

Wu G, Bazer FW, Burghardt RC, Johnson GA, Kim SW, Knabe DA, Li P, Li XL, McKnight JR, Satterfield MC, Spencer TE (2011) Proline and hydroxyproline metabolism: implications for animal and human nutrition. Amino Acids 40:1053–1063

Wu Z, Hou Y, Hu S, Bazer FW, Meininger Cynthia J, McNeal Catherine J, Wu G (2016) Catabolism and safety of supplemental l-arginine in animals. Amino Acids 48:1541–1552

Wu G, Bazer FW, Johnson GA, Hou YQ (2018) Arginine nutrition and metabolism in growing, gestating and lactating swine. J Anim Sci 96:5035–5051

Wu ZL, Hou YQ, Dai ZL, Hu CA, Wu G (2019) Metabolism, nutrition and redox signaling of hydroxyproline. Antioxid Redox Signal 30:674–682

Wu G, Bazer FW, Lamb GC (2020a) Introduction: Significance, challenges and strategies of animal production. In: Lamb GC, Wu G (eds) Animal Agriculture: Challenges, Innovations, and Sustainability (Bazer FW. Elsevier, New York, pp 1–20

Wu CS, Wei Q, Wang H, Kim DM, Balderas M, Wu G, Lawler J, Safe S, Guo S, Devaraj S, Chen Z, Sun Y (2020b) Protective effects of ghrelin on fasting-induced muscle atrophy in aging mice. J Gerontol A 75:621–630

Wu G, Meininger CJ, McNeal CJ, Bazer FW, Rhoads JM (2021) Role of L-arginine in nitric oxide synthesis and health in humans. Adv Exp Med Biol 1332:167–188

Xue GP, Snoswell AM, Fishlock RC (1988) Quantitative study on creatine metabolism in sheep tissues. Biochem Int 16:623–628

Wyss M, Kaddurah-Daouk R (2000) Creatine and creatinine metabolism. Physiol Rev 80:1107–1213

Zeitoun M, Al-Ghoneim A, Al-Sobayil K, Al-Dobaib S (2016) L-Arginine modulates maternal hormonal profiles and neonatal traits during two stages of pregnancy in sheep. Open J Anim Sci 6:95–104

Zhang H, Sun LW, Wang ZY, Deng MT, Zhang GM, Guo RH, Ma TW, Wang F (2016) Dietary N-carbamylglutamate and rumen-protected L-arginine supplementation ameliorate fetal growth restriction in undernourished ewes. J Anim Sci 94:2072–2085

Zhang H, Zhao F, Nie H, Ma T, Wang Z, Wang F, Loor JJ (2018) Dietary N-carbamylglutamate and rumen-protected L-arginine supplementation during intrauterine growth restriction in undernourished ewes improve fetal thymus development and immune function. Reprod Fert Dev 30:1522–1531

Zhang H, Peng A, Guo S, Wang M, Loor JJ, Wang H (2019) Dietary *N*-carbamylglutamate and L-arginine supplementation improves intestinal energy status in intrauterine-growth-retarded suckling lambs. Food Funct 10:1903–1914

Zhang Q, Hou YQ, Bazer FW, He WL, Posey EA, Wu G (2021) Amino acids in swine nutrition and production. Adv Exp Med Biol 1285:81–107

Hepatic Glucose Metabolism and Its Disorders in Fish

11

Xinyu Li, Tao Han, Shixuan Zheng, and Guoyao Wu

Abstract

Carbohydrate, which is the most abundant nutrient in plant-sourced feedstuffs, is an economically indispensable component in commercial compound feeds for fish. This nutrient can enhance the physical quality of diets and allow for pellet expansion during extrusion. There is compelling evidence that an excess dietary intake of starch causes hepatic disorders, thereby further reducing the overall food consumption and growth performance of fish species. Among the severe metabolic disturbances are glycogenic hepatopathy (hepatomegaly caused by the excessive accumulation of glycogen in hepatocytes) and hepatic steatosis (the accumulation of large vacuoles of triacylglycerols in hepatocytes). The development of those disorders is mainly due to the limited ability of fish to oxidize glucose and control blood glucose concentration. The prolonged elevations of blood glucose increase glucose intake by the liver, and excess glucose is stored either as glycogen through glycogenesis in hepatocytes or as triglycerides via lipogenesis in tissues, depending on the species. In some fish species (e.g., largemouth bass), the liver has a low ability to regulate glycolysis, gluconeogenesis, and glycogen breakdown in response to high starch intake. For most species of fish, the liver size increases with lipid or glycogen accumulation when they have a high starch intake. It is a challenge to develop the same set of diagnostic criteria for all fish species as their physiology or metabolic patterns differ. Although glycogenic hepatopathy appears to be a common disease in carnivorous fish, it has been under-recognized in many studies. As a result, understanding these diseases and their pathogeneses in different fish species is crucial for manufacturing cost-effective pellet diets to promote the health, growth, survival, and feed efficiency of fish in future.

Keywords

Fish · Carbohydrate metabolism · Glucose · Liver diseases · Disorders

Abbreviations

ACC	Acetyl-CoA carboxylase
FAS	FA synthase complex
GH	Glycogenic hepatopathy
GLUT	Glucose transporter

X. Li · T. Han · G. Wu (✉)
Department of Animal Science, Texas A&M University, College Station, TX 77843, USA
e-mail: g-wu@tamu.edu

T. Han
Department of Aquaculture, Zhejiang Ocean University, Zhoushan 316022, Zhejiang, China

S. Zheng
Guangdong Yuehai Feeds Group Co., Ltd., Zhanjiang 524017, Guangdong, China

G. Wu (ed.), *Recent Advances in Animal Nutrition and Metabolism*, Advances in Experimental Medicine and Biology 1354,
https://doi.org/10.1007/978-3-030-85686-1_11

HIS	Hepatosomatic index
LDH	Lactate dehydrogenase
NAFLD	Nonalcoholic fatty liver disease
NASH	Nonalcoholic steatohepatitis
NRC	National Research Council
PK	Pyruvate kinase
TAGs	Triacylglycerols

11.1 Introduction

The global aquaculture production has increased over the past decade at an average rate of about 3.5% per year and provides more than 50% of fish fillet for human consumption (FAO 2020). About 70% of global aquaculture relies on commercial compound feeds and, therefore, their improvement plays an important role in sustaining aquaculture development worldwide. Aquafeeds are formulated with different ingredients to meet the nutritional requirements of aquatic animals. In order to enhance the production of farmed fish per unit of land and water, fish are usually fed high-energy diets. It is also recommended to provide as much as no-protein energy sources, such as starch, in fish feeds to minimize the use of high-priced protein sources. Moreover, extruded feeds are the dominant feed type in many commercial fish cultures because they have better chemical and physical properties than expansion-compression feeds (Welker et al. 2018). The recommended starch levels in extruded aquatic feeds are usually more than 20% in order to obtain good floating properties (Riaz et al. 2011). Under intensive farming conditions, although some of the non-carnivorous fish fed such diets can grow at a desirable rate to a market weight (Li et al. 2021c), carnivorous fish (e.g., largemouth bass) often have a lower growth performance, a lower ability to resist stresses, and a higher rate of mortality when fed diets containing $\geq 20\%$ starch, as compared with diets containing <10% starch (Li et al. 2021b, c).

Dietary carbohydrate is not required for fish species (NRC 2011; Wilson 1994) but has significant role in animal metabolism such as the production of NADPH and ribose-5-phosphate via the pentose cycle (Wu 2018). In addition, dietary fiber is beneficial for the intestinal health of herbivorous and omnivorous fish by stimulating intestinal motility, inhibiting intestinal inflammation, and increasing the availability of metabolic fuels (e.g., butyrate and acetate) for the distal intestine (Wu 2018). Thus, an adequate provision of carbohydrate contributes to better growth performance and a protein-sparing effect in some fish species (Li et al. 2013; Wang et al. 2005). Dietary carbohydrates are the least expensive form of energy, but their utilization varies greatly among different fish species (Wilson 1994). For example, omnivorous and herbivorous fish can utilize dietary starch as an energy source at high efficiencies even when their diets contain as much as 40% starch (Stone 2003; Wilson 1994). By contrast, a dietary level of 15–25% digestible carbohydrate has been recommended for carnivorous fish (NRC 2011). However, excess dietary starch can produce fatty fish, reduce feed consumption, and impair the proper utilization of other dietary nutrients (Mohanta et al. 2009). Of particular note, some species of fish (e.g., largemouth bass) have a low ability to oxidize glucose to CO_2 (Li et al. 2020d, e) and develop metabolic disorders [e.g., glycogenic hepatopathy (GH)] when fed diets containing $\geq 10\%$ starch (Li et al. 2020b, c).

Recent reviews have summarized the functional ability of farmed fish species to utilize dietary starch as an energy source (Kamalam et al. 2017). However, most of the previous studies only focused on the effects of dietary starch levels and sources on fish growth and feed utilization within a relatively short period of feeding trials (mostly in 8 weeks). In practical fish-production settings, dietary starch-induced hepatic diseases present a major problem that can last for months or even years, depending on species. As a result, it is very important to pay more attention to the relationship between hepatic disorders and dietary starch in fish species. In this article, we will highlight current concepts about glucose metabolism in the liver of fish and their hepatic disorders (including hepatic diseases) of dietary origin.

11.2 Structure of the Liver in Fish

In fish, the liver plays an important role in nutrition, metabolism (both anabolism and catabolism), and physiology (Brusle et al. 1996). This organ is crucial for nutrient digestion by synthesizing and secreting bile salts that solubilize dietary fats and fat-soluble vitamins in the lumen of the intestine (Hou et al. 2020). Bile salts also exert antimicrobial effects in the intestine, serve as olfactory stimuli, and regulate whole-body cholesterol homeostasis (Buchinger et al. 2014). Bile salts are formed from the covalent conjugation of bile acids with taurine, glycine, or both, depending on animal species (Hofmann et al. 2010; Russell 2003; Wu 2018). As in dogs (Oberbauer and Larsen 2021) and cats (Che et al. 2021), all bile acids are conjugated with taurine rather than glycine in teleost fish (including hybrid-striped bass, largemouth, and zebrafish) with a few exceptions (Hagey et al. 2010; Hofmann et al. 2010; Tammar 1974). Besides producing bile salts, the liver serves as a storage site for fats and glycogen (Faccioli et al. 2014). In addition, the liver is important for the destruction of old blood cells and maintaining proper blood chemistry. Furthermore, the liver plays a central role in nitrogen (waste) disposal and excretion (Brusle et al. 1996).

Generally, the liver can be divided into two or three lobes in fish, depending on the species (Brusle et al. 1996; Faccioli et al. 2014). The main cell type of the liver is the hepatocyte. The hepatocytes of fish contain lesser amounts of organelles than those of mammals, indicating a lower synthetic activity and a lower rate of secretion in the former (Brusle et al. 1996). In contrast to mammals, the hepatic parenchyma of fish has no distinct lobules. The hepatocytes of fish are arranged as cords forming cell plates, each of which separates several lacunae (interconnected spaces between hepatic sinusoids) to form the vascular (sinusoids) and biliary (canaliculi) network (Brusle et al. 1996; Faccioli et al. 2014; Wilkins et al. 2013). According to their arrangements, there are three patterns of hepatic parenchyma in fish. In the first pattern as reported for largemouth bass, the pike, and rainbow trout, hepatocytes are radially arranged around the central vein. In the second pattern as reported for the hagfish, hepatocytes lie in the form of tubules with a bile canaliculus running through the center of this structure and with sinusoids forming an extended network around the tubule. In the third pattern as reported for some freshwater fish and marine fish, hepatocytes lie in anastomosing laminae around the central vein (Mokhtar 2017).

The vascular organization of the liver have two afferent blood vessels (hepatic artery and portal vein) and an efferent (hepatic or central vein) vessel located at the hilum (Truslé et al. 1996). Sinusoids can receive blood directly from portal vein radicles with which they are contiguous, along with hepatic artery radicles (Wilkins et al. 2013; Akiyoshi and Inoue 2004). Generally, sinusoids can be classified as three categories (Akiyoshi and Inoue 2004): (a) the cord-like form, in which the hepatocyte lining is single-layered, and the hepatic sinusoids are enlarged with straight capillaries; (b) the tubular form, in which sinusoidal capillaries are narrow, with irregularly shaped sinusoids appearing throughout the interstice between the hepatic plates as shown in Fig. 11.1e; and (c) the solid form, in which the hepatocyte lining is multilayered, with the hepatic sinusoids being narrow and short tortuous capillaries. The structure of hepatic sinusoids is physiologically important, and is essential for hepatic function (Akiyoshi and Inoue 2004).

The liver structure and functions of fish are affected by many factors, including sex, temperature, feeding, and environment (Faccioli et al. 2014). The liver is a target organ for many nutritional regulation (including factors such as food quality and quantity) that can alter its structure and metabolism (Brusle et al. 1996). Normally, the liver shows reddish brown color due to the rich vascularity, indicating that the animals are in a good nutritional status and in normal health (Труslé et al. 1996). Some cases of yellowish, pale, or light pink color have been recorded in fish fed high-energy feeds

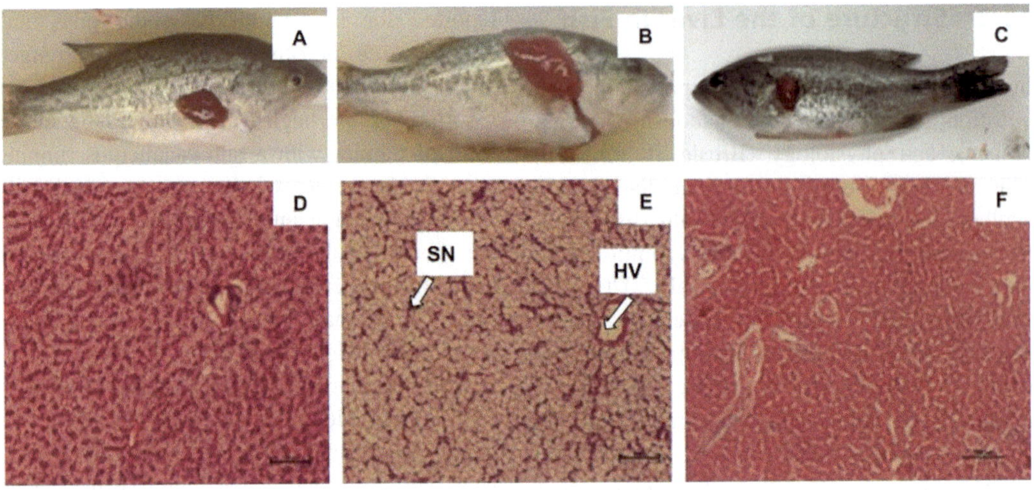

Fig. 11.1 Representative pictures of the whole body, liver, and hepatic histology of juvenile largemouth bass fed a diet containing 10 or 28% dextrinized starch for 8 weeks. Composition of the diet was the same as previously described (Li et al. 2020e). The initial and final body weights of the fish were ~18 and ~50 g, respectively. A & D (normal liver): The liver from fish fed a diet with 10% starch showed relative regular shape with reddish color and normal hepatocytes. B & E (liver hypertrophy): The liver from fish fed a diet with 28% starch showed light pink color and larger hepatocytes with peripherally located nuclei. C & F (liver atrophy): The necrosis in the liver from unhealthy fish fed a diet with low (7.26%) fishmeal and 4.3% dextrinized starch; the abnormal liver showed an unclear nucleus and unclear sinusoids (SN). HV, hepatic vein

(Fig. 11.1b). The size, shape, and volume of the liver are adapted to the anatomical space available between other visceral organs. Hepatosomatic index (HSI = liver weight/body weight × 100) is regarded as one of the most important indicators of the relative liver size in fish (Li et al. 2021a, b). HSI is also highly sensitive to nutritional status in many fishes, which can be directly affected by the quantity and quality of ingested food (Table 11.1). In a study of red spotted grouper (*Epinephelus akaara*), HSI values can range from 2.37 to 4.47% when fish are fed diets with different levels of protein and starch (Wang et al. 2016a). Generally, high HSI values are often related to low growth performance and poor health in fish species (Deng et al. 2011; Li et al. 2014; Ren et al. 2011). However, an overly low HSI may be an indicator of certain hepatic diseases. As shown in Fig. 11.1 A and C, the size of the liver with necrosis from unhealthy fish was approximately only half of the size of a normal liver.

11.3 Pathophysiology of Liver Metabolic Disease

Glycogenic hepatopathy (hepatomegaly caused by the excessive accumulation of glycogen in hepatocytes) and hepatic steatosis (the accumulation of large vacuoles of triglycerides in hepatocytes) are the common hepatic disorders in fish fed high-starch diets. In humans, the diagnosis of nonalcoholic fatty liver disease (NAFLD) has been well developed. The hallmark feature of NAFLD is steatosis, as shown in Fig. 11.2a. NAFLD also represents a histopathologic spectrum ranging from steatosis alone to necroinflammation, summarized as nonalcoholic steatohepatitis (NASH), with progression to advanced fibrosis and cirrhosis (Fig. 11.2b, c). Generally, there are two main reasons for hepatic steatosis (fatty liver): (1) alterations in hepatic uptake, synthesis, metabolism, and secretion of fatty acids; and (2) inflammation and oxidative

Table 11.1 Hepatic parameters, blood glucose concentrations, and their responses to high dietary starch intake for different farmed fish species

Fish species	Main starch sources, min–max levels of starch in the diet (%)	Feeding period (days)	Hepatic glycogen (mg/g of wet tissue)	Hepatic lipid (mg/g of wet tissue)	HSI (%)	Blood glucose (mM)	References
Herbivore							
Grass carp (*Ctenopharyngodon idellus*)	Maize starch, 6–38	62	6.76–8.14	28.2–65.7 ↑	1.13–1.68 ↑	7.49–8.73	Li et al. (2014)
	Wheat starch, 20–47	56	82.4–106	204–319 ↑	2.81–3.85 ↑	6.30–7.01	Tian et al. (2012)
	Corn starch, 12–42	56		47.3–79.4 ↑	2.63–3.69 ↑	5.81–7.01 ↑	Cai et al. (2018)
	Wheat starch, 17–26	56		107–204 ↑	2.36–2.78 ↑		Chen et al. (2012a)
Blunt snout bream (*Megalobrama amblycephala*)	Dextrin and plant meals, 19–42[d]	70	23.8–58.4 ↑	57.3–69.1	1.74–2.24 ↑	2.64–3.79 ↓	Li et al. (2013)
Omnivore							
Gibel carp (*Carassius auratus gibelio*)	Maize starch, 6–38	62	11.7–15.0 ↑	26.33–46.3 ↑	3.49–5.12 ↑	6.33–7.53 ↓	Li et al. (2014)
Nile tilapia (*Oreochromis niloticus*)	Maize grain and wheat bran, 18–40[d]	63			1.30–1.80 ↑		Ali and Al-Asgah (2001)
	Dextrin, 11–54[d]	56		↑[a]	3.70–4.54 ↑	2.65–3.05 ↑[b]	Xie et al. (2017)
Hybrid tilapia (*Oreochromis niloticus* × *O. aureus*)	Maize starch, 6–46	56			1.96–2.43 ↑		Wang et al. (2005)
Silver barb (*Puntius gonionotus*)	Dextrin, 22–38	90			1.78–2.05 ↑		Mohanta et al. (2009)
Hybrid lemon fin barb (*Barbonymus gonionotus* ♀ × *Hypsibarbus wetmorei male* ♂)	Tapioca starch, 29–44[d]	60	↑[a]	↑[a]	1.74–1.82 ↑		Sulaiman et al. (2020)
African catfish (*Clarias gariepinus*)	Maize starch and wheat flour, 15.4–37.7[d]	56	6.0–7.3	41.2–76.3	0.93–1.28 ↑[b]		Ali and Jauncey, (2004)
Yellowfin seabream (*Sparus latus*)	Raw maize starch, 10–30[d]	70	47–92 ↑	221–228 ↑	1.59–2.42 ↑	4.6–5.7 ↑[b]	Wu et al. (2007)
Lebranche mullet (*Mugilliza Valenciennes*)	Maize starch and dextrin, 32–46	34	18.2–48.3 ↑	2.80–3.84 ↑[c]	1.39–1.68 ↑[b]		Zamora-Sillero et al. (2013)

(continued)

Table 11.1 (continued)

Fish species	Main starch sources, min–max levels of starch in the diet (%)	Feeding period (days)	Hepatic glycogen (mg/g of wet tissue)	Hepatic lipid (mg/g of wet tissue)	HSI (%)	Blood glucose (mM)	References
Carnivore							
Rainbow trout (*Oncorhynchus mykiss*)	Gelatinized potato starch, 9–36	56	20–58 ↑[b]	19–34 ↓	0.90–1.30 ↑[b]		Yamamoto et al. (2001)
Cobia (*Rachycentron canadum* L)	Gelatinized maize starch, 0–30	63	4.1–32.8 ↑	202–325 ↑	2.41–4.50 ↑	1.2–5.5 ↑	Ren et al. (2011)
Golden pompano (*Trachinotus ovatus*)	Raw maize starch 0–28	56	157–287 ↑		1.19–1.46 ↑	3.29–6.29 ↑	Zhou et al. (2015)
Obscure puffer (*Takifugu obscurus*)	Gelatinized maize starch, 10–30	60	24.6–36.3 ↑	345–435 ↑	10.1–11.3 ↑	3.22–3.91 ↑	Liu et al. (2015)
Giant croaker (*Nibea japonica*)	Maize starch, 4–33[d]	56		240–269	1.39–1.79 ↑		Li et al. (2015)
Large yellow croaker (*Larmichthys crocea*)	Wheat starch, 7–29[d]	56	18.4–27.2 ↑		0.62–0.83	3.08–4.58 ↑	Zhou et al. (2016)
European sea bass (*Dicentrarchus labrax*)	Gelatinized starch, 12–30	70	40.9–105 ↑		1.59–3.11↑	6.65–7.49	Moreira et al. (2008)
Largemouth bass (*Micropterus salmoides*)	Wheat starch, 5–20	84	54.0–70.4 ↑	45.7–46.0		2.10–3.73 ↑	Lin et al. (2018)
	Maize starch, 0–25	60	58.4–81.8 ↑	34.4–66.4 ↑			Ma et al. (2019)
	Wheat starch and rice mid, 13–25[d]	148	273–322 ↑	37–51 ↑	1.20–2.10 ↑	3.39–4.83 ↑	Amoah et al. (2008)
Red-spotted grouper (*Epinephelus akaara*)	Maize starch, 0–30	56	169 - 235 ↑	9.2–14.7 ↓[b]			Wang et al. (2016a)
Hybrid grouper (*Epinephelus fuscoguttatus*♀ × *E. lanceolatus*♂)	Cassava starch, 0–18	70	127–175 ↑	44.4–67.6 ↓[b]	1.59–2.35 ↑		Li et al. (2019a)
Hybrid grouper (♂ *Epinephelus lanceolatus* × ♀ *E. fuscoguttatus*)	Maize starch, 0–28	56	117–181 ↑	63.2–74.7 ↓[b]	2.09–2.98 ↑		Luo et al. (2016)

HSI, Hepatosomatic index = liver weight/body weight × 100

[a]They are determined by the histological examination

[b]There is no significant difference ($P > 0.05$), but with a decrease or increase trend

[c]Triglycerides content in the liver (mg/g)

[d]Nitrogen-free extract (dry matter — crude protein — crude lipids — ash— crude fiber)

↑ Indicating an increase with dietary starch level; ↓ Indicating a number decrease with dietary starch level injury. The adverse outcome is hepatic damage that progresses to fibrosis and liver failure (Siddiqui et al. 2015). The definition of NAFLD requires that there is evidence for hepatic steatosis, either by imaging or by histology and there are no causes for secondary hepatic fat accumulation. In humans, NAFLD is defined by the presence of macrovesicular fat accumulation in more than 5% of hepatocytes in the liver when individuals consume less than 20 g of alcohol per day (Yuan and Bambha 2015). However, currently there are no common criteria for the diagnosis of the fatty liver among different species of fish, because different species have huge differences in physiology and metabolism as shown in Table 11.1. The content of lipids can be as much as 700 mg/g of wet weight in the liver of healthy Atlantic cod (*Gadus morhua*), while the liver of healthy Atlantic salmon (*Salmo salar*) only has nearly 100 mg lipids/g of wet weight (NRC 2011). As a result, it is imperative to develop specific criteria for the definition of the fatty liver in different fish species.

Glycogenic hepatopathy can be identified as the hepatocyte cytoplasm being markedly expanded by excess glycogen, imparting a "glassy" appearance (as shown in Fig. 11.3) in many species of animals. In humans, GH is associated with hepatic disorders in type 1 diabetes mellitus (Maharaj et al. 2017) and with NAFLD in type 2 diabetes mellitus (Chandel et al. 2017). Our recent findings have revealed that long-term GH will lead to cirrhosis and fibrosis in the liver of largemouth bass (Fig. 11.3c and d). In the GH of humans, glycogen loading, hepatomegaly, and abnormal liver enzymes are associated with other symptoms, including growth retardation and/or dwarfism, delayed puberty, cushingoid features, and hypercholesterolemia (Maharaj et al. 2017). In the GH of fish species, an enlarged liver size results from the excess glycogen accumulation in hepatocytes. However, GH in fish appears to be underrecognized by many researchers and farmers, even though this metabolic disorder has been described several times over the years for different species. This may be due to both the rarity of this lesion among terrestrial animals and a greater emphasis on fatty liver disease than GH. In humans, the histology of GH reveals these key features: (1) marked glycogen accumulation, leading to pale and swollen hepatocytes, (2) no or mild fatty change, (3) no or minimal inflammation, (4) no or minimal spotty lobular necrosis, and (5) intact architecture with no significant fibrosis (Torbenson et al. 2006). Hepatic histology remains a useful tool to confirm the diagnosis and exclude other diseases or helps to discover the concomitant chronic liver disease. Periodic acid-Schiff (PAS) stainings have been developed to identify excessive cytoplasmic glycogen in the liver of fish with GH (Fig. 11.3a and b). In histological examinations of the liver, glycogen can be removed after digestion with diastase as a means to diagnose GH (Torbenson et al. 2006; Saxena et al. 2010). As noted previously, unified diagnostic criteria of GH in fish species are difficult to establish due to the large variations of hepatic glycogen content among different fish species

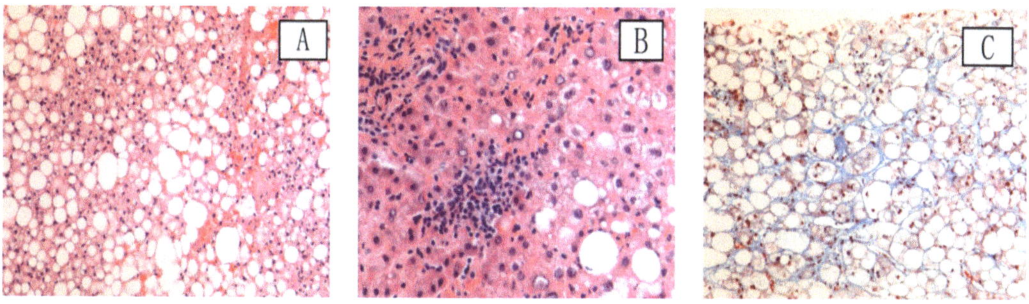

Fig. 11.2 Liver histology of human patients with nonalcoholic fatty liver disease. **a** Steatosis: Hematoxylin–eosin stain. **b** Mononuclear inflammatory infiltration: hematoxylin–eosin stain. **c** Fibrosis pattern around hepatocytes. The liver samples were examined by using the Masson's trichrome staining. Taken from free available figures in the open-access article of Cabezas et al. (2012)

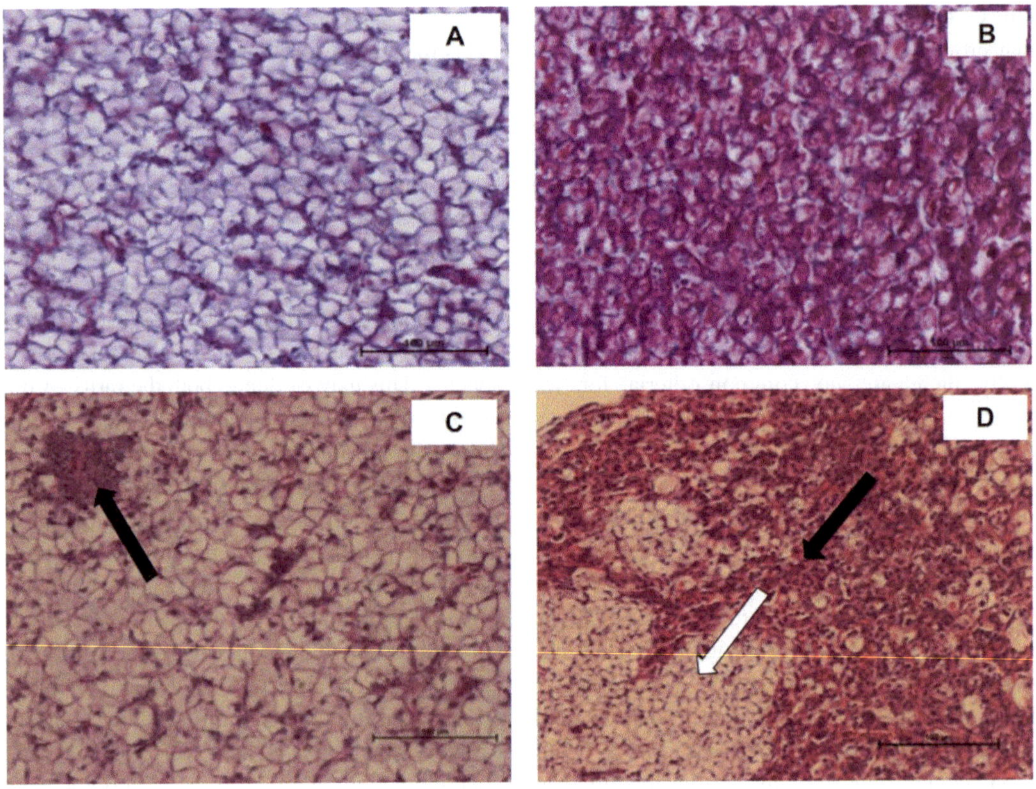

Fig. 11.3 Presence of excess glycogen in the liver of largemouth bass fed a diet containing 15% dextrinized starch, 45% crude protein, and 10% lipids (dry matter basis) for 8 weeks. Composition of the diet was the same as previously described (Li et al. 2020d). The initial and final body weights of the fish were ~8 and ~50 g, respectively. **a** Periodic Acid-Schiff stain with diastase showed enlarged and swollen hepatocytes in largemouth bass with glycogenic hepatopathy (GH). There was no steatosis or inflammation in the liver. **b** Periodic acid-Schiff staining of the liver tissue showed excess glycogen accumulation in the hepatocytes of largemouth bass. **c** H&E staining of the liver tissue from largemouth bass showed glycogenic hepatopathy (GH), with the presence of inflammation (black arrow). **d** Some largemouth bass (not all of them) with long-term GH had cirrhosis (white arrow) and fibrosis (black arrow) in the liver

(Table 11.1). For example, the hepatic glycogen content in healthy grouper is about 10 times higher than that in healthy grass carp. However, these hepatic parameters for healthy fish are very important for future diagnoses of GH in a given species.

11.4 Digestion and Metabolism of Dietary Carbohydrates

As the least expensive form of dietary energy, carbohydrates are widely included in aquafeeds (Jia et al. 2022; Li et al. 2021). Carbohydrates are polyhydroxyaldehydes (aldoses) or polyhydroxyketones (ketoses) composed of carbon, hydrogen, and oxygen (Wu 2018). Based on their structure and degree of their polymerization, carbohydrates can be classified as: monosaccharides, disaccharides, oligosaccharides, and polysaccharides. The susceptibility of carbohydrate to enzymatic degradation or bacterial fermentation and their effects on animal physiology is dependent on the composition and molecular structure of the specific carbohydrate. Generally, the apparent digestibility and intestinal uptake of carbohydrates decrease with increasing complexities (glucose <

dextrin < starch) (Kamalam et al. 2017). Moreover, compared to the raw starches, gelatinized starches generally have better digestibility for fish (NRC 2011). It should be noted that the non-starch polysaccharides in plant feedstuffs are often considered to be indigestible and thus of no nutritional value for fish (Maas et al. 2020). Carnivorous fish do not have a need for dietary fiber but can tolerate some dietary fiber (e.g., <10% and at least 20% in the diets of rainbow trout and largemouth bass, respectively; Hilton et al. 2011; Li et al. 2021c). However, dietary fiber (usually $\leq 7\%$ in compound aquafeeds) is essential for the health of many species of fish (particularly herbivores and omnivores) and may be beneficial for intestinal motility and health in some carnivorous fish (Davies 1985; Li et al. 2021c).

11.4.1 The Digestion of Dietary Starch and the Absorption of Glucose into the Portal Vein

In spite of large variations in the structural and functional anatomy of their gastrointestinal tract, almost all fish species possess the enzymatic apparatus for the hydrolysis and absorption of simple and more complex carbohydrates (Krogdahl et al. 2005). Digestion is the primary limiting step in the efficient utilization of carbohydrate (e.g., starch) for growth. The pancreatic α-amylase is an important enzyme for hydrolyzing starch into maltose. Maltose and short-chain dextrin can be further hydrolyzed by various brush border enzymes (disaccharidases or glucosidases) to produce glucose. The α-amylase and disaccharidases activities in different fish species have been well reported, which are often lower in carnivorous fish species than in herbivores and omnivores. Glucose is a polar (water-soluble) molecule that cannot readily diffuse across the hydrophobic cell membrane. Transport of glucose from the intestinal lumen to the blood stream is performed by specific transporters in enterocyte membranes, namely, the electrogenic, Na^+-dependent glucose symporter (SGLT1) in the brush border/apical membrane and the facilitative, Na^+-independent glucose transporter (GLUT2) in the basolateral membrane (Kamalam et al. 2017). Glucose concentrations in the liver are maintained at equilibrium with glucose in the blood (Castillo et al. 2009; Terova et al. 2009). In rainbow trout, GLUT2 is expressed at high abundance in the liver and at lower abundance in the intestine and kidneys, and is undetectable in other tissues (Hall et al. 2006). Increasing the dietary intake of starch augments the concentration of glucose in the blood and glycogen synthesis in the liver of fish, such as sea bass (Viegas et al. 2015) and largemouth bass (Li et al. 2020b).

11.4.2 Blood Glucose Concentrations

In fish, the absorption of the diet-derived glucose by the intestine is very efficient, and contributes to an increase in blood glucose (Hilton et al. 1987; NRC 2011). Insulin and glucagon are the two major pancreatic endocrine hormones that regulate whole-body glucose metabolism and blood glucose concentrations in higher vertebrates. In the fed state, pancreatic β cells secrete insulin in response to an increase in blood glucose concentration. Although the basal concentrations of blood glucose range from 3 to 7 mM depending on fish species, the peak concentrations vary from 5 to 30 mM among different fish species after oral feeding or the administration of glucose (Kamalam et al. 2017). In fish species, blood glucose clearance after the ingestion of a high-starch meal is used as an invasive but non-destructive criterion for assessing the ability of fish to metabolize glucose (Kamalam et al. 2017). As shown in Table 11.1, blood glucose concentration generally increases with increasing starch intake in most of carnivorous fish species, including largemouth bass (Fig. 11.4). However, herbivorous and omnivorous fish species can control their blood glucose concentrations within narrow physiological ranges even when they are fed diets with high carbohydrate levels. By contrast, carnivorous fish are largely insulin- and

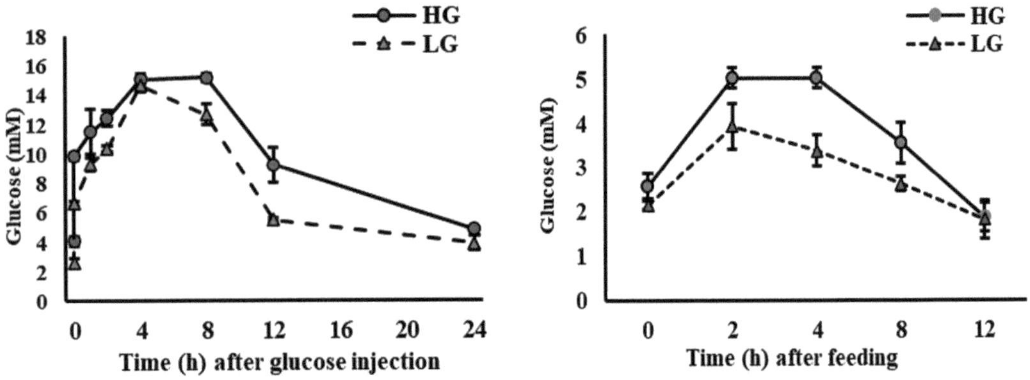

Fig. 11.4 Concentrations of glucose in the serum of largemouth bass after either satiety feeding or a single intraperitoneal injection of glucose (0.5 mol/100 g body weight). Values are means ± SEM, n = 4 fish at each sampling time. Largemouth bass (the initial body weight of ~80 g/fish) were fed, for 4 weeks, a high-starch diet containing 20% dextrinized starch (dry matter basis; Li et al. 2020d) or a low-starch diet containing 5% dextrinized starch plus 15% cellulose (dry matter basis; Li et al. 2020d). The composition of other ingredients in both the high- and low-starch diets was the same. All diets contained 45% crude protein and 10% lipids. At the end of the 4-week feeding, fish fed the 20%-starch diet exhibited glycogenic hepatopathy and had a body weight of ~140 g, whereas fish fed the 5%-starch diet had a normal liver and a body weight of ~160 g. Thereafter, all the fish were starved for 24 h and then either were re-fed the 5%-starch diet at 3% of the body weight or received a single intraperitoneal injection of glucose (0.5 mol/100 g body weight), followed by the collection of blood samples for glucose analysis at the indicated time points, as we described previously (Li et al. 2020b). HG: Fish fed the 20%-starch diet exhibited a high degree of glycogenic hepatopathy. LG: Fish fed the 5%-starch diet lacked glycogenic hepatopathy. In response to the single intraperitoneal injection of glucose, the LG fish had lower concentrations of glucose in the plasma than the HG fish at 8 and 12 h post administration ($P < 0.05$). In response to the refeeding, the LG fish had lower concentrations of glucose in the plasma than the HG fish at 2, 4, and 8 h post refeeding ($P < 0.05$)

glucose-resistant with regard to carbohydrate metabolism, and have a lower ability to regulate blood glucose concentration than herbivorous and omnivorous fish species (Kamalam et al. 2017). Food deprivation generally leads to decreased plasma glucose concentrations in many fish species without altering glucose uptake into the brain (Blasco et al. 1996).

The insulin- and glucose-resistance in fish may be related to the low expression of glucose transporter-4 (GLUT4) in their insulin-responsive peripheral tissues such as skeletal muscle and heart (Kamalam et al. 2017). As in mammals, GLUT4 is regarded as an important factor for the regulation of glucose homeostasis in lower vertebrates (Marín-Juez et al. 2013). In fish, skeletal muscle is one of many tissues that express GLUT4 at both protein and mRNA levels, and is the major site for glucose disposal (Capilla et al. 2004; Díaz et al. 2007). Studies with the Atlantic cod showed that among the tissues examined, GLUT4 was expressed most abundantly in the heart, strongly in red and white skeletal muscles, and at lower levels in the gills, gonads, intestine, and kidneys (Hall et al. 2006). However, the actual rates of glucose uptake by these tissues or whole-body glucose utilization were not determined by the authors (Hall et al. 2006). In hybrid-striped bass, skeletal muscle has a much lower rate of glucose transport, as compared with the proximal intestine, liver, and kidneys (Jia et al. 2021).

GLUT4 mRNA or protein levels in the skeletal muscle of teleost fish can be increased by hormonal stimuli (i.e., insulin and IGF-1), and swimming can stimulate AMP-activated protein kinase for glucose catabolism (Marín-Juez et al. 2013, 2014). Although fish GLUT4 is a structural and functional homolog of mammalian GLUT4, fish GLUT4 has a lower affinity for glucose and a broader substrate specificity than mammalian GLUT4 (Marín-Juez et al. 2014).

Moreover, the intracellular trafficking characteristics of fish GLUT4 are different than those of mammalian GLUT4 (Marín-Juez et al. 2013, 2014; Díaz et al. 2007). This is consistent with our recent results that skeletal muscle of largemouth bass has a lower activity to take up glucose and convert extracellular glucose into glycogen and lipids than the liver (Fig. 11.5).

11.4.3 Glucose Metabolism

In vertebrates (including fish), the liver plays a central role in controlling glucose homeostasis by serving as a consumer and/or a producer of glucose, depending on nutritional and physiological states, thereby keeping blood glucose concentration and other metabolic fuels in a

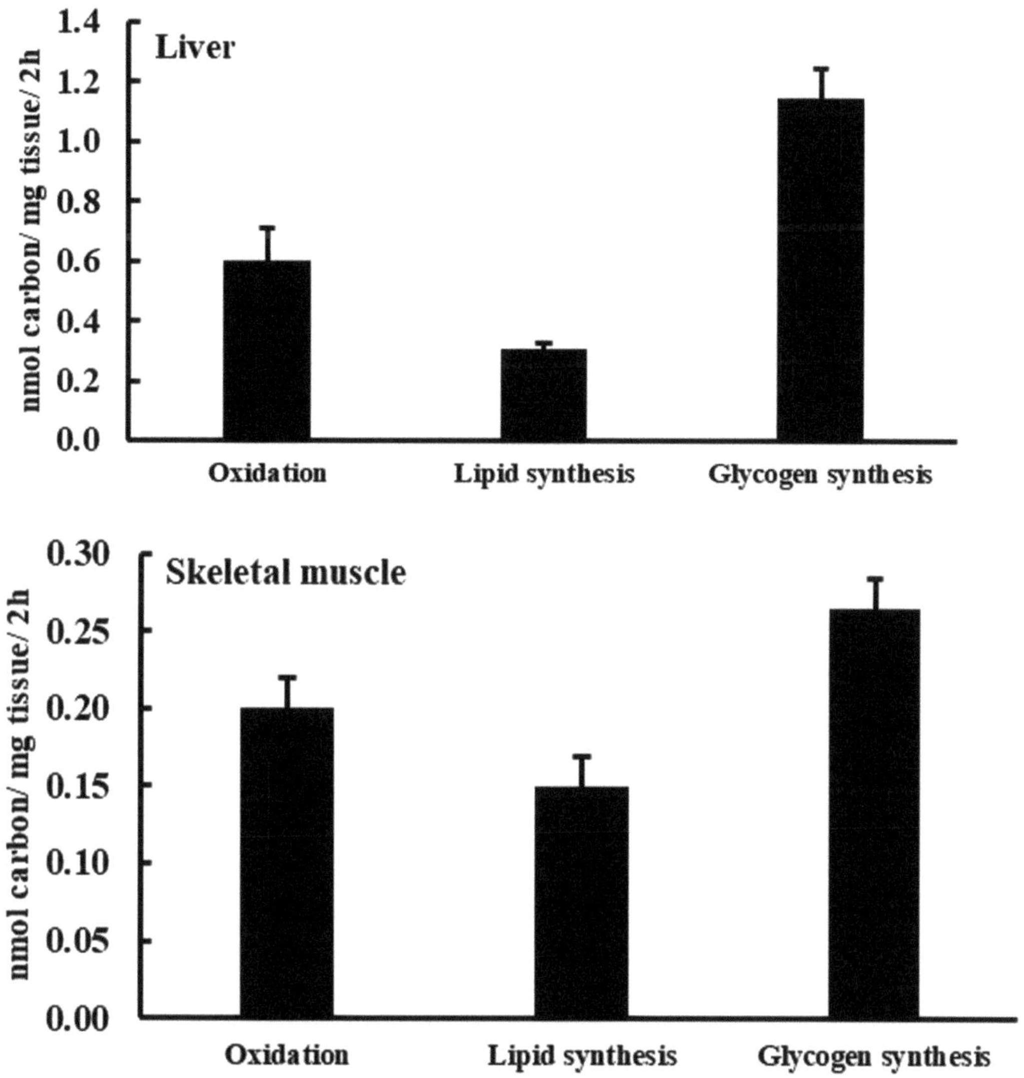

Fig. 11.5 The rates of glucose oxidation, as well as lipid and glycogen syntheses from glucose in the liver and skeletal muscle of largemouth bass (the initial body weight = ~5 g/fish) fed for 8 weeks a diet containing 10% dextrinized starch, 45% crude protein, and 10% lipids (dry matter basis; Li et al. 2020d). The final body weight of the fish was 25–30 g. Slices of the liver (~20 mg) or skeletal muscle (~50 mg) was incubated in an oxygenated (95% O_2/5% CO_2) Krebs bicarbonate medium with 5 mM glucose and [U-^{14}C]glucose (Li et al. 2020b). Data are expressed as means ± SEM, n = 10

steady state (Kamalam et al. 2017; Enes et al. 2009). Blood glucose enters the liver via the hepatic portal vein. In fish, prolonged elevations of blood glucose increase glucose intake by their liver. Glucose is metabolized by the liver via four main pathways: (1) the pentose phosphate pathway (pentose cycle) to produce NADPH and ribose 5-phosphate; (2) glycogenesis to store excess glucose as glycogen (3) lipogenesis to store excess glucose as triglycerides; (4) the glycolysis and the Krebs cycle to oxidize glucose to CO_2 and water, as well as produce ATP and GTP (Fig. 11.6). However, excess starch intake by fish increases their blood glucose concentrations, leading to liver insulin resistance, the excess accumulation of glycogen or lipids in hepatocytes, and the impairment of liver function (Prisingkorn et al. 2017). Hyperglycemia also causes glucose autooxidation and the production of reactive oxygen species, as well as protein glycosylation (Fang et al. 2002).

11.4.3.1 Glycolysis

In fish, hepatic glycolysis is a critical metabolic pathway to maintain glucose homeostasis. Glycolysis provides some energy and intermediates for the body (Jia et al. 2021). The pathway of glycolysis is a sequence of enzyme-catalyzed reactions in which glucose is converted to pyruvate. This metabolic pathway exists in all

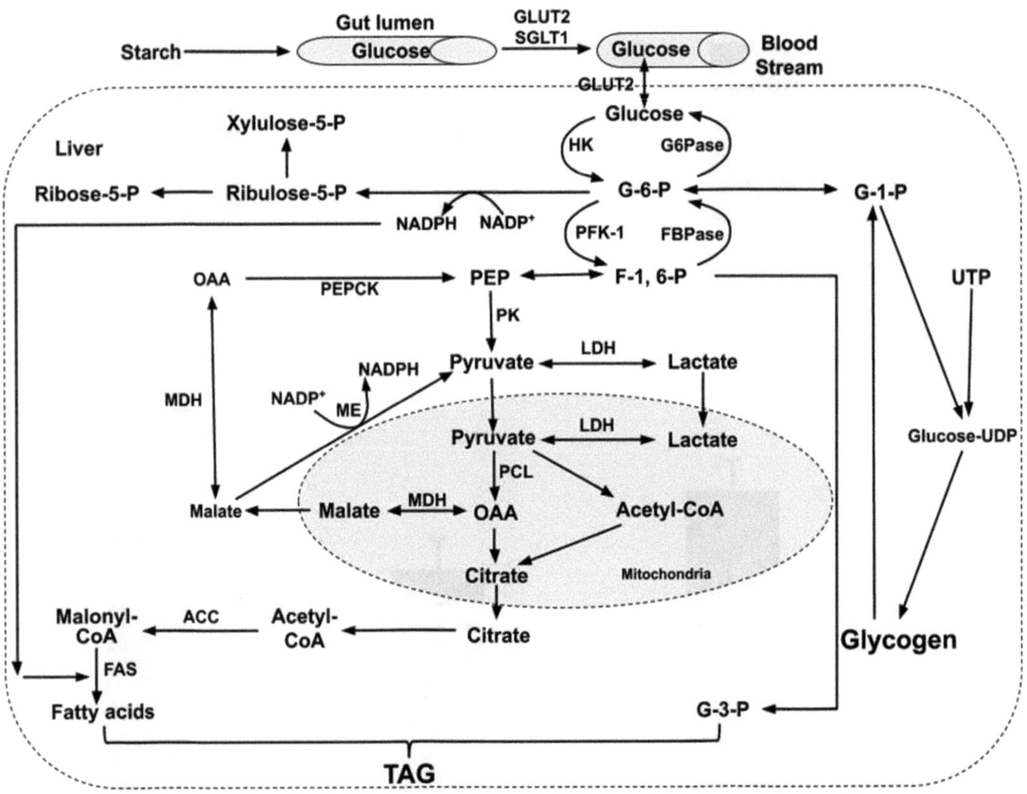

Fig. 11.6 Utilization of dietary carbohydrate in fish through the processes of digestion, intestinal glucose transport, and glucose metabolism in the liver and other tissues. SGLT1, Na^+-dependent glucose symporter; HK, hexokinases; LDH, lactate dehydrogenase; PFK-1, phosphofructokinase-1; PEPCK, phosphoenolpyruvate carboxykinase; PCL, pyruvate carboxylase; G6Pase, glucose-6-phosphatase; G-6-P, glucose 6-phosphate; G-3-P, glucose 3-phosphate; F-6-P, fructose-1,6-bisphosphate; PEP, phosphoenolpyruvate; PK, pyruvate kinase; G-1-P, glucose 1-phosphate; UTP, uridine triphosphate; UDP, uridine diphosphate; Ac-CoA, Acetyl-CoA; OAA, Oxaloacetic acid; FAS, fatty acid synthase; α-KG, α-ketoglutaric acid; ACC, acetyl-CoA carboxylase; TAG, triglycerides

cell types of fish (Enes et al. 2009; Hemre et al. 2002) and produces ATP in the absence of oxygen. Glycolysis in cells is regulated by not only glucose uptake but also three rate-controlling reactions: (1) the phosphorylation of glucose to glucose 6-phosphate by hexokinases (HK) in the presence of Mg^{2+}; (2) the conversion of fructose 6-phosphate into fructose-1,6-bisphosphate by phosphofructokinase-1 (PFK-1); and (3) the conversion of phosphoenolpyruvate (PEP) into pyruvate by pyruvate kinase (PK). It is generally accepted that the activities of these glycolytic enzymes play an important role in glucose utilization and metabolism by fish (Kamalam et al. 2017). Similar to mammals, there are four closely related hexokinase isozymes in fish, named as HK I-III and glucokinase (HK IV). Among them, glucokinase has a low affinity for glucose (high Km), and is important for regulating blood glucose concentrations after a meal. Hepatic glucokinase expression and activity are strongly related to changes in blood glucose concentrations in fish, including rainbow trout, common carp, and gilthead seabream (Panserat et al. 2014).

In cells with mitochondria, the pyruvate produced via glycolysis can be further catabolized to acetyl-CoA, which subsequently enters the Krebs cycle for oxidation to CO_2 with the production of NADH and $FADH_2$ (Wu 2018). Both NADH and $FADH_2$ are oxidized via the mitochondrial respiratory chain to produce water and ATP (30 mol ATP/mol glucose). Pyruvate dehydrogenase (PDH), a very large and multienzyme complex in the mitochondrial matrix, converts pyruvate into acetyl-CoA. This enzyme connects glycolysis with the Krebs cycle and, therefore, represents a key regulatory step in glucose metabolism (Saunier et al. 2016). Note that in contrast to terrestrial mammals, the red blood cells of fish and birds as well as almost all amphibians and reptiles retain a nucleus and functional mitochondria (Martos-Sitcha et al. 2017; Stier et al. 2013). In cells without mitochondria and under anaerobic or hypoxic conditions, pyruvate is converted into lactate by lactate dehydrogenase (LDH). As a result, the total production of ATP or CO_2 from glucose oxidation is dependent on the flux of pyruvate into the Krebs cycle or the LDH. Interestingly, a high activity of LDH has been observed in some tissues of fish species (Moon and Foster 1995), indicating a rapid flux of pyruvate through this enzyme. It should be kept in mind that lactate can be oxidized to CO_2 or utilized for glucose synthesis in the liver and kidneys, depending on nutritional and physiological states (Wu 2018). Additionally, other tissues (e.g., skeletal muscle, brain, heart, and intestine) may oxidize lactate via the cooperation of enzymes in the cytosol and mitochondria (Skilleter and Kun 1972; Szczesna-Kaczmarek 1990). The shuttling of lactate between producer and consumer cells can fulfill the physiological requirements for ATP production, gluconeogenesis, and the hormonal regulation of the metabolic pathways (Adeva-Andany et al. 2014; Brooks 2018). For example, in the Cori cycle, lactate produced by anaerobic glycolysis in skeletal muscle, red blood cells, and immunocytes enters the liver where lactate is converted into glucose or glycogen, depending on nutritional and physiological states. However, it has been reported that the Cori cycle is less important in certain fish species than other animals (Jobling 1994; Milligan and Pagnotta 1991; Tang and Boutilier 1991), possibly due to a limited synthesis of glucose and glycogen in the liver. There is a suggestion that lactate may be an important substance for the maintenance and functions of teleost tissues (Moon and Foster, 1995). It remains to be defined whether this may be valid for other fish species.

Generally, an enhancement of glycolysis is indicative of increased glucose catabolism and an attempt to reduce blood glucose concentration. There are reports that high dietary carbohydrate can enhance hepatic glucokinase expression and activity in herbivorous fish, such as *Megalobrama amblycephala* (Li et al. 2016). However, this ability is usually limited in carnivorous fish (Viegas et al. 2015). Although hepatic glucokinase activity is increased with increasing the dietary starch level from 10 to 20% in European sea bass and gilthead sea bream (Enes et al. 2006; Moreira et al. 2008), the enzymatic activity does not further increase when the dietary starch level is higher than 20% (Moreira et al. 2008). Similar results have also been reported for large yellow croaker (Zhou et al. 2016). Growing evidence shows that a low glucokinase or PK activity in tissues limits the ability of fish to catabolize glucose (Li et al. 2020d, e; Tranulis et al. 1991; Moon and Foster, 1995). These findings clearly indicate that fish (especially carnivorous fish) can regulate glycolysis only in response to a narrow range of dietary carbohydrate levels. While dietary protein has long been known to be a major source of energy for fish (Wilson and Halver 1986), our recent studies have shown that hybrid-striped bass (Jia et al. 2017), zebrafish (Jia et al. 2017), largemouth bass (Li et al. 2020d, e, f, g), and crustaceans (Li et al. et al. 2021d) use amino acids (glutamate, glutamine, and aspartate) as primary metabolic fuels in the intestine, liver, skeletal muscle, and kidneys. In all these tissues, glucose and fatty acids make only a minor contribution to ATP production. Thus, the oxidation on glucose plus palmitate contributes to only 18% of total ATP production in the whole body of juvenile largemouth bass, with 82% of the needed ATP being generated from the oxidation of amino acids (Li and Wu, 2019; Li et al. 2020d). This is a major reason why fish, like crustaceans (Li and Wu 2022; Li et al. 2021d), have a particularly high requirement for dietary amino acids.

11.4.3.2 Gluconeogenesis

Gluconeogenesis is the principal metabolic pathway for glucose production, and its main precursors include lactate, pyruvate, glycerol, and gluconeogenic amino acids. This metabolic pathway occurs in the liver and kidneys. A major function of gluconeogenesis is to meet the needs of the body for glucose when dietary carbohydrate is inefficient. Although the conversion of pyruvate into glucose occurs via the same intermediates as those in glycolysis, different enzymes are used for catalyzing four irreversible reactions: pyruvate carboxylase, phosphoenolpyruvate carboxykinase (PEPCK), fructose-1,6-bisphosphatase (FBPase), and glucose-6-phosphatase. In some fish species (e.g., carnivores), endogenous glucose synthesis plays an important role in maintaining glucose homeostasis in the body (Skiba-Cassy et al. 2013). For example, in European seabass, the contribution of endogenously derived glucose to the total blood glucose is 85% and 54%, respectively, when fed a diet without starch or containing 30% digestible starch (Viegas et al. 2012, 2015).

In higher vertebrates, glucagon and starvation are known to enhance the expression of gluconeogenic enzymes, whereas insulin, refeeding, and high dietary starch intake have the opposite effect (Wu 2018). Similar results were observed in omnivorous fish (Kamalam et al. 2017). For example, endogenous glucose synthesis is depressed by the high-starch diet in carp (Shimeno et al. 1995). It is generally accepted that carnivorous species have a limited ability to regulate glucogenic enzymes in response to nutritional status and dietary starch intake (Kamalam et al. 2017). Thus, in rainbow trout, no inhibition of gluconeogenesis was observed regardless of whether they consumed carbohydrates (Panserat et al. 2001, 2002). Clearly, species differences exist in the regulation of hepatic gluconeogenesis among fish species. There are suggestions that a futile cycle based on the phosphorylation and immediate dephosphorylation of glucose may contribute to persistent postprandial hyper-glycemia in carnivorous fish, such as rainbow trout fed high-starch diets (Kamalam et al. 2017; Skiba-Cassy et al. 2013). On the contrary, gluconeogenesis is decreased by a high-starch diet in largemouth bass (Lin et al. 2018; Ma et al. 2019) and silver sea bream (Leung et al. 2012).

11.4.3.3 The Pentose Phosphate Pathway (Pentose Cycle)

The pentose cycle is an alternative route for glucose oxidation, which is also the second most important pathway for glucose metabolism in animal cells (Wu 2018). The major function of this cycle is to provide NADPH and ribose 5-phosphate for biosynthetic processes in animals (Wu 2018). The beginning molecule for the pentose phosphate pathway is glucose-6-P, which is the second intermediate metabolite of the glycolysis pathway. Glucose-6-P is oxidized to ribulose-5-phosphate via a series of reactions in the presence of $NADP^+$. The production of NADPH by glucose-6-P dehydrogenase is irreversible and this enzyme is strongly inhibited by NADPH and fatty acyl-CoA. Because NADPH is required by anti-oxidative reactions in cells, the dependence of the organisms on the oxidation of fatty acids (that generates fatty acyl-CoA) for the major energy source is not beneficial for their health. For completing the pentose cycle, ribulose-5-phosphate is isomerized by phosphopentose isomerase to produce ribose-5-phosphate from ribulose-5-P. The latter is one of the main building blocks of nucleic acids. Furthermore, phosphopentose epimerase catalyzes a chiralty rearrangement about the center carbon of the pentose, yielding xylulose-5-phosphate. Both ribose-5-phosphate and xylulose-5-phosphate can then be rearranged via transketolase to produce glyceraldehyde-3-P and sedoheptulose-7-P. Subsequently, transaldolase converts glyceraldehyde-3-P and sedoheptulose-7-P into erythrose-4-P and fructose-6-P.

An appropriate intake of dietary carbohydrate can exert its protective role against oxidative stress in the liver by direct roles of lactate (Groussard et al. 2020) and pyruvate (Wang et al. 2007) in scavenging free radicals and reactive oxygen species and also through an increase in the activity of glucose-6-phosphate dehydrogenase to generate NADPH for the regeneration of glutathione (GSH) (Castro et al. 2016; Wu 2018). A high level of dietary starch may contribute to protection from oxidative stress in common dentex (*Dentex dentex;* Pérez-Jiménez et al. 2017). Therefore, starch is a better energy source for fish because of their low oxidative stress compared to lipids, especially fish oil and saturated fats (Castro et al. 2016; Torfi Mozanzadeh et al. 2017). This effect of dietary starch has been previously described in some fish species, such as rainbow trout (*Oncorhynchus mykiss*), sole (*Solea senegalensis*), and yellow catfish (*Pelteobagrus fulvidraco*), in which oxidative damage was lower when diets had high starch levels (Álvarez et al. 1999; Castro et al. 2012; Wang et al. 2014). Consistent with this notion, the activities of some serum enzymes, such as ALT and AST, increased with decreasing the ratio of dietary carbohydrate to lipids in silvery-black porgy (*Sparidentex hasta*), possibly due to the damage of the liver (Torfi Mozanzadeh et al. 2017). Moreover, the NADPH produced via the pentose cycle plays a vital role in fatty acid synthesis in cells (Fig. 11.6). As a result, those cells (e.g., hepatocytes, mammary epithelial cells, and activated macrophages) that have a high activity of the pentose cycle are usually active in the synthesis of fatty acids. However, as noted previously, hyper-glycemia induced by excess starch intake can cause oxidative stress, resulting in the damage to organs in the body (Wu 2020a). Furthermore, high dietary starch is harmful to fish species (e.g., largemouth bass) that are very sensitive to dietary starch intake (Li et al. 2020b, c; Li et al. 2021c). Thus, the diets of fish should contain an appropriate level of starch to achieve its physiological benefits, while preventing GH.

11.4.3.4 Glycogenesis and Glycogenolysis

Similar to higher vertebrates, fish can store excess glucose in the form of glycogen via glycogenesis and degrade glycogen to glucose via glycogenolysis, depending on nutritional states and physiological needs. These two metabolic pathways are present in many tissues of fish but occur primarily in their liver and skeletal muscle (Wu 2018). Bruslé et al. (1996) reported that the hyaloplasm (the clear, fluid portion of the cytoplasm) of fish hepatocytes contained a variable amount of stored products, such as glycogen in a typical rosette pattern and lipid globules of different electron densities. Hepatic glycogen content is extremely variable in fish, and can reach up to 320 mg/g (Table 11.1). Glycogen is present in the cytosol in the form of granules, ranging in diameter from 10 to 40 nm. In the liver, glycogen synthesis and degradation are regulated to maintain blood-glucose levels to prevent hypo- and hyper-glycemia in healthy animals. For glycogen synthesis, glucose is first phosphorylated to glucose 6-P, which then is isomerized to glucose 1-P. Glucose 1-P reacts with uridine triphosphate (UTP) to form uridine diphosphate glucose (UDPG) by UDP-Glucose pyrophosphorylase. Catalyzed by glycogen synthase, the glucose residues from the UDPG donor add to a growing glycogen molecule, with glucose residues being linked in α-1,4 bonds. Both blood-borne glucose and the glucose synthesized via gluconeogenesis from lactate and amino acids are used for hepatic glycogen synthesis.

Glycogen degradation consists of three steps: (1) the release of glucose 1-P from glycogen, (2) the remodeling of the glycogen substrate via debranching to permit further degradation, and (3) the conversion of glucose 1-P into glucose 6-P for further metabolism. The glucose 6-P derived from the breakdown of glycogen is initial substrate for glycolysis and the pentose phosphate pathway to yield NADPH and ribose 5-P. Glucose 6-P also can be converted into free glucose in the liver and kidneys (Viegas 2012).

Glycogen metabolism in animal tissues is regulated via the cAMP-dependent cell signaling (Wu 2018). As in mammals, glycogen synthesis (glycogenesis) and breakdown (glycogenolysis) in the liver of fish are catalyzed by glycogen synthase (GSase) and glycogen phosphorylase (GPase), respectively. These two enzymes are sensitive to cAMP-dependent phosphorylation and dephosphorylation reactions. Protein phosphorylation, which occurs during fasting and exercise with an increase in blood glucagon concentration, inhibits GSase but activates GPase to favor glycogen breakdown. By contrast, protein dephosphorylation, which occurs during starch and protein feeding with an increase in blood insulin concentration, activates GSase but inhibits GPase to favor glycogen synthesis from glucose. Thus, a change in the ratio of glucagon to insulin, the circulating levels of catecholamine hormones (e.g., epinephrine and norepinephrine), and other factors that affect cAMP availability plays an important role in regulating hepatic glycogen metabolism.

A significant concern over feeding carnivorous fish is the dietary starch level. In general, plant-sourced feedstuffs contain much more digestible carbohydrates than animal-sourced feedstuffs (Hou et al. 2019; Li et al. 2021e; Li and Wu 2020, 2022). Excess dietary starch may be utilized for the synthesis of glycogen and fatty acids in the liver. In some carnivorous fish, such as largemouth bass, excess glucose is preferentially converted into glycogen instead of fats or CO_2 plus water. As shown in Table 11.1, hepatic glycogen content and the HSI of three grouper species were increased with increasing dietary starch levels, whereas hepatic lipid content tended to decrease possibly due to a more hydrophilic cytosol. In our recent studies with largemouth bass, we found that blood glucose clearance was related to the degree of glycogen accumulation in liver (Fig. 11.4). When the rate of glycogen synthesis from glucose is much higher than the rate of lipid synthesis from glucose, hepatic glycogen is an important sink for excess blood glucose in this species. In humans, glycogen storage disorders (GSD) are well defined as a group of inborn errors of metabolism with abnormal storage or utilization of glycogen (Maharaj et al. 2017). These metabolic disease result from various genetic deficiencies of enzymes for glycogen breakdown or synthesis, or from mutations of proteins regulating glycogen metabolism. Many fish species have similar problems, mainly due to their limited ability to oxidize glucose in the whole body and their preference to converting glucose into glycogen in the liver when fed a high-starch diet (Palmer et al. 1972). As shown in Table 11.1, some carnivorous fish species have high glycogen content in the liver (expressed per fresh organ weight), such as 115 mg/g in European sea bass, 287 mg/g in Golden pompano (*Trachinotus ovatus*), and 227 mg/g in grouper (*Epinephelus akaara*). In most of these species, an increase in the dietary content of digestible starch will increase the deposition of glycogen in the liver. For example, in *E. akaara* (a carnivorous fish), hepatic glycogen content (expressed per fresh organ weight) was increased from 169 to 227 mg/g with increasing dietary starch levels (Wang et al. 2016a). Similar results were observed in some herbivorous fish species, such as Blunt snout bream (*Megalobrama amblycephala*) (Li et al. 2013). Using [U-^{14}C]glucose as a tracer, we found that increasing the dietary starch level from 5 to 15% increased the synthesis of glycogen from glucose in the whole liver of largemouth bass (Li et al. 2020b).

In humans, an excess and irreversible accumulation of glycogen in hepatocytes will cause glycogenic hepatomegaly, leading to liver dysfunction and hepatomegaly (Lin and Kao 2012). The negative effects of high dietary starch on the growth, liver function, non-specific immune responses, and hepatic anti-oxidative status of fish have been well documented (Zhou et al. 2013a, b; Waagbø et al. 1994) due to glycogen deposition in the liver. For example, the hypertrophy of hepatocytes in hybrid catfish occurs with increasing dietary starch levels as a consequence of glycogen accumulation (Bernardes et al. 2016). A high level of dietary starch results in high HSI values in many fish species, including silver barb (*Puntius gonionotus*; Mohanta et al. 2009), cobia (*Rachycentron canadum*; Ren et al. 2011), and giant croaker (*Nibea japonica*; Li et al. 2014), as detailed in Table 11.1. High dietary starch also causes pathological changes in the hepatic appearance of fish and their hepatic histology (Tan et al. 2007), which may result from hepatic structural abnormalities and dysfunction. Furthermore, a decrease of growth accompanied by an increase of hepatic glycogen occurs in some fish species fed diets containing high dietary starch levels (Wang et al. 2016a; Fynn-Aikins et al, 1992).

11.4.3.5 Lipogenesis from Glucose

Over-ingestion of starch is a major cause of nonalcoholic fatty liver disease (NAFLD) in both humans and farm animals (Zivkovic et al. 2007). As mentioned above, the catabolism of glucose produces acetyl-CoA or NADPH via glycolysis plus PDH and the pentose cycle, which can increase fatty acid (FA) synthesis in the liver. Long-chain FAs (cytotoxic molecules at elevated concentrations) can be esterified to TAGs for storage. The process during which the liver synthesizes FAs and TAGs from glucose- or amino acid-derived acetyl-CoA is called de novo lipogenesis (DNL). FA synthesis occurs in the cytosol, and this metabolic pathway includes a complex cytosolic polymerization in which glucose is converted to acetyl-CoA by glycolysis and the oxidation of pyruvate. Acetyl-CoA carboxylase (ACC) then converts acetyl-CoA and bicarbonate into malonyl-CoA. The production of malonyl-CoA is the initial and controlling step in FA synthesis. Then, malonyl-CoA and acetyl-CoA form a C_4 fatty acid chain by the FA synthase complex (FAS) (Fig. 11.6), a multi-functional enzyme complex that can catalyze the subsequent reactions of decarboxylation, reduction by NADPH, dehydration, and reduction by NADPH. The major source of NADPH is the pentose cycle but malic enzyme also generates NADPH (Fig. 11.6). A C_4 fatty acid chain further elongates to palmitic acid (C_{16}) by the addition of 6 C_2 units via 6 malonyl-CoA. Depending on the metabolic state, FAs can be incorporated into TAGs, which consist of the glycerol backbone to which three fatty acid molecules are esterified. Much evidence shows that the amount of de novo synthesis of lipids from glucose is limited in the liver of some fish species, particularly carnivorous fish such as largemouth bass and rainbow trout (Brauge et al. 1995; Rawles et al. 2008; NRC 2011; Li et al. 2020b). Nonetheless, there is evidence that in rainbow trout, hepatic lipogenesis from [U-^{14}C] glucose is increased with increasing dietary starch levels, such that the production of phospholipids, TAGs, and fatty acids is elevated (Brauge et al. 1995). Likewise, using [U-^{14}C] glucose as a tracer, we found that increasing the dietary starch level from 5 to 15% moderately increased the synthesis of lipids from glucose in the whole liver of largemouth bass (Li et al. 2020b). However, hepatic lipid content was paradoxically 39% lower in largemouth bass fed with 15%-starch diet than fish fed with 5%-starch diet (Li et al. 2020b), indicating that the export of lipids from the liver was greater in the former than in the latter.

It has been hypothesized that glucose may increase lipid deposition in the liver of fish by reducing the use of lipids for metabolic fuels (NRC 2011). This is partly because the glucose-derived malonyl-CoA (the carboxylation product of the glucose-derived acetyl-CoA) inhibits the activity of carnitine palmitoyltransferase I, a mitochondrial enzyme that is responsible for the transport of long-chain acyl-CoA from the cytosol into the mitochondria for β-oxidation (Wu 2018). In the liver and skeletal muscle of largemouth bass, the oxidation of glucose into acetyl-CoA is limited and, therefore, increasing the dietary starch level from 5 to 15% slightly increased intramuscular lipid content by 14% over an 8-week period (Li et al. 2020b). This is likely true for many carnivorous fish.

Lipogenesis may play an important role in controlling glucose homeostasis in some fish (Kamalam et al. 2017). The stimulation of lipogenesis by dietary glucose is related to a high insulin secretion when fish are fed a high-starch diet. The expression of key lipogenic enzymes is controlled at the transcriptional level through the combined actions of several factors such as sterol regulatory element-binding protein-1c (SREBP-1c), liver X receptor (LXR), and the carbohydrate response element-binding protein (ChREBP) (Kamalam et al. 2017). In mammals, hepatic lipogenesis is regulated independently by insulin and glucose, through the activation of SREBP-1c and ChREBP, which can further cause hepatic steatosis (Fabbrini et al. 2010; Dentin et al. 2005). Previous studies have shown that hepatic lipogenesis in rainbow trout is stimulated by insulin, as indicated by increases in hepatic mRNA and protein abundances as well as the enzymatic activity of ATP-citrate lyase, acetyl-CoA carboxylase, and fatty acid synthase (Polakof et al. 2011). Thus, elevated blood insulin concentrations are positively correlated with the improvement in the postprandial clearance of blood glucose in rainbow trout that have been genetically selected for high intramuscular fat content (Skiba-Cassy et al. 2009).

Increased activities or expression of hepatic lipogenic enzymes by high dietary starch levels have been observed in some fish species (NRC 2011), including juvenile tilapia (*Oreochromis niloticus*) (Jiang et al. 2013), juvenile white sturgeon (Fynn-Aikins et al. 1992), and European seabass (Dias et al. 2005). Juvenile tilapia (*Oreochromis niloticus* × *O. aureus*) fed diets with $\geq$ 30% starch had significantly higher hepatic lipid content than fish fed diets with 6 or 14% starch (Wang et al. 2005). Hepatic lipid content increased with an increasing dietary starch intake in juvenile cobia (*Rachycentron canadum*; Ren et al. 2011), rainbow trout (*Oncorhynchus mykiss*; Brauge et al. 1994), and Wuchang bream (*Megalobrama amblycephala*; Zhou et al. 2013b), as summarized in Table 11.1. In juvenile tilapia, the liver displayed a large number of vacuoles in hepatocytes when the dietary starch level is over 40% (Jiang et al. 2013). All of those results indicate that high or excess starch uptake may increase lipogenesis in the liver of fish, which may further lead to a fatty liver and related metabolic diseases. However, this notion does not apply to largemouth bass (as noted previously) and possibly some other fish species. Consistent with this view, Dai et al. (2016) have indicated that, unlike rodents or humans, the hepatic expression of fatty acid biosynthetic enzymes in rainbow trout is more responsive to dietary protein or amino acid intake than dietary starch or glucose intake. This discrepancy represents another important metabolic difference among carnivores, omnivores, and herbivores. As shown in Table 11.1, carnivores, such as grouper species, prefer to deposit glucose as glycogen in the liver rather than lipids.

11.5 Inflammation, Oxidative Stress, and Fibrosis

Hepatic inflammation, fibrosis, and cirrhosis can occur at the middle or late stage of fatty liver disease when the liver is damaged to a large extent (Farrell and Larter 2006; Marchesini et al. 2003; Cabezas et al. 2012). Hepatic fibrosis is the inappropriate repair of the liver in response to its chronic injuries such as NAFLD, whereas inflammation is the most common and important feature of liver fibrosis (Szabo et al. 2018). Excess fat accumulation induces lipotoxicity and the release of cell damage-associated substances, which further activate both Kupffer cells and hepatic stellate cells to promote inflammation and fibrosis, respectively. Activated Kupffer cells produce inflammatory cytokines and chemokines such as tumor necrosis factor-alpha (TNFα), interleukin-1β (IL-1β), interleukin-6 (IL-6), and

C–C motif ligand s2 and 5 (CCL2 and CCL5), which contribute to injury and inflammatory necrosis in hepatocytes (Byrne 2010; Arrese et al. 2016). During the development of NAFLD, mitochondria also produce ROS that damage hepatocytes, trigger inflammation, and contribute to insulin resistance (Satapati et al. 2015). Fortunately, hepatic inflammation and fibrosis can be reversed if they are detected in the early stages and effective steps are taken to prevent further damage and repair the injured tissues. Cirrhosis has generally been regarded as a permanent condition due to severe, prolonged damage to the liver by inflammation/fibrosis, and is irreversible; therefore, this metabolic disease must be controlled.

It has been suggested that the features of the GH include the absence of significant inflammation, fibrosis, and cirrhosis in humans (Torbenson et al. 2006). Our recent study indicated that under the long-term stress of hepatic glycogen accumulation, the liver of fish could further exhibit abnormal conditions such as inflammation, fibrosis, and cirrhosis (Fig. 11.7). This is also the first evidence that hepatic inflammation, fibrosis, and cirrhosis can be developed from GH in aquatic animals. Therefore, metabolic disorders in the liver of farmed fish fed commercial pellet diets with high starch levels may be attributed to the GH. It has been suggested that there is link between liver glycogen content and antioxidant capacity in salmon (Lygren and Hemre 2001). For example, excessive dietary starch leads to oxidative stress and impaired innate immunity, thereby negatively affecting the health of fish, as reported for *M. amblycephala* (Zhou et al. 2013b) and *M. salmoides* (Lin et al. 2018; Ma et al. 2019; Guo et al. 2020). In juvenile grouper (*E. akaara*), high dietary starch levels can result in poor growth and inflammatory immune responses (Yang et al. 2018). Future studies are warranted to better define the development of GH, inflammation, oxidative stress, and fibrosis in the liver of fish, as well as terrestrial mammals (including humans) and birds (Wu 2020a). Largemouth bass is an excellent animal model to address this important issue.

11.6 Treatment and Prevention of GH and Hepatic Steatosis in Fish

As highlighted previously, in many fish species (particularly carnivores), excess starch intake can induce hepatic disorders because of abnormal hepatic glycogenosis and/or lipogenesis resulting from prolonged elevations of blood glucose. The simple, best way to prevent this metabolic problem is to control the dietary starch level and intake. Due to its good water stability, flavor, and improved utilization efficiency, pelleted feeds (especially floating) are now commonly manufactured through extrusion (a process used to create objects of a fixed cross-sectional area) for the intensive production of aquafeeds (Watanabe 2002). The recommended starch levels for floating and sinking extrusion feeds are 20% and 10%, respectively, to create sufficient expansion and low densities (Riaz et al. 2011). As many carnivorous fish cannot tolerate more than 10% of starch, the excess starch intake may be a common challenge for feeding these fish (e.g., largemouth bass and grouper) in practical production settings. Moreover, the sources of starch have a significant impact on the extrusion process (Riaz et al. 2011). For example, the content of amylose in starch improves binding and assists in expansion at the extruder die, and its level varies with starch sources (Chinnaswamy and Hanna 1984; Riaz et al. 2011). The amylose content is 55%, 20%, 22%, and 28% in high amylose corn, potato, tapioca, and wheat, respectively (Riaz et al. 2011). In the future, low-starch pelleted aquafeeds, with floating or sinking, should be developed.

Except for the control of dietary starch levels, some nutrients (e.g., highly unsaturated fatty acid, phospholipids, choline, betaine, and carnitine) may be used to prevent the development of fatty livers in animals (Wu 2020b). To date, agents for NASH treatment include insulin sensitizers (e.g., thiazolidinediones, metformin, and incretin mimetics), antioxidants (e.g., vitamin E, vitamin C, and betaine), and cytoprotective agents (e.g., ursodeoxycholic acid and pentoxifylline), and

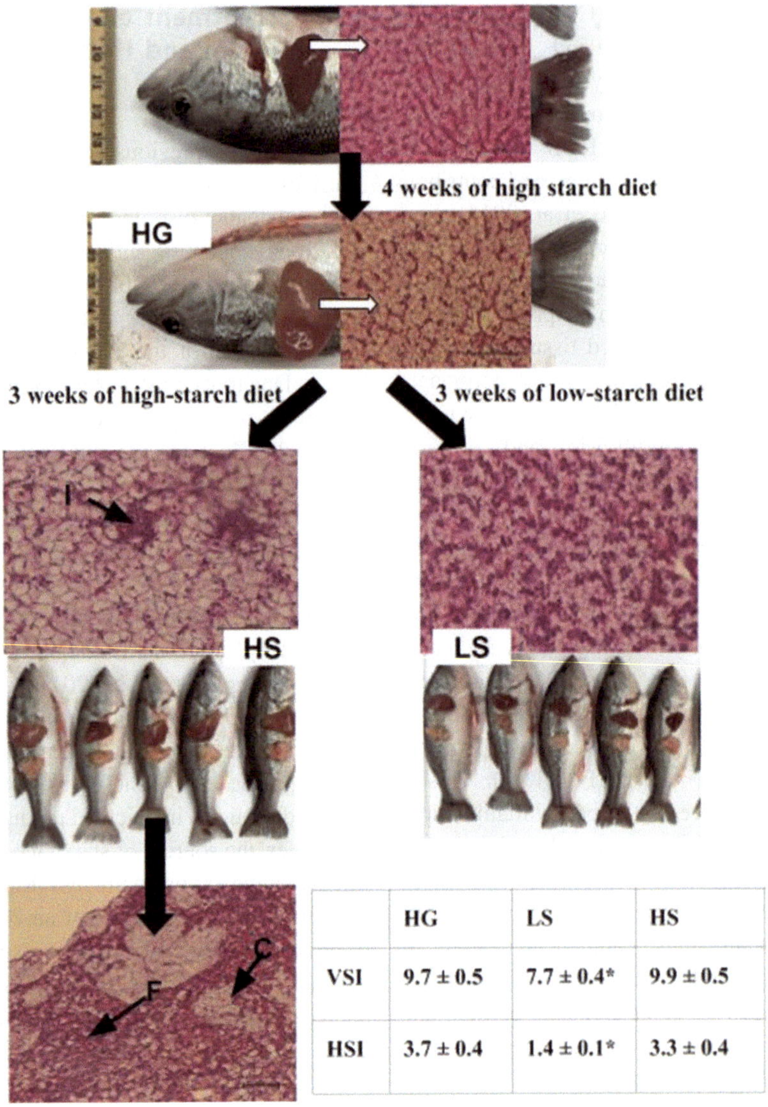

	HG	LS	HS
VSI	9.7 ± 0.5	7.7 ± 0.4*	9.9 ± 0.5
HSI	3.7 ± 0.4	1.4 ± 0.1*	3.3 ± 0.4

Fig. 11.7 Histological analysis of the liver of juvenile largemouth bass fed a high-starch (HS) diet containing 20% dextrinized starch (dry matter basis; Li et al. 2020d) for 4 weeks, followed by either the HS diet or a low-starch (LS) diet containing 5% dextrinized starch plus 15% cellulose (dry matter basis; Li et al. 2020d) for 3 weeks. The composition of other ingredients in both the HS and LS diets was the same. All diets contained 45% crude protein and 10% lipids. HG: Fish with a high degree of glycogenic hepatopathy. HS to HG: Fish fed the HS diet exhibited an abnormal structure, glycogenic hepatopathy, and inflammation (I) in the liver after the additional 3-week feeding. Some fish with glycogenic hepatopathy further developed cirrhosis (C) and fibrosis (F) in the liver. LS to HG: Fish fed the LS diet had a normal structure in the liver and no glycogenic hepatopathy after the additional 3-week feeding. The final body weight of the fish was ~95 g. HSI, Hepatosomatic index (%) = liver weight/body weight × 100. VSI, viscerosomatic index (%) = visceral weight/body weight × 100

others (e.g., bile acids, vitamin D) (Torres et al. 2012). However, an ideal treatment for NASH patients has not yet been established, and most of these therapeutic trials lack a sufficient power to definitively show a benefit in humans (Torres et al. 2012). Interestingly, the findings of recent studies revealed that dietary additive, such as L-carnitine (Jin et al. 2019a), choline (Jin et al. 2019b), fenofibrate (Jin et al. 2020), bile acids (Liao et al. 2020), butylated hydroxytoluene (Yu et al. 2018), can partially prevent or treat the fatty liver symptom. Because arginine has been reported to effectively reduce hepatic fats in rats (Fu et al. 2005; Jobgen et al. 2009) and tilapia (Li et al. 2020a), this amino acid is a promising nutrient to improve liver function in farmed animals, including fish (Li et al. 2009; Wu 2021; Wu et al. 2021).

Glycogenic hepatopathy without fibrosis and cirrhosis is reversible as glycogen can be degraded to glucose in response to low dietary starch intake. In humans, improved glycemic control is the mainstay of the management for GH (Sherigar et al. 2018). We found that GH could be eliminated when largemouth bass with the existing hepatic disorder were fed a low-starch diet for at most 3 weeks (Fig. 11.7). Additionally, the growth, protein utilization efficiency, feed utilization efficiency, and feed intake were also improved in the largemouth bass fed a low-starch diet. To date, there has not yet been research to develop pharmacological treatment strictly targeting or preventing GH in humans and animals (Sherigar et al. 2018). In largemouth bass consuming a high amount of starch, dietary supplementation with bile acids has been reported to enhance liver function and immune responses, and to reduce oxidative stress, thereby improving the growth performance and feed efficiency of the fish (Guo et al. 2020). However, the underlying mechanisms are unknown.

11.7 Concerns Over GH and Hepatic Steatosis in Fish

The potential negative effects of excess dietary starch have been under-recognized in some previous studies with fish. Thus, in nutrition research, starch has often been used as an isoenergetic substitute for dietary protein and lipids. For example, in a study to determine the requirement of Mexican silverside (*Menidia estor* Jordan 1879) for dietary protein, seven isoenergetic diets were prepared to contain 25 to 55% protein by decreasing the dietary starch level from 49.5 to 24.5% (Martínez-Palacios et al. 2007). Although no hepatic variables were measured by these authors, the feed intake of the fish fed the 25%-protein diet was only about 60% of that for the fish fed the 40%-protein diet. In another study involving the requirement of *E. akaara* for dietary protein, an enlarged liver size and increased blood glucose concentrations were observed when fish were fed the lowest protein diet that contained the highest starch level (Wang et al. 2016b). Similar results were reported in other fish species such as juvenile Spinibarbus hollandi (Yang et al. 2003) and juvenile black sea bream (Zhang et al. 2010), in which the liver size, feed intake, and hepatic glycogen were inversely correlated with dietary protein levels but positively correlated with dietary starch levels. As a result, the high dietary starch level itself may induce negative effects on the growth, feed utilization, and hepatosis of the fish, which could influence the accuracy of their recommended requirements for dietary protein or lipids. As a carnivore species, the recommended dietary starch level for largemouth bass is less than 10% (Li et al. 2020d, e; Lin et al. 2018; Ma et al. 2019). We found that this fish developed severe GH and exhibited poor growth performance when fed diets containing $\geq$ 15% starch (Li et al. 2020d, e). However, as shown in Table 11.2, the starch inclusion levels in many nutrition studies with this species varied from 10 to 30%, and this issue must be considered when determining the dietary requirements of fish for nutrients and the physiological functions of the nutrients.

Much evidence shows that excess starch intake results in an enlarged liver size in most fish species, and fish can further develop either GH or a fatty liver, depending on their species, dietary nutrient composition, and experimental conditions (Table 11.1). Alternations in hepatic variables may not necessarily indicate the onset of diseases, because there are normal ranges or fluctuations in physiological variables under different conditions and at different stages of life (Wu 2022). Although high dietary starch

Table 11.2 Starch inclusion levels in diets used for nutritional studies of largemouth bass (*Micropterus salmoides*)

Objectives of studies	Initial body weight (g/fish)	Dietary protein levels (%)	Dietary lipid levels (%)	Dietary starch levels (%), sources	References
Protein requirement	11.28	31–68	10.00	0–12, dextrin	Anderson et al. (1981)
Lipid requirement	16.2	40.0	7–16	15.6, wheat flour	Bright et al. (2005)
	8.7	48–51	12.00	11–26, wheat flour	
Vitamin C requirement	6.67	45.1–45.5	12.5–12.8	20, wheat flour	Chen et al. (2015)
Oxidized fish oil effects	5.09–5.12	45.0	13	20, wheat flour	Chen et al. (2012b)
Fat types	15.7	40.0	7.3–8.1	26, wheat flour	Tidwell et al. (2007)
Arginine requirement	24.7–25.2	45.3–46.4	12.2–12.4	11, α-starch	Zhou et al. (2012)
Lipid sources	5	46.8–47.8	13.7–13.9	23, wheat flour	Subhadra et al. (2006)
Protein sources	3.1–6.9	37.8–42.9	6.8–8.5	20, wheat flour	Tidwell et al. (2005)
Selenium supplementation	4.9	48.0	9.21	16.8, wheat flour	Zhu et al. (2012)
Vitamin E requirements	7.54	48.4	13.4	13, α-starch	Li et al. (2018)
Lysine Requirements	1.29	43.0–43.3	9.12–9.37	30, dextrin	Dairiki et al. (2007)
Vitamin E and selenium supplementation	6.35	44.5–45.5	13.0–13.5	20, wheat flour	Chen et al. (2013)
Butylated hydroxytoluene supplementation	6.20	50.3–51.3	13.4–15.1	21, wheat flour	Yu et al. (2018)
Oxidized fish oil effects	31.46	55	12.3–12.8	12, wheat flour	Yin et al. (2019)
Starch sources	36.3	52.3–52.4	11.2–12.3	10, different sources[a]	Li et al. (2019b)
Lipid sources	33.8	43.3–44.2	8.3–8.7	24, maize starch	Zhang et al. (2019)
Synbiotics[b] supplementation	4.5	47.0–47.3	12.1–12.3	16–17, wheat flour	Gong et al. (2019)
Brewer's yeast hydrolysate supplementation	34	44.3	7.5	20, wheat flour	Zhou et al. (2018)

[a]10% of cassava starch, potato starch, pea starch, or dextrin in each treatment; [b] Mixtures of probiotics

intake can result in oxidative stress, suppress innate immunity, and affect the health status of fish species (Lin et al. 2018; Zhou et al. 2013b), it is not clear whether these alternations are common in GH or the fatty liver. As a result, knowledge about the pathogenesis and biomarkers of metabolic diseases are very important for improving the utilization of carbohydrate and maintaining the health of farmed fish. Both GH and the fatty liver should be considered chronic liver diseases that can last over months and even years, eventually leading to cirrhosis and the end-stage liver failure. The duration of production for farmed fish species is usually more than 6 months. In grouper culture, it usually takes about 1–2 years to raise them to a market size (Tucker 1999). However, the experimental period of most nutritional studies with fish is usually only about 8 weeks (Table 11.1). As a result, the negative impact of the hepatic diseases induced by an excess starch intake on fish production could be underestimated if researchers and farmers rely only on results from short-term experiments.

11.8 Conclusions and Perspectives

Although many studies have been conducted to understand carbohydrate nutrition and metabolism in different fish species, the relationship between dietary starch and hepatic diseases in all aquatic animals is still not clear. Glycogenic and steatosis hepatopathy are the common liver diseases in fish fed high-starch diets. For most species of fish, the size of the liver is increased with the accumulation of lipids or glycogen in this organ when the animals are fed a high-starch diet. Although the fatty liver syndrome is more commonly recognized than GH in most animal species, GH should be considered in the diagnosis of hepatic disorders when fish exhibit an unusually enlarged liver and excess glycogen accumulation. For carnivorous fish, high dietary starch levels generally result in high blood glucose concentrations. An augmented influx of glucose from the diet leads to an excess accumulation of glycogen or lipids in the liver, depending on the species of fish. Based on results of current studies, carnivorous fish appear to deposit the diet-derived glucose as glycogen rather than TAGs in the liver. For some species, GH is often misdiagnosed as hepatic steatosis solely based on the appearance and size of the liver. It is a challenge to develop unified diagnostic criteria of GH and hepatic steatosis for all species of fish because they can differ substantially in both physiology and metabolism. As a result, understanding these hepatic diseases and their pathogenesis in different fish species is key to developing species- and developmental stage-specific pelleted feeds to ensure high efficiency, high productivity, and sustainability of aquaculture in the future.

Acknowledgements This work was supported by Texas A&M AgriLife Research (H-8200) and Guangdong Yeuhai Feeds Group Co., Ltd. We thank students and research assistants in our laboratory for helpful discussions.

References

Adeva-Andany M, López-Ojén M, Funcasta-Calderón R, Ameneiros-Rodríguez E, Donapetry-García C, Vila-Altesor M, Rodríguez-Seijas J (2014) Comprehensive review on lactate metabolism in human health. Mitochondrion 17:76–100

Akiyoshi H, Inoue A (2004) Comparative histological study of teleost livers in relation to phylogeny. Zool Sci 21:841–850

Ali A, Al-Asgah NA (2001) Effect of feeding different carbohydrate to lipid ratios on the growth performance and body composition of nile tilapia (*Oreochromis niloticus*) fingerlings. Anim Res 50:91–100

Ali MZ, Jauncey K (2004) Optimal dietary carbohydrate to lipid ratio in African catfish *Clarias gariepinus* (Burchell 1822). Aquacult Int 12:169–180

Alvarez MJ, Lopez-Bote CJ, Diez A, Corraze G, Arzel J, Dias J, Kaushik SJ, Bautista JM (1999) The partial substitution of digestible protein with gelatinized starch as an energy source reduces susceptibility to lipid oxidation in rainbow trout (*Oncorhynchus mykiss*) and sea bass (*Dicentrarchus labrax*) muscle. J Anim Sci 77:3322–3329

Amoah A, Coyle SD, Webster CD, Durborow RM, Bright LA, Tidwell JH (2008) Effects of graded levels of carbohydrate on growth and survival of largemouth bass, *Micropterus salmoides*. J World Aquacult Soc 39:397–405

Anderson RJ, Kienholz EW, Flickinger SA (1981) Protein requirements of smallmouth bass and largemouth bass. Nutr J 111:1085–1097

Arrese M, Cabrera D, Kalergis AM, Feldstein AE (2016) Innate immunity and inflammation in NAFLD/NASH. Digest Dis Sci 61:1294–1303

Bernardes CL, Navarro RD, Guerra-Santos B, Fortes-Silva R (2016) Effects of dietary carbohydrate/lipid ratios on growth, body composition, and nutrient

utilization of hybrid catfish (*Pseudoplatystoma reticulatum x Leiarius marmoratus*). Rev Colomb Cienc Pec 29:58–65
Blasco Ferna´ndez-Borra´s J, Marimon I, Requena A, J (1996) Plasma glucose kinetics and tissue uptake in brown trout in vivo: effect of an intravascular glucose load. J Comp Physiol B 165:534–541
Brauge C, Medale F, Corraze G (1994) Effect of dietary carbohydrate levels on growth, body composition and glycaemia in rainbow trout, *Oncorhynchus mykiss,* reared in seawater. Aquaculture 123:109–120
Brauge C, Corraze G, Médale F (1995) Effects of dietary levels of carbohydrate and lipid on glucose oxidation and lipogenesis from glucose in rainbow trout, *Oncorhynchus mykiss*, reared in freshwater or in seawater. Comp Biochem Physiol A 111:117–124
Bright LA, Coyle SD, Tidwell JH (2005) Effect of dietary lipid level and protein energy ratio on growth and body composition of largemouth bass *Micropterus salmoides*. J World Aquacult Soc 36:129–134
Brooks GA (2018) The science and translation of lactate shuttle theory. Cell Metab 27:757–785
Brusle J, Anadon GG (1996) The structure and function of fish liver. In: Munshi JSD, Dutta HM (eds) Fish Morphology, CRC Press, Boca Raton. pp 77–93
Buchinger TJ, Li W, Johnson NS (2014) Bile salts as semiochemicals in fish. Chem Senses 39:647–654
Byrne CD (2010) Fatty liver: role of inflammation and fatty acid nutrition. Prostaglandins Leukot Ess Fat Acids 82:265–271
Cabezas J, Mayorga M, Crespo J (2012) Nonalcoholic fatty liver disease: a pathological view. In: Tagaya N (ed) Liver biopsy—indications, procedures, results. InTechOpen, Croatia, pp 161–188
Cai W, Liang XF, Yuan X, Liu L, He S, Li J, Li B, Xue M (2018) Different strategies of grass carp (*Ctenopharyngodon idella*) responding to insufficient or excessive dietary carbohydrate. Aquaculture 497:292–298
Capilla E, Díaz M, Albalat A, Navarro I, Pessin JE, Keller K, Planas JV (2004) Functional characterization of an insulin-responsive glucose transporter (GLUT4) from fish adipose tissue. Am J Physiol 287:E348–E357
Castillo J, Crespo D, Capilla E, Diaz M, Chauvigne F, Cerda J, Planas JV (2009) Evolutionary structural and functional conservation of an ortholog of the GLUT2 glucose transporter gene (SLC2A2) in zebrafish. Am J Physiol 297:R1570–R1581
Castro C, Diógenes AF, Coutinho F, Panserat S, Corraze G, Pérez-Jiménez A, Peres H, Oliva-Teles A (2016) Liver and intestine oxidative status of gilthead sea bream fed vegetable oil and carbohydrate rich diets. Aquaculture 464:665–672
Castro C, Pérez-Jiménez A, Guerreiro I, Peres H, Castro-Cunha M, Oliva-Teles A (2012) Effects of temperature and dietary protein level on hepatic oxidative status of Senegalese sole juveniles (*Solea senegalensis*). Comp Biochem Physiol A 163:372–378
Chandel A, Scarpato B, Camacho J, McFarland M, Mok S (2017) Glycogenic hepatopathy: resolution with minimal glucose control. Case Rep Hepatol 2017:7651387
Che DS, Nyingwa PS, Ralinala KM, Maswanganye GMT, Wu G (2021) Amino acids in the nutrition, metabolism, and health of domestic cats. Adv Exp Med Biol 1285:217–231
Chen YJ, Tian LX, Yang HJ, Chen PF, Yuan Y, Liu YJ, Liang GY (2012a) Effect of protein and starch level in practical extruded diets on growth, feed utilization, body composition, and hepatic transaminases of juvenile grass carp, *Ctenopharyngodon idella*. J World Aquacult Soc 43:187–197
Chen YJ, Liu YJ, Yang HJ, Yuan Y, Liu FJ, Tian LX, Liang GY, Yuan RM (2012b) Effect of dietary oxidized fish oil on growth performance, body composition, antioxidant defence mechanism and liver histology of juvenile largemouth bass *Micropterus salmoides*. Aquac Nutr 18:321–331
Chen YJ, Liu YJ, Tian LX, Niu J, Liang GY, Yang HJ, Yuan Y, Zhang YQ (2013) Effect of dietary vitamin E and selenium supplementation on growth, body composition, and antioxidant defense mechanism in juvenile largemouth bass (*Micropterus salmoide*) fed oxidized fish oil. Fish Physiol Biochem 39:593–604
Chen YJ, Yuan RM, Liu YJ, Yang HJ, Liang GY, Tian LX (2015) Dietary vitamin C requirement and its effects on tissue antioxidant capacity of juvenile largemouth bass, *Micropterus salmoides*. Aquaculture 435:431–436
Chinnaswamy R, Bhattacharya KR (1984) Relationship between amylose content and expansion characteristics of parboiled rice. J Cereal Sci 2:273–279
Dai W, Panserat S, Kaushik S, Terrier F, Plagnes-Juan E, Seiliez I, Skiba-Cassy S (2016) Hepatic fatty acid biosynthesis is more responsive to protein than carbohydrate in rainbow trout during acute stimulations. Am J Physiol 310:R74–86
Dairiki JK, Dias CTD, Cyrino JEP (2007) Lysine requirements of largemouth bass, *Micropterus salmoides*: a comparison of methods of analysis of dose-response trials data. J Appl Aquacult 19:1–27
Davies SJ (1985) The role of dietary fibre in fish nutrition. In: Roberts RJ (ed) Recent advances in aquaculture (Muir JF. Croom Helm, London, United Kingdom, pp 219–249
Díaz M, Antonescu CN, Capilla E, Klip A, Planas JV (2007) Fish glucose transporter (GLUT)-4 differs from rat GLUT4 in its traffic characteristics but can translocate to the cell surface in response to insulin in skeletal muscle cells. Endocrinology 148:5248–5257
Deng DF, Ju ZY, Dominy W, Murashige R, Wilson RP (2011) Optimal dietary protein levels for juvenile Pacific threadfin (*Polydactylus sexfilis*) fed diets with two levels of lipid. Aquaculture 316:25–30
Dentin R, Girard J, Postic C (2005) Carbohydrate responsive element binding protein (ChREBP) and sterol regulatory element binding protein-1c (SREBP-

1c): two key regulators of glucose metabolism and lipid synthesis in liver. Biochimie 87:81–86

Dias J, Alvarez MJ, Arzel J, Corraze G, Diez A, Bautista JM, Kaushik SJ (2005) Dietary protein source affects lipid metabolism in the European seabass (*Dicentrarchus labrax*). Comp Biochem Physiol A 142:19–31

Enes P, Panserat S, Kaushik S, Oliva-Teles A (2006) Effect of normal and waxy maize starch on growth, food utilization and hepatic glucose metabolism in European sea bass (*Dicentrarchus labrax*) juveniles. Comp Biochem Physiol A 143:89–96

Enes P, Panserat S, Kaushik S, Oliva-Teles A (2009) Nutritional regulation of hepatic glucose metabolism in fish. Fish Physiol Biochem 35:519–539

Farrell GC, Larter CZ (2006) Nonalcoholic fatty liver disease: from steatosis to cirrhosis. Hepatology 43 (S1):S99-112

Fabbrini E, Sullivan S, Klein S (2010) Obesity and nonalcoholic fatty liver disease: biochemical, metabolic, and clinical implications. Hepatology 51:679–689

Faccioli CK, Chedid RA, Bombonato MTS, Vicentini CA, Vicentini IBF (2014) Morphology and histochemistry of the liver of carnivorous fish *Hemisorubim platyrhynchos*. Int J Morphol 715–720

Fang YZ, Yang S, Wu G (2002) Free radicals, antioxidants, and nutrition. Nutrition 18:872–879

FAO (2020) The state of world fisheries and aquaculture 2020. Food and Agriculture Organization of the United Nations, Rome

Fu WJ, Haynes TE, Kohli R, Hu J, Shi W, Spencer TE, Carroll RJ, Meininger CJ, Wu G (2005) Dietary L-arginine supplementation reduces fat mass in Zucker diabetic fatty rats. J Nutr 135:714–721

Fynn-Aikins K, Hung SS, Liu W, Li H (1992) Growth, lipogenesis and liver composition of juvenile white sturgeon fed different levels of D-glucose. Aquaculture 105:61–72

Gong Y, Yang F, Hu J, Liu C, Liu H, Han D, Jin J, Yang Y, Zhu X, Yi J, Xie S (2019) Effects of dietary yeast hydrolysate on the growth, antioxidant response, immune response and disease resistance of largemouth bass (*Micropterus salmoides*). Fish Shellfish Immun 94:548–557

Groussard C, Morel I, Chevanne M, Monnier M, Cillard J, Delamarche A (2020) Free radical scavenging and antioxidant effects of lactate ion: an in vitro study. J Appl Physiol 89:169–175

Guo JL, Kuang WM, Zhong YF, Zhou YL, Chen YJ, Lin SM (2020) Effects of supplemental dietary bile acids on growth, liver function and immunity of juvenile largemouth bass (*Micropterus salmoides*) fed high-starch diet. Fish Shellfish Immun 97:602–607

Hagey LR, Møller PR, Hofmann AF, Krasowski MD (2010) Diversity of bile salts in fish and amphibians: evolution of a complex biochemical pathway. Physiol Biochem Zool 83:308–321

Hall JR, Short CE, Driedzic WR (2006) Sequence of Atlantic cod (*Gadus morhua*) GLUT4, GLUT2 and GPDH: developmental stage expression, tissue expression and relationship to starvation-induced changes in blood glucose. J Exp Biol 209:4490–4502

Hemre GI, Mommsen TP, Krogdahl A (2002) Carbohydrates in fish nutrition: effects on growth, glucose metabolism and hepatic enzymes. Aquac Nutr 8:175–194

Hilton JW, Plisetskaya EM, Leatherland JF (1987) Does oral 3,5,3¢-triiodo-L-thyronine affect dietary glucose utilization and plasma insulin levels in rainbow trout (*Salmo gairdneri*). Fish Physiol Biochem 4:113–120

Hilton JW, Atkinson JL, Slinger SJ (2011) Effect of increased dietary fiber on the growth of rainbow trout (*Salmo gairdneri*). Can J Fish Aquatic Sci 40:81–85

Hofmann AF, Hagey LR, Krasowski MD (2010) Bile salts of vertebrates: structural variation and possible evolutionary significance. J Lipid Res 51:226–246

Hou Y, He W, Hu S, Wu G (2019) Composition of polyamines and amino acids in plant-source foods for human consumption. Amino Acids 51:1153–1165

Hou YQ, Hu SD, Li XY, He WL, Wu G (2020) Amino acid metabolism in the liver: nutritional and physiological significance. Adv Exp Med Biol 1265:21–37

Jia S, Li X, Zheng S, Wu G (2017) Amino acids are major energy substrates for tissues of hybrid striped bass and zebrafish. Amino Acids 49:2053–2063

Jia SC, Li XY, He WL, Wu G (2021) Oxidation of energy substrates in tissues of fish: metabolic significance and implications for gene expression and carcinogenesis. Adv Exp Med Biol 1332:67–83

Jia SC, Li XY, He WL, Wu G (2022) Protein-sourced feedstuffs for aquatic animals in nutrition research and aquaculture. Adv Exp Med Biol 1354:237–261

Jiang L, Wu H, Huang K, Ma Y, Yang Q, Yu D, Zhong L (2013) Effects of dietary carbohydrate levels on growth performance and liver metabolism functions of juvenile tilapia (*Oreochromis niloticus*). J Fish China 37:245–255

Jin M, Pan T, Cheng X, Zhu TT, Sun P, Zhou F, Ding X, Zhou Q (2019a) Effects of supplemental dietary L-carnitine and bile acids on growth performance, antioxidant and immune ability, histopathological changes and inflammatory response in juvenile black seabream (*Acanthopagrus schlegelii*) fed high-fat diet. Aquaculture 504:199–209

Jin M, Pan T, Tocher DR, Betancor MB, Monroig Ó, Shen Y, Zhu T, Sun P, Jiao L, Zhou Q (2019b) Dietary choline supplementation attenuated high-fat diet-induced inflammation through regulation of lipid metabolism and suppression of NFκB activation in juvenile black seabream (*Acanthopagrus schlegelii*). J Nutr Sci 8:e38

Jin M, Zhu T, Tocher DR, Luo J, Shen Y, Li X, Pan T, Yuan Y, Betancor MB, Jiao L, Sun P (2020) Dietary fenofibrate attenuated high-fat-diet-induced lipid accumulation and inflammation response partly through regulation of pparα and sirt1 in juvenile black seabream (*Acanthopagrus schlegelii*). Dev Comp Immunol 109:103691

Jobgen WJ, Meininger CJ, Jobgen SC, Li P, Lee MJ, Smith SB, Spencer TE, Fried SK, Wu G (2009) Dietary L-arginine supplementation reduces white-fat gain and enhances skeletal muscle and brown fat masses in diet-induced obese rats. J Nutr 139:230–237

Jobling M (1994) Book of fish bioenergetics- fish and fisheries series 13, 1st edn. Chapman & Hall, London, United Kingdom

Kamalam BS, Medale F, Panserat S (2017) Utilization of dietary carbohydrates in farmed fishes: new insights on influencing factors, biological limitations and future strategies. Aquaculture 467:3–27

Krogdahl Å, HEMRE GI, Mommsen TP, (2005) Carbohydrates in fish nutrition: digestion and absorption in postlarval stages. Aquac Nutr 11:103–122

Leung LY, Woo NY (2012) Influence of dietary carbohydrate level on endocrine status and hepatic carbohydrate metabolism in the marine fish *Sparus sarba*. Fish Physiol Biochem 38:543–554

Li XF, Wang Y, Liu WB, Jiang GZ, Zhu J (2013) Effects of dietary carbohydrate/lipid ratios on growth performance, body composition and glucose metabolism of fingerling blunt snout bream *Megalobrama amblycephala*. Aquac Nutr 19:701–708

Li X, Zhu X, Han D, Yang Y, Jin J, Xie S (2014) Carbohydrate utilization by herbivorous and omnivorous freshwater fish species: a comparative study on gibel carp (*Carassius auratus gibelio*. var CAS III) and grass carp (*Ctenopharyngodon idellus*). Aquac Res 47:128–139

Li XY, Wang JT, Han T, Hu SX, Jiang YD (2015) Effects of dietary carbohydrate level on growth and body composition of juvenile giant croaker *Nibea japonica*. Aquac Res 46:2851–2858

Li XY, Wu G (2019) Oxidation of energy substrates in tissues of largemouth bass (*Micropterus salmoides*). J Anim Sci 97(Suppl 3):68–69

Li XF, Xu C, Zhang DD, Jiang GZ, Liu WB (2016) Molecular characterization and expression analysis of glucokinase from herbivorous fish *Megalobrama amblycephala* subjected to a glucose load after the adaption to dietary carbohydrate levels. Aquaculture 459:89–98

Li S, Lian X, Chen N, Wang M, Sang C (2018) Effects of dietary vitamin E level on growth performance, feed utilization, antioxidant capacity and nonspecific immunity of largemouth bass, *Micropterus salmoides*. Aquac Nutr 24:1679–1688

Li S, Li Z, Zhang J, Sang C, Chen N (2019a) The impacts of dietary carbohydrate levels on growth performance, feed utilization, glycogen accumulation and hepatic glucose metabolism in hybrid grouper (*Epinephelus fuscoguttatus*♀× *E. lanceolatus*♂). Aquaculture 512:734351

Li S, Sang C, Wang A, Zhang J, Chen N (2019b) Effects of dietary carbohydrate sources on growth performance, glycogen accumulation, insulin signaling pathway and hepatic glucose metabolism in largemouth bass, *Micropterus salmoides*. Aquaculture 513:734391

Li P, He WL, Wu G (2021e) Composition of amino acids in foodstuffs for humans and animals. Adv Exp Med Biol 1332:189–210

Li P, Wu G (2020) Composition of amino acids and related nitrogenous nutrients in feedstuffs for animal diets. Amino Acids 52:523–542

Li P, Wu G (2022) Functional molecules of intestinal mucosal products in animal nutrition and health. Adv Exp Med Biol 1354:263–277

Li P, Mai K, Trushenski J, Wu G (2009) New developments in fish amino acid nutrition: towards functional and environmentally oriented aquafeeds. Amino Acids 37:43–53

Li SL, Zhang YC, Liu N, Chen JQ, Guo LN, Dai ZL, Wang C, Wu ZL, Wu G (2020a) Dietary L-arginine supplementation reduces lipid accretion by regulating fatty acid metabolism in Nile tilapia (*Oreochromis niloticus*). J Anim Sci Biotechnol 11:82

Li XY, Zheng SX, Ma XK, Cheng KM, Wu G (2020b) Effects of dietary starch and lipid levels on the protein retention and growth of largemouth bass (*Micropterus salmoides*). Amino Acids 52:999–1016

Li XY, Zheng SX, Ma XK, Cheng KM, Wu G (2020c) Effects of dietary protein and lipid levels on growth performance, feed utilization, and liver histology of largemouth bass (*Micropterus salmoides*). Amino Acids 52:1043–1061

Li XL, Zheng SX, Jia SC, Song F, Zhou CP, Wu G (2020d) Oxidation of energy substrates in tissues of largemouth bass (*Micropterus salmoides*). Amino Acids 52:1017–1032

Li XY, Zheng SX, Han T, Song F, Wu G (2020) Effects of dietary protein intake on the oxidation of glutamate, glutamine, glucose and palmitate in tissues of largemouth bass (*Micropterus salmoides*). Amino Acids 52:1491–1503

Li XL, Zheng SX, Wu G (2020f) Nutrition and metabolism of glutamate and glutamine in fish. Amino Acids 52:671–691

Li XY, Zheng SX, Wu G (2020g) Amino acid metabolism in the kidneys: nutritional and physiological significance. Adv Exp Med Biol 1265:71–95

Li XY, Zheng SX, Ma XK, Cheng KM, Wu G (2021a) Use of alternative protein sources for fishmeal replacement in the diet of largemouth bass (*Micropterus salmoides*). Part I: effects of poultry by-product meal and soybean meal on growth, feed utilization, and health. Amino Acids 53:33–47

Li XY, Zheng SX, Cheng KM, Ma XK, Wu G (2021b) Use of alternative protein sources for fishmeal replacement in the diet of largemouth bass (*Micropterus salmoides*). Part II: effects of supplementation with methionine or taurine on growth, feed utilization, and health. Amino Acids 53:49–62

Li XY, Zheng SX, Wu G (2021c) Nutrition and functions of amino acids in fish. Adv Exp Med Biol 1285:133–168

Li XY, Han T, Zheng SX, Wu G (2021d) Nutrition and functions of amino acids in aquatic crustaceans. Adv Exp Med Biol 1285:169–198

Liao Z, Sun B, Zhang Q, Jia L, Wei Y, Liang M, Xu H (2020) Dietary bile acids regulate the hepatic lipid homeostasis in tiger puffer fed normal or high-lipid diets. Aquaculture 519:734935

Lin YC, Kao CH (2012) Glycogenic hepatopathy in type 1 diabetes mellitus. Liver Int 32:1294

Lin SM, Shi CM, Mu MM, Chen YJ, Luo L (2018) Effect of high dietary starch levels on growth, hepatic glucose metabolism, oxidative status and immune response of juvenile largemouth bass, *Micropterus salmoides*. Fish Shellfish Immun 78:121–126

Liu XH, Ye CX, Zheng LM, Ou CC, Wang AL, Ye JD, Kong JH (2015) Dietary maize starch influences growth performance, apparent digestibility coefficient, and hepatic enzyme activities of carbohydrate metabolism in obscure puffer, Takifugu obscurus (Abe). J World Aquac Soc 46:102–113

Luo Y, Wu X, Li W, Jiang S, Lu S, Wu M (2016) Effects of different corn starch levels on growth, protein input, and feed utilization of juvenile hybrid grouper (*male Epinephelus lanceolatus× female E. fuscoguttatus*). N Am J Aquacult 78:168–173

Lygren B, Hemre GI (2001) Influence of dietary carbohydrate on antioxidant enzyme activities in liver of Atlantic salmon (*Salmo salar* L.). Aquacult Int 9:421–427

Ma HJ, Mou MM, Pu DC, Lin SM, Chen YJ, Luo L (2019) Effect of dietary starch level on growth, metabolism enzyme and oxidative status of juvenile largemouth bass, *Micropterus salmoides*. Aquaculture 498:482–487

Maharaj V, Fitz M, Ding X (2017) Drug-induced liver injury in the setting of glycogenic hepatopathy. J Gen Intern Med 32:714–717

Marchesini G, Bugianesi E, Forlani G, Cerrelli F, Lenzi M, Manini R, Natale S, Vanni E, Villanova N, Melchionda N, Rizzetto M (2003) Nonalcoholic fatty liver, steatohepatitis, and the metabolic syndrome. Hepatology 37:917–923

Marín-Juez R, Diaz M, Morata J, Planas JV (2013) Mechanisms regulating GLUT4 transcription in skeletal muscle cells are highly conserved across vertebrates. PLoS One 8:e80628

Marín-Juez R, Capilla E, Carvalho-Simoes F, Camps M, Planas JV (2014) Structural and functional evolution of glucose transporter 4 (GLUT4): a look at GLUT4 in fish. In: Szablewski L (ed) Glucose homeostasis. IntechOpen, London, pp 37–67

Martínez-Palacios CA, Ríos-Durán MG, Ambriz-Cervantes L, Jauncey KJ, Ross LG (2007) Dietary protein requirement of juvenile Mexican Silverside (*Menidia estor Jordan* 1879), a stomachless zooplanktophagous fish. Aquac Nutr 13:304–310

Martos-Sitcha JA, Bermejo-Nogales A, Calduch-Giner JA, Jaume Pérez-Sánchez J (2017) Gene expression profiling of whole blood cells supports a more efficient mitochondrial respiration in hypoxia-challenged gilthead sea bream (Sparus aurata). Front Zool 14:34

Maas RM, Verdegem MC, Wiegertjes GF, Schrama JW (2020) Carbohydrate utilisation by tilapia: a meta-analytical approach. Rev Aquacult 12:1851–1866

Milligan LC, Pagnotta A (1991) The role of blood glucose in the restoration of muscle glycogen during recovery from exhaustive exercise in rainbow trout (*Oncorhynchus mykiss*) and winter flounder (*Pseudopleuronectes americanus*). J Exp Biol 161:489–508

Mohanta KN, Mohanty SN, Jena J, Sahu NP, Patro B (2009) Carbohydrate level in the diet of silver barb, *Puntius gonionotus* (Bleeker) fingerlings: effect on growth, nutrient utilization and whole body composition. Aquac Res 40:927–937

Mokhtar DM (2017) Fish histology: from cells to organs. Apple Academic Press, Palm Bay, Florida

Moreira IS, Peres H, Couto A, Enes P, Oliva-Teles A (2008) Temperature and dietary carbohydrate level effects on performance and metabolic utilisation of diets in European sea bass (*Dicentrarchus labrax*) juveniles. Aquaculture 274:153–160

Moon TW, Foster GD (1995) Tissue carbohydrate metabolism, gluconeogenesis and hormonal and environmental influences. Biochem Mol Biol Fish 4:65–100

National Research Council (NRC) (2011) Nutrient requirements of fish and shrimp. National Academies Press, Washington DC

Oberbauer AM, Larsen JA (2021) Amino acids in dog nutrition and health. Adv Exp Med Biol 1285:199–216

Palmer TN, Ryman BE (1972) Studies on oral glucose intolerance in fish. J Fish Biol 4:311–319

Panserat S, Plagnes-Juan E, Kaushik S (2001a) Nutritional regulation and tissue specificity of gene expression for proteins involved in hepatic glucose metabolism in rainbow trout (*Oncorhynchus mykiss*). J Exp Biol 204:2351–2360

Panserat S, Plagnes-Juan E, Breque J, Kaushik S (2001b) Hepatic phosphoenolpyruvate carboxykinase gene expression is not repressed by dietary carbohydrates in rainbow trout (*Oncorhynchus mykiss*). J Exp Biol 204:359–365

Panserat S, Plagnes-Juan E, Kaushik S (2002) Gluconeogenic enzyme gene expression is decreased by dietary carbohydrates in common carp (*Cyprinus carpio*) and gilthead seabream (*Sparus aurata*). Biochim Biophys Acta 1579:35–42

Panserat S, Rideau N, Polakof S (2014) Nutritional regulation of glucokinase: a cross-species story. NutrRes Rev 27:21–47

Pérez-Jiménez A, Abellán E, Arizcun M, Cardenete G, Morales AE, Hidalgo MC (2017) Dietary carbohydrates improve oxidative status of common dentex (*Dentex dentex*) juveniles, a carnivorous fish species. Comp Biochem Physiol A 203:17–23

Prisingkorn W, Prathomya P, Liu H, Deng FY, Zhao YH, Wang WM (2017) Transcriptomic and metabolomics analyses show that high carbohydrate induces fatty liver in blunt snout bream (*Megalobrama amblycephala*). Preprints 2017:2017010117

Polakof S, Medale F, Larroquet L, Vachot C, Corraze G, Panserat S (2011) Regulation of de novo hepatic lipogenesis by insulin infusion in rainbow trout fed a high-carbohydrate diet. J Anim Sci 89:3079–3088

Rawles SD, Smith SB, Gatlin DM (2008) Hepatic glucose utilization and lipogenesis of hybrid striped bass (*Morone chrysops*× *Morone saxatilis*) in response to dietary carbohydrate level and complexity. Aquac Nutr 14:40–50

Ren MC, Ai QH, Mai K, Ma HM, Wang XJ (2011) Effect of dietary carbohydrate level on growth performance, body composition, apparent digestibility coefficient and digestive enzyme activities of juvenile cobia, *Rachycentron canadum* L. Aquac Res 42:1467–1475

Riaz MN, Rokey GJ (2011) Extrusion problems solved: food, pet food and feed. Elsevier, New York, NY

Russell DW (2003) The enzymes, regulation, and genetics of bile acid synthesis. Annu Rev Biochem 72:137–174

Satapati S, Kucejova B, Duarte JA, Fletcher JA, Reynolds L, Sunny NE, He T, Nair LA, Livingston K, Fu X, Merritt ME (2015) Mitochondrial metabolism mediates oxidative stress and inflammation in fatty liver. J Clin Invest 125:4447–4462

Saunier E, Benelli C, Bortoli S (2016) The pyruvate dehydrogenase complex in cancer: an old metabolic gatekeeper regulated by new pathways and pharmacological agents. Int J Cancer 138:809–817

Saxena P, Turner I, McIndoe R (2010) Hepatobiliary and pancreatic: glycogenic hepatopathy: a reversible condition. J Gastroenterol Hepatol 25:646–646

Sherigar JM, De Castro J, Yin YM, Guss D, Mohanty SR (2018) Glycogenic hepatopathy: a narrative review. World J Hepatol 10:172

Shimeno S, Kheyyali D, Shikata T (1995) Metabolic response to dietary carbohydrate to protein rations in carp. Fish Sci 61:277–281

Siddiqui RA, Xu Z, Harvey KA, Pavlina TM, Becker MJ, Zaloga GP (2015) Comparative study of the modulation of fructose/sucrose-induced hepatic steatosis by mixed lipid formulations varying in unsaturated fatty acid content. Nutr Metab 12:41

Skilleter DN, Kun E (1972) The oxidation of L-lactate by liver mitochondria. Arch Biochem Biophys 152:92–104

Skiba-Cassy S, Lansard M, Panserat S, Médale F (2009) Rainbow trout genetically selected for greater muscle fat content display increased activation of liver TOR signaling and lipogenic gene expression. Am J Phys 297:R1421–R1429

Skiba-Cassy S, Panserat S, Larquier M, Dias K, Surget A, Plagnes-Juan E, Kaushik S, Seiliez I (2013) Apparent low ability of liver and muscle to adapt to variation of dietary carbohydrate: protein ratio in rainbow trout (*Oncorhynchus mykiss*). Br J Nutr 109:1359–1372

Stier A, Bize P, Schull Q, Zoll J, Singh F, Geny B, Gros F, Royer C, Massemin S, Criscuolo F (2013) Avian erythrocytes have functional mitochondria, opening novel perspectives for birds as animal models in the study of ageing. Front Zool 10:33

Stone D, Allan G, Anderson A (2003) Carbohydrate utilization by juvenile silver perch, *Bidyanus bidyanus* (Mitchell). III. The protein-sparing effect of wheat starch-based carbohydrates. Aquac Res 34:123–134

Subhadra B, Lochmann R, Rawles S, Chen R (2006) Effect of dietary lipid source on the growth, tissue composition and hematological parameters of largemouth bass (*Micropterus salmoides*). Aquaculture 255:210–222

Sulaiman MA, Kamarudin MS, Romano N, Syukri F (2020) Effects of increasing dietary carbohydrate level on feed utilisation, body composition, liver glycogen, and intestinal short chain fatty acids of hybrid lemon fin barb (*Barbonymus gonionotus*♀ *Hypsibarbus wetmorei male*♂). Aquacul Rep 16:100250

Szabo G, Saha B, Ambade A (2018) The Liver as an immune organ. In: Boyer TD, Terrault NA, Lindor KD (eds) Zakim and Boyer's Hepatology (Sanyal AJ. Elsevier, New York, pp 66–76

Szczesna-Kaczmarek A (1990) L-lactate oxidation by skeletal muscle mitochondria. Int J Biochem 22:617–620

Tammar AR (1974) Bile salts in fishes. Chem Zool 8:595–612

Tan Q, Xie S, Zhu X, Lei W, Yang Y (2007) Effect of dietary carbohydrate-to-lipid ratios on growth and feed utilization in Chinese longsnout catfish (*Leiocassis longirostris* Günther). J Appl Ich 23:605–610

Terova G, Rimoldi S, Brambilla F, Gornati R, Bernardini G, Saroglia M (2009) In vivo regulation of GLUT2 mRNA in sea bass (Dicentrarchus labrax) in response to acute and chronic hypoxia. Comp Biochem Physiol B 152:306–316

Tang Y, Boutilier RG (1991) White muscle intracellular acid base and lactate status following exhaustive exercise: a comparison between freshwater- and seawater-adapted rainbow trout. J Exp Biol 156:153–171

Tian LX, Liu YJ, Yang HJ, Liang GY, Niu J (2012) Effects of different dietary wheat starch levels on growth, feed efficiency and digestibility in grass carp (*Ctenopharyngodon idella*). Aquacult Int 20:283–293

Tidwell JH, Coyle SD, Bright LA, Yasharian D (2005) Evaluation of plant and animal source proteins for replacement of fish meal in practical diets for the largemouth bass *Micropterus salmoides*. J World Aquacult Soc 36:454–463

Tidwell JH, Coyle S, Bright LA (2007) Effects of different types of dietary lipids on growth and fatty acid composition of largemouth bass. North Am. J Aquacult 69:257–264

Torbenson M, Chen YY, Brunt E, Cummings OW, Gottfried M, Jakate S, Liu YC, Yeh MM, Ferrell L (2006) Glycogenic hepatopathy: an underrecognized hepatic complication of diabetes mellitus. Am J Surg Pathol 30:508–513

Torfi Mozanzadeh M, Yavari V, Marammazi JG, Agh N, Gisbert E (2017) Optimal dietary carbohydrate-to-lipid ratios for silvery-black porgy (*Sparidentex hasta*) juveniles. Aquac Nutr 23:470–483

Torres DM, Williams CD, Harrison SA (2012) Features, diagnosis, and treatment of nonalcoholic fatty liver disease. Clin Gastroenterol Hepatol 10:837–858

Tranulis MA, Christophersen B, Blom AK, Borrebaek B (1991) Glucose dehydrogenase, glucose-6-phosphate dehydrogenase and hexokinase in liver of rainbow trout (*Salmo gairdneri*). effects of starvation and temperature variations. Comp Biochem Physiol B 99:687–691

Tucker JW Jr (1999) Grouper aquaculture. South Reg Aquac Cent Publ 721:1–11

Viegas IDDSM (2012) Sources of blood glucose and liver glycogen in the seabass (*Dicentrarchus labrax* L.): implications to carbohydrate metabolism in fish. PhD Dissertation, Universidade de Coimbra, Portugal

Viegas I, Rito J, Jarak I, Leston S, Caballero-Solares A, Metón I, Pardal MA, Baanante IV, Jones JG (2015) Contribution of dietary starch to hepatic and systemic carbohydrate fluxes in European seabass (*Dicentrarchus labrax* L.). Br J Nutr 113:1345–1354

Waagbø R, Glette J, Sandnes K, Hemre GI (1994) Influence of dietary carbohydrate on blood chemistry, immunity and disease resistance in Atlantic salmon, *Salmo salar* L. J Fish Dis 17:245–258

Wang XF, Perez E, Liu R, Yan LJ, Mallet RT, Yang SU (2007) Pyruvate protects mitochondria from oxidative stress in human neuroblastoma SK-N-SH cells. Brain Res 1132:1–9

Wang Y, Liu YJ, Tian LX, Du ZY, Wang JT, Wang S, Xiao WP (2005) Effects of dietary carbohydrate level on growth and body composition of juvenile tilapia, *Oreochromis niloticus*× *O. aureus*. Aquac Res 36:1408–1413

Wang LN, Liu WB, Lu KL, Xu WN, Cai DS, Zhang CN, Qian Y (2014) Effects of dietary carbohydrate/lipid ratios on non-specific immune responses, oxidative status and liver histology of juvenile yellow catfish *Pelteobagrus fulvidraco*. Aquaculture 426:41–48

Wang J, Li X, Han T, Yang Y, Jiang Y, Yang M, Xu Y, Harpaz S (2016a) Effects of different dietary carbohydrate levels on growth, feed utilization and body composition of juvenile grouper *Epinephelus akaara*. Aquaculture 459:143–147

Wang J, Jiang Y, Li X, Han T, Yang Y, Hu S, Yang M (2016b) Dietary protein requirement of juvenile red spotted grouper (*Epinephelus akaara*). Aquaculture 450:289–294

Watanabe T (2002) Strategies for further development of aquatic feeds. Fish Sci 68:242–252

Welker TL, Overturf K, Snyder S, Liu K, Abernathy J, Frost J, Barrows FT (2018) Effects of feed processing method (extrusion and expansion-compression pelleting) on water quality and growth of rainbow trout in a commercial setting. J Appl Aquac 30:97–124

Wilkins Benjamin J, Pack M (2013) Zebrafish models of human liver development and disease. Compr Physiol 3:1213–1230

Wilson RP (1994) Utilization of dietary carbohydrate by fish. Aquaculture 124:67–80

Wilson RP, Halver JE (1986) Protein and amino acid requirements of fishes. Annu Rev Nutr 6:225–244

Wu G, Meininger CJ, McNeal CJ, Bazer FW, Rhoads JM (2021) Role of L-arginine in nitric oxide synthesis and health in humans. Adv Exp Med Biol 1332:167–187

Wu XY, Liu YJ, Tian LX, Mai KS, Yang HJ, Liang GY (2007) Effects of raw corn starch levels on growth, feed utilization, plasma chemical indices and enzyme activities in juvenile yellowfin seabream *Sparus latus* Houttuyn. Aquac Res 38:1330–1338

Wu G (2018) Principles of animal nutrition. CRC Press, Boca Raton, Florida

Wu G (2020a) Management of metabolic disorders (including metabolic diseases) in ruminant and nonruminant animals. In: Lamb GC, Wu G (eds) Animal agriculture: challenges, innovations, and sustainability (Bazer FW). Elsevier, New York, pp 471–492

Wu G (2020b) Important roles of dietary taurine, creatine, carnosine, anserine and hydroxyproline in human nutrition and health. Amino Acids 52:329–360

Wu G (2021) Amino Acids: biochemistry and Nutrition, 2nd edition. CRC Press, Boca Raton, Florida

Wu G (2022) Nutrition and metabolism: Foundations for animal growth, development, reproduction, and health. Adv Exp Med Biol 1354:1–24

Xie D, Yang L, Yu R, Chen F, Lu R, Qin C, Nie G (2017) Effects of dietary carbohydrate and lipid levels on growth and hepatic lipid deposition of juvenile tilapia, *Oreochromis niloticus*. Aquaculture 479:696–703

Yamamoto T, Konishi K, Shima T, Furuita H, Suzuki N, Tabata M (2001) Influence of dietary fat and carbohydrate levels on growth and body composition of rainbow trout *Oncorhynchus mykiss* under selffeeding conditions. Fish Sci 67:221–227

Yang SD, Lin TS, Liou CH, Peng HK (2003) Influence of dietary protein levels on growth performance, carcass composition and liver lipid classes of juvenile *Spinibarbus hollandi* (Oshima). Aquac Res 34:661–666

Yang Y, Han T, Xiao J, Li X, Wang J (2018) Transcriptome analysis reveals carbohydrate-mediated liver immune responses in *Epinephelus akaara*. Sci Rep 8:1

Yin P, Xie S, Huo Y, Guo T, Fang H, Zhang Y, Liu Y, Tian L, Niu J (2019) Effects of dietary oxidized fish oil on growth performance, antioxidant defense system, apoptosis and mitochondrial function of juvenile largemouth bass (*Micropterus salmoides*). Aquaculture 500:347–358

Yu LL, Yu HH, Liang XF, Li N, Wang X, Li FH, Wu XF, Zheng YH, Xue M (2018) Dietary butylated hydroxytoluene improves lipid metabolism, antioxidant and anti-apoptotic response of largemouth bass (*Micropterus salmoides*). Fish Shellfish Immun 72:220–229

Yuan L, Bambha K (2015) Bile acid receptors and nonalcoholic fatty liver disease. World J Hepatol 7:2811

Zamora-Sillero J, Ramos LRV, Romano LA, Monserrat JM, Tesser MB (2013) Effect of dietary dextrin levels on the growth performance, blood chemistry, body composition, hepatic triglicerides and glycogen of Lebranche mullet juveniles (*Mugilliza Valenciennes* 1836, Mugilidae). J Appl Ichthyol 29:1342–1347

Zhang J, Zhou F, Wang LL, Shao Q, Xu Z, Xu J (2010) Dietary protein requirement of juvenile black sea bream, *Sparus macrocephalus*. J World Aquac Soc 41:151–164

Zhang W, Tan B, Liu K, Dong X, Yang Q, Chi S, Liu H, Zhang S, Wang H (2019) Effects of different dietary lipids on growth, body composition and lipid metabolism-related enzymes and genes in juvenile largemouth bass, *Micropterus salmoides*. Aquac Nutr 25:1318–1326

Zhou H, Chen N, Qiu X, Zhao M, Jin L (2012) Arginine requirement and effect of arginine intake on immunity in largemouth bass, *Micropterus salmoides*. Aquac Nutr 18:107–116

Zhou CP, Liu B, Xie J, Ge XP, Xu P, Zhou QL, Pan LK, Chen RL (2013a) Effect of dietary carbohydrate level on growth performance, blood chemistry, hepatic enzyme activity, and growth hormone gene expression in Wuchang bream (*Megalobrama amblycephala*). Isr J Aquac 65:882–890

Zhou C, Liu B, Ge X, Xie J, Xu P (2013b) Effect of dietary carbohydrate on the growth performance, immune response, hepatic antioxidant abilities and heat shock protein 70 expression of Wuchang bream, *Megalobrama amblycephala*. J Appl Ich 29:1348–1356

Zhou C, Ge X, Niu J, Lin H, Huang Z, Tan X (2015) Effect of dietary carbohydrate levels on growth performance, body composition, intestinal and hepatic enzyme activities, and growth hormone gene expression of juvenile golden pompano, *Trachinotus ovatus*. Aquaculture 437:390–397

Zhou M, Liang R, Mo J, Yang S, Gu N, Wu Z, Babu VS, Li J, Huang Y, Lin L (2018) Effects of brewer's yeast hydrolysate on the growth performance and the intestinal bacterial diversity of largemouth bass (*Micropterus salmoides*). Aquaculture 484:139–144

Zhou P, Wang M, Xie F, Deng DF, Zhou Q (2016) Effects of dietary carbohydrate to lipid ratios on growth performance, digestive enzyme and hepatic carbohydrate metabolic enzyme activities of large yellow croaker (*Larmichthys crocea*). Aquaculture 452:45–51

Zhu Y, Che Y, Liu Y, Yang H, Liang G, Tian L (2012) Effect of dietary selenium level on growth performance, body composition and hepatic glutathione peroxidase activities of largemouth bass *Micropterus salmoide*. Aquac Res 43:1660–1668

Zivkovic AM, German JB, Sanyal AJ (2007) Comparative review of diets for the metabolic syndrome: implications for nonalcoholic fatty liver disease. Am J Clin Nutr 86:285–300

Protein-Sourced Feedstuffs for Aquatic Animals in Nutrition Research and Aquaculture

12

Sichao Jia, Xinyu Li, Wenliang He, and Guoyao Wu

Abstract

Aquatic animals have particularly high requirements for dietary amino acids (AAs) for health, survival, growth, development, and reproduction. These nutrients are usually provided from ingested proteins and may also be derived from supplemental crystalline AA. AAs are the building blocks of protein (a major component of tissue growth) and, therefore, are the determinants of the growth performance and feed efficiency of farmed fish. Because protein is generally the most expensive ingredient in aqua feeds, much attention has been directed to ensure that dietary protein feedstuff is of high quality and cost-effective for feeding fish, crustaceans, and other aquatic animals worldwide. Due to the rapid development of aquaculture worldwide and a limited source of fishmeal (the traditionally sole or primary source of AAs for aquatic animals), alternative protein sources must be identified to feed aquatic animals. Plant-sourced feedstuffs for aquatic animals include soybean meal, extruded soybean meal, fermented soybean meal, soybean protein concentrates, soybean protein isolates, leaf meal, hydrolyzed plant protein, wheat, wheat hydrolyzed protein, canola meal, cottonseed meal, peanut meal, sunflower meal, peas, rice, dried brewers grains, and dried distillers grains. Animal-sourced feedstuffs include fishmeal, fish paste, bone meal, meat and bone meal, poultry by-product meal, chicken by-product meal, chicken visceral digest, spray-dried poultry plasma, spray-dried egg product, hydrolyzed feather meal, intestine-mucosa product, peptones, blood meal (bovine or poultry), whey powder with high protein content, cheese powder, and insect meal. Microbial sources of protein feedstuffs include yeast protein and single-cell microbial protein (e.g., algae); they have more balanced AA profiles than most plant proteins for animal feeding. Animal-sourced ingredients can be used as a single source of dietary protein or in complementary combinations with plant and microbial sources of proteins. All protein feedstuffs must adequately provide functional AAs for aquatic animals.

Keywords

Protein · Amino acids · Function · Fish · Shrimp · Health · Aquaculture

Abbreviations

AA	Amino acid
CP	Crude protein
EAA	Nutritionally essential amino acid

Jia S and Li X contributed equally to this work.

S. Jia · X. Li · W. He · G. Wu (✉)
Department of Animal Science, Texas A&M University, College Station, TX 77843, USA
e-mail: g-wu@tamu.edu

G. Wu (ed.), *Recent Advances in Animal Nutrition and Metabolism*, Advances in Experimental Medicine and Biology 1354,
https://doi.org/10.1007/978-3-030-85686-1_12

GDH	Glutamate dehydrogenase
GOT	Glutamate–oxaloacetate transaminase
GPT	Glutamate–pyruvate transaminase
HSB	Hybrid-striped bass
LMB	Largemouth bass
NEAA	Nutritionally nonessential amino acid
NRC	National Research Council
RAS	Recirculating aquaculture system

12.1 Introduction

Protein is the major dry matter component of growth in aquatic animals (Li et al. 2021c, d). The deposition of 1 g protein in tissues is associated with the retention of 3 g water in the body (Wu 2018). Thus, lean-tissue gain in fish, shrimp and crabs is a primary determinant of their growth rates and feed efficiencies. Both plant- and animal-sourced ingredients are the regular sources of dietary protein. Different proteins in feedstuffs have different compositions of all physiologically and nutritionally essential amino acids (AAs). On a dry matter basis, plant-sourced feedstuffs generally contain a lower content of crude protein (CP) and AAs than animal-sourced feedstuffs (Hou et al. 2019; Li and Wu 2020). For example, while corn grain and sorghum are excellent sources of energy for swine and poultry, they contain only 9–10% CP, 0.21–0.25% lysine, and 0.07–0.1% tryptophan, as well as imbalanced proportions of AAs. For comparison, fishmeal contains 63.4% CP, 5.28% lysine, and 0.7% tryptophan (Li et al. 2011). Also, poultry by-product meal contains 64.3% CP, 3.44% lysine, and 0.49% tryptophan.

Fishmeal has traditionally been included at high concentrations in the diets of carnivorous fish (Hardy 2010). However, fishmeal is an expensive ingredient and its cost has continued to increase due to heightened demand associated with the expansion of world aquaculture (Tacon et al. 2011). In addition, fishmeal is a finite global resource. Therefore, the use of fishmeal as the sole source of dietary protein for fish production is not sustainable in the long term (Kaushik and Seiliez 2010; Turchini et al. 2019). To solve this practical problem, extensive research efforts have been made in recent years to identify suitable alternatives to fishmeal for use in the diets of various aquatic animal species (Table 12.1), including hybrid-striped bass (HSB) (Berge et al. 1999; Brown et al. 1997; Day et al. 2000; Gatlin et al. 2007; Perez-Velasquez et al. 2019; Rawles et al. 2006, 2009; Rossi et al. 2015; Tacon and Metian 2008). The main objective of this article is to highlight protein source feedstuffs for aquatic animals.

12.2 Basic Concepts in the AA Nutrition of Aquatic Animals

12.2.1 Definitions of Nutritionally Essential and Nonessential AAs, and Functional AAs

Since 1912, an AA has been classified as nutritionally "essential" (EAA) or "nonessential" (NEAA) based on the growth and nitrogen balance of mammals (Wu et al. 2013a). AAs that are not formed by animals had traditionally been considered as nutritionally essential for animals, including fish and crustaceans (NRC 2011). By contrast, AAs that are formed by animals had traditionally been considered as nutritionally nonessential for animals, including fish and crustaceans (NRC 2011). Historically, much emphasis had been placed on the content of EAAs in aquafeed ingredients (Table 12.2) and the requirements of aquatic animals for EAAs (Mambrini and Kaushik 1995; NRC 2011; Twibell et al. 2003), whereas NEAAs had been previously thought to be dispensable in diets because they are synthesized de novo in animals (Li et al. 2021c). However, over the past 25 years, there has been growing interest in the physiological roles of AAs for roles other than protein synthesis in humans and other animals (Andersen et al. 2016; Wu 2013a). The syntheses of many low-molecular-weight substances (e.g., nitric oxide, polyamines, glutathione, creatine, melanin, and heme) require AAs, including those

Table 12.1 Estimated use of commercial feed and fishmeal in aquaculture and feed conversion ratio between 1995 and 2020[a]

Species group	Use of commercial feed in aquaculture (%)	Average FCR[b]	% Fishmeal in feed
Marine shrimps			
1995	75	2.0	28
2005	89	1.8	24
2010	95	1.6	16
2015	97	1.5	12
2020	100	1.4	8
Marine fish			
1995	50	2.0	50
2005	70	1.9	38
2010	73	1.9	26
2015	75	1.8	18
2020	80	1.8	12
Salmon			
1995	100	1.5	45
2005	100	1.3	35
2010	100	1.3	22
2015	100	1.3	16
2020	100	1.3	12
Carps[c]			
1995	20	2.0	10
2005	45	1.8	8
2010	50	1.8	2
2015	55	1.7	1
2020	60	1.6	1
Tilapias			
1995	70	2.0	10
2005	80	1.8	8
2010	85	1.7	3
2015	90	1.6	2
2020	95	1.6	1

[a]Adapted from Tacon et al. (2011)
[b]Feed conversion ratio (feed/gain ratio)
[c]Excluding the silver carp, bighead carp, and Indian major carps

that are synthesized by animal cells. Much evidence has also shown that the amount of AAs synthesized in the body may not be sufficient to meet the metabolic needs of animals, such as for maximal growth in the young and optimal health in the life cycle (Hou et al. 2015; Wu et al. 2013b). Additionally, nitrogen balance is not a sensitive indicator of optimal dietary requirements of animals for all AAs (Wu et al. 2014).

In recent years, a new nutritional concept of functional AAs has been proposed to formulate new generations of improved diets that incorporate NEAAs (Wu 2010). Functional AAs are defined as those AAs that regulate key metabolic

Table 12.2 Profiles of nutritionally essential amino acids and limiting amino acids in fishmeal and selected alternative protein sources

	Nutritionally essential amino acids														Limiting amino acids		
	Protein (%)	Arg	Cys	His	Ile	Leu	Lys	Met	Phe	Thr	Tyr	Trp	Val	Met + Cys	Phe + Tyr	1°	2°	3°
Species		*Requirement (% of protein)*																
Atlantic salmon[a]		5.0		2.2	3.1	4.2	6.7	1.9	2.5	3.1		0.8	3.3	3.1	5.0			
Trout[a]		3.9		2.1	2.9	3.9	6.3	1.8	2.4	2.9		0.8	3.2	2.9	4.7			
Carp[a]		5.3		1.6	3.1	4.4	6.9	2.2	4.1	4.7		0.9	4.4	3.1	6.3			
Tilapia[a]		4.1		3.4	3.4	6.6	5.5	2.4	3.8	3.8		1.0	5.2	3.4	5.5			
Catfish[a]		4.1		2.1	2.8	4.5	5.5	2.1	2.4	2.4		0.7	2.8	3.1	5.5			
Average[a]		4.5		2.3	3.1	4.7	6.2	2.1	3.0	3.4		0.9	3.8	3.1	5.4			
Feedstuffs		*Content (% of protein)*																
Maize distillers wet grains and solubles[b]	31.8	3.9	1.9	2.3	3.1	10.1	3.0	1.8	4.1	3.3	—	0.4	4.3	3.7	—	Lys	Trp	Arg
Maize distillers dried grains and solubles[b]	44	3.4	2	2.4	3.5	12	2.6	1.9	4.6	3.2	4.1	0.5	4.4	3.9	8.7	Lys	Trp	–
Brewer's yeast, dehydrated[b]	48.9	4.4	0.9	2.0	4.6	6.2	6.3	1.5	3.6	4.4	2.7	1.1	4.9	2.4	6.3	Met + Cys	His	–
Earthworm, dehydrated[b]	57.9	6.8	1.1	2.6	4.5	7.8	7	1.8	3.9	4.1	3.3	1	5.1	3	7.2	Met + Cys	–	–
Maize gluten meal[b]	67.2	3.0	1.8	2.0	4.0	15.9	1.7	2.4	6.1	3.3	4.8	0.5	4.5	4.2	10.9	Lys	Trp	Arg
Wheat grain[b]	12.6	4.7	2.2	2.3	3.4	6.5	2.9	1.6	4.5	2.9	2.7	1.2	4.3	3.8	7.2	Lys	Thr	–
Faba bean[b]	29	9.0	1.2	2.6	4.1	7.1	6.2	0.8	4.0	3.5	2.8	0.8	4.6	2.0	6.8	Met + Cys	Trp	–
Lupin, blue, seeds[b]	33.8	11.0	1.5	2.7	4.2	6.9	4.7	0.7	4.0	3.4	3.6	0.8	3.9	2.2	7.6	Met + Cys	Lys	–
Pea seeds[b]	23.9	8.4	1.4	2.5	4.2	7.1	7.2	1.0	4.7	3.8	3.1	0.9	4.8	2.4	7.8	Met + Cys	–	–
Linseed meal, expeller-extracted[b]	33.7	10	1.8	2.3	4.7	7.4	5	0.9	3.9	3.9	4.6	0.7	4.4	2.6	8.5	Lys	Met + Cys	–
Sunflower meal, solvent-extracted, dehulled and partially dehulled[b]	37.7	8.5	1.7	2.5	4.1	6.2	3.5	2.3	4.4	3.6	2.4	1.2	4.9	4.0	6.8	Lys	–	–

(continued)

Table 12.2 (continued)

	Nutritionally essential amino acids															Limiting amino acids		
	Protein (%)	Arg	Cys	His	Ile	Leu	Lys	Met	Phe	Thr	Tyr	Trp	Val	Met + Cys	Phe + Tyr	1°	2°	3°
Canola (Rapeseed) meal[c]	38.1	5.8	2.4	2.7	4.0	6.8	5.3	2.0	3.9	4.3	2.8	1.2	5.1	4.4	6.7	Lys	–	–
Blood meal[c]	89.6	5.5	2.1	6.2	2.8	12.7	9.2	1.3	6.5	4.4	3.2	1.5	9.2	3.4	9.7	Ile	–	–
Cookie meal[c]	12.3	4.7	1.5	1.8	4.1	7.2	3.3	1.5	4.1	3.4	4.5	1.2	4.3	3.0	8.5	Lys	–	–
Feedstuffs	*Content (% of protein)*																	
Canola (Rapeseed) meal, solvent-extracted, low erucic, low glucosinolates[b]	38.3	6.1	2.3	2.6	4.0	6.7	5.5	2.1	3.9	4.4	3.1	1.3	5.1	4.4	7.0	Lys	–	–
Corn grain[c]	9.3	4.1	2.2	2.5	3.7	12.2	2.7	2.3	4.9	3.3	4.6	0.8	4.7	4.4	9.6	Lys	Trp	–
Cottenseed meal[c]	40.3	11.3	1.7	2.7	3.0	5.6	4.1	1.6	5.0	3.1	2.7	1.1	4.2	3.4	7.7	Lys	Thr	Ile
Feather meal[c]	82.1	7.0	5.1	1.1	4.6	8.2	2.6	0.9	4.8	4.8	2.5	1.0	7.0	6.0	7.3	Lys	His	–
Fish meal[c]	63.4	7.6	1.1	2.4	5.1	8.3	8.3	3.2	4.4	6.5	3.7	1.1	6.0	4.2	8.1	–	–	–
Meat and bone meal[c]	52	7.1	0.9	2.3	3.7	6.8	6.1	2.1	3.6	4.7	2.8	0.8	4.3	3.1	6.3	Trp	Met + Cys	Lys
Peanut meal[c]	43.9	12.9	1.5	2.2	3.2	5.6	3.1	1.1	4.4	3.8	3.2	0.9	3.9	2.6	7.6	Lys	Met + Cys	–
Poultry by-product meal[c]	64.3	7.2	1.6	2.0	3.6	6.5	5.3	2.2	3.7	4.4	2.9	0.8	4.5	3.8	6.5	Lys	Thr	Trp
Soybean meal[c]	43.6	7.3	1.6	2.6	4.7	7.9	6.4	1.4	5.1	4.0	3.8	1.4	4.8	3.0	8.9	Met + Cys	–	–
Soybean meal(Dehulled) [c]	51.8	6.0	1.3	2.2	4.1	7.1	5.5	1.2	4.7	3.9	3.3	1.2	4.3	2.6	8.0	Met + Cys	Lys	His
Sorghum grain[c]	10.1	4.1	1.9	2.3	3.8	12.0	2.2	2.0	5.0	3.2	4.5	1.0	5.0	3.9	9.5	Lys	Arg	Thr

[a]Data from NRC (2011)

[b]Data from Feedpedia: http://www.feedipedia.org/. Values were calculated as (amount of amino acid x 100)/amount of crude protein. Protein content refers to crude protein content and is expressed on the as-fed basis

[c]Data from Li et al. (2011). Values were calculated as (amount of amino acid x 100)/amount of crude protein. Protein content refers to crude protein content and is expressed on the as-fed basis

pathways to improve health, survival, growth, development, reproduction, and productivity of organisms (Wu 2010). Functional AAs (e.g., arginine, cysteine, glutamate, glutamine, glycine, methionine, proline, and tryptophan) can be EAAs and the traditionally classified NEAAs (Li and Wu 2018; Wu et al. 2014). A deficiency or imbalance of functional AAs may impair protein synthesis, metabolism, and homeostasis in the whole body of animals. For example, arginine may increase resistance to *Edwardsiela ictaluri* in channel catfish through the production of its metabolite (i.e., nitric oxide) and to induce intestinal maturation via its another metabolite, spermine, in sea bass (Andersen et al. 2016; Buentello and Gatlin 2001; Costas et al. 2011). In addition, glutamine may affect the secretion of pituitary hormones in rainbow trout (Andersen et al. 2016), whereas glycine and taurine can help to maintain osmolality in tissues of aquatic animals (Li et al. 2021c) and may enhance their growth (Chatzifotis et al. 2008; Li et al. 2009; Lunger et al. 2007; Qiyou et al. 2011; Salze and Davis 2015). Furthermore, supplementing glutamine, glutamate, and aspartate to a conventional diet may improve the integrity of the intestinal epithelium by providing additional metabolic energy and substrates of synthetic processes for optimal intestinal growth and maintenance in pigs and rats (Hou et al. 2015; Wu et al. 2014). Also, an inadequate supply of arginine from the maternal diet impairs fetal development and growth in gestating swine (Liu et al. 2012; Mateo et al. 2008; Wu et al. 2018) and rats (Wu et al. 2013a). Similar effects of glutamate (Caballero-Solares et al. 2015), glutamine (Caballero-Solares et al. 2015; Qiyou et al. 2011), and arginine (Buentello and Gatlin 2001) have been reported for fish.

12.2.2 High Requirements of Fish for Dietary Protein

Dietary requirements of fish for protein range from 30 to 60% of dietary dry matter based on their species, age, size, and feeding habits (Wilson 2002). Such requirements are much greater than those for mammals and birds such as swine (12–20%), chickens (14–22%), and cattle (10–18%) (Hopkins 1992; Kaushik and Seiliez 2010; NRC 2000, 2011, 2012; Wu et al. 2014). However, protein content (14–18%) and AA composition in the whole body of fish (Li et al. 2021c, d) are generally similar to those in terrestrial animals, such as pigs, cattle, rats, and chickens (Latshaw and Bishop 2001; Li et al. 2021c; Lobley et al. 1980; Smits et al. 1988).

Several reasons have been postulated to explain the high requirements of fish for dietary protein. First, the basal energy needs of fish are less than those of terrestrial animals due to their poikilothermic and ammoniotelic life mode (Kaushik and Seiliez 2010). Thus, the need for dietary substances (e.g., lipids and carbohydrates) as substrates for ATP production is lower for fish, which results in the higher content of protein in fish diets than the diets of land animals. However, this explanation is not satisfactory because the oxidation of AAs, like fatty acids and glucose, can also produce ATP in fish (Li et al. 2020a, b, c, f). Second, fish may have a lower ability to oxidize glucose and fatty acids (van den Thillart 1986), and therefore these nutrients may be easily stored in the body as glycogen and triacylglycerols, respectively. Their excessive accumulation impairs the functions of tissues and cells. For this reason, a higher content of dietary protein may be necessary to prevent metabolic dysfunction in fish, particularly carnivorous fish (Li et al. 2020d, e; Li et al. 2021a, b).

Third, fish have a remarkably high ability to use dietary protein as an energy source (Van Waarde 1983; Weber and Haman 1996). Among the three types of macronutrients (carbohydrates, protein, and lipids), most fish do not use carbohydrates as a major energy source (Cowey and Walton 1988). However, a high rate of AA catabolism in the whole body of fish has been observed (Jürss and Bastrop 1995; Wilson 2002). It has been estimated that up to 85% of the energy requirement of teleost fish is provided by AAs, depending on the fish's developmental stage (Van Waarde 1983). The livers and kidneys of fish generally have high rates of AA oxidation (Ballantyne 2001; Li et al. 2020f), similar to mammal and bird livers and kidneys (Wu 2013b). Furthermore, AAs are the major metabolic fuels for marine fish embryos and yolk-sac larvae (Cowey and Walton 1988). Available evidence shows that the oxidation of AAs as an entity contributes to 50–70% of total energy needs of marine fish embryos and yolk-sac larvae (Rønnestad and Fyhn 1993; Rønnestad et al. 1999). The contribution from AAs to meeting energy requirements is higher and is metabolically more efficient in fish than in mammals and birds (Wu 2021) because ammonia is directly excreted from fish into the surrounding environment without a direct need for ATP (Kaushik and Seiliez 2010). For the carbon skeletons of AAs to enter the Krebs cycle and generate ATP, the amino group of AAs is ultimately liberated as ammonia through many metabolic pathways. As in terrestrial mammals (Zhang et al. 2021) and birds (He et al. 2021), AA transaminases and glutamate dehydrogenases play critical roles in AA metabolism and ammonia production by fish (Hughes et al. 1983; Li et al. 2021c). Of particular note, we have recently reported that glutamate, glutamine, and aspartate are the major metabolic fuels for the proximal intestine, liver, kidneys, and skeletal muscle of HSB and zebrafish (Jia et al. 2017), as well as largemouth bass (LMB; Li et al. 2020d) and crustaceans (Li et al. 2021d). The use of these three AAs as the major energy sources for fish is a reasonable explanation for the particularly high requirements of fish for dietary protein. Therefore, dietary protein plays an important role in the growth of fish (primarily protein synthesis) and their ATP production (mainly via AA catabolism). As with mammals, AAs also have other functions in fish (Li et al. 2009).

12.3 Functional Amino Acids

There are more than 700 AAs in nature, but the number of canonical amino acids that serve as precursors for the synthesis of proteins in aquatic animals are 20 and they are called proteinogenic AAs. Numerous studies showed that proteinogenic AAs not only serve as building blocks for protein synthesis but also play many other crucial roles in the metabolism and physiology of animals (Wu et al. 2013a), including the regulation of fatty acid oxidation and synthesis to reduce excess white fat accretion (Jobgen et al. 2006; Li et al. 2020g). The traditional classification of EAAs and NEAAs based on whether they can be synthesized de novo is not adequate to address the importance of their functions in animals, including fish (Hou et al. 2015; Li et al. 2021c). Therefore, the new concept of "functional amino acids" is now widely accepted in the international community of nutritional science and has been exploited to improve the growth, health, and productivity of animals (Wu et al. 2014), including fish (Andersen et al. 2016; Caballero-Solares et al. 2015).

Besides terrestrial mammals and birds (Wu 2010), the roles of functional AAs in fish nutrition and health have also been documented in an increasing number of studies (Li et al. 2021c; Wu et al. 2011). These roles include the regulation of gene expression, reproduction, osmoregulation, and metamorphosis; provision of the bulk of ATP for the small intestine; activation of protein synthesis; control of appetite and body composition; modulation of immune response; and prevention of infectious disease (Andersen et al. 2016). When an appropriate amount of a functional AA is supplemented to animals, the pattern of all other AAs in diets may not need to be adjusted (Wu et al. 2014).

Over the past 10 years, there has been active research on the nutrition of glutamate and glutamine as functional AAs in fish (Caballero-Solares et al. 2015; Li et al. 2020c). Glutamate is one of the most abundant AAs in aqua feeds (Li et al. 2011). It is also a physiologically important AA to constitute proteins because of its chemical structure, and the negative charge of glutamate helps to stabilize the structure of protein by forming ionic bonds (Brosnan and Brosnan 2013). Glutamate was previously thought to be an NEAA because it is synthesized de novo. There are several metabolic pathways to synthesize glutamate from other AAs and their metabolites in vivo (Hou and Wu 2018). For example, glutamate can be formed from α-ketoglutarate and ammonia by glutamate dehydrogenase or from α-ketoacids by

Table 12.3 Activities of glutamate dehydrogenase (GDH), glutamate–pyruvate transaminase (GPT), and glutamate–oxaloacetate transaminase (GOT) in the liver of animals

Animal	Enzyme	Activity	References
Pig	GDH	385	Bush et al. (2002)
Chicken	GDH	50	Lee et al. (1972)
Largemouth bass	GDH	52	Li et al. (2020a)
Largemouth bass	GPT	95	Li et al. (2020a)
Largemouth bass	GOT	46	Li et al. (2020a)
Hybrid-striped bass	GDH	446	Jia et al. (2021)
Hybrid-striped bass	GPT	502	Jia et al. (2021)
Hybrid-striped bass	GOT	518	Jia et al. (2021)
Red drum	GDH	200	Chan (2016)
Tilapia	GDH	462	Bhaskar (1994)
Tilapia	GPT	463	Abdel-Tawwab et al. (2010)
Tilapia	GOT	300	Abdel-Tawwab et al. (2010)
Atlantic salmon	GDH	185	Rossignol et al. (2011)
Atlantic salmon	GPT	250	Fynn-Aikins et al. (1995)
Atlantic salmon	GOT	230	Fynn-Aikins et al. (1995)

Values are expressed as nmol/mg protein/min

aminotransferases (Table 12.3). Glutamate can also be produced from glutamine by glutaminase, which is a mitochondrial enzyme in fish (Li et al. 2021c) and terrestrial animals (He et al. 2021; Zhang et al. 2021). Furthermore, the metabolism of most AAs (including arginine, proline, and histidine) via a series of enzymes yields glutamate (Wu et al. 2013a). Of note, in cells lacking mitochondria, glutamine cannot replace glutamate due to the absence of glutaminase.

Many studies across different animal species have shown that a large amount of dietary glutamate is metabolized within the small intestine (foregut), primarily by enterocytes. For example, 95–97% of dietary glutamate is metabolized by the small intestine of young pigs in the first pass (Wu et al. 1998), and 74% of dietary glutamate was metabolized in the first pass by premature human infants on enteral feeding (Haÿs et al. 2007). Glutamate, glutamine, and aspartate are the major energy fuels for mammalian enterocytes (Wu et al. 1998). Thus, because of the extensive first-pass catabolism of glutamate in the small intestine of mammals, the concentration of glutamate in plasma is usually low and is not affected substantially by dietary glutamate intake (Brosnan and Brosnan 2013; Wu 2021). Similar results have been reported for fish (Jia et al. 2017; Li et al. 2020a) and crustaceans (Li et al. 2021d). In addition, the gastrointestinal tract receives a glutamate signal for the presence of protein digestion by activating taste receptors in the tongue, stomach, and small intestine. Dietary glutamate can also activate umami taste receptors and further increase their appetite (Wu et al. 2013a). Furthermore, glutamate serves as an excitatory neurotransmitter in the central nervous system to induce food intake, while promoting nucleotide synthesis in the gut and improving the availability of soybean meal-based feed in fish such as rainbow trout (Yoshida et al. 2016).

Glutamate dehydrogenase (GDH) of the liver and white muscle of goldfish showed much greater activity than other dehydrogenases and the enzymes of the purine nucleotide cycle (Van Waarde and Kesbeke 1982). Notably, the activity of GDH is particularly high in the intestine of fish (Li et al. 2020a). Besides GDH, the glutamate–oxaloacetate transaminase (GOT) and glutamate–pyruvate transaminase (GPT), whose activities are generally high in animal (including fish) tissues, also participate in AA degradation (Campbell et al. 1983; French et al. 1983; Watford and Wu 2005). The activities of GDH, GOT,

Table 12.4 Phosphate-activated glutaminase and glutamine synthetase (GS) activities in tissues of fish[a]

Tissue	Bowfin[b]		Lake char[b]		Largemouth bass[c]	Hybrid-striped bass[d]	
	Glutaminase	GS	Glutaminase	GS	Glutaminase	Glutaminase	GS
Red muscle	2.2	0.86	2.4	0.09	–	–	–
White muscle	1.2	1.37	1.2	0.11	5.21	0.18	0.014
Brain	26.8	72.0	22.2	118	–	–	–
Gill	1.4	1.80	3.0	1.42	–	–	–
Heart	4.4	1.08	7.6	0.28	–	–	–
Liver	2.6	0.78	2.2	0.80	15.1	1.23	0.23
Kidney	3.0	1.45	3.2	0.80	62.0	1.79	0.29
Intestine	1.4	1.08	1.8	0.81	15.9	3.42	0.17

[a]Values expressed as nmol/mg protein/min are calculated based on the content of 10% protein in tissues
[b]Adapted from Chamberlin et al. (1991)
[c]Adapted from Li et al. (2020a)
[d]Adapted from Jia (2019)
GS, glutamine synthetase

and GPT in the liver of some fish, swine, and chickens are summarized in Table 12.3. In contrast to fish (Li et al. 2020c), the GDH activity is negligible in the small intestine of terrestrial mammals and birds (Wu et al. 2013a).

Like glutamate, glutamine is another AA that is abundantly present in all vertebrates (Wu et al. 2013a). Glutamine is endogenously synthesized from glutamate and ammonia by glutamine synthetase. It is deaminated into glutamate plus ammonia by phosphate-activated glutaminase. Due to their chemical structures and abundances, glutamine and glutamate play an important role in ammonia detoxification (Albrecht and Norenberg 2006). However, compared with GDH, GPT, and GOT, the activities of glutaminase and glutamine synthetase in fish tissues are much lower except for the brain (Chamberlin et al. 1991). Therefore, the synthesis of glutamine from glutamate may play a more important role in the detoxification of ammonia in the ammonotelic teleost (Ip and Chew 2010). Of note, glutamine regulates protein turnover to favor protein accretion in terrestrial animals (Wu 2021). For example, in chicken skeletal muscles, intracellular glutamine levels are positively related to protein synthesis (Watford and Wu 2005). High rates of protein synthesis have also been reported for rat skeletal muscle perfused with high levels of glutamine (MacLennan et al. 1987) and for chick skeletal muscles incubated with elevated levels of glutamine (Wu and Thompson 1990). To date, little is known about the role of glutamine in regulating protein synthesis in fish. Nonetheless, there is also unequivocal evidence that dietary supplementation with glutamine enhances antioxidative capacities, the growth of fish, and protein content in their intestine (Cheng et al. 2011, 2012; Yan and Zhou 2006). In some tissues, such as the small intestine of pigs, glutamine can replace glutamate due to the presence of glutaminase, but glutamate cannot replace glutamine because of the very low activity of glutamine synthetase (Haynes et al. 2009). Many tissues of fish possess both glutamine synthetase and glutaminase (Table 12.4) and, therefore, a glutamine–glutamate cycle to regulate the intracellular concentrations and release of glutamine.

Great impediments to the productivity of aquatic animals are: (1) the inadequate provision of nutrients, particularly AAs, which are the most predominant components of diets and the building blocks of tissue proteins (Li et al. 2021c, d); and (2) high rates of mortality in production settings [e.g., 33% of mortality in HSB even when fed diets containing adequate protein levels (36.3–41.5% CP) that meet the NRC-recommended requirements of the fish for protein (D'Abramo et al. 2000)]. This problem may be alleviated or prevented by the adequate provision of functional AAs for fish.

Besides the use of zebrafish as an animal model for biomedical research (Laale 1977), both HSB and LMB are agriculturally important species to produce high-quality protein for human consumption. Therefore, zebrafish, HSB, and LMB are useful models to assess AA requirements of fish, the quality of feedstuff proteins for farmed fish, as well as the nutrition and metabolism of functional AAs in fish.

12.4 Hybrid-Striped Bass

The HSB (*Morone saxatilis* ♂ x white bass *M. chrysops* ♀) are the offspring of cross-breeding striped bass and white bass and are a very popular sportfish throughout the United States, particularly in large reservoirs (Quagrainie 2015). HSB, also known as a wiper or whiterock bass, is a cross between the striped bass (*Morone saxatilis*) and the white bass (*Morone chrysops*). The HSB was first produced in South Carolina in the mid-1960s by fertilizing the eggs of the striped bass with the sperm of the white bass (Bulak et al. 2013). This hybrid generally refers to the original cross, namely the palmetto bass. The reciprocal cross between the female white bass and the male striped bass produced in subsequent years is called the sunshine bass. The HSB not only gains some superior traits inherited from its parental stocks but also shows outbreeding characteristics for the enhancement of growth performance. For example, the HSB grows faster and has a better survival rate than the striped bass and the white bass. Moreover, the HSB has a greater ability to resist disease and can tolerate various water conditions, such as salinities of 0–30 ppt, pH of 6–10, temperatures of 4–29 °C, and the dissolved oxygen of 6–12 mg/L; the maximum growth of the HSB occurs at salinities of 3–7 ppt, pH of 7.0–8.5, and temperatures of 25–27 °C (D'Abramo and Frinsko 2008; Harrell 2016; Hodson 1989; McGinty and Hodson 2008). Thus, the HSB is widely cultured in most regions of the United States as both a sportfish and a foodfish in all production systems such as ponds, tanks (e.g., closed or semi-closed recirculating aquaculture systems), or cage systems (Lougheed and Nelson 2001). For example, in the United States, HSB production is now a major aquaculture industry, ranking #3 in sales ($100 million) after catfish and trout (USDA 2019a). In many states, including Texas, HSB farming is currently the second-largest aquacultural enterprise behind only catfish farming (Treece 2017). The HSB has also been introduced to several other countries and regions in Europe (Ende et al. 2018) and Asia (Liu and Liao 1999).

Publications regarding HSB aquaculture started to increase substantially in the 1990s. Successes at using fishmeal-based diets to feed HSB and determining nutrient requirements for HSB have helped the aquaculture industry to expand rapidly in the southern regions of the United States (Quagrainie 2015). As the fishmeal price continues to increase due to the limited resources of fish in oceans, studies on the replacement of fishmeal by alternative sources of proteins for different species of fish have emerged over the past 25 years. For HSB, fishmeal could be partially replaced by poultry by-product meal (50%) and soybean meal (75%) (Gallagher 1994; Rawles et al. 2006). Some benefits of AA supplementation to HSB diets have also been confirmed. For example, dietary supplementation with arginine and/or glutamine can improve the growth performance, immune responses, and intestinal morphology of HSB (Cheng et al. 2011).

As noted previously, the HSB, like other fish species, have a much higher requirement for dietary protein than terrestrial mammals and poultry (NRC 2011), even though the content of protein or AAs in their bodies is largely similar among all animals (Wu et al. 2013a). Based on the content of carbohydrates (~20%), lipids (~10%), and protein (~50%) in the diet of HSB, dietary protein may provide substantial amounts of energy for the growth of the fish. Dietary glutamine and glutamate, traditionally classified as NEAAs, are abundant in proteins of animal and plant origins, such as fishmeal, poultry by-product meal, and soybean meal, which are widely used as protein sources for aqua feeds (Li et al. 2011). Glutamine is a major energy source for many types of mammalian cells, including Hela cells, enterocytes, and tumor cells. There are suggestions that glutamine may serve as a major energy substrate for leukocytes and enterocytes in fish (Cheng et al. 2011; Li et al. 2009) as in mammals (Wu et al. 1998). Thus, dietary glutamine supplementation could increase the growth of HSB (Cheng et al. 2012) and half-smooth tongue sole (Liu et al. 2015), as reported for young pigs (Wu et al. 1996) and chickens (Bartell and Batal 2007), indicating that the regular diet may not provide sufficient glutamine to either mammals and poultry or fish. This further extends the concept of functional AAs broadly to farm animals of both agricultural and biomedical importance. Furthermore, our results reveal the tissue-specific metabolism of glutamate, glutamine, and aspartate in their utilization by the HSB (Jia et al. 2017). This foundational knowledge can guide the formulation of new cost-effective diets for feeding fish, as well as the development of alternatives of protein sources to fishmeal in aquaculture.

12.5 Largemouth Bass

The largemouth bass (*Micropterus salmoides*) is native to North America. Because of its popularity as a sportfish and its high market value as a food, the intensive culture of LMB in the United States began in the 1960s (Brecka et al. 1996), and this fish is now reared in many other countries (including China) worldwide (Bai and Li 2018; Tidwell et al. 2019). This species has several desirable characteristics for aquaculture, such as easy adaptation to freshwater, tolerance to a wide range of temperatures, good flesh quality, and strong disease resistance. The LMB can tolerate various water conditions, such as salinities of 2–5 ppt, pH of 6–10, temperatures of 18–32 °C, and the dissolved oxygen of 8–12 mg/L; the maximum growth of the LMB occurs at salinities of 2.5–3.5 ppt, pH of 6.5–8.5, and temperatures of 24–30 °C (Stuber et al. 1982). In the United States, besides its production as a food, the LMB is the top sportfish, representing about 74% of the farms that provide sportfish (USDA 2019b).

The LMB is a typical carnivorous fish (eating fish and invertebrates) with a short gastrointestinal segment and is traditionally fed directly low-value (trash) fish on farms (Tidwell et al. 2019). However, low-value fish feeding in fish cultivation will increase risks for infectious diseases and environmental pollution. As a carnivorous fish, the dietary protein requirement of LMB is more than 40% (Table 12.5), and an optimal level of CP in the diet is about 45% (Li et al. 2020d, e). Moreover, 50–70% of dietary protein for LMB feeding is typically provided by fishmeal. Obviously, the expansion of aquaculture for LMB production cannot be supported only by trash fish or high inclusion levels of fishmeal in diets. To address this critical issue, it is imperative to determine the relative importance of AAs, fatty acids, and glucose as metabolic fuels for LMB, and to reduce the use of fishmeal in its diet. Li et al. (2020d, e) reported that the LMB fed a diet containing $\geq$10% starch (dry matter basis) developed glycogenic hepatopathy but not a fatty liver and exhibited metabolic disorders in the liver. Of note, the LMB fed a low-fishmeal, soybean meal-based diet exhibited the black skin syndrome (Li et al. 2021a) that can be largely prevented by dietary supplementation with 0.5% methionine (Li et al. 2021b). This fish has different metabolic patterns than the HSB.

12.6 Aquaculture for the Provision of Food with High-Quality Protein

Fish is an important food that provides humans with high-quality protein and highly bioavailable minerals, especially in developing countries (Merino et al. 2012). The cultivation of fish and shellfish in terrestrial freshwater and marine systems of aquaculture has been rapidly growing at an annual rate of 7.8% worldwide between 1990 and 2010 and continues to grow, exceeding the growth of other food sectors including poultry,

Table 12.5 Reported requirements of largemouth bass (LMB) for dietary protein and lipids

Body weight of LMB (g)	Study period (days)	Crude protein requirement (% of dry diet)	Lipid requirement (% of dry diet)	References
1.8–6.7	36–50	40–41	10	Anderson et al. (1981)
5–15	26–68	40–41	10	Anderson et al. (1981)
14–23	64	43.6	10	Portz et al. (2001)
122–436	365	47	3.7	Tidwell et al. (1996)
16–79	84	43.4	7.6–17.4	Bright et al. (2005)
10–87	88	46–49	11.5–14	Chen et al. (2012)
8.7–52	56	55–58	13.6	Huang et al. (2017)
10–36	56	51.6	12.2	Cai et al. (2020)
100–320	56	50.5	12.2	Cai et al. (2020)
200–526	84	47.8	12.2	Cai et al. (2020)
13–42	56	47.3	10	Zhou et al. (2020)
4.8–26.7	56	45	10	Li et al. (2020d)
18.4–52.5	56	45	10	Li et al. (2020e)

pork, dairy, and grains during the same period (Troell et al. 2014; FAO 2020). In 2012, farmed fish production reached a record of 66 million tons globally, passing the beef production of 63 million tons for the first time (FAO 2018). Aquaculture provides not only high-quality protein for improving human nutrition and health but also an economic income for farmers worldwide. For example, US aquaculture sales in 2018 amounted to $1.5 billion, which was an increase of $144.0 million or 10.5% from 2013 (USDA 2019b). In contrast to the rapid growth of aquaculture, the capture of fish in the oceans and rivers reached a plateau of 90 million tons in the mid-1990s and stayed stable thereafter.

The current state of protein production from aquaculture and capture fisheries is directly a combined result of the expansion of the world population, economic development, climate and ecosystem changes, applications of scientific research, and many other factors (Merino et al. 2012; FAO 2020). Accordingly, the yield of aquaculture production is predicted to be close to that of captured fish by 2030 or sooner (Brander 2007). Currently, aqua feeds have heavily relied on fishmeal and fish oil, which are mainly proceeded from wild-caught small pelagics such as anchovies, sardines, and menhaden (Li et al. 2021c, d). Aquaculture will become even more important to supply high-quality animal protein in countries and regions where livestock and poultry species have high rates of morbidity and mortality due to widespread infectious diseases in terrestrial animals.

Fishmeal has traditionally been considered as an ingredient with highly digestible protein for aquafeed (Tacon et al. 2011). This feedstuff contains high levels of protein, vitamins, and minerals, as well as a balanced profile of AAs. However, the quantity of marine fisheries has a maximum potential of around 80 million tons per year (FAO 2018). Given the rapid expansion of aquaculture and the limited natural resources of fishmeal in association with the high requirements of dietary protein for many farmed carnivorous/piscivorous fish species, global fishmeal and fish oil supplies cannot meet the growing demand. Thus, the inclusion levels of fishmeal in aqua feeds will have to be reduced. Clearly, it is not sustainable to feed fish with fish.

12.7 Replacement of Fishmeal with Alternative Protein Sources

Feed cost constitutes more than half of the operating cost in intensive aquaculture, and ingredients of protein sources are the most expensive part in aquafeed. Replacing fishmeal in aqua feeds with alternative protein resources is a promising solution to reduce the use of fishmeal and the cost of feed for producing many fish species (Li et al. 2021e). Other than fishmeal, possible protein sources to meet the dietary requirement of farmed fish for high-quality protein include plant meals, domestic animal products, single-cell proteins, and insect proteins (Trushenski and Gause 2013). Because of extensive scientific research in this field, fishmeal levels in feeds for most fish species have decreased nearly by half in the past two decades. Indeed, fishmeal-free feeds have been successfully developed for many freshwater species such as cyprinids and tilapia (Tacon et al. 2011). However, fishmeal-free aqua feeds have not been achieved for carnivorous fish, such as HSB and LMB, without compromising their food intake, growth, skin color, or health.

It appears that many species, especially carnivorous and marine fish, show significantly lower growth performance when fed a low-fishmeal experimental diet, even though the experimental diets appeared to be nutritionally adequate in the provision of carbohydrates, lipids, vitamins, minerals, and EAAs (Oliva-Teles et al. 2015). Deleterious effects on nutrient utilization and fish health originally reported from those fishmeal replacement studies have been confirmed by the results of additional studies (Li et al. 2021c). Adverse effects include lower feed conversion rate, lower digestibility, intestinal inflammation, or enteritis (Andersen et al. 2016). For example, sub-acute enteritis in the distal intestine of Atlantic salmon was developed in a six-week feeding experiment using soybean meal-based feed (Baeverfjord and Krogdahl 1996). Moreover, some studies showed reductions in both feed intake by fish and the apparent digestibility of protein due to changes in the palatability and physical properties of feeds (Kaushik et al. 1995). Another emerging concern is the effect of fishmeal substitution on nutrient composition in the body of fish and their fillets quality. Numerous studies have shown some adverse effects of plant protein sources on the quality of fish fillets, with their color being most negatively affected by dietary plant-source proteins (Gaylord et al. 2010; Oliva-Teles et al. 2015). The following sections describe fishmeal and its major alternative feedstuffs.

12.7.1 Fishmeal

Fishmeal is the coarse flour made from fresh raw fish or fish parts by cooking, pressing, drying, and milling. The quality of fishmeal is affected by many factors, including source species, processing and storage conditions, shelf life, and adulteration with other ingredients of no or lower nutritional quality (e.g., urea and feather meal). The major source of fishmeal is some harvested small marine fish, such as anchovies, mackerel, sardines, menhaden, and herring (FAO 2018). As noted previously, fishmeal is a highly digestible feed ingredient for farmed animals and an excellent source of high-quality protein and fatty acids as well as highly bioavailable minerals and vitamins (Li et al. 2021e). Some fatty acids in fishmeal are essential for animal growth, such as long-chain omega-3 fatty acids, eicosapentaenoic acid, and docosahexaenoic acid (Wu 2018). The CP content of a high-quality fishmeal normally ranges from 60 to 72% by dry weight. The properly balanced profile of EAAs in fishmeal makes it a highly valued protein supplement for young growing terrestrial animals. An inclusion rate of less than 10% fishmeal in diets is beneficial for improving the nutrition and growth of starters and weaned pigs (Kats et al. 1992; Stoner et al. 1990). The use of fishmeal in diets can also increase the body weight, daily weight gain, and feed intake of broilers (Nixey 2010). Because the absolute amounts of feed intake by swine and poultry are high, the quantity of fishmeal used to feed land animals is tremendous.

Globally, the majority of fishmeal is used to feed fish due to the increased production of farmed fish and the wide use of compounded (formulated) feed for feeding them. For example, the percentage of commercial feed used for farmed marine fish has been estimated to gradually increase from 50 to 80% between 1995 and 2020 (FAO 2020). Currently, fishmeal is the major protein source in formulated feed for marine and carnivorous fish (Olsen and Hasan 2012). Some herbivorous and omnivorous fish such as Nile tilapia (*Oreochromis niloticus*), common carp (*Cyprinus carpio*), and crucian carp (*Carassius carassius*) also need a relatively high level of fishmeal in compounded feed (Olsen and Hasan 2012). Besides the abundance of AAs such as arginine, taurine, methionine, and lysine, fishmeal has an attractive odor to fish (possibly due to the presence of trimethylamine) that other protein sources (such as plants, meat and bone meal, poultry by-products, and insects) lack. This is a major reason why the substantial or complete replacement of fishmeal in diets for some farmed fish is difficult even though the provision of conventional nutrients (including EAAs) from the experimental diet appears to be adequate. In addition, fishmeal may contain a higher content of bioactive substances such as glutathione than other sources of protein ingredients (Li et al. 2011).

Over the past three decades, numerous studies have been carried out by research institutions and the aquaculture feed industry to generate detailed knowledge on the digestive processes and nutritional requirements of many farmed fish species (NRC 2011). Thus, since 1995, the dependency of aquaculture on fishmeal to feed many fish species, including carnivorous, marine and salmon, has been dramatically reduced but feed efficiencies in the fish have been considerably improved (Table 12.1). For example, the FAO (2018) reported that the feed conversion ratio (FCR, feed/gain ratio) of tilapias that were fed commercial compounded feed was 2.0 in 1995 and this value has been predicted to be reduced to 1.6 by 2020 (Tacon et al. 2011).

12.7.2 Alternative Protein Sources to Replace Fishmeal

The quality of a protein feedstuff depends upon many factors, including (a) its ability to supply sufficient amounts of all proteinogenic AAs; (b) its protein digestibility; and (c) the presence of anti-nutritional or toxic substances (Blaufuss and Trushenski 2012; Brown et al. 1997; Glencross et al. 2007). While fishmeal has traditionally been the major AA source for aquatic animals (Table 12.1), its alternatives have attracted more and more attention in aquafeed production (FAO 2020). Plant-sourced feedstuffs for aquatic animals include soybean meal, extruded soybean meal, fermented soybean meal, soybean protein concentrates, soybean protein isolates, leaf meal, hydrolyzed plant protein, wheat, wheat hydrolyzed protein, canola meal, cottonseed meal, peanut meal, sunflower meal, peas, rice, dried brewers grains, and dried distillers grains. Some of these products contain many anti-nutritional factors (e.g., trypsin inhibitors, gossypol, oxalates, goitrogens, tannins, cyanogenic glycosides, chlorogenic acids, toxic AAs) (Glencross et al. 2007; Wu 2018). To date, animal-sourced feedstuffs for aquatic animals are manufactured from the by-products of fish, poultry, pork, beef, and insects. These products include fishmeal, fish paste, bone meal, meat & bone meal, poultry by-product meal, chicken by-product meal, chicken visceral digest, spray-dried poultry plasma, spray-dried egg product, hydrolyzed feather meal, intestine-mucosa product, peptones (partial protein hydrolysates), blood meal (bovine or poultry), whey powder with high protein content, cheese powder, and insect meal (Li et al. 2020g; Li and Wu 2022). Microbial sources of protein feedstuffs include yeast protein and single-cell microbial protein (e.g., algae); they have more balanced AA profiles than most plant proteins for animal feeding (Li and Wu 2020). Of note, animal-sourced ingredients contain little or no anti-nutritional factors and can be used as a single source of dietary protein or in complementary combinations with plant and microbial sources of proteins (Li et al. 2021e).

12.7.2.1 Plant Protein Sources

Various plant feedstuffs are commonly used as alternative protein sources for the diets of farmed fish, including meals from soybean, wheat, and peas (Blaufuss and Trushenski 2012; Fournier et al. 2004; Oliva-Teles et al. 2015). Energy density, AA content, fiber, anti-nutritional factors, and nutrient digestibilities are the main factors to be considered when replacing fishmeal with plant protein sources in compounded feeds. Plant meals with high energy density, such as wheat meal, usually have high carbohydrate content, but carnivorous species cannot utilize carbohydrates well (Stone 2003). Plant meals are deficient in some EAAs such as lysine, methionine, and tryptophan, and contain no taurine or creatine (Li et al. 2011). Note that only animal products provide taurine and creatine (Wu et al. 2013a). Table 12.2 lists the nutritional requirements of five common farmed fish species (Atlantic salmon, rainbow trout, common carp, tilapia, and catfish) for AAs, the AA composition of protein in various feedstuffs, as well as the first, second, and third limiting AAs in these feedstuffs. As shown in Table 12.2, the most common first limiting AA in plant-source protein is usually methionine, lysine, or tryptophan. Interestingly, all plant feedstuffs, with the possible exception of cottonseed meal, are deficient in methionine, cysteine, lysine, and tryptophan relative to animal growth. A deficiency of one EAA will limit protein synthesis in tissues, leading to an increase in the oxidation of all other proteinogenic AAs (Wu 2021). Furthermore, dietary fiber increases satiety, gastric emptying, the transit of chime through the gastrointestinal tract, digestive passage rate, gastrointestinal tract weight, endogenous fluid secretion by the gut, pancreatic lipase secretion, enterocyte proliferation, maintenance energy requirement, the availability of metabolic fuels (e.g., butyrate) for the distal intestine, and fecal bulk, while reducing the pancreatic secretion of some digestive enzymes (α-amylase, elastase-1 and chymotrypsin), nutrient digestibilities, constipation, intestinal inflammation, and risk for colon cancer (Davies 1985; Lin et al. 2020; Perry and Ying 2016).

In addition to the above-mentioned nutritional drawbacks of plant protein sources, they also contain many anti-nutritional factors, including protease inhibitors, lectins, saponins, and phytate (Oliva-Teles et al. 2015). These anti-nutritional factors reduce the digestion or absorption of nutrients and may antagonize the function of AAs and vitamins in the gastrointestinal tract. To alleviate the adverse impacts of anti-nutritional factors present in plant feedstuffs, these ingredients can be chemically, mechanically, and biologically processed through methods such as heat processing, solvent extraction, dehulling, pelleting, extrusion, micronizing, autoclaving, or the use of exogenous enzymes (Glencross et al. 2007; Jobling et al. 2001; Krogdahl et al. 2010). For instance, fiber (a nonstarch polysaccharide) in many plant feedstuffs can be significantly reduced through the treatment with β-glucanases, pentosanases (arabinose and xylanase), and other enzymes (e.g., β-1,4-mannanase, β-1,6-galactosidase, and β-1,4-mannosidase) to increase the relative content of protein in the feedstuffs (Wu 2018). Phytate, which is covalently bound to phosphorus, reduces the bioavailability of minerals in soybean meal and can be treated by adding phytase to feeds to increase the release of nutrients from the feed matrix (Gatlin et al. 2007).

Some promising alternate plant protein sources to replace fishmeal in aquafeed are protein concentrates produced from soy, wheat, and other grains, as well as oilseeds and algae, because of their high protein content (60–80%, dry matter basis) and their low content of anti-nutritional factors (Blaufuss and Trushenski 2012; Hardy 2010). For example, salmonid species of fish could be fed a diet containing up to 75% of soy protein concentrate without developing intestinal enteritis (Kaushik et al. 1995; Refstie et al. 2001; Stickney et al. 1996). Their first limiting AAs are usually lysine, threonine, tryptophan, and methionine. However, owing to the high cost of their manufacturing, plant protein concentrates are currently not yet economically feasible or used as feed ingredients in the aquaculture industry. This challenge can be overcome by the use of animal by-products (Li et al. 2021e). Considering the limited resource of marine fish, relatively unstable fishmeal price, and improved processing technologies, plant protein concentrates may have great potential as aqua feeds on a large scale. Of all plant concentrate meals, soybean protein concentrate is most commonly used for laboratory research.

Soybean protein concentrate is a good source of many AAs for fish feeding (Berge et al. 1999; Li and Wu 2020; USSEC 2008). Based on the definition of The Association of American Feed Control Officials (Berk 1992), "soy protein concentrate is prepared from high-quality, sound, clean, dehulled soybean seeds by removing most of the oil and water-soluble nonprotein constituents and must contain not less than 70% protein on a moisture-free basis." The content of most AAs in soy protein concentrate is equal to, or greater than, that of menhaden fishmeal, but soy protein concentrate has a lower content of methionine and lysine than fishmeal and contains no taurine or creatine (Li and Wu 2020). Of all plant concentrate meals, soybean protein concentrate has partially or completely replaced fishmeal in experimental diets for many farmed fish species without compromising their growth performance (Hansen et al. 2007; Kaushik and Seiliez 2010). Atlantic salmon that were fed diets with 75% of total protein being replaced by soybean protein concentrate may achieve rapid growth, compared with a fishmeal-based diet (Storebakken et al. 2000). The growth rate of rainbow trout was not affected when they were fed a fishmeal-free diet containing soybean protein concentrate as the sole protein source (Kaushik et al. 1995).

Most plant-sourced protein feedstuffs contain a high content of starch (NRC 2011). Dietary starch is a major source of energy for human and terrestrial animals, and may help to lower fishmeal content in the diets of some fish species by sparing some dietary AAs (Enes et al. 2006; Li et al. 2021c). It has been reported that fish do not have requirements for dietary starch, but some evidence suggests that an appropriate amount of dietary starch can confer a protein-sparing effect in many species of fish (Hemre et al. 2002; NRC 2011). However, the ability of fish to utilize dietary starch varies greatly among different species. In general, the maximum inclusion of starch in diets has been suggested to be 15–25% for marine or carnivorous fish but can be up to 50% for herbivorous and omnivorous species (NRC 2011). However, we found that the dietary content of $\geq 10\%$ starch (dry matter basis) caused hepatic structural abnormalities and dysfunction in the LMB (Li et al. 2020d, e; Li et al. 2021c). Thus, care must be exercised when using plant-sourced feedstuffs as protein sources in the diets of fish.

12.7.2.2 Animal By-Products

Apart from plant proteins, terrestrial animal by-product meals are considered good substitutes of fishmeal based on their nutritional quality and competitively low prices (Gaylord and Rawles 2007; Li et al. 2021e). Animal by-product meals are made from a variety of animal organs or tissues that are left over after the principal food components have been obtained. These processed animal protein ingredients include, but are not limited to, blood meal, intestinal mucosa, feather meal, meat and bone meal, and poultry by-product meal (Li et al. 2021e). Compared with fishmeal, animal by-product meals have a profile of AAs more similar to that in the animal body than plant-source proteins. Notably, the content of some AAs, such as lysine, methionine, cysteine, and tryptophan, in plant-source proteins is relatively low (Li and Wu 2020). However, the proximate composition of animal feedstuffs is highly variable depending on their raw materials, particularly those of poultry by-products and meat and bone meals. Animal by-product meals have good palatability and no anti-nutritional factors. Inclusion of animal by-product meals as protein sources in fish feeds can range up to 20–40% without compromising fish growth in many studies (Oliva-Teles et al. 20152015). For biosafety reasons, the use of animal by-product meals in fish diets is regulated in many nations and regions.

Poultry by-product meal is now commonly used as a protein source for fish feeding (Hill et al. 2019; Li et al. 2021c; Mambrini et al. 1999; Nengas et al. 1999; Pine et al. 2008). According to the definition of AAFCO, poultry by-product meal is "the ground, rendered, clean parts of the carcass of slaughtered poultry such as necks, heads, feet, undeveloped eggs, gizzards and intestines (provided their content is removed), exclusive of feathers (except in such amounts as might occur unavoidably in good processing practices)." The nutrient composition of poultry by-product meals is highly variable, depending on the raw materials and manufacturers. However, typical high-quality poultry by-product meals can have protein content between 75 and 90% (dry matter basis) with a relatively low content of ash and fat. Therefore, the feedstuff's proximate composition should be carefully evaluated before use. Like the poultry by-product meal, other poultry-derived by-products with excellent AA profiles and digestibilities include chicken by-product meal, chicken visceral digest, spray-dried egg product, and spray-dried poultry plasma (Li et al. 2020g).

Results of studies over the past decade reveal that soybean protein and poultry by-products contain high levels of specific AAs such as glutamate and glutamine (Li et al. 2011, 2020g). This allows the nutritional possibility to replace fishmeal in feed in fish diets with a combination of plant- and animal-sourced feedstuffs. After dietary requirements of fish for EAAs are met, some NEAAs (e.g., glutamate and glutamine) are likely to play an important role in sparing the need for high fishmeal content in diets and maintaining intestinal health (Li et al. 2020c). It should be borne in mind that most non-fishmeal sourced ingredients contain one or more AAs that are relatively low compared with fishmeal. Therefore, the insufficiency of particular AAs in the feeds with fishmeal being replaced by alternative protein sources may be corrected by supplementing the feed with crystalline AAs or a complementary mixture of animal by-products (Li et al. 2021e).

12.7.2.3 Microbial Sources of Protein Feedstuffs

Single-cell proteins (also known as microbial proteins) are produced from edible unicellular microorganisms. They include certain species of yeast (*Saccharomyces cerevisiae*, *Pichia pastoris*, *Candida utilis*, *Torulopsis coralline, and Geotrichum candidum*), algae (*Spirulina* and *Chlorella*), fungi (*Aspergillus oryzae*, *Fusarium venenatum*, *Sclerotium rolfsii*, *Polyporus, and Trichoderma*), and bacteria (*Rhodobacter capsulatus*) (Ritala et al. 2017). In cultures, these microbes can rapidly convert animal wastes (e.g., ammonia, other nitrogenous metabolites, and sulfur) and indigestible carbohydrates that are harmful to the environment into usable nutrients (protein, fatty acids, and nucleic acids) for animal feeding. This is of great significance for both environmental protection and sustainable resource utilization (Wu 2022).

Single-cell proteins contain high amounts of proteins. For example, the content of total AAs in algae spirulina meal and the marine yeast product is 77.5% (Li and Wu 2020) and 61–70% (Ritala et al. 2017), respectively. Glutamate is the most abundant AA in the peptides plus the free AA pool in algae spirulina meal, followed by leucine, alanine, arginine, valine, and serine in descending order (Li and Wu 2020). Compared with soy protein concentration, algae spirulina meal contains 70% and 85% more alanine and methionine, respectively, but 35% and 38% less cysteine and histidine, respectively (Li and Wu 2020). Thus, a combination of algae spirulina meal and soy protein concentration may be nutritionally advantageous for aquatic animals than either feedstuff alone. To date, little is known about the content of glutamate, glutamine, aspartate, and asparagine in single-cell protein feedstuffs other than algae. In addition, yeast is an abundant source of glutathione (a major antioxidant; Wu et al. 2013a), whereas microbial nucleic acids may be beneficial for intestinal nutrition and health of aquatic animals (Li et al. 2007). Furthermore, chitin and glucan from fungal cell walls contribute indigestible fiber to the diet.

Some microorganisms can be fed directly as whole-cell preparations to animals as a source of single-cell proteins, whereas other microorganisms must be lysed before their use for animal feeding to release proteins. For example, *Euglena* (a motile, single-celled organism that is commonly found in aquatic habitats) does not require disruption before feeding to animals because this cell has proteinaceous pellicles rather than a cell wall (Chae et al. 2006). In addition, intact bacteria or yeast can be included in the diets of ruminants because these animals contain lysozymes in their abomasum (Wu 2018). In contrast, most microorganisms must be disrupted before their feeding to nonruminant and aquatic animals. The cell wall of most microorganisms can be broken down by mechanical forces (crushing, crumbling, grinding, pressure homogenization, or ultrasonication), hydrolytic enzymes (lysozymes), chemical disruption with detergents, or combinations of these methods (Ritala et al. 2017).

12.8 Culturing of Fish for Nutrition Research

Current fish nutrition studies are mainly conducted via two systems: the indoor recirculating aquaculture system (RAS) and outdoor ponds. The RAS is a system in which water is partially reused after undergoing treatment (Ebeling and Timmons 2012). This system is used to rear fish in indoor tanks because it provides fish with a controllable and stable living environment. In order to maintain healthy fish, the RAS needs a continuous supply of clean incoming water with an optimal temperature and an optimal level of dissolved oxygen. Therefore, the RAS essentially consists of those sub-systems: tanks, a filtering system, a temperature control unit (water heater and temperature monitor), an air supply system (air pump, air tubing, and air-stone), and a water recirculating system (water pump and pipes). Such an indoor system was developed in our aquaculture nutrition research involving the HSB, LMB, and zebrafish, as shown in Fig. 12.1. The filtering system itself contains mechanical filters, bio-filters, and UV lights to remove particles (e.g., feces and leftover feed), detoxify harmful waste products (e.g., ammonia and nitrite), and kill pathogens, respectively. Clean water is added to each tank only when the accumulated waste materials need to be removed or when the RAS's water volume is low due to evaporation and splashing. About one-third of the water in a tank is replaced with fresh water when the water is changed (usually every day) to maintain sufficient oxygen in the remaining water. Water conditions of our RAS are as follows: temperature, 24–27 °C; salinity, 2–5 ppt; pH, 6.5–7.5; NH_4^+, <2 mg/L; NO_2^-, <1 ppm; NO_3^-, <20 ppm; and dissolved O_2, 7–9 ppm. To minimize the concentrations of excreted metabolites in water, a tank holding 55 L water can house up to fifteen 65-g fish (e.g., hybrid-striped bass and largemouth bass).

Compared with the open ponds, using the RAS for fish nutrition studies provides various benefits. First, the conditions of the water are easily controlled and stable. Thus, the RAS minimizes the impact of irrelevant factors, and the living environment of fish will not confound their response to dietary treatments. Second, the indoor RAS prevents intruders (e.g., predators and prey of experimental fish) because the facility is maintained in a closed system. Third, the RAS allows for ease of management, harvest, and feeding because the system rears fish at a high density without compromising their health. Fourth, the RAS system can reduce the risks of inclement weather, natural water pollution, and infectious diseases. Thus, the RAS is useful for nutrition research and promising for the future farming of aquatic animals. Using our RAS, we have successfully reared HSB, LMB [e.g., 12 fish (205 g/fish) per tank with 55 L of water], and zebrafish (Jia et al. 2017; Li et al. 2020a, b, c, d, e, f), as well as shrimp and crabs (Li et al. 2021d) to conduct well-controlled nutrition research. Our work has identified glycogenic hepatopathy and hepatic steatosis in the LMB fed a high-starch diet (Li et al. 2021e) and discovered the black skin syndrome in the LMB fed diets containing inadequate methionine (Li et al. 2021a, b). However, the RAS does have disadvantages compared with studies conducted in open ponds. For example, it is labor-intensive and expensive

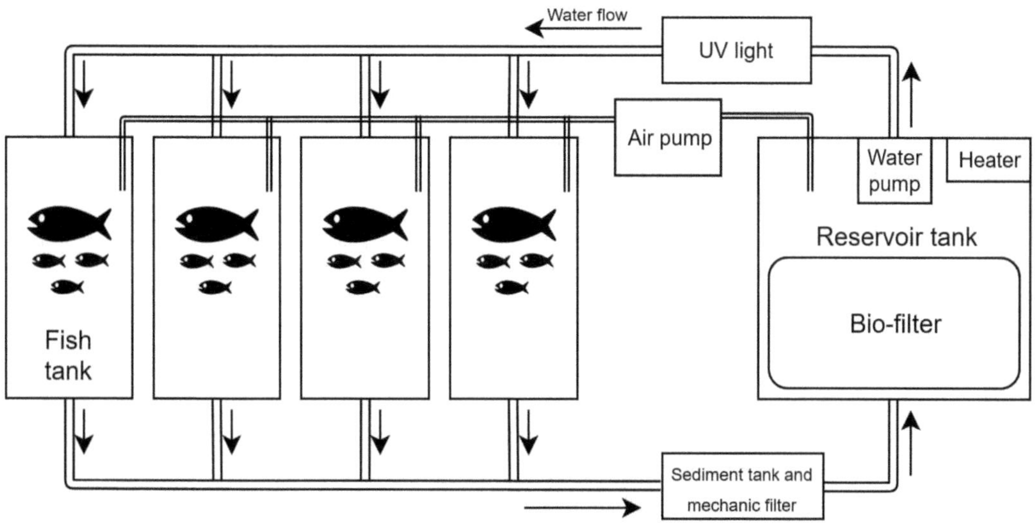

Fig. 12.1 Flowchart of a recirculating aquaculture system to rear fish. The water for housing fish is prepared by mixing fresh deionized water with sea salt (1.0–1.5 mg/l). The salty water is added into a large reservoir tank and pumped into individual tanks through pipes. The outflowing water from each tank [length (53.8 cm) × height (28.0 cm) × width (34.4 cm); holding 55 L of water] is collected into the sediment tank and filtered by a mechanical filter before returning to the reservoir tank. Any solids, feces, or uneaten food from the fish tanks are filtered in the sediment tank through a mechanical filter. A bio-filter is used to convert ammonia in the water into nitrite and nitrate by nitrifying bacteria. The water is then recirculated back to the fish tanks through an in-line UV light that kills pathogens, fungi, or other micro-organisms. The air pump is used to aerate water in the fish tanks and the reservoir tank via tubes connected to an air-stone. A submersible water heater in the reservoir tank is used to maintain a desired temperature of water in the fish tanks. Each tank can house up to fifteen 65-g fish (e.g., hybrid-striped bass and largemouth bass)

to maintain daily tanks with clean water. In addition, the RAS limits the number of large fish or sub-adult fish in a tank. Furthermore, energy consumption and greenhouse gas emissions are the two most stringent limiting factors for the RAS (Ahmed and Turchini 2021). These shortcomings of the RAS can be alleviated by technological innovations, including (1) the automation of the system; (2) improvements in the formulation of aquafeeds to minimize excess ammonia accumulation; and (3) the development of artificial intelligence techniques to teach the system how to recognize and predict patterns, identify problematic processes, and generate early warnings of imminent malfunction.

Nutrition research often involves the periodic weighing of fish (e.g., every 2, 4, or 8 weeks). To calm down fish during the weighing process, they are usually transferred from their rearing tank to a container (e.g., a round-shape plastic container with an internal diameter of 30 cm for juvenile fish) with an oxygenated neutral solution [50 ppm MS222 (tricaine methanesulfonate, the recommended anesthetic agent for aquatic animals) buffered with 100 ppm sodium bicarbonate] (Bowker et al. 2012). This solution (e.g., 4 L for up to fifteen 50-g fish) with 0.4% salinity (for fish such as hybrid-striped bass and largemouth bass) is prepared by adding 50 mg MS-222, 100 mg sodium bicarbonate, and 4 g sea salt to 1 L of deionized water, followed by 5-min aeration via an air-stone connected to an air pump. After being weighed in a balance, fish should be immediately placed back to their original tank.

12.9 Summary

Fish generally require a very high level of dietary protein (e.g., 30–60% of dry matter, depending on species) for maintenance and growth. Amino acids provide the bulk of energy for fish in a tissue-specific manner. Results of recent studies have shown that the proximal intestine, liver, kidneys, and skeletal muscle of fish use glutamate, glutamine, and aspartate as major metabolic fuels (Li et al. 2021c). These AAs and other proteinogenic AAs must be provided in adequate amounts and ratios in the diets of all fish species. Carnivorous fish also need sufficient taurine from their diets. Although fishmeal has traditionally been used as the sole or primary source of AAs in compound aqua feeds for fish, there is an urgent need to identify alternative sources of protein feedstuffs so as to sustain the global aquaculture (Wu 2022). Growing evidence shows that plant- and animal-sourced feedstuffs can be successfully used to replace all or most fishmeal in aqua feeds, depending on fish species. Functional AAs must be considered when formulating the diets of aquatic animals (e.g., fish, shrimp, and crabs) to improve their growth, development, health, and productivity.

Acknowledgments This work was supported by Texas A&M AgriLife Research (H-8200). The authors thank students and research assistants in our laboratory for helpful discussions.

References

Abdel-Tawwab M, Ahmad MH, Khattab YAE, Shalaby AME (2010) Effect of dietary protein level, initial body weight, and their interaction on the growth, feed utilization, and physiological alterations of Nile tilapia, *Oreochromis niloticus* (L.). Aquaculture 298:267–274

Ahmed N, Turchini GM (2021) Recirculating aquaculture systems (RAS): environmental solution and climate change adaptation. J Cleaner Prod 297:126604

Albrecht J, Norenberg MD (2006) Glutamine: a Trojan horse in ammonia neurotoxicity. Hepatology 44:788–794

Anderson RJ, Kienholz EW, Flickinger SA (1981) Protein requirements of smallmouth bass and largemouth bass. J Nutr 111:1085–1097

Andersen S, Waagbø R, Espe M (2016) Functional amino acids in fish health and welfare. Front Biosci 8:143–169

Baeverfjord G, Krogdahl A (1996) Development and regression of soybean meal induced enteritis in Atlantic salmon, *Salmo salar* L., distal intestine: a comparison with the intestines of fasted fish. J Fish Dis 19:375–387

Bai J, Li S (2018) Development of largemouth bass (Micropterus salmoides) culture. In: Gui JF, Tang Q, Li Z, Liu Z, De Silva SS (eds) Aqualculture in China: success stories and modern trends. Wiley, Oxford, pp 421–429

Ballantyne JS (2001) Amino acid metabolism. Fish Physiol 20:77–107

Bartell SM, Batal AB (2007) The effect of supplemental glutamine on growth performance, development of the gastrointestinal tract, and humoral immune response of broilers. Poult Sci 86:1940–1947

Berge GM, Grisdale-Helland B, Helland SJ (1999) Soy protein concentrate in diets for Atlantic halibut (*Hippoglossus hippoglossus*). Aquaculture 178:139–148

Berk Z (1992) Technology of production of edible flours and protein products from soybeans. Food and Agriculture Organization of the United Nations, Rome, Italy

Bhaskar M (1994) Changes in the liver protein fractions of Tilapia mossambica (Peters) during acclimation to low and high pH media. Fish Res 19:179–186

Blaufuss P, Trushenski J (2012) Exploring soy-derived alternatives to fish meal: using soy protein concentrate and soy protein isolate in hybrid striped bass feeds. N Am J Aquac 74:8–19

Bowker JD, Trushenski JT, Gaikowski MP, Straus DL, Editors (2012) Guide to Using Drugs, Biologics, and Other Chemicals inAquaculture. American Fisheries Society Fish Culture Section, Bethesda, MD

Brander KM (2007) Global fish production and climate change. Proc Natl Acad Sci USA 104:19709–19714

Brecka BJ, Wahl DH, Hooe ML (1996) Growth, survival, and body composition of largemouth bass fed various commercial diets and protein concentrations. ProgIve Fish-Cult 58:104–110

Bright LA, Coyle SD, Tidwell JH (2005) Effect of dietary lipid level and protein energy ratio on growth and body composition of largemouth bass *Micropterus salmoides*. J World Aquac Soc 36:129–134

Brosnan JT, Brosnan ME (2013) Glutamate: a truly functional amino acid. Amino Acids 45:413–418

Brown PB, Twibell R, Jonker Y, Wilson KA (1997) Evaluation of three soybean products in diets fed to juvenile hybrid striped bass *Morone saxatilis* × *M. chrysops*. J World Aquac Soc 28:215–223

Buentello JA, Gatlin DM (2001) Effects of elevated dietary arginine on resistance of channel catfish to exposure to *Edwardsiella ictaluri*. J Aquat Anim Health 13:194–201

Bulak JS, Coutant CC, Rice JA (2013) Biology and management of inland striped bass and hybrid striped bass. American Fisheries Society, Bethesda, MD

Bush JA, Wu G, Suryawan A et al (2002) Somatotropin-induced amino acid conservation in pigs involves differential regulation of liver and gut urea cycle enzyme activity. J Nutr 132:59–67

Caballero-Solares A, Viegas I, Salgado MC et al (2015) Diets supplemented with glutamate or glutamine improve protein retention and modulate gene expression of key enzymes of hepatic metabolism in gilthead seabream (*Sparus aurata*) juveniles. Aquaculture 444:79–87

Cai ZN, Qian XQ, Xie SQ (2020) Optimal dietary protein concentrations for largemouth bass (*Micropterus salmoides*) of different sizes (10–500 g). Aquacult Int 28:831–840

Campbell JW, Aster PL, Vorhaben JE (1983) Mitochondrial ammoniagenesis in liver of the channel catfish *Ictalurus punctatus*. Am J Physiol 244:R709–R717

Chae SR, Hwang EJ, Shin HS (2006) Single cell protein production of Euglena gracilis and carbon dioxide fixation in an innovative photo-bioreactor. Bioresour Technol 97:322–329

Chamberlin ME, Glemet HC, Ballantyne JS (1991) Glutamine metabolism in a holostean (Amia calva) and teleost fish (*Salvelinus namaycush*). Am J Physiol 260:R159–R166

Chan J (2016) Investigating hepatic glutamate dehydrogenase activity as a cause for the significant loss of amino acid utilization efficiency in growing rainbow trout (*Oncorhynchus mykiss*). The University of Guelph, Guelph, Canada

Chatzifotis S, Polemitou I, Divanach P, Antonopoulou E (2008) Effect of dietary taurine supplementation on growth performance and bile salt activated lipase activity of common dentex, *Dentex dentex*, fed a fish meal/soy protein concentrate-based diet. Aquaculture 275:201–208

Chen NS, Xiao WW, Liang QL, Zhou HY, Ma XL, Zhao M (2012) Effect of dietary lipid to protein ratios on growth performance, body composition and non-specific immunity of largemouth bass *Micropterus salmoides*. J Fish China 36:1270–1280

Cheng Z, Buentello A, Gatlin DM (2011) Effects of dietary arginine and glutamine on growth performance, immune responses and intestinal structure of red drum, *Sciaenops ocellatus*. Aquaculture 319:247–252

Cheng Z, Gatlin DM, Buentello A (2012) Dietary supplementation of arginine and/or glutamine influences growth performance, immune responses and intestinal morphology of hybrid striped bass (*Morone chrysops* × *Morone saxatilis*). Aquaculture 362–363:39–43

Costas B, Conceição LEC, Dias J et al (2011) Dietary arginine and repeated handling increase disease resistance and modulate innate immune mechanisms of Senegalese sole (*Solea senegalensis Kaup*, 1858). Fish Shellfish Immunol 31:838–847

Cowey CB, Walton MJ (1988) Studies on the uptake of (^{14}C) amino acids derived from both dietary (^{14}C) protein and dietary (^{14}C) amino acids by rainbow trout, *Salmo gairdneri Richardson*. J Fish Biol 33:293–305

D'Abramo LR, Frinsko MO (2008) Hybrid striped bass: pond production of food fish. Southern Regional Aquaculture Center, Publication No. 303, pp 1–4

D'Abramo LR, Ohs CL, Taylor JB (2000) Effects of reduced levels of dietary protein and menhaden fish meal on production, dressout, and biochemical composition of phase III sunshine bass (*Morone chrysops* x *Morone saxatilis*) cultured in earthen ponds. J World Aquac Soc 31:316–325

Davies SJ (1985) The role of dietary fibre in fish nutrition. In: Muir JF, Roberts RJ (eds) Recent advances in aquaculture. Springer, Boston, MA, pp 219–249

Day OJ, GonzAlez HGP (2000) Soybean protein concentrate as a protein source for turbot *Scophthalmus maximus* L. Aquac Nutr 6:221–228

Ebeling JM, Timmons MB (2012) Recirculating aquaculture systems. In: Tidwell JH (ed) Aquaculture production systems. Wiley, Oxford, pp 245–277

Ende SSW, Fuchs V, Schuhn A, von der Marwitz C, Wirtz A, Henjes J, Slater M (2018) Growth performance of hybrid striped bass (*Morone chrysops* × *M. saxatilis*) fed with commercial pike perch and trout diets. Int Aquat Res 10:57–63

Enes P, Panserat S, Kaushik S, Oliva-Teles A (2006) Effect of normal and waxy maize starch on growth, food utilization and hepatic glucose metabolism in European sea bass (*Dicentrarchus labrax*) juveniles. Comp Biochem Physiol A 143:89–96

FAO (2018) The state of world fisheries and aquaculture 2018—meeting the sustainable development goals. Italy, Rome

FAO (2020) The state of world fisheries and aquaculture 2020. Food and Agriculture Organization of the United Nations, Rome, Italy

Fournier V, Huelvan C, Desbruyeres E (2004) Incorporation of a mixture of plant feedstuffs as substitute for fish meal in diets of juvenile turbot (*Psetta maxima*). Aquaculture 236:451–465

French CJ, Mommsen TP, Hochachka PW (1981) Amino acid utilisation in isolated hepatocytes from rainbow trout. Eur J Biochem 113:311–317

Fynn-Aikins K, Hughes SG, Vandenberg GW (1995) Protein retention and liver aminotransferase activities in Atlantic salmon fed diets containing different energy sources. Comp Biochem Physiol A 111:163–170

Gallagher ML (1994) The use of soybean meal as a replacement for fish meal in diets for hybrid striped bass (*Morone saxatilis* × *M. chrysops*). Aquaculture 126:119–127

Gatlin DM III, Barrows FT, Brown P, Dabrowski K, Gaylord TG, Hardy RW et al (2007) Expanding the utilization of sustainable plant products in aquafeeds—a review. Aquaculture Res 38:551–579

Gaylord TG, Barrows FT, Overturf K et al (2010) An overview of progress toward developing an all plant-based diet for rainbow trout. Bull Fish Res Agency 31:9–14

Gaylord TG, Rawles SD (2007) The Modification of poultry by-product meal for use in hybrid striped bass *Morone chrysops* × *M. saxatilis* diets. J World Aquac Soc 36:363–374

Glencross B, Rutherford N, Jones B (2011) Evaluating options for fishmeal replacement in diets for juvenile barramundi (*Lates calcarifer*). Aquac Nutr 17:e722–e732

Glencross BD, Booth M, Allan GL (2007) A feed is only as good as its ingredients ? a review of ingredient evaluation strategies for aquaculture feeds. Aquac Nutr 13:17–34

Hansen AC, Rosenlund G, Karlsen Ø et al (2007) Total replacement of fish meal with plant proteins in diets for Atlantic cod (*Gadus morhua L.*) I—effects on growth and protein retention. Aquaculture 272:599–611

Hardy RW (2010) Utilization of plant proteins in fish diets: effects of global demand and supplies of fishmeal. Aquac Res 41:770–776

Harrell R (2016) Cultured aquatic species information programme—Morone hybrid (genus Morone, hybrids). In: Fisheries and aquaculture. FAO, Rome, Italy

Haynes TE, Li P, Li X et al (2009) L-glutamine or L-alanyl-L-glutamine prevents oxidant- or endotoxin-induced death of neonatal enterocytes. Amino Acids 37:131–142

Haÿs SP, Ordonez JM, Burrin DG, Sunehag AL (2007) Dietary glutamate is almost entirely removed in its first pass through the splanchnic bed in premature infants. Pediatr Res 62:353–356

He WL, Li P, Wu G (2021) Amino acid nutrition and metabolism in chickens. Adv Exp Med Biol 1285:109–131

Hemre GI, Mommsen TP, Krogdahl Å (2002) Carbohydrates in fish nutrition: effects on growth, glucose metabolism and hepatic enzymes. Aquac Nutr 8:175–194

Hill JC, Alam MS, Watanabe WO et al (2019) Replacement of Menhaden fish meal by poultry by-product meal in the diet of juvenile red porgy. N Am J Aquac 81:81–93

Hodson RG (1989) Hybrid striped bass: biology and history. e-Southern Regional Aquaculture Center, Publication No. 300, pp 1–4

Hopkins KD (1992) Reporting fish growth: a review of the basics. J World Aquac Soc 23:173–179

Hou Y, Wu G (2018) L-glutamate nutrition and metabolism in swine. Amino Acids 50:1497–1510

Hou Y, Yin Y, Wu G (2015) Dietary essentiality of nutritionally non-essential amino acids for animals and humans. Exp Biol Med 240:997–1007

Hou YQ, He WL, Hu SD, Wu G (2019) Composition of polyamines and amino acids in plant-source foods for human consumption. Amino Acids 51:1153–1165

Huang D, Wu Y, Lin Y, Chen J, Karrow N, Ren X, Wang Y (2017) Dietary protein and lipid requirements for juvenile largemouth bass, *Micropterus salmoides*. J World Aquac Soc 48:782–790

Hughes SG, Rumsey GL, Nesheim MC (1983) Branched-chain amino acid aminotransferase activity in the tissues of lake trout. Comp Biochem Physiol 76B:429–431

Ip YK, Chew SF (2010) Ammonia production, excretion, toxicity, and defense in fish: a review. Front Physiol 1:134

Jia S (2019) Nutritional roles of glutamate and glutamine in the growth of juvenile hybrid striped bass. M.S. thesis. Texas A&M University, College Station, TX, USA

Jia S, Li X, Zheng S, Wu G (2017) Amino acids are major energy substrates for tissues of hybrid striped bass and zebrafish. Amino Acids 49:2053–2063

Jia S, Li XY, He WL, Wu G (2021) Oxidation of energy substrates in tissues of fish: metabolic significance and implications for gene expression and carcinogenesis. Adv Exp Med Biol 1332:67–83

Jobgen WS, Fried SK, Fu WJ et al (2006) Regulatory role for the arginine–nitric oxide pathway in metabolism of energy substrates. J Nutr Biochem 17:571–588

Jobling M, Gomes E, Dias J (2001) Feed types, manufacture and ingredients. In: Boujard T, Jobling M (eds) Food intake in fish (Houlihan D. Blackwell Science, Oxford, UK, pp 25–48

Jürss K, Bastrop R (1995) Amino acid metabolism in fish. Amino acid metabolism in fish. Biochem Mol Biol Fishes 4:159–189

Kats LJ, Laurin JL, Tokach MD et al (1992) Comparison of spray-dried blood meal and fish by-products in the phase II starter pig diet. Kansas Agric Exp Stn Res Reports 37–40

Kaushik SJ, Cravedi JP, Lalles JP et al (1995) Partial or total replacement of fish meal by soybean protein on growth, protein utilization, potential estrogenic or antigenic effects, cholesterolemia and flesh quality in rainbow trout, *Oncorhynchus mykiss*. Aquaculture 133:257–274

Kaushik SJ, Seiliez I (2010) Protein and amino acid nutrition and metabolism in fish: current knowledge and future needs. Aquac Res 41:322–332

Krogdahl Å, Penn M, Thorsen J et al (2010) Important antinutrients in plant feedstuffs for aquaculture: an update on recent findings regarding responses in salmonids. Aquac Res 41:333–344

Laale HW (1977) The biology and use of zebrafish, *Brachydanio rerio* in fisheries research a literature review. J Fish Biol 10:121–173

Latshaw JD, Bishop BL (2001) Estimating body weight and body composition of chickens by using noninvasive measurements. Poult Sci 80:868–873

Lee DJW, McNab JM, Shannon DWF, Blair R (1972) Enzyme studies with the livers of chicks fed semi-synthetic diets containing crystalline amino acids and diammonium citrate. Br Poult Sci 13:229–235

Li P, Wu G (2018) Roles of dietary glycine, proline and hydroxyproline in collagen synthesis and animal growth. Amino Acids 50:29–38

Li P, Wu G (2020) Composition of amino acids and related nitrogenous nutrients in feedstuffs for animal diets. Amino Acids 52:523–542

Li P, Wu G, Gatlin DM (2007) Nucleotide supplementation may offer health benefits in cultured fish, but more study needed. Glob Aquac Advocate 10:74–75

Li P, Mai K, Trushenski J, Wu G (2009) New developments in fish amino acid nutrition: towards functional and environmentally oriented aquafeeds. Amino Acids 37:43–53

Li X, Rezaei R, Li P, Wu G (2011) Composition of amino acids in feed ingredients for animal diets. Amino Acids 40:1159–1168

Li XL, Zheng SX, Jia SC, Song F, Zhou CP, Wu G (2020a) Oxidation of energy substrates in tissues of largemouth bass (*Micropterus salmoides*). Amino Acids 52:1017–1032

Li XY, Zheng SX, Han T, Song F, Wu G (2020b) Effects of dietary protein intake on the oxidation of glutamate, glutamine, glucose and palmitate in tissues of largemouth bass (*Micropterus salmoides*) Amino Acids 52:1491–1503

Li XL, Zheng SX, Wu G (2020c) Nutrition and metabolism of glutamate and glutamine in fish. Amino Acids 52:671–691

Li XY, Zheng SX, Ma XK, Cheng KM, Wu G (2020d) Effects of dietary starch and lipid levels on the protein retention and growth of largemouth bass (*Micropterus salmoides*). Amino Acids 52:999–1016

Li XY, Zheng SX, Ma XK, Cheng KM, Wu G (2020e) Effects of dietary protein and lipid levels on growth performance, feed utilization, and liver histology of largemouth bass (*Micropterus salmoides*). Amino Acids 52:1043–1061

Li XY, Zheng SX, Wu G (2020f) Amino acid metabolism in the kidneys: nutritional and physiological significance. Adv Exp Med Biol 1265:71–95

Li SL, Zhang YC, Liu N, Chen JQ, Guo LN, Dai ZL, Wang C, Wu ZL, Wu G (2020g) Dietary L-arginine supplementation reduces lipid accretion by regulating fatty acid metabolism in Nile tilapia (*Oreochromis niloticus*). J Anim Sci Biotechnol 11:82

Li XY, Zheng SX, Ma XK, Cheng KM, Wu G (2021a) Use of alternative protein sources for fishmeal replacement in the diet of largemouth bass (*Micropterus salmoides*). Part I: effects of poultry by-product meal and soybean meal on growth, feed utilization, and health. Amino Acids 53:33–47

Li XY, Zheng SX, Cheng KM, Ma XK, Wu G (2021b) Use of alternative protein sources for fishmeal replacement in the diet of largemouth bass (*Micropterus salmoides*). Part II: effects of supplementation with methionine or taurine on growth, feed utilization, and health. Amino Acids 53:49–62

Li XY, Zheng SX, Wu G (2021c) Nutrition and functions of amino acids in fish. Adv Exp Med Biol 1285:133–168

Li XY, Han T, Zheng SX, Wu G (2021d) Nutrition and functions of amino acids in aquatic crustaceans. Adv Exp Med Biol 1285:169–198

Li P, He WL, Wu G (2021e) Composition of amino acids in foodstuffs for humans and animals. Adv Exp Med Biol 1332:189–209

Li P, Wu G (2022) Functional molecules of intestinal mucosal products in animal nutrition and health. Adv Exp Med Biol1354:263–277

Lin S-M, Zhou X-M, Zhou Y-L, Kuang W-M, Chen Y-J (2020) Intestinal morphology, immunity and microbiota response to dietary fibers in largemouth bass, Micropterus salmoide. Fish Shellfish Immunol 103:135–142

Liu FG, Liao IC (1999) Effect of feeding regimen on the food consumption, growth, and body composition in hybrid striped bass *Morone saxatilis* × *M. chrysops*. Fish Sci 65:513–519

Liu J, Mai K, Xu W et al (2015) Effects of dietary glutamine on survival, growth performance, activities of digestive enzyme, antioxidant status and hypoxia stress resistance of half-smooth tongue sole (*Cynoglossus semilaevis Günther*) post larvae. Aquaculture 446:48–56

Liu XD, Wu X, Yin YL et al (2012) Effects of dietary l-arginine or N-carbamylglutamate supplementation during late gestation of sows on the miR-15b/16, miR-221/222, VEGFA and eNOS expression in umbilical vein. Amino Acids 42:2111–2119

Lobley GE, Milne V, Lovie JM et al (1980) Whole body and tissue protein synthesis in cattle. Br J Nutr 43:491–502

Lougheed M, Nelson B (2001) Hybrid striped bass production. Michigan State University, East Lansing, Michigan, Markets and Marketing

Lunger AN, McLean E, Gaylord TG et al (2007) Taurine supplementation to alternative dietary proteins used in fish meal replacement enhances growth of juvenile cobia (*Rachycentron canadum*). Aquaculture 271:401–410

MacLennan PA, Brown RA, Rennie MJ (1987) A positive relationship between protein synthetic rate and intracellular glutamine concentration in perfused rat skeletal muscle. FEBS Lett 215:187–191

Mambrini M, Kaushik SJ (1995) Indispensable amino acid requirements of fish: correspondence between quantitative data and amino acid profiles of tissue proteins. J Appl Ichthyol 11:240–247

Mambrini M, Roem AJ, Carvèdi JP et al (1999) Effects of replacing fish meal with soy protein concentrate and of DL-methionine supplementation in high-energy, extruded diets on the growth and nutrient utilization of rainbow trout, *Oncorhynchus mykiss*. J Anim Sci 77:2990–2999

Mateo RD, Wu G, Moon HK et al (2008) Effects of dietary arginine supplementation during gestation and lactation on the performance of lactating primiparous sows and nursing piglets1. J Anim Sci 86:827–835

McGinty AS, Hodson RG (2008) Hybrid striped bass: hatchery phase. Southern Regional Aquaculture Center. Publication No. 300, pp 1–6

Merino G, Barange M, Blanchard JL et al (2012) Can marine fisheries and aquaculture meet fish demand from a growing human population in a changing climate? Glob Environ Chang 22:795–806

National Research Council (NRC) (2000) Nutrient requirements of beef cattle. National Academies Press, Washington DC

National Research Council (NRC) (2011) Nutrient requirements of fish and shrimp. National Academies Press, Washington DC

National Research Council (NRC) (2012) Nutrient requirements of swine. National Academies Press, Washington DC

Nengas I, Alexis MN, Davies SJ (1999) High inclusion levels of poultry meals and related byproducts in diets for gilthead seabream Sparus aurata L. Aquaculture 179:13–23

Nixey C (2010) Nutrition and feeding of organic poultry by robert blair. Br Poult Sci 51:309–309

Oliva-Teles A, Enes P, Peres H (2015) Replacing fishmeal and fish oil in industrial aquafeeds for carnivorous fish. In: Feed and feeding practices in aquaculture. Elsevier, pp 203–233

Olsen RL, Hasan MR (2012) A limited supply of fishmeal: impact on future increases in global aquaculture production. Trends Food Sci Technol 27:120–128

Perez-Velazquez M, Gatlin DM III, González-Félixa ML, García-Ortega A, de Cruz CR, Juárez-Gómez ML, Chen K (2019) Effect of fishmeal and fish oil replacement by algal meals on biological performance and fatty acid profile of hybrid striped bass (*Morone chrysops* ♀ × *M. saxatilis* ♂). Aquaculture 507:83–90

Perry JR, Ying W (2016) A review of physiological effects of soluble and insoluble dietary fibers. J Nutr Food Sci 6:476

Pine HJ, Daniels WH, Davis DA et al (2008) Replacement of fish meal with poultry by-product meal as a protein source in pond-raised sunshine bass, *Morone chrysops* ♀ × *M. saxatlis* ♂ diets. J World Aquac Soc 39:586–597

Portz L, Cyrino JEP, Martino RC (2001) Growth and body composition of juvenile largemouth bass *Micropterus salmoides* in response to dietary protein and energy levels. Aquac Nutr 7:247–254

Qiyou X, Qing Z, Hong X et al (2011) Dietary glutamine supplementation improves growth performance and intestinal digestion/absorption ability in young hybrid sturgeon (Acipenser schrenckii ♀ × Huso dauricus♂). J Appl Ichthyol 27:721–726

Quagrainie K (2015) Profitability of hybrid striped bass cage aquaculture in the midwest. Purdue University Extension, West Lafayette, IN

Rawles SD, Gaylord TG, McEntire ME, Freeman DW (2009) Evaluation of poultry by-product meal in commercial diets for hybrid striped bass, *Morone chrysops* ♀ × *Morone saxatilis* ♂, in pond production. J World Aquac Soc 40:141–156

Rawles SD, Riche M, Gaylord TG et al (2006) Evaluation of poultry by-product meal in commercial diets for hybrid striped bass (*Morone chrysops* ♀× *M. saxatilis* ♂) in recirculated tank production. Aquaculture 259:377–389

Refstie S, Storebakken T, Baeverfjord G, Roem AJ (2001) Long-term protein and lipid growth of Atlantic salmon (*Salmo salar*) fed diets with partial replacement of fish meal by soy protein products at medium or high lipid level. Aquaculture 193:91–106

Ritala A, Häkkinen ST, Toivari M, Wiebe MG (2017) Single cell protein—state-of-the-art, industrial landscape and patents 2001–2016. Front Microbiol 8:2009

Rønnestad I, Fyhn HJ (1993) Metabolic aspects of free amino acids in developing marine fish eggs and larvae. Rev Fish Sci 1:239–259

Rønnestad I, Thorsen A, Finn RN (1999) Fish larval nutrition: a review of recent advances in the roles of amino acids. Aquaculture 177:201–216

Rossi W, Tomasso JT, Gatlin DM III (2015) Performance of cage-raised, overwintered hybrid striped bass fed fishmeal- or soybean-based diets. North Am J Aquac 77:178–185

Rossignol O, Dodson JJ, Guderley H (2011) Relationship between metabolism, sex and reproductive tactics in young Atlantic salmon (*Salmo salar L.*). Comp Biochem Physiol A 159:82–91

Salze GP, Davis DA (2015) Taurine: a critical nutrient for future fish feeds. Aquaculture 437:215–229

Smits CHM, Moughan PJ, Smith WC (1988) Chemical whole-body composition of the 20 kg liveweight growing pig. New Zeal J Agric Res 31:155–157

Stickney RR, Hardy RW, Koch K et al (1996) The effects of substituting selected oilseed protein concentrates for fish meal in rainbow trout *Oncorhynchus mykiss* diets. J World Aquac Soc 27:57–63

Stone DAJJ (2003) Dietary carbohydrate utilization by fish. Rev Fish Sci 11:337–369

Stoner GR, Allee GL, Nelssen JL et al (1990) Effect of select menhaden fish meal in starter diets for pigs. J Anim Sci 68:2729–2735

Storebakken S, Roem (2000) Growth, uptake and retention of nitrogen and phosphorus, and absorption of other minerals in Atlantic salmon *Salmo salar* fed diets with fish meal and soy-protein concentrate as the main sources of protein. Aquac Nutr 6:103–108

Stuber RJ, Gebhart G, Maughan OE (1982) Habit suitability index models: largemouth bass. U.S. Fish and Wildlife Service, Washington, DC. Publication FWS/OBS-82/10.16

Tacon AGJ, Metian M (2008) Global overview on the use of fish meal and fish oil in industrially compounded aquafeeds: trends and future prospects. Aquaculture 285:146–158

Tacon AGJ, Hasan MR, Metian M (2011) Demand and supply of feed ingredients for farmed fish and crustaceans: trends and prospects. FAO Fisheries and Aquaculture Technical Paper No. 564, Rome, Italy

Tidwell JH, Webster CD, Coyle SD (1996) Effects of dietary protein level on second year growth and water quality for largemouth bass (*Micropterus salmoides*) raised in ponds. Aquaculture 145:213–223

Tidwell JH, Coyle SD, Bright LA (2019) Largemouth bass aquaculture Largemouth bass production in China. 5M Published Ltd., Sheffield, pp 37–47

Treece GD (2017) The Texas aquaculture industry–2017. Texas Aquaculture Association, Austin, Texas

Troell M, Naylor RL, Metian M et al (2014) Does aquaculture add resilience to the global food system? Proc Natl Acad Sci USA 111:13257–13263

Trushenski J, Gause B (2013) Comparative value of fish meal alternatives as protein sources in feeds for hybrid striped bass. N Am J Aquac 75:329–341

Turchini GM, Trushenski JT, Glencross BD (2019) Thoughts for the future of aquaculture nutrition: realigning perspectives to reflect contemporary issues related to judicious use of marine resources in aquafeeds. North Am J Aquaculture 81:13–39
Twibell RG, Griffin ME, Martin B et al (2003) Predicting dietary essential amino acid requirements for hybrid striped bass. Aquac Nutr 9:373–381
USDA (2019a) USDA/ARS national program 106 aquaculture action plan 2020–2024. www.ars.usda.gov/ARSUserFiles/np106/NP106%20Aquaculture
USDA (2019b) Aquaculture: results from the 2018 census of aquaculture. https://www.nass.usda.gov/Publications/Highlights/2019/2017Census_Aquaculture_in_2018.pdf
U.S. Soybean Export Councel (USSEC 2008) soy protein concentrate technical bulletin. Chesterfield, MO
van den Thillart G (1986) Energy metabolism of swimming trout (*Salmo gairdneri*)—Oxidation rates of palmitate, glucose, lactate, alanine, leucine and glutamate. J Comp Physiol B 156:511–520
Van Waarde A (1983) Aerobic and anaerobic ammonia production by fish. Comp Biochem Physiol 74b:675–684
Van Waarde A, Kesbeke F (1982) Nitrogen metabolism in goldfish, *Carassius auratus L.* activities of amidases and amide synthetases in goldfish tissues. Comp Biochem Physiol B 71:599–603
Watford M, Wu G (2005) Glutamine metabolism in uricotelic species: variation in skeletal muscle glutamine synthetase, glutaminase, glutamine levels and rates of protein synthesis. Comp Biochem Physiol B 140:607–614
Weber JM, Haman F (1996) Pathways for metabolic fuels and oxygen in high performance fish. Comp Biochem Physiol A 113:33–38
Wilson RP (2002) Amino acids and proteins. In: Halver JE, Hardy RW (eds) Fish nutrition. Academic Press, New York, pp 143–179
Wu G (1998) Intestinal mucosal amino acid catabolism. J Nutr 128:1249–1252
Wu G (2010) Functional amino acids in growth, reproduction, and health. Adv Nutr 1:31–37
Wu G (2013a) Amino acids: biochemistry and nutrition. CRC Press, Boca Raton, Florida,
Wu G (2013b) Functional amino acids in nutrition and health. Amino Acids 45:407–411
Wu G (2014) Dietary requirements of synthesizable amino acids by animals: a paradigm shift in protein nutrition. J Anim Sci Biotechnol 5:34
Wu G (2018) Principles of animal nutrition. CRC Press, Boca Raton, Florida
Wu G (2021) Amino acids: biochemistry and nutrition, 2nd edn. CRC Press, Boca Raton, Florida
Wu G (2022) Nutrition and metabolism: Foundations for animal growth, development, reproduction, and health. Adv Exp Med Biol 1354:1–24
Wu G, Bazer FW, Burghardt RC et al (2011) Proline and hydroxyproline metabolism: implications for animal and human nutrition. Amino Acids 40:1053–1063
Wu G, Bazer FW, Dai Z et al (2014) Amino acid nutrition in animals: protein synthesis and beyond. Annu Rev Anim Biosci 2:387–417
Wu G, Bazer FW, Johnson GA, Hou Y (2018) Arginine nutrition and metabolism in growing, gestating, and lactating swine. J Anim Sci 96:5035–5051
Wu G, Bazer FW, Satterfield MC et al (2013a) Impacts of arginine nutrition on embryonic and fetal development in mammals. Amino Acids 45:241–256
Wu G, Meier SA, Knabe DA (1996) Dietary glutamine supplementation prevents jejunal atrophy in weaned pigs. J Nutr126:2578–2584
Wu G, Wu Z, Dai Z et al (2013b) Dietary requirements of "nutritionally non-essential amino acids" by animals and humans. Amino Acids 44:1107–1113
Wu G, Thompson JR (1990) The effect of ketone bodies on protein turnover in isolated skeletal muscle from the fed and fasted chick. Int J Biochem 22:263–268
Yan L, Zhou XQ (2006) Dietary glutamine supplementation improves structure and function of intestine of juvenile Jian carp (*Cyprinus carpio var. Jian*). Aquaculture 256:389–394
Yoshida C, Maekawa M, Bannai M, Yamamoto T (2016) Glutamate promotes nucleotide synthesis in the gut and improves availability of soybean meal feed in rainbow trout. SpringerPlus 5:1021
Zhang Q, Hou YQ, Bazer FW, He WL, Posey EA, Wu G (2021) Amino acids in swine nutrition and production. Adv Exp Med Biol 1285:81–107
Zhou YL, Guo JL, Tang RJ, Ma HJ, Chen YJ, Lin SM (2020) High dietary lipid level alters the growth, hepatic metabolism enzyme, and anti-oxidative capacity in juvenile largemouth bass *Micropterus salmoides*. Fish Physiol Biochem 46:125–134

13 Functional Molecules of Intestinal Mucosal Products and Peptones in Animal Nutrition and Health

Peng Li and Guoyao Wu

Abstract

There is growing interest in the use of intestinal mucosal products and peptones (partial protein hydrolysates) to enhance the food intake, growth, development, and health of animals. The mucosa of the small intestine consists of the epithelium, the lamina propria, and the muscularis mucosa. The diverse population of cells (epithelial, immune, endocrine, neuronal, vascular, and elastic cells) in the intestinal mucosa contains not only high-quality food protein (e.g., collagen) but also a wide array of low-, medium-, and high-molecular-weight functional molecules with enormous nutritional, physiological, and immunological importance. Available evidence shows that intestinal mucosal products and peptones provide functional substances, including growth factors, enzymes, hormones, large peptides, small peptides, antimicrobials, cytokines, bioamines, regulators of nutrient metabolism, unique amino acids (e.g., taurine and 4-hydroxyproline), and other bioactive substances (e.g., creatine and glutathione). Therefore, dietary supplementation with intestinal mucosal products and peptones can cost-effectively improve feed intake, immunity, health (the intestine and the whole body), well-being, wound healing, growth performance, and feed efficiency in livestock, poultry, fish, and crustaceans. In feeding practices, an inclusion level of an intestinal mucosal product or a mucosal peptone product at up to 5% (as-fed basis) is appropriate in the diets of these animals, as well as companion and zoo animals.

Keywords

Mucosal product · Nutrition · Health · Animals

13.1 Introduction

The small intestine and its mucosa account for approximately 3% and 0.6% of the body weight in market-weight livestock, respectively (Table 13.1). After an animal is slaughtered, the empty small intestine without luminal content is often used to manufacture intestinal mucosal products and peptones (partial protein hydrolysates) as ingredients for animal diets (Wilkinson and Meeker 2021). These palatable products provide large amounts of highly digestible protein (~60% on dry matter basis) with balanced ratios of amino acids relative to lysine (which is

P. Li
North American Renderers Association, Alexandria, VA, USA

G. Wu (✉)
Department of Animal Science, Texas A&M University, College Station, TX, USA
e-mail: g-wu@tamu.edu

G. Wu (ed.), *Recent Advances in Animal Nutrition and Metabolism*, Advances in Experimental Medicine and Biology 1354,
https://doi.org/10.1007/978-3-030-85686-1_13

Table 13.1 Weights of the empty small intestine and its mucosa in growing pigs

Body weight (kg)	Weight of the small intestine[a] (% of BW)	Weight of the mucosa[b] (% of BW)	Ratio of the mucosal weight to the small-intestinal weight (g/g)
10	3.36 ± 0.14	0.789 ± 0.03[c]	0.236 ± 0.007 [c, d]
50	3.20 ± 0.13	0.693 ± 0.02[d]	0.217 ± 0.006[d, e]
100	3.08 ± 0.17	0.619 ± 0.04[e]	0.201 ± 0.008[e]

Values are means ± SEM, n = 6. Pigs were fed typical corn- and soybean meal-based diets and were slaughtered at 6 h after the last feeding
[a]Including the duodenum, jejunum, and ileum without intestinal content
[b]Including the mucosa of the duodenum, jejunum, and ileum
[c-e]Within a column, means not sharing the same superscript letter differ ($P < 0.05$), as analyzed by one-way analysis of variance and the Student–Newman–Keuls multiple comparison (Assaad et al. 2014)
BW = body weight

often the first limiting amino acid) for animal utilization (Table 13.2). Besides being an abundant source of the building blocks for tissue protein, the small-intestinal mucosa also contains many functional molecules, including hormones, large peptides, small peptides, antimicrobials, cytokines, bioamines, as well as a variety of other nitrogenous substances (e.g., taurine, creatine, serotonin, and glutathione) with regulatory, anti-oxidative, anti-inflammatory, endocrine modulatory, and immunity-enhancing effects (Beaumont and Blachier 2020; Flynn et al. 2020; Halloran et al. 2021; Hansen et al. 2008; He and Wu 2020; Li and Wu 2018; Li et al. 2021e; Madara 1991; 2020; Ren et al. 2020; Rezaei et al. 2013; Wu 2020a, b; 2021). Much evidence shows that the health of the intestine is crucial for the maximum growth performance and feed efficiencies in farm animals, including cattle (Gilbreath et al. 2021), swine (Zhang et al. 2021a), poultry (He et al. 2021a, b), fish (Li et al. 2021a, c, d), and crustaceans (Li et al. 2021b).

The major objective of this article is to summarize a wide variety of functional molecules in porcine intestinal mucosal tissues and research performed with various animals showing unique roles of the mucosal products and peptones in animal nutrition and health. Future research on those functional molecules will provide much-needed scientific evidence to support the uses of porcine intestinal mucosal products and peptones in the animal feed industry as well as the direct optimization of processes for manufacturing intestinal mucosal products, including enzyme treatment and mechanical drying, towards a higher retention of those highly bioactive components and the maximum efficacy of intestinal mucosal products in animal nutrition and health.

13.2 The Mucosa of the Small Intestine

The small intestine is defined as that portion of the digestive tract between the pylorus and the ileocecal valve and consists of the duodenum, jejunum, and ileum. The jejunum and ileum constitute approximately 40% and 60%, respectively, of the small intestine below the duodenum (Wu 2018). The wall of the small intestine consists of four main layers: the mucosa, the submucosa, the muscularis externa, and the serosa (Madara 1991). The mucosa consists of the epithelium, the lamina propria, and the muscularis mucosa. The epithelium rests on the underlying lamina propria covered with the basal lamina. The lamina propria is supported by the underlying muscularis mucosa (two thin layers of smooth muscle together with varying amounts of elastic tissue). These four layers of tissues are rich in structural proteins (including membrane proteins and extracellular collagens) and nutrients (including free amino acids, minerals, and vitamins) that are of high quality as food for animals (Wu 2018).

Table 13.2 Composition of total amino acids and other functional substances in intestinal-mucosa product, corn grain, and soybean meal[a,b]

Nutrient	Porcine intestinal-mucosal product[c]	Corn grain[d]	Soybean meal[e]
Proteinogenic Amino Acids (% of feedstuff; as-fed basis)			
Alanine	3.69	0.71	1.97
Arginine	3.76	0.38	3.20
Asparagine	1.57	0.35	2.10
Aspartate	3.70	0.43	3.12
Cysteine	0.90	0.20	0.70
Glutamine	2.96	1.02	3.78
Glutamate	5.97	0.64	4.22
Glycine	4.32	0.40	2.30
Histidine	1.29	0.23	1.15
4-Hydroxyproline	1.03	0.00	0.09
Isoleucine	2.56	0.34	2.03
Leucine	4.21	1.13	3.45
Lysine	4.10	0.25	2.81
Methionine	1.18	0.21	0.60
Phenylalanine	2.32	0.46	2.20
Proline	3.15	1.06	2.44
Serine	3.04	0.45	2.14
Tryptophan	0.68	0.07	0.62
Threonine	2.83	0.31	1.78
Tyrosine	2.12	0.43	1.67
Valine	3.36	0.44	2.08
Other Functional Substances (mg/kg feedstuff; as-fed basis)			
Taurine	1652	0.0	0.0
β-Alanine	83	8.8	8.2
Carnosine	938	0.0	0.0
Anserine	64	0.0	0.0
Creatine	38	0.0	0.0
Creatinine	2915	0.0	0.0
Creatine phosphate	1183	0.0	0.0
Putrescine[f]	522	568[g]	16[g]
Spermidine	1289	216[g]	194[g]
Spermine	1376	17[g]	74[g]
Total glutathione	172	115	169

[a]Dry matter content of the feedstuffs was 89.0%
[b]Molecular weights of intact AA were used to calculate the content of peptide-bound AA in feed ingredients
[c]Analyzed by high-performance liquid chromatography as we described (Li and Wu 2020)
[d]Adapted from Li et al. (2011)
[e]Li and Wu (2020)
[f]Analyzed by high-performance liquid chromatography (Hou et al. 2018)
[g]Hou et al. (2019)

Further knowledge of cell composition in the small-intestinal mucosa is important to understand its functional molecules. The epithelium of the small intestine consists of two compartments: the villus and the crypt of Lieberkühn (Madari 1991). Enterocytes constitute more than 80% of the mucosal epithelial cell population in the small intestine. Other cell types in the villus include goblet cells, intraepithelial cells, and enteroendocrine cells. The crypt contains stem cells, Paneth cells, and enteroendocrine cells. The lamina propria is a connective tissue layer, which carries both blood and lymphatic vessels, and contains a variety of mononuclear cells (e.g., T-lymphocytes, B-lymphocytes, plasma cells, and macrophages), polymorphonuclear granulocytes (e.g., mast cells, neutrophils, and eosinophils), and small nerve fibers (Ren et al. 2020). In addition, the lamina propria has numerous glands, with the ducts opening to the mucosal epithelium. These different cell types produce various kinds of soluble proteins (including growth factors), peptides (including those for stimulating food intake and killing bacteria), anti-oxidative substances, and regulators of nutrient metabolism.

13.3 Functions of Molecules Present in the Mucosa of the Small Intestine

The mucosa of the small intestine contains large amounts of nutrients (Hansen et al. 2008; Lussier et al. 2001; Madara 1991) and regulatory factors [including functional amino acids, their bioactive metabolites (e.g., serotonin, nitric oxide, and polyamines), growth factors, enzymes, and mediators of immune responses; Wu 2021] (Table 13.3 and Table 13.4). Of note, the intestine contains about 95% of the total serotonin (a neurotransmitter) in the whole body (Yabut et al. 2019), and polyamines [putrescine, spermidine, and spermine (antioxidants and growth promoters)] are particularly abundant in the gut mucosa (Wu 2021). At physiological concentrations, the beneficial roles of proteins, peptides, amino acids, and their metabolites in the nutrition and function of the intestine and the whole body of animals are discussed in the following sections.

13.3.1 Provision of High-Quality Protein and Functional Amino Acids

Protein is the major component of dry matter in the mucosa of the small intestine. The amounts of amino acids and their ratios to lysine are greater in the mucosal product than in corn and soybean meal (Li et al. 2021e). Specifically, the mucosal protein contains high proportions of glutamate, glutamine, and aspartate (Table 13.2), which serve as major energy sources for enterocytes to sustain the structure and function of the small intestine of mammals [including humans (Wu 1998), pigs (Beaumont and Blachier 2020; Hou and Wu 2018; Zhang et al. 2021a), cattle, and sheep (Wu et al. 2021], fish (Li et al. 2020a, b; Jia et al. 2021, 2022), and crustaceans (Li et al. 2021b). In addition, glutamate and aspartate provide the bulk of energy for the small intestine of poultry (He et al. 2021a, b). Intestinal mucosal products and peptones are also rich in conditionally essential amino acids (e.g., arginine, glycine, proline, and 4-hydroxyproline) and nutritionally essential amino acids (e.g., lysine, methionine, threonine, and branched-chain amino acids) (Table 13.2), as compared with plant-sourced ingredients (Hou et al. 2019; Li and Wu 2020; Li et al. 2011; Li et al. 2021e). These functional amino acids are both substrates and activators of tissue protein synthesis as well as cell signaling molecules for anabolic metabolism, thereby enhancing the growth and health of the small intestine and other tissues

Table 13.3 Composition of total, free, protein-bound, and peptide-bound amino acids in a spray-dried porcine intestinal-mucosal peptone product[a]

	Total AA	Free AA or substance	AA in Proteins + Peptides	AA in proteins	AA in peptides
Proteinogenic amino acids					
Alanine	35.09	16.96	18.13	9.32	8.81
Arginine	35.92	12.82	23.10	14.56	8.54
Asparagine	15.18	3.362	11.82	2.08	9.74
Aspartate	37.17	15.21	21.96	8.00	13.96
Cysteine	9.85	4.14	5.71	4.96	0.75
Glutamine	27.84	0.0114	27.83	15.26	12.57
Glutamate	57.68	14.58	43.11	9.81	33.3
Glycine	49.5	7.12	42.37	21.80	20.57
Histidine	12.85	5.80	7.05	4.90	2.15
4-Hydroxyproline	7.89	2.54	5.35	1.99	3.36
Isoleucine	24.43	12.42	12.01	5.51	6.50
Leucine	43.01	22.16	20.86	15.52	5.34
Lysine	40.67	19.50	21.17	15.59	5.58
Methionine	12.01	5.45	6.56	6.12	0.44
Phenylalanine	23.31	12.79	10.52	5.28	5.24
Proline	31.63	14.42	17.22	6.54	10.68
Serine	33.72	12.06	21.65	10.09	11.56
Threonine	28.52	12.76	15.75	6.00	9.75
Tryptophan	6.28	4.193	2.09	1.62	0.47
Tyrosine	22.37	12.47	9.90	6.11	3.79
Valine	30.48	16.82	13.65	6.65	7.00
Other functional substances					
Citrulline	1.40	1.40	–	–	–
Ornithine	1.54	1.54	–	–	–
Taurine	1.64	1.64	–	–	–
β-Alanine	0.081	0.081	–	–	–
Carnosine	–	0.955	–	–	–
Anserine	–	0.067	–	–	–
Creatine	–	0.042	–	–	–
Creatinine	–	2.72	–	–	–
Creatine phosphate	–	1.43	–	–	–
Putrescine	–	0.53	–	–	–
Spermidine	–	1.36	–	–	–
Spermine	–	1.43	–	–	–
Total glutathione	–	0.175	–	–	–

[a] Enzymes-treated porcine mucosal tissues. Values are g/kg product (as-fed basis). The content of dry matter in this product was 88.74%. Proteins and peptides are perchloric acid-insoluble and soluble molecules, respectively. Adapted from Li and Wu (2020).

Table 13.4 Presence of functional molecules in the small-intestinal mucosa of animals

Molecules	Classification	Sources	Major functions
N-Acetylserotonin	Trp metabolite	Enteroendocrine cells	Inhibit production of inflammatory Cytokines and serve as an antioxidant
S-Adenosylmethionine	Met metabolite	Multiple mucosal cells	Stimulate growth of epithelial cells
β-Alanine	Nutrient	All cell types	Synthesis of carnosine, anserine, and related dipeptides; a component of pantothenic acid and coenzyme A
α-Amino acids	Nutrients	All cell types	Antioxidants, DNA and protein synthesis, cell signaling and growth; nutrient metabolism and transport; immunity
γ-Aminobutyrate	Glu metabolite	Enteroendocrine cells	Increase food intake by animals
Anserine	β-Ala-1MeHis	All cell types of most animals	Antioxidant and buffering
Anthranilic acid	Trp metabolite	IEL and mucosal T-cells	Inhibit production of inflammatory cytokines and enhance immunity
Antibodies (e.g., IgA)	Proteins	Plasma cells (B-cell)	Cell-mediated immunity
Azurocidin	Protein	Neutrophils	Antimicrobial and innate immunity
Belanine	β-Ala-3MeHis	All cell types of most animals	Antioxidant and buffering
Carnosine	β-Ala-Histidine	All cell types of most animals	Antioxidant and buffering support innate immunity
Catalase	Protein	Multiple mucosal cells	Antimicrobials and innate immunity
Cathelicidins	Small peptides	Paneth cells	Antimicrobials and innate immunity
Cathepsin G	Protease	Neutrophils	Antimicrobial and innate immunity
Cathepsin K	Protease	Goblet cells & enterocytes	Antimicrobial and innate immunity
Cholecystokinin (CCK)	Polypeptide	Enteroendocrine cells	Stimulate pancreatic secretions
Chymase	Protease	Mast cells	Antimicrobial and regulation of intestinal mucosal homeostasis
Creatine and creatine-P	Arg, Gly and	All cell types Met metabolite	Energy metabolism, antioxidant and buffering
α-Defensins (cryptdins)	Small peptides	Paneth cells	Antimicrobials and innate immunity
β-Defensins	Small peptides	Enterocytes, plasma cells and granulocytes	Antimicrobials and innate immunity
Diamine oxidases	Enzymes	Multiple mucosal cells	Antimicrobial and innate immunity
Digestive enzymes[a]	Proteins	Enterocytes	Enhance digestion of protein, carbohydrate, and fat in diets

(continued)

Table 13.4 (continued)

Molecules	Classification	Sources	Major functions
Dopamine	Bioamine	Mucosal neurons	Increase intestinal motility and Food intake
Dopuin	62-AA peptide	Mucosa of pig small intestine	Antimicrobial
EGF	Protein	Mucosal Brunner's glands	Promote growth of epithelial cells and restitution of the epithelium
Fibroblast growth factors	Proteins	Fibroblasts	Stimulate growth of epithelial cells
Gastrin	Small peptide	Enteroendocrine cells in the duodenum	Stimulate gastric acid secretion to enhance digestion and kill bacteria, while promoting gut maturation
Ghrelin	Polypeptide	Enteroendocrine cells	Stimulate food intake by animals
Glucagon-like peptide-1	Polypeptide	Enteroendocrine cells	Regulate glucose metabolism
Glucagon-like peptide-2	Polypeptide	Enteroendocrine cells	Promote intestinal mucosal integrity
Glucosamine-6-phosphate	Aminosugar	Multiple mucosal cells	Enhance synthesis of glycoproteins
Glutathione	γ-Glu-Cys-Gly	Multiple mucosal cells	A major antioxidant
Glutathione reductase	Protein	Multiple mucosal cells	An antioxidant enzyme
Histadins[b]	Large peptides	Paneth cells	Antimicrobials and innate immunity
Histamine	His metabolite	Mast cells	Stimulate gastric acid secretion to enhance digestion and kill bacteria
IGF-I	Large peptide	enterocytes and goblet cells	Stimulate growth of epithelial cells
Interferon-γ	Cytokine	IEL and mucosal T-cells	Regulates epithelial homeostasis
Interleukin-2	Cytokine	IEL	Protect the intestinal epithelium
Interleukin-4	Cytokine	IEL	Protect the intestinal epithelium
Interleukin-10	Protein	Fibroblasts and IEL	Increase survival of mucosal T-cells
Lysozymes	Glycoside hydrolases	Paneth cells, enterocytes and neutrophils	Antimicrobials, innate immunity and maintain extracellular matrix
Melatonin	Bioamine	Enteroendocrine cells	regulate food intake and digestion
Minerals	Nutrients	All cell types	Antioxidants, DNA and protein synthesis, cell growth; coenzymes; nutrient metabolism and transport; immunity
Mucins	Glycoproteins	Goblet cells	Maintain intestinal mucosal integrity
Myeloperoxidase	Protein	Macrophages and lymphocytes	Kill pathogenic organisms and support innate immunity

(continued)

Table 13.4 (continued)

Molecules	Classification	Sources	Major functions
Nucleotides	Metabolites	Multiple mucosal cells	Stimulate growth of epithelial cells
Oleoylethanolamide	Lipid derivative	Enterocytes	Stimulate uptake of fatty acids
Osteopontin	Glycoprotein	Enterocytes	Increase ion and nutrient transport
PEC-60	60-AA peptide	Mucosa of pig small intestine	Antimicrobial
Peroxidase	Enzyme	Multiple mucosal cells	Kill pathogenic organisms and support innate immunity
Phospholipase A2	Enzyme	Multiple mucosal cells	Antimicrobial and innate immunity
Polyamines[c]	Bioamines	Multiple mucosal cells	Stimulate DNA and protein synthesis; ion transport; intestinal cell proliferation, migration, maturation, and repair
Peptide YY	Small peptide	Enteroendocrine cells	Regulate gut development & motility
PR-39 peptide[d]	Small peptide	Paneth cells	Antibacterial and innate immunity
Serotonin	Bioamine	Enteroendocrine cells	Regulate luminal sensing of nutrients, mucosal secretions and inflammation, nutrient absorption, and inflammation, nutrient absorption, gut motility, food intake, and energy balance
Superoxide dismutase	Enzyme	Multiple mucosal cells	Kill pathogenic organisms and support innate immunity
Taurine	Cys metabolite	All cell types	Antioxidant and osmoregulation
TGF-α	Protein	Enterocytes and macrophages	Enhance growth of epithelial cells and restitution of the epithelium
TGF-β	Protein	T-cells and goblet cells	Enhance growth of epithelial cells and restitution of the epithelium
VEGF	Protein	Endothelial cells	Maintain microvasculature structure to support nutrient transport
Vitamins	Nutrients	All cell types	Antioxidants, DNA and protein synthesis, cell growth; coenzymes; nutrient metabolism and transport; immunity

[a]Including disaccharidases (lactase-phlorizin hydrolase and sucrase-isomaltase) and peptidases
[b]Histine-rich peptides
[c]Putrescine + spermidine + spermine
[d]A proline-arginine-rich peptide
AA = amino acid; β-Ala = β-alanine; Arg = L-arginine; Cys = cysteine; EGF = epidermal growth factors; Glu, glutamate; Gly = glycine; IEL = intraepithelial lymphocytes; 1MeHis = L-1-methylhistidine; 3MeHis = L-1-methylhistidine; Met = L-methionine; TGF = transforming growth factor; VEGF = vascular endothelial growth factor

(including skeletal muscle) in animals (Bazer et al. 2021; Paudel et al. 2021, Sah et al. 2021, Shen et al. 2021, Wang et al. 2014; Wu et al. 2021; Yang et al. 2021).

Intestinal mucosal peptone products contain relatively large amounts of small peptides and free amino acids, as analyzed using acid, alkaline, and enzymatic hydrolyses (Li and Wu 2020). The content of total, free, protein-bound, and peptide-bound amino acids in a spray-dried porcine peptone product (enzymes-treated porcine mucosal tissues) is shown in Table 13.3. The near absence of free glutamine in this peptone product (that contained water) resulted from the hydrolysis of this amino acid into glutamate spontaneously at elevated temperature (Wu 2021) and via the action of glutaminase. By contrast, porcine peptone products contain large amounts of free ornithine and citrulline, which can be generated from arginine via arginase and arginine deiminase, respectively. Free, protein-bound, and peptide-bound amino acids account for 39, 30, and 31% of the total amino acids in the spray-dried porcine peptone product (Li and Wu 2020). Spray-dried porcine peptone products also contain β-alanine, small peptides [carnosine (β-alanyl-histidine), anserine (β-alanyl-1-methylhistidine), and glutathione (γ-Glu-Cys-Gly)], creatine, creatine phosphate, polyamines, and glutathione (Table 13.3).

As abundant sources of amino acids, the use of animal-derived products may substantially reduce the inclusion level of synthetic DL-methionine or its analogs as well as crystalline lysine, threonine and tryptophan in the diets of swine, poultry, fish, crustaceans, and other animals (Li et al. 2021e). Growing evidence shows that arginine, glutamate, glutamine, aspartate, glycine, and proline are inadequately synthesized by swine fish fed the regular diets with minimal protein content (Hou and Wu 2018; Wu 2009; Wu et al. 2018). Similarly, results of feeding trials have revealed that poultry (He et al. 2021a, b), fish (Li et al. 2009, 2021a), and crustaceans (Li et al. 2021b) do not synthesize sufficient amounts of glutamate, glutamine, glycine, and proline when fed regular diets. Because animals have dietary requirements for biosynthesizable amino acids (Hou et al. 2015, 2016), these nutrients or their sources from animal-derived feedstuffs can be included in diets for their maximum growth performance and optimum health (Hou and Wu 2017).

13.3.2 Provision of Factors for Regulation of Food Intake

Amino acids affect gastrointestinal motility and the neurological network for the control of feeding behavior (Wu 2021). Thus, a deficiency of dietary amino acids (e.g., glycine, lysine, methionine, and tryptophan) reduces food consumption by animals (Harper et al. 1970; He and Wu 2020). By contrast, the provision of adequate amounts of amino acids in proper ratios from intestinal mucosal products or peptones plays an important role in stimulating food intake by animals through both enhancing the release of neuropeptide Y from the arcuate nucleus in the hypothalamus and inhibiting general control nonderepressing kinase 2 in the anterior piriform cortex of the brain (Gietzen et al. 2007). Similarly, ghrelin enhances food intake by activating the arcuate nucleus and the nucleus tractus solitarii of the brainstem (Sartin et al. 2011). Additionally, γ-aminobutyrate (Pu et al. 1999), dopamine (Volkow et al. 2011) and melatonin (Angers et al. 2003) in the gut mucosa promote intestinal motility and food intake by animals. Likewise, there is evidence that serotonin activates overall feeding in *C. elegans* (Song and Avery 2012) and that serotonin in the gut enhances intestinal motility, food intake, and anti-oxidative responses, while inhibiting intestinal inflammation (Yabut et al. 2019). Furthermore, gastrin and histamine augment gastric acid secretion to aid in digestion, thereby increasing gastrointestinal emptying and meal consumption (Wu 2018).

13.3.3 Provision of Factors for the Stimulation of Nutrient Absorption

Both the areas of microvilli and the expression of transporters in absorptive enterocytes are enhanced by growth factors and amino acids , as well as their metabolites (e.g., polyamines and nitric oxide) (Ferraris 1994; Wu et al. 2021). Thus, epidermal growth factor and serotonin increase the absorption of water, Na^+, Cl^-, and glucose from the jejunum (Yabut et al. 2019; Opleta-Madsen et al. 1991). Similar results have been reported for transforming growth factor-α (Hardin and Gall 1992). In addition, vitamin D in intestinal mucosal products and peptones can stimulate the absorption of dietary calcium and phosphorus (Khanal and Nemere 2008). Interestingly, oleoylethanolamide (a lipid derivative) has been found to enhance the uptake of fatty acids by the small intestine (Yang et al. 2007). Most recently, we discovered that osteopontin, which is present in the mucosa, increases the transport of ions, amino acids, and glucose by porcine placental cells (Johnson et al. 2011) and porcine enterocytes (our unpublished data). Of particular note, intestinal mucosal products and peptones contain a large amount of taurine (Table 13.2), which promotes the intestinal absorption of lipids and fat-soluble vitamins and protects the gut and other tissues from oxidative damage in animals (Oberbauer and Larsen 2021; Wu 2020a). For animals that do not synthesize taurine (e.g., cats, tigers, and most carnivorous fish), intestinal mucosal products and peptones provide an abundant source of this amino acid (Che et al. 2021; Herring et al. 2021; Li and Wu 2020). For comparison, plant-sourced foods lack taurine (Hou et al. 2019).

13.3.4 Provision of Growth Factors and Enhancers

The mucosa of the small intestine provides relative high amounts of growth-promoting molecules, such as growth factors, cytokines, polyamines, *S*-adenosylmethionine, and nucleotides (Table 13.4). The protein and polypeptide growth factors are all virtually absent from processed grain ingredients such as corn grain and soybean meal (Li et al. 2011). Growth factors in intestinal mucosal products and peptones include epidermal growth factor (Drozdowski and Thomson), fibroblast growth factors (Dignass and Sturm 2001), insulin-like growth factor-I (Dvorák et al. 1996), glucagon-like growth factor-II (Burrin et al. 2001), vascular endothelial growth factor (Jones et al. 1999), and interleukin-12 (Al-Sadi et al. 2009). These growth factors and enhancers support polyamines, DNA, and protein syntheses in intestinal epithelial cells, as well as maintain the normal structure and function of the intestinal epithelium and vasculature, thereby sustaining the mucosal homeostasis and the absorption of nutrients into the blood circulation. An increase in the availability of dietary nutrients can promote the growth, development, wound healing, and health of the whole body (including the small intestine, liver, skeletal muscle and bones).

13.3.5 Provision of Natural Antimicrobial Agents

Mucosal products are a rich source of antimicrobial peptides and antimicrobial proteins against Gram-positive and Gram-negative bacteria, as well as fungi and enveloped viruses (Muniz et al. 2012; Thacker 2013). These antimicrobial substances include cathelicidins, defensins, and enzymes (e.g., lysozymes and myeloperoxidase) (Table 13.4), and some of them are particularly rich in proline and arginine (Hou et al. 2017). Therefore, mucosal antimicrobials protect animals from infection, while improving their health and well-being (Bevins and Salzman 2011). Additionally, by reducing the number of microorganisms in the lumen of the small intestine, which degrade 20%–95% of dietary amino acids (Wu 1998), intestinal mucosal products, and mucosal peptones can enhance the entry of these nutrients into the portal circulation (Dai et al. 2011), thereby promoting tissue protein synthesis, intestinal and whole-body growth, as well as the efficiency of nutrient utilization and the sustainability of animal agriculture (Wu 2022).

13.3.6 Provision of Antioxidants and Their Precursors

The mucosa of the small intestine contains high concentrations (e.g., 5–10 mM for many substances) of antioxidants (e.g., taurine, β-alanine, carnosine, anserine, creatine, creatine phosphate, and glutathione) and their precursors, including arginine, glutamate, glycine, cysteine, and methionine (Rezaei et al. 2013; Li and Wu 2020; Wang et al. 2014; Wu 2020a; Wu et al. 2004)] and enzymes (e.g., proline oxidase, catalase, peroxidase, and superoxide dismutase) (Dillon and Wu 2021; Fang et al. 2002). Notably, taurine, carnosine, anserine, and creatine are absent from plant-sourced ingredients (Hou et al. 2019; Li et al. 2011; Li and Wu 2020; Wu 2021). Furthermore, β-alanine [a substrate for the syntheses of functional dipeptides (e.g., anserine, balenine, carcinine, and carnosine) as well as pantothenic acid and coenzyme A that are essential for cell metabolism in animals (Wu 2018)] is present in intestinal mucosal products and peptones. The content of this amino acid in the animal products is usually much greater than that in plant-sourced feedstuffs such as corn grains and soybean meal (Table 13.2). In addition, intestinal mucosal products and peptones are good sources of water- and lipid-soluble vitamins (e.g., vitamins of the B complex, vitamin E, and vitamin D), as well as trace elements (e.g., Se, Fe, Zn, Cu, and Mn) (McLean et al. 2005). These micronutrients support metabolic pathways (including anti-oxidative reactions, ATP production, and protein synthesis) in the small intestine and other tissues of the body (Fang et al. 2002).

13.3.7 Provision of Factors for Supporting Immune Function

The mucosa of the small intestine provides numerous factors produced by its constitutive immunocytes and other mucosal cells that participate directly and indirectly in immune responses (Table 13.4). Specifically, in both the gut and other tissues, innate and acquired immune systems are regulated by a cooperative network of immunoglobulins (primarily IgA), cytokines (e.g., interferon-γ, interleukins 2, 4, and 10), as well as the availability of amino acids for the synthesis of these regulatory proteins and peptides (Li et al. 2007). In addition, many proteins (e.g., catalase, lysozymes, cathelicidins, and myeloperoxidase) contribute to innate immunity (McNabb and Tomasi 1981). Furthermore, the oxidation of 4-hydroxyproline and proline by the intestinal mucosal 4-hydroxyproline oxidase and proline oxidase, respectively, as well as the degradation of polyamines by mucosal diamine oxidases, yields hydrogen peroxide, thereby killing pathogens in the lumen of the gut (Hu et al. 2021; Wu 2013; Wu et al. 2019). Finally, amino acids (e.g., arginine, cysteine, glycine, and tryptophan) positively affect immune responses either directly or indirectly through their metabolites (e.g., NO, H_2S, CO, and anthranilic acid; Li et al. 2007). Improving the functions of both innate and acquired immune systems is crucial for preventing infectious diseases in animals, particularly pigs, ruminants, poultry, fish, and crustaceans (Wu 2021).

13.3.8 Provision of Factors for Supporting Intestinal Mucosal Barrier Function

The intestinal mucosa contains a relatively large amount of mucins (7.6%), which consists of 22% aminosugars and as much as 30% threonine (Lien et al. 1997; Satchithanandam et al. 1990). These hexosamines are utilized by goblet cells of the gut to synthesize mucins and other glycoproteins, which are crucial for protecting the intestinal epithelium from physical injury by ingested food and luminal pathogens (McGuckin et al. 2011). Thus, animals have high requirements for dietary threonine for optimum intestinal health (Wang et al. 2010). In addition, epidermal growth factor, insulin-like growth factor, and transforming growth factor-β stimulate the growth of epithelial cells and restitution of the epithelium (Dignass and Sturm 2001; Drozdowski and Thomson 2009). Moreover, glutamine, glutamate and glycine (abundant amino acids in the intestinal mucosa) increase the expression of zonula occludens, claudins, and other tight-junction proteins (Chen et al. 2021; Li et al. 2016; Rhoads and Wu 2009), whereas 4-hydroxyproline protects the gut from oxidative injury and inflammation (Ji et al. 2018; Zhang et al. 2021b). Such physiological changes, which are triggered by proteins and amino acids, are required for sustaining intestinal integrity and mucosal-barrier function (Turner 2009).

13.4 Inclusion Levels of Intestinal Mucosal Products and Peptones

Intestinal mucosal products and peptones contain highly digestible proteins (Moughan 2003), balanced amino acids (Li and Wu 2020), and functional nutrients (Table 13.4). Only a relatively small amount of these feedstuffs is added to animal diets to improve their composition of AAs and confer nutritional and physiological functions (e.g., the provision of antioxidant molecules and appetite enhancers) in farm and captive animals. For example, the inclusion level of an intestinal mucosal product or a mucosal peptone product in the diets of weanling pigs can be 2–5% (as-fed basis; dry matter content = 90%; Cromwell 2006; Li et al. 2021e). In addition, 5% intestinal mucosal product (e.g., spray-dried porcine mucosa tissue products) can be used to replace 5% fishmeal (as-fed basis) in the diets of the rice field eel, high-value carnivorous fish (Tang et al. 2021). Based on the principles of animal nutrition (Wu 2018), such an inclusion level of an intestinal mucosal product or a mucosal peptone product is also appropriate for the diets of other animals, including poultry, fish, crustaceans, and companion and zoo animals. Much evidence shows that dietary supplementation with intestinal mucosal products is effective in improving intestinal morphology, health, and function, as well as growth performance and feed efficiencies in farm animals (Li et al. 2021e; Hou et al. 2017).

13.5 Conclusion

In summary, molecules present in small-intestinal mucosal products and peptones fulfill many nutritional and protective functions in the gut and other organs of animals. These animal-sourced feedstuffs, mainly out of porcine origin, provide high-quality protein, as well as enzymes, secretory proteins, large peptides, small peptides, amino acids, and other bioactive substances (e.g., carnosine, serotonin, polyamines, creatine, and glutathione) with versatile antioxidative, anti-inflammatory, and immuno-modulatory functions. Furthermore, intestinal mucosal products and peptones exert beneficial effects on: (1) physiological (e.g., enterocytes, goblet cells, Paneth cells, endocrine cells, intestinal motility, regeneration of epithelial cells, and tight junctions), (2) biochemical (e.g., low pH of gastric juice, mucus, antimicrobial peptides, antimicrobial proteins, and antibodies), (3) innate and acquired immune (e.g., lymphocytes, phagocytic cells, mast cells, neutrophils, and gut-associated lymphoid tissues), and (4) endocrine, neuronal, and intestinal-microbial systems. Through positively affecting these physiological systems, dietary supplementation with intestinal mucosal products, mucosal peptones, or their combinations is expected to improve food intake, immunity, wound healing, health (both intestinal and whole-body), well-being, growth performance, and food efficiency in livestock, poultry, fish, and crustaceans. Consumption of intestinal mucosal products and peptones can also be beneficial for the health of companion and zoo animals.

Acknowledgements This work was supported by Texas A&M AgriLife Research (H-8200). We thank research associates and students in our laboratory for helpful discussions.

References

Al-Sadi R, Boivin M, Ma T (2009) Mechanism of cytokine modulation of epithelial tight junction barrier. Front Biosci 14:2765–2778

Angers K, Haddad N, Selmaoui B, Thibault L (2003) Effect of melatonin on total food intake and macronutrient choice in rats. Physiol Behav 80:9–18

Assaad H, Zhou L, Carroll RJ, Wu G (2014) Rapid publication-ready MS-Word tables for one-way ANOVA. SpringerPlus 3:474

Bazer FW, Seo H, Johnson GA, Wu G (2021) One-carbon metabolism and development of the conceptus during pregnancy: lessons from studies with sheep and pigs. Adv Exp Med Biol 1285:1–15

Beaumont M, Blachier F (2020) Amino acids in intestinal physiology and health. Adv Exp Med Biol 1265:1–20

Bevins CL, Salzman NH (2011) Paneth cells, antimicrobial peptides and maintenance of intestinal homeostasis. Nature Rev Microbiol 9:356–368

Burrin DG, Petersen Y, Stoll B, Sangild P (2001) Glucagon-like peptide 2: a nutrient-responsive gut growth factor. J Nutr 131:709–712

Che DS, Nyingwa PS, Ralinala KM, Maswanganye GMT, Wu G (2021) Amino acids in the nutrition, metabolism, and health of domestic cats. Adv Exp Med Biol 1285:217–231

Chen JQ, Yang YC, Yang C, Dai ZL, Kim IH, Wu G, Wu ZL (2021) Dietary supplementation with glycine enhances intestinal mucosal integrity and ameliorates inflammation in C57BL/6J mice with high-fat diet-induced obesity. J Nutr 151:1769–1778

Cromwell GL (2006) Rendered products in swine nutrition. In: Meeker DL and Hamilton CR (eds) Essential Rendering National renderers association, Alexandria, VA, USA. pp 141–157

Dai ZL, Wu G, Zhu WY (2011) Amino acid metabolism in intestinal bacteria: links between gut ecology and host health. Front Biosci 16:1768–1786

Dignass AU, Sturm A (2001) Peptide growth factors in the intestine. Eur J Gastroenterol Hepatol 13:763–770

Dillon EL, Wu G (2021) Cortisol enhances ctrulline synthesis from proline in enterocytes of suckling piglets. Amino Acids. https://doi.org/10.1007/s00726-021-03039-y

Drozdowski L, Thomson AB (2009) Intestinal hormones and growth factors: effects on the small intestine. World J Gastroenterol 15:385–406

Dvorák B, Stephana AL, Holubec H, Williams CS, Philipps AF, Koldovskoý O (1996) Insulin-like growth factor-I (IGF-I) mRNA in the small intestine of suckling and adult rats. FEBS Lett 388:155–160

Fang YZ, Yang S, Wu G (2002) Free radicals, antioxidants, and nutrition. Nutrition 8:872–879

Ferraris RP (1994) Regulation of intestinal nutrient transport. In: Johnson LR (ed) Physiology of the gastrointestinal tract. Raven Press, New York, pp 1821–1844

Flynn NE, Shaw MH, Becker JT (2020) Amino acids in health and endocrine function. Adv Exp Med Biol 1265:97–109

Gietzen DW, Hao S, Anthony TG (2007) Mechanisms of food intake repression in indispensable amino acid deficiency. Annu Rev Nutr 27:63–78

Gilbreath KR, Bazer FW, Satterfield MC, Wu G (2021) Amino acid nutrition and reproductive performance in ruminants. Adv Exp Med Biol 1285:43–61

Halloran KM, Stenhouse C, Wu G, Bazer FW (2021) Arginine, agmatine and polyamines: key regulators of conceptus development in mammals. Adv Exp Med Biol 1332:85–105

Hansen MB, Witte AB (2008) The role of serotonin in intestinal luminal sensing and secretion. Acta Physiol (oxf) 193:311–323

Hardin JA, Gall DG (1992) The effect of TGF alpha on intestinal solute transport. Regul Pept 39:169–176

Harper AE, Benevenga NJ, Wohlhueter RM (1970) Effects of ingestion of disproportionate amounts of amino acids. Physiol Rev 50:428–558

He WL, Wu G (2020) Metabolism of amino acids in the brain and their roles in regulating food intake. Adv Exp Med Biol 1265:167–185

He WL, Li P, Wu G (2021a) Amino acid nutrition and metabolism in chickens. Adv Exp Med Biol 1285:109–131

He WL, Furukawa K, Toyomizu M, Nochi T, Bailey CA, Wu G (2021b) Interorgan metabolism, nutritional impacts, and safety of dietary L-glutamate and L-glutamine in poultry. Adv Exp Med Biol 1332:107–128

Herring CM, Bazer FW, Wu G (2021) Amino acid nutrition for optimum growth, development, reproduction, and health of zoo animals. Adv Exp Med Biol 1285:233–253

Hou YQ, Wu G (2017) Nutritionally nonessential amino acids: a misnomer in nutritional sciences. Adv Nutr 8:137–139

Hou YQ, Wu G (2018) L-Glutamate nutrition and metabolism in swine. Amino Acids 50:1497–1510

Hou YQ, Yin YL, Wu G (2015) Dietary essentiality of "nutritionally nonessential amino acids" for animals and humans. Exp Biol Med 240:997–1007

Hou YQ, Yao K, Yin YL, Wu G (2016) Endogenous synthesis of amino acids limits growth, lactation and reproduction of animals. Adv Nutr 7:331–342

Hou YQ, Wu ZL, Dai ZL, Wang GH, Wu G (2017) Protein hydrolysates in animal nutrition: industrial production, bioactive peptides, and functional significance. J Anim Sci Biotechnol 8:24

Hou YQ, Li XL, Dai ZL, Wu ZL, Bazer FW, Wu G (2018) Analysis of glutathione in biological samples by HPLC involving pre-column derivatization with *o*-phthalaldehyde. Methods Mol Biol 1694:105–115

Hou YQ, He WL, Hu SD, Wu G (2019) Composition of polyamines and amino acids in plant-source foods for human consumption. Amino Acids 51:1153–1165

Hu SD, He WL, Wu G (2021) Hydroxyproline in animal metabolism, nutrition, and cell signaling. Amino Acids. https://doi.org/10.1007/s00726-021-03056-x

Ji Y, Dai ZL, Sun S, Ma X, Yang Y, Tso O, Wu G, Wu ZL (2018) Hydroxyproline attenuates dextran sulfate sodium-induced colitis in mice: Involvment of the NF-kB signaling and oxidative stress. Mol Nutr Food Res 62:e1800494

Jia SC, Li XY, He WL, Wu G (2021) Oxidation of energy substrates in tissues of fish: Metabolic significance and implications for gene expression and carcinogenesis. Adv Exp Med Biol 1332:67–83

Jia SC, Li XY, He WL, Wu G (2022) Protein-sourced feedstuffs for aquatic animals in nutrition research and aquaculture. AdvExp Med Biol 1354:237–261

Johnson GA, Frank JW, Li XL, Bayless KJ, Burghardt RC, Bazer FW, Wu G (2011) Osteopontin expressed at the uterine-placental interface increases ion transport across the pig placenta. Placenta 32:A53

Jones MK, Tomikawa M, Mohajer B, Tarnawski AS (1999) Gastrointestinal mucosal regeneration: role of growth factors. Front Biosci 4:D303–D309

Khanal RC, Nemere I (2008) Regulation of intestinal calcium transport. Annu Rev Nutr 28:179–196

Li P, Yin YL, Li DF, Kim SW, Wu G (2007) Amino acids and immune function. Br J Nutr 98:237–252

Li P, Mai KS, Trushenski J, Wu G (2009) New developments in fish amino acid nutrition: towards functional and environmentally oriented aquafeeds. Amino Acids 37:43–53

Li XL, Rezaei R, Li P, Wu G (2011) Composition of amino acids in feed ingredients for animal diets. Amino Acids 40:1159–1168

Li W, Sun KJ, Ji Y, Wu ZL, Wang W, Dai Z, Wu G (2016) Glycine regulates expression and distribution of claudin-7 and ZO-3 proteins in porcine intestinal epithelial cells. J Nutr 146:964–969

Li P, Wu G (2018) Roles of dietary glycine, proline and hydroxyproline in collagen synthesis and animal growth. Amino Acids 50:29–38

Li P, Wu G (2020) Composition of amino acids and related nitrogenous nutrients in feedstuffs for animal diets. Amino Acids 52:523–542

Li XL, Zheng SX, Wu G (2020a) Nutrition and metabolism of glutamate and glutamine in fish. Amino Acids 52:671–691

Li XY, Zheng SX, Han T, Song F, Wu G (2020b) Effects of dietary protein intake on the oxidation of glutamate, glutamine, glucose and palmitate in tissues of largemouth bass (*Micropterus salmoides*) Amino Acids 52:1491–1503

Li XY, Zheng SX, Wu G (2021a) Nutrition and functions of amino acids in fish. Adv Exp Med Biol 1285:133–168

Li XY, Han T, Zheng SX, Wu G (2021b) Nutrition and functions of amino acids in aquatic crustaceans. Adv Exp Med Biol 1285:169–198

Li XY, Zheng SX, Ma XK, Cheng KM, Wu G (2021c) Use of alternative protein sources for fishmeal replacement in the diet of largemouth bass (*Micropterus salmoides*). Part I: effects of poultry by-product meal and soybean meal on growth, feed utilization, and health. Amino Acids 53:33–47

Li XY, Zheng SX, Cheng KM, Ma XK, Wu G (2021d) Use of alternative protein sources for fishmeal replacement in the diet of largemouth bass (*Micropterus salmoides*). Part II: effects of supplementation with methionine or taurine on growth, feed utilization, and health. Amino Acids 53:49–62

Li P, He WL, Wu G (2021e) Composition of amino acids in foodstuffs for humans and animals. Adv Exp Med Biol 1332:189–209

Lien KA, Sauer WC, Fenton M (1997) Mucin output in ileal digesta of pigs fed a protein-free diet. Z. Ernährungswiss 36:182–190

Lussier C, Sodek J, Beaulieu JF (2001) Expression of SPARC/osteonectin/BM4O in the human gut: predominance in the stroma of the remodeling distal intestine. J Cell Biochem 81:463–476

Madara JL (1991) Functional morphology of epithelium of the small intestine. In: Physiological A (ed) Handbook of Physiology: The Gastrointestinal System (Shultz SG. Society, Bethesda, pp 83–120

McGuckin MA, Lindén SK, Sutton P, Florin TH (2011) Mucin dynamics and enteric pathogens. Nat Rev Microbiol 9:265–278

McLean JA, Karadas F, Surai PF, McDevitt RM, Speake BK (2005) Lipid-soluble and water-soluble antioxidant activities of the avian intestinal mucosa at different sites along the intestinal tract. Comp Biochem Physiol B 141:366–372

McNabb PC, Tomasi TB (1981) Host defense mechanisms at mucosal surfaces. Annu Rev Microbiol 35:477–496

Moughan PJ (2003) Amino acid availability—aspects of chemical analysis and bioassay methodology. Nutr Res Rev 16:127–141

Muniz LR, Knosp C, Yeretssian G (2012) Intestinal antimicrobial peptides during homeostasis, infection, and disease. Front Immunol 3:310

Oberbauer AM, Larsen JA (2021) Amino acids in dog nutrition and health. Adv Exp Med Biol 1285:199–216

Opleta-Madsen K, Hardin J, Gall DG (1991) Epidermal growth factor upregulates intestinal electrolyte and nutrient transport. Am J Physiol 260:G807-814

Paudel S, Wu G, Wang XQ (2021) Amino acids in cell signaling: regulation and function. Adv Exp Med Biol 1332:17–33

Pu S, Jain MR, Horvath TL, Diano S, Kalra PS, Kalra SP (1999) Interactions between neuropeptide Y and γ-aminobutyric acid in stimulation of feeding: a morphological and pharmacological analysis. Endocrinology 140:933–940

Ren WK, Bin P, Yin YL, Wu G (2020) Impacts of amino acids on the intestinal defensive system. Adv Exp Med Biol 1265:133–151

Rezaei R, Knabe DA, Tekwe CD, Dahanayaka S, Ficken MD, Fielder SE, Eide SJ, Lovering SL, Wu G (2013) Dietary supplementation with monosodium glutamate is safe and improves growth performance in postweaning pigs. Amino Acids 44:911–923

Rhoads JM, Wu G (2009) Glutamine, arginine, and leucine signaling in the intestine. Amino Acids 37:111–122

Sah N, Wu G, Bazer FW (2021) Regulation of gene expression by amino acids in animal cells. Adv Exp Med Biol 1332:1–15

Sartin JL, Whitlock BK, Daniel JA (2011) Triennial Growth Symposium: neural regulation of feed intake: modification by hormones, fasting, and disease. J Anim Sci 89:1991–2003

Satchithanandam S, Vargofcak-Apker M, Valvert RJ, Leeds AR, Sassidy MM (1990) Alteration of gastrointestinal mucin by fiber feeding in rats. J Nutr 120:1179–1184

Shen JZ, Wu G, Guo SD (2021) Amino acids in autophagy: Regulation and function. Adv Exp Med Biol 1332:51–66

Song BM, Avery L (2012) Serotonin activates overall feeding by activating two separate neural pathways in Caenorhabditis elegans. J Neurosci 32:1920–1931

Tang T, Zhong L, Yu D, Li P, Hu Y (2021) Evaluation of dried porcine solubles in diets of rice field eel (*Monopterus albus*). Aquaculture 531:735897

Thacker PA (2013) Alternatives to antibiotics as growth promoters for use in swine production: a review. J Anim Sci Biotechnol 4:35

Turner JR (2009) Intestinal mucosal barrier function in health and disease. Nat Rev Immunol 9:799–809

Volkow ND, Wang GJ, Baler RD (2011) Reward, dopamine and the control of food intake: implications for obesity. Trends Cogn Sci 15:37–46

Wang WW, Zeng XF, Mao XB, Wu G, Qiao SY (2010) Optimal dietary true ileal digestible threonine for supporting mucosal barrier in the small intestine of weanling pigs. J Nutr 140:981–986

Wang WW, Dai ZL, Wu ZL, Lin G, Jia SC, Hu SD, Dahanayaka S, Wu G (2014) Glycine is a nutritionally essential amino acid for maximal growth of milk-fed young pigs. Amino Acids 46:2037–2045

Wilkinson AD, Meeker DL (2021) How agricultural rendering supports sustainability and assists livestock's ability to contribute more than just food. Anim Front 11:24–34

Wu G (1998) Intestinal mucosal amino acid catabolism. J Nutr 128:1249–1252

Wu G (2009) Amino acids: metabolism, functions, and nutrition. Amino Acids 37:1–17

Wu G (2018) Principles of animal nutrition. CRC Press, Boca Raton, Florida

Wu G (2020a) Important roles of dietary taurine, creatine, carnosine, anserine and hydroxyproline in human nutrition and health. Amino Acids 52:329–360

Wu G (2020b) Metabolism and functions of amino acids in sense organs. Adv Exp Med Biol 1265:201–217

Wu G (2021) Amino acids: biochemistry and nutrition, 2nd ed. CRC Press, Boca Raton, Florida

Wu G (2022) Nutrition and metabolism: Foundations for animal growth, development, reproduction, and health. Adv Exp Med Biol 1354:1–24

Wu G, Fang YZ, Yang S, Lupton JR, Turner ND (2004) Glutathione metabolism and its implications for health. J Nutr 134:489–492

Wu G, Bazer FW, Johnson GA, Hou YQ (2018) Arginine nutrition and metabolism in growing, gestating and lactating swine. J Anim Sci 96:5035–5051

Wu ZL, Hou YQ, Dai ZL, Hu CA, Wu G (2019) Metabolism, nutrition and redox signaling of hydroxyproline. Antioxid Redox Signal 30:674–682

Wu G, Meininger CJ, McNeal CJ, Bazer FW, Rhoads JM (2021) Role of L-arginine in nitric oxide synthesis and health in humans. Adv Exp Med Biol 1332:167–188

Yabut JM, Crane JD, Green AE, Keating DJ, Khan WI, Steinberg GR (2019) Emerging roles for serotonin in regulating metabolism: New implications for an ancient molecule. Endocrine Rev 40:1092–1107

Yang YK, Chen M, Georgeson KE, Harmon CM (2007) Mechanism of oleoylethanolamide on fatty acid uptake in small intestine after food intake and body weight reduction. Am J Physiol 292:R235–R241

Yang Y, He Y, Jin Y, Wu G, Wu ZL (2021) Amino acids in endoplasmic reticulum stress and redox signaling. Adv Exp Med Biol 1332:35–49

Zhang Q, Hou YQ, Bazer FW, He WL, Posey EA, Wu G (2021a) Amino acids in swine nutrition and production. Adv Exp Med Biol 1285:81–107

Zhang Y, Mu T, Jia H et al. (2021b) Protective effects of glycine against lipopolysaccharide-induced intestinal apoptosis and inflammation. Amino Acids. https://doi.org/10.1007/s00726-021-03011-w

Use of Genome Editing Techniques to Produce Transgenic Farm Animals

14

Alayna N. Hay, Kayla Farrell,
Caroline M. Leeth, and Kiho Lee

Abstract

Recombinant proteins are essential for the treatment and diagnosis of clinical human ailments. The availability and biological activity of recombinant proteins is heavily influenced by production platforms. Conventional production platforms such as yeast, bacteria, and mammalian cells have biological and economical challenges. Transgenic livestock species have been explored as an alternative production platform for recombinant proteins, predominantly through milk secretion; the strategy has been demonstrated to produce large quantities of biologically active proteins. The major limitation of utilizing livestock species as bioreactors has been efforts required to alter the genome of livestock. Advancements in the genome editing field have drastically improved the ability to genetically engineer livestock species. Specifically, genome editing tools such as the CRISPR/Cas9 system have lowered efforts required to generate genetically engineered livestock, thus minimizing restrictions on the type of genetic modification in livestock. In this review, we discuss characteristics of transgenic animal bioreactors and how the use of genome editing systems enhances design and availability of the animal models.

Keywords

Genome editing · Livestock · Animal Bioreactor

Abbreviations

Cas9	CRISPR-associated 9
CHO	Chinese hamster ovary
CRISPR	Clustered regularly interspaced short palindromic repeats
DSB	Double strand break
ES	Embryonic Stem cells
HAC	Human artificial chromosomes
HDR	Homology-directed repair
hEPO	Human erythropoietin
hG-CSF	Human granulocyte-colony stimulating factor
hLF	Human lactoferrin
hPA	Human plasminogen activator
HSA	Human serum albumin
Ig	Immunoglobulin
NHEJ	Non-homologous end joining

A. N. Hay · K. Farrell ·
CarolineM. Leeth
Department of Animal and Poultry Sciences, Virginia Tech, Blacksburg, VA, USA

K. Lee (✉)
Division of Animal Sciences, University of Missouri, Columbia, MO, USA
e-mail: kiholee@missouri.edu

G. Wu (ed.), *Recent Advances in Animal Nutrition and Metabolism*, Advances in Experimental Medicine and Biology 1354,
https://doi.org/10.1007/978-3-030-85686-1_14

PGC	Primordial germ cell
SCNT	Somatic cell nuclear transfer
sgRNA	Single guide RNA
TALEN	Transcription activator-like effector nucleases
ZFN	Zinc finger nucleases

14.1 Introduction

Recombinant proteins are essential in the biopharmaceutical industry for the treatment and diagnosis of clinical human diseases, including recombinant insulin for the treatment of Type I Diabetes (Vajo et al. 2001), clotting factors for blood clotting disorders (Powell 2014), cystic fibrosis treatment (Cutting 2005), and other disease treatments. The pharmaceutical demand for recombinant proteins promotes the development of production platforms capable of effectively supplying substantial quantities of recombinant proteins. Conventionally, therapeutic recombinant proteins are produced by bacteria (mainly *Escherichia coli*), yeast, and mammalian cell lines. However, these conventional platforms have biological and economical limitations creating a demand for alternative routes to produce recombinant proteins. Since the 1980s, transgenic animals have been explored for their potential to serve as bioreactors, namely, through the secretion of recombinant proteins into milk which can then be readily purified (Shepelev et al. 2018; Simons et al. 1987; Clark et al. 1989). Unlike their bacteria and yeast counterparts, the transgenic animal bioreactor is capable of carrying out the necessary post translational modifications for production of biologically active and functional proteins (Shepelev et al. 2018; Wang et al. 2013). Currently, there are three FDA approved recombinant proteins produced by transgenic animals for clinical use: ATryn (rEVO Biologics Inc), Ruconest (Pharming) and Kanuma (Alexion Pharmaceuticals, Inc). Livestock species are a suitable organism to serve as bioreactors due to the amount of daily milk production compared to traditional laboratory animals (Wu 2022). The production of recombinant proteins through transgenic animals is achieved by inserting an exogenous gene(s) coding for the protein(s) of interest into the selected species genome. Unfortunately, technologies to modify the genome of livestock is suboptimal, compared to rodents, and often costly and time consuming. Transgenic livestock expressing exogenous genes can be produced by pronuclear injection (Clark et al. 1989; Simons et al. 1987) or by utilizing somatic cell nuclear transfer (SCNT) technology (Wilmut et al. 1997). However, the efficiency of these approaches is poor and establishing bioreactors that can stably secrete proteins of interest is often challenging. Recent developments of genome editing systems, such as the CRISPR/Cas9 system, now offer advanced methods to alter the genome of livestock species and integrate a gene of interest into a specific locus of the genome.

In this review we discuss how the development of genome editing systems impacts the use of livestock animals as bioreactors. Attributes of past and current genetic engineering techniques employed to produce transgenic animal bioreactors will be described. Furthermore, benefits of utilizing livestock as bioreactors and available livestock models will be introduced to forecast the use of livestock species as bioreactors in the era of genome editing.

14.2 Conventional Production of Recombinant Proteins

Conventional production methods for recombinant proteins include bacteria, yeast, and mammalian cell culture systems. While these systems are capable of producing recombinant proteins, the cellular machinery necessary for post translational modifications is lacking in bacteria platforms, therefore, limiting their ability to produce complex recombinant proteins (Sanchez-Garcia et al. 2016). Similarly, post translational modifications can occur in yeast platforms; however, glycosylation and sialyation patterns do not always resemble those of humans, potentially rendering the yeast produced proteins immunogenic to humans and inappropriate for biotherapeutic use (Maksimenko et al. 2013; Dyck et al. 2003). Although the use of mammalian cells, primarily Chinese hamster ovary (CHO) cells, mitigates improper post translational modification issues, this platform is a timely and expensive.

To overcome issues associated with conventional recombinant protein production methods, the use of transgenic animals to produce recombinant proteins through milk secretion has been explored. The estimated expense to establish a large scale recombinant protein production platform using mammalian cell culture is $250–500 million, whereas only $80 million is required to establish a system producing a similar amount of recombinant protein using an animal-based bioreactor system (Dyck et al. 2003). Cell culture-based platforms, such as using CHO cells, often require the optimization of cell culture conditions which can be timely and costly (Kim et al. 2012). The longevity of mammalian cell-based platforms and resulting protein production is also often not infinite. However, once established, a transgenic animal can be bred and the transgene passed on to offspring, continuing the stable production of the desired recombinant protein (Wang et al. 2013; Zhao et al. 2015; He et al. 2018). The CHO cell culture produces approximately 10 g/L of recombinant protein in 10–12 days (Kim et al. 2012), whereas the transgenic cow mammary gland can produce 1 to 14 g of recombinant protein per liter of milk and 50 L or more of milk can be collected from the cow daily (Monzani et al. 2016). The mammalian cell bioreactor platforms have significantly contributed to the production of recombinant proteins in biopharmaceutical industry. However, the transgenic animal-based bioreactor system offers an innovative opportunity to overcome the shortcomings associated with conventional production platforms and expand the pharmaceutical industry. Unfortunately, available genetic engineering technologies have been the main obstacle to utilizing livestock species as bioreactors.

14.3 Difficulty in Establishing the Transgenic Animal Bioreactor Through Conventional Approaches

The use of animals to produce a recombinant protein(s) requires the expression of an exogenous gene(s) coding for the protein of interest. The ability to integrate an exogenous gene into the genome of the host animal was first explored via pronuclear injection. Pronuclear injection involves the injection of exogenous DNA fragments into the pronucleus of developing embryos, and the exogenous DNA fragments can potentially integrate into the host genome (Fig. 14.1a). In 1987, the first transgenic mouse producing sheep β-lactoglobulin in milk was generated through pronuclear injection (Simons et al. 1987). Shortly after this discovery, the first transgenic sheep producing human blood clotting factor IX was generated, paving the basis for the use of transgenic livestock bioreactors for biopharmaceutical protein production (Clark et al. 1989).

The major drawback of pronuclear injection is random integration of exogenous gene into the genome. Random integration of an exogenous gene impedes the ability to predict the expression level of the gene of interest. Likewise, inactivation of an exogenous gene can also occur as a result of random integration, if the gene is integrated into a closed chromatin structure (Swift et al. 1984; Schilit et al. 2016; West and Gill 2016). Furthermore, when utilizing the pronuclear injection method, the copy number of integrated exogenous genes cannot be controlled, and consequently an essential endogenous gene(s) may be potentially disrupted or alterations in epigenetic marks controlling transgene expression level may occur (Schilit et al. 2016). Transgenic animals generated through pronuclear injection primarily carry a mosaic genotype and multiple rounds of breeding is required to establish founder animals, properly expressing the gene of interest. Although it is possible to generate transgenic animals through pronuclear injection, this technique often results in complications, therefore limiting its use for the production of transgenic livestock models.

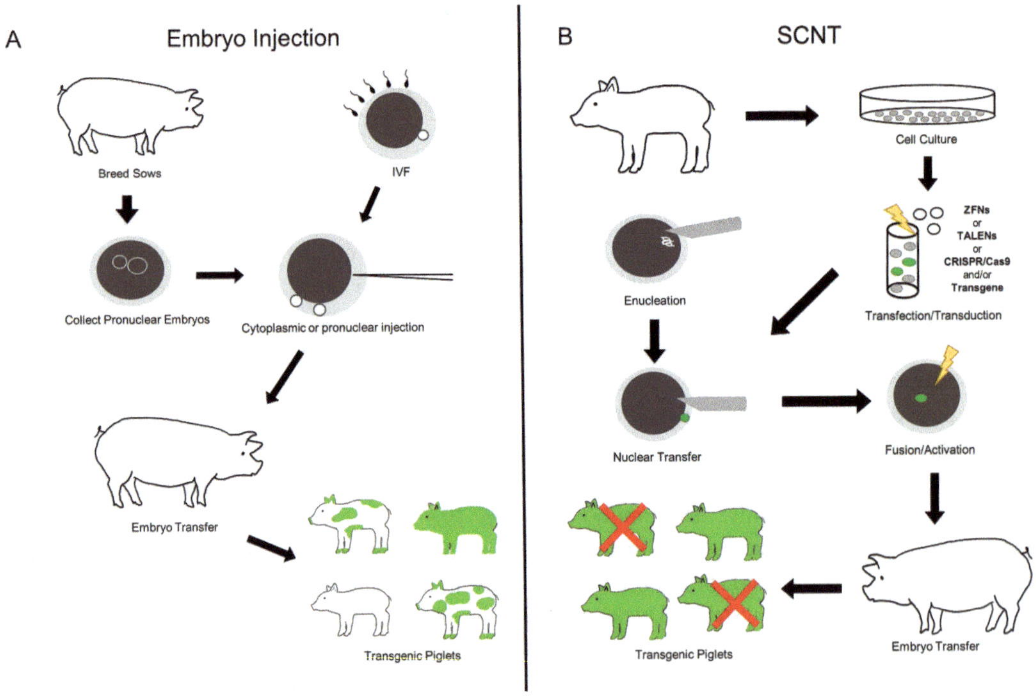

Fig. 14.1 Schematic to produce genetically engineered livestock. **a** Pronuclear stage embryos are injected with a transgene construct, either into the pronucleus or cytoplasm. Embryo transfer of the embryos into a recipient female results in transgenic animals expressing different degree of the desired mutation (GFP expression) due to efficacy of the transgenesis and mosaic genotype. **b** For somatic cell nuclear transfer (SCNT), donor cells are genetically modified to contain the gene of interest. Genome editing systems can be used to facilitate integration of gene of interest into a safe harbor location. Then, a genetically modified cell is placed in the perivitelline space of an enucleated oocyte and fused into the oocyte. The reconstructed SCNT embryos are transferred into recipient females. All resulting piglets should contain the transgene; however, SCNT-derived piglets often suffer from lethal health complications

Embryonic stem (ES) cells offer an alternative route to establish animals that carry an exogenous DNA fragment. Genetically modified ES cells can be expanded through colonial cell propagation, genotyped to ensure genetic modification, and then introduced into developing embryos (Thomas and Capecchi 1987). In order to efficiently manipulate ES cells for genetic engineering, pluripotency characteristics of ES cells should be continuously maintained in-vitro (Blomberg and Telugu 2012). Mouse ES cells are well characterized and maintain its pluripotency in-vitro, an ideal source to engineer gene knock-in and knockout mouse models (Jin et al. 2000). However, livestock ES cells remain to be fully characterized and their pluripotency is not well preserved in-vitro (Soto and Ross 2016). Additionally, founder animals generated through ES cell-mediated manipulation are chimeric and therefore, extensive breeding is required to establish a population of transgenic animals. Extensive breeding is manageable in rodents but becomes laborious when generating animals with lengthy gestation periods and sexual maturity time such as livestock species.

To combat the lack of ES cells and avoid issues related to pronuclear injection, transgenic livestock has been generated through incorporating SCNT (Park et al. 2006; Simons et al. 1987; Wall et al. 1991; Wright et al. 1991; Krimpenfort et al. 1991; Schnieke et al. 1997). The ability to perform nuclear transfer with somatic cells was first demonstrated when Dolly the sheep was born in 1997 (Wilmut et al. 1997). The development of SCNT allows researchers to generate transgenic animals with intended modifications as somatic cells can be manipulated in-vitro and screened for proper genetic modification before being used for SCNT (Fig. 14.1b) (McCreath et al. 2000). Targeted genetic modifications in somatic cells followed by SCNT technique offer a mechanism to generate livestock models carrying an exogenous gene on a specific locus, which helps to overcome issues of the random integration of exogenous genes following pronuclear injection. SCNT technology has been successfully employed for the generation of transgenic animal bioreactors. For instance, goats secreting human glycoprotein α-fetoprotein in their milk were generated with SCNT technology (Parker et al. 2004) and more recently, cows designed to secrete human serum albumin (HSA) in milk (Luo et al. 2016). SCNT enables the production of transgenic animals with targeted modifications. However, the efficiency of generating viable and healthy animals through SCNT is low due to developmental complications associated with the technology (Bertolini et al. 2016; Lai et al. 2002; Dai et al. 2002; Young 1998; Park et al. 2006). Recent advancements in genome editing systems facilitate efficient production of genetically modified livestock with targeted modifications by increasing the efficiency of targeted modifications in somatic cells. Genome editing systems also provide a route for transgenesis without applying SCNT technology, thereby offering opportunities to avoid limitations associated with conventional genetic engineering approaches.

14.4 Application of Genome Editing Systems

The ability to engineer the genome has been immensely improved by genome editing systems in the form of site-specific endonucleases. The basic principle of genome editing systems is recognizing a specific sequence of the genome and introducing a double strand DNA break (DSB) at a specified locus (Cong et al. 2013; Horvath and Barrangou 2010; Ruan et al. 2015). When the DSB occurs, it is repaired through one of two repair mechanisms: non-homologous end joining (NHEJ) or homology-directed repair (HDR). If NHEJ occurs, random base pairs can be inserted or deleted (indels) potentially resulting in a gene knockout by the introduction of a frameshift mutation leading to the generation of premature stop codon, consequently inactivating the targeted gene. Exogenous gene sequences can be introduced into a specific locus of the genome through the HDR pathway (Christian et al. 2010). In livestock, utilization of the NHEJ pathways allows researchers to modify multiple alleles, thus reducing the need of breeding (Hauschild et al. 2011), and targeted insertions are now practical by stimulating the HDR pathway. Conventional targeted modification of the genome in somatic cells by relying on endogenous homologous recombination is extremely poor and in some cases one out of million cells is modified (Mir and Piedrahita 2004).

There are three main types of genome editing systems: Zinc finger nucleases (ZFN), transcription activator-like effector nucleases (TALENs) or Clustered regularly interspaced short palindromic repeats (CRISPR)/CRISPR-associated 9 (Cas9) and its derivatives. The mechanism of recognizing a specific sequence on the genome and introducing a DSB varies amongst the genome editing systems, but the functional steps of the systems are comparable. The application of genome editing systems has significantly lowered efforts to modify the genome of livestock. In particular, the development of the CRISPR/Cas9 system considerably impacted the production of genetically engineered livestock. The CRISPR/Cas9 system comprises the adaptive immune system of bacteria cells and targets exogenous DNA. The system utilizes a single-guide RNA (sgRNA), which contains an RNA sequence complementary to the target DNA and trans-activating CRISPR RNA (tracr-RNA). By utilizing the sgRNA sequences which are designed to bind to a specified location in the genome, the Cas9 endonuclease is directed to a target location in the genome. The Cas9 will induce a DSB at the target site if the protospacer adjacent motif (PAM) sequence is present at the target location (Jinek et al. 2012). The CRISPR/Cas9 system is preferred over ZFNs or TALENs because of its simplicity as it requires only a sgRNA that contains the 20 bp target sequence inserted into a targeting vector (Cong et al. 2013) or synthesized in vitro (Wang et al. 2013). On the contrary, ZFNs and TALENs utilize a multicomponent array which requires a number of ligation reactions for assembly, increasing the effort and time required by the users compared to CRISPR/Cas9 system (Ellis et al. 2013; Holkers et al. 2013).

Genome editing systems allow for efficient targeted gene insertion (gene knock-in) via HDR pathway into a safe harbor locus such as *ROSA26,* which permits the expression of exogenous genes under the control of a tissue specific promoter (Fig. 14.2). In livestock, multiple safe harbor locations within the genome have been proposed as "safe" locations to insert exogenous DNA fragment because targeted modifications on the loci typically result in no harmful impacts to the animal; genes such as *AAVS2*, *CLYBL*, *Rosa26*, and *pH11* have all been proposed as a safe harbor location (Ruan et al. 2015; Cerbini et al. 2015). Genome editing systems can be utilized in genetically engineered livestock production by efficiently modifying the genome of somatic cells for SCNT. Previously, using ZFNs, we were able to introduce targeted modifications on *CMAH* gene at over 40% efficiency in pig fibroblast cells by utilizing the HDR pathway (Kwon et al. 2013), demonstrating that the HDR pathway is active and targeted insertion can effectively occur using genome editing systems in pig somatic cells.

A novel route, independent of pronuclear injection and SCNT, has also been developed by directly injecting genome editing systems into a developing embryo to modify the genome of livestock. In our previous study, two genes (*RAG2* and *IL2RG*) were simultaneously disrupted during embryogenesis at 100% targeting efficiency by injecting RNA form of CRISPR/Cas9 systems into developing embryos (Lei et al. 2016). Since this approach does not rely on SCNT, no piglets were presented with any health complications associated with the SCNT. Gene knock-in can also occur during embryogenesis if a genome editing system targeting a safe harbor locus, and DNA fragments containing an exogenous gene and homology sequence to the safe harbor locus are injected into developing embryos. Although the efficiency of knock-in events during embryogenesis in livestock varies depending on the size of the exogenous gene and target locus, reported efficiencies can be as high as 100% with the CRISPR/Cas9 system (Peng et al. 2015). The success of this approach is higher than typical success rates from pronuclear injection (1–12%) (Hammer et al. 1985; Wheeler and Walters 2001), and the knock-in events in embryos avoid issues associated with SCNT and random integration. However, since the knock-in events are introduced as embryo develop, it is possible that mosaic genotypes may be introduced, similar to pronuclear injection (Fig. 14.1a).

While genome editing systems have vastly improved the livestock genome editing field, there are limitations that should be mentioned. Not all sites on the genome are appropriate to target as the efficiency of genome editing can be significantly impacted by the nucleotide composition of the target loci (Gaj 2013). In addition,

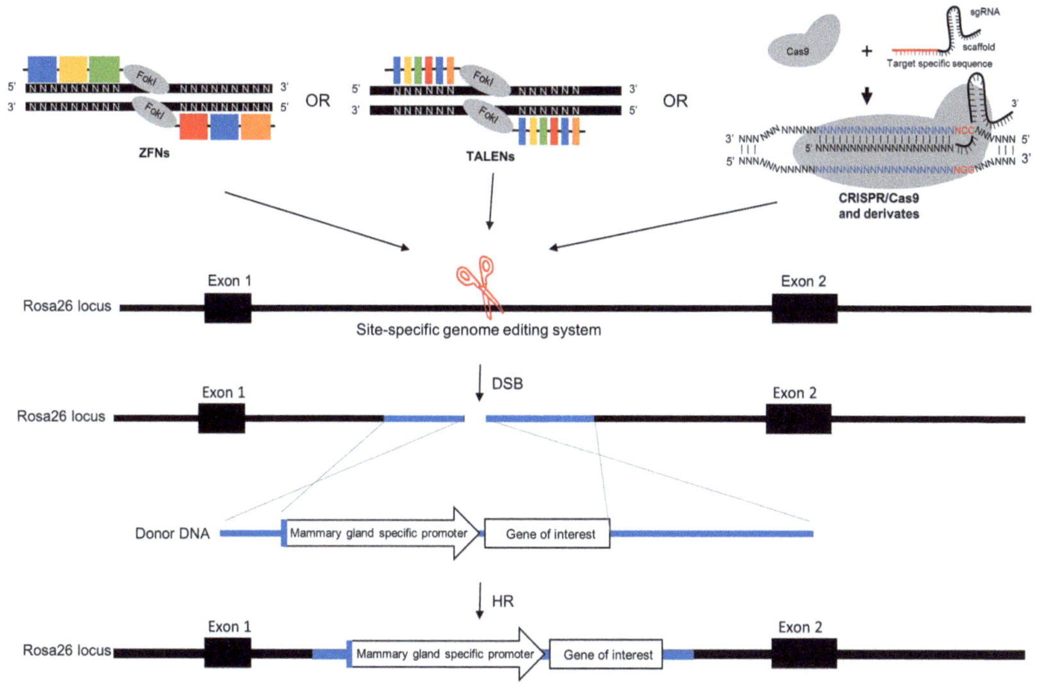

Fig. 14.2 Schematic of knock-in approach for targeted insertion of milk specific transgene into the safe harbor gene, *ROSA26*. Genome editing systems can be used to target a specified location on *ROSA26* for the desired knock-in. A double strand break introduced by the genome editing system facilitates the introduction of donor DNA, containing a mammary gland specific promoter, gene of interest sequence, and homology arms. The construct containing the gene of interest is integrated via homologous-directed repair

inappropriate targeting can result in negative health impacts due to target location. Specifically, unintended mutations, a phenomenon known as off-targeting, may lead to compromised health or an unexpected phenotype (Doench et al. 2016; Zhang et al. 2015; Carey et al. 2019). In our recent study, we identified potential off-targeting events in one out of four CRISPR/Cas9 systems when injected into developing pig embryos (Carey et al., 2019), indicating off-targeting events can be introduced by genome editing systems in livestock. To address the side effects and limitations of genome editing systems, systems recognizing more diverse sequences (Zetsche et al. 2017) and carrying enhanced safety have been proposed (Komor et al. 2018; Cong et al. 2013). Advancements in genome editing technologies are a continuous effort and offer novel approaches to design animal bioreactors.

14.5 Examples of Transgenic Animal Bioreactors

Mammalian species including the cow, goat, sheep, rabbit, and pig can serve as bioreactors because of their ability to synthesize and secrete proteins in milk during lactation periods. Although the context of this review is focused on animal bioreactors secreting target proteins through the mammary gland, other secretory tissues and fluids such as urine, blood and seminal fluid have also been utilized (Wang et al. 2013). Avian species, such as the chicken, can also serve as a transgenic animal bioreactor because recombinant proteins can be supplied daily through egg production (Lee et al. 2020). Selection of the animal for recombinant protein production is determined by the amount of protein needed, the size of the production and animal housing facilities, and the time allotted for production. In general, the larger the animal species, the greater the milk yield and subsequent recombinant protein production. However, the required facility size and maintenance expenses also increase with animal size, along with animal sexual maturity age and gestation period, which collectively increase animal production time. Furthermore, physiological factors such as glycosylation and sialylation patterns should also be considered, as certain species are able to mimic patterns observed in human proteins better than others.

When engineering a transgenic animal bioreactor, the design of gene constructs to produce target proteins typically includes the sequence for the gene of interest, a milk protein/mammary gland promoter to enhance tissue expression specificity, and often times cis-regulatory elements to express multiple genes from a single construct (Clark et al. 1989; Shepelev et al. 2018). While traditional transgenesis methodology permits the production of transgenic animal bioreactors, the targeted knock-in efficiency from the use of genome editing systems is remarkably more efficient than traditional methods, which further enables the use of livestock as bioreactors. Currently utilized bioreactor models, discussed in this review, can be found in Table 14.1.

- *Rabbit*

The rabbit is an appealing mammary gland bioreactor due to its physical and physiological characteristics. Due to their size, they are relatively inexpensive to maintain and can be housed under specific pathogen free (SPF) conditions if necessary. The gestation period of rabbits is only 35 days with sexual maturity achieved at the young age of 4–5 months old, and they have the ability to produce 4–7 L per year. The protein content of their milk is higher than the cow, 14% compared to 5%, respectively (Fan and Watanabe 2003). The transgenic rabbit bioreactor has been designed to produce recombinant human α-glucosidase (Van den Hout et al. 2001), human plasminogen activator (hPA) (Song et al. 2016), and the FDA approved C1 esterase inhibitor, Ruconest (Pharming) (van Veen et al. 2012).

Pronuclear injection of a gene construct containing a milk protein promoter and the gene sequence for a target human protein has been the leading method to produce transgenic rabbits. While the use of a milk protein promoter increases protein yield in a bioreactor, the appropriate integration of the gene seldomly occurs (Van den Hout et al. 2001; van Veen et al. 2012; Song et al. 2016). The rabbit is a challenging species in which to perform SCNT; therefore, the pronuclear injection method has been a predominant route to produce transgenic rabbits (Zakhartchenko et al. 2011). Recently, the application of genome editing systems has expanded on the techniques for genetic engineering in the rabbit and increased the efficiency to integrate the gene of interest into a specific locus by employing HDR machinery during embryogenesis (Bosze et al. 2003; Wang et al. 2014; Yang et al. 2014, 2019). Advancements in rabbit genome editing have led to the characterization of safe harbor genes, such as *ROSA26*, providing optimal locations for exogenous gene expression (Yang et al. 2016). For example, Yang et al. (2016), generated an rbRosa26- Cre reporter knock-in rabbits with a resulting knock-in efficiency of 35%. The ability to perform targeted gene integration in rabbits will undoubtedly expand the production of transgenic rabbit bioreactors.

- *Goat*

As a traditional dairy species, the goat yields a relatively high amount of milk (600–800 L/year) that can potentially contain up to 5 g/L of target recombinant protein. Furthermore, the goat does not require as much housing space like larger dairy species, has a relatively short gestation period (5 months), and reaches sexual maturity within a year (Wang et al. 2013). The transgenic goat has been used to produce numerous recombinant proteins including human granulocyte- colony stimulating factor (hG-CSF) (Ko et al. 2000), human glycoprotein α-fetoprotein

(Parker et al. 2004), CuZn-SOD and EC-SOD (Lu et al. 2018), human lactoferrin (Cui et al. 2015), and FDA approved ATryn an antithrombin (Maksimenko et al. 2013).

Unlike rabbits, the transgenic goat bioreactor has been generated through both pronuclear injection and SCNT. For example, ATryn secreting goats were generated by pronuclear injection into goat embryos of the human antithrombin cDNA sequence fused with the regulatory elements of the goat β-casein gene (Adiguzel et al. 2009). Utilizing SCNT technology has allowed for the generation of transgenic goat bioreactors that produce human α-fetoprotein (Parker et al. 2004) or two forms of human superoxide dismutase (SOD), CuZn and EC (Lu et al. 2018), in milk. Recently, the use of genome editing systems has enhanced the ability to modify somatic cells for SCNT, which enables the introduction of complicated targeted modifications. For instance, TALENs were utilized to precisely integrate the human lactoferrin (hLF) gene into the β-lactoglobulin (BLG) endogenous gene site in the genome of primary goat fibroblasts at 13% efficiency, thereby expressing hLF while inactivating goat BLG. Goats derived from the modified cells lacked the gene for the milk allergen BLG; therefore, the allergenicity of the produced milk was decreased and the *BLG* gene was proved as an efficient gene integration site for mammary gland specific expression. This study exemplifies the beneficial impact of genome editing systems as multiple genetic modifications can be expedited by using this technology (Cui et al. 2015).

- *Cow*

The cow is the largest sized livestock species utilized as a bioreactor and comparatively produces a large quantity of milk, classifying them as ideal for large-scale production of recombinant proteins. However, due to its size, gestation period (9 months), and time to reach sexual maturity (15 months), producing a herd of cows is a timely upfront investment (Wang et al. 2013). Despite these disadvantages, the cow has been used to produce recombinant human proteins such as lactoferrin (hLF) (van Berkel et al. 2002; Wang et al. 2017) and human serum albumin (HSA) (Luo et al. 2016).

Pronuclear injection was initially utilized to engineer a transgenic cow bioreactor to produced functional hLF in milk secretions (van Berkel et al. 2002). Another hLF cow was generated through SCNT and produced functional hLF but animal health complications associated with SCNT limited the number of animals produced (Wang et al. 2017). The application of genome editing systems offers a methodology to proficiently introduce targeted modifications and allows for efficient gene integration in transgenic cow bioreactors. For example, transgenic cows were engineered to secrete human serum albumin (HSA) by targeting the bovine β-lactoglobulin locus and inserting the *HSA* gene into the genome of fibroblast cells with TALEN nickases, a derivative of TALENs, followed by SCNT. The TALEN nickases proved an efficient way to genetically manipulate the bovine fibroblast cells; 4.8% of cell clones contained the desired modifications and cows generated from the cells resulted in functional HSA production in their milk (Luo et al. 2016). Efficient production of transgenic cows using genome editing systems and SCNT promotes the use of cows as bioreactors, despite the prolonged time required for establishing a transgenic herd through breeding (Carvalho et al. 2019; Kaiser et al. 2017; Clarissa Varajão Cardoso 2019).

- *Pig*

The pig is a valuable animal model in biomedical research because of its physiological similarities with the human. Due to the pig's value as a biomedical research model, extensive efforts have been invested into improving genetic engineering in the pig. The expanded genetic engineering strategies (Wu and Bazer 2019) in the pig are beneficial for the design of pig bioreactors. The pig has a shorter gestation period (approximately 4 months) than the goat or cow and is a litter bearing species, producing 10–15 piglets per gestation cycle (Walters et al. 2011). The ability to rapidly propagate a herd of pigs is an appealing property for transgenic animal bioreactors.

One of the first transgenic pig bioreactors was generated through pronuclear injection to generate a pig that secreted human blood clotting factor VIII (Paleyanda et al. 1997). Likewise, transgenic pigs secreting human erythropoietin (hEPO) in their milk were generated by pronuclear microinjection of a transgene containing the human DNA sequence of hEPO and the mouse WAP promoter (Park et al. 2006). Bitransgenic pigs expressing genes for the human Furin enzyme and clotting factor IX (pro-peptide form) were generated by the genetic engineering of somatic cells followed by SCNT technology to serve as a means to expand treatment options for hemophilia B patients (Zhao et al. 2015). Since precise genome modifications can be carried out at higher efficiency with genome editing systems, such as the CRISPR/Cas9 system (Ruan et al. 2015; Wang et al. 2016; Zheng et al. 2017), more diverse transgenic pig bioreactors are expected to be introduced into the biopharmaceutical industry.

- *The Avian bioreactor: The Chicken Egg*

Avian eggs are a compelling bioreactor platform because birds such as the chicken are relatively inexpensive to maintain, can be housed in a small space, and can produce on average 300 eggs/year, which provides an abundant source of protein at a relatively low cost (Woodfint et al. 2018). Additionally, recombinant proteins are easier to purify from eggs than milk and the protein glycosylation pattern in the chicken egg closely relates to the human, potentially decreasing immunogenicity of chicken egg derived recombinant proteins (Lillico et al. 2005). Despite these advantages, there are limited avian models designed as a bioreactor due to low efficiency in transgenesis. Unlike mammals, technologies such as pronuclear injection and SCNT cannot be applied in avian species. Alternative methods such as direct injection of viral vectors (Salter et al. 1987) or primordial germ cell (PGC) mediated transgenesis (Oishi et al. 2018) have been utilized. Currently, Kanuma (Alexion Pharmaceuticals, Inc), a human lysosomal acid lipase, is the only FDA approved recombinant protein derived from transgenic chicken eggs (Sheridan 2016).

Chickens producing biologically active human interferon α-2b (IFNα-2b) were one of the first transgenic chickens to express a human protein in their eggs (Rapp et al. 2003). The transgenic chicken eggs were produced by the injection of retroviral vector containing the human (Ifnα-2b) gene sequence under the control of the cytomegalovirus (CMV) promoter into the subgerminal cavity of windowed eggs (Rapp et al. 2003). Recombinant proteins derived from chicken eggs displayed increased biological activity over proteins produced from *E. Coli* (Kwon et al. 2008) or comparable activity to CHO cell produced proteins (Kodama et al. 2008), demonstrating the effectivity of chickens as bioreactors. Similar to the use of mammary gland specific promoters in mammals, the ovalbumin promoter and corresponding regulatory elements have been used in gene constructs for targeted expression of exogenous genes, which results in increased protein production in the egg (Zhu et al. 2005). For example, transgenic chickens were engineered to express hEPO under the control of ovabulimin promoter, which resulted in restricted expression of the transgene to the oviduct. The egg derived hEPO recombinant protein was equalivalent in function to hEPO generated through CHO cell culture, further supporting the use of the transgenic chicken as a bioreactor (Kwon et al. 2018).

The implementation of targeted genome editing techniques, such as the CRISPR/Cas9 system, has provided a pivotal opporutnity for the expansion of avian bioreactors. Through the use of the CRISPR/Cas9 system, the *hIFNB* gene sequence was integrated into the translation initiation site of the chicken *OVA* locus via HDR in PGCs. Then, PGCs carrying the intended modification were injected into chicken embryos to generate four chimeric founder roosters. The chimeric roosters were bred to wild type hens and 14–22.5% of G1 founder birds expressed *hIFNB* gene and produced eggs containing functional hIFNβ (Oishi et al. 2018). The chicken eggs expressing the *hIFNB* gene as a consequence of the knock-in event contained a significantly higher level of hIFNβ (1.86–4.42 mg of hIFNβ/mL) compared to eggs from lentiviral derived transgenic hens (3.5–426 μg of hIFNβ/mL), which relied on the random integration of the *hIFNB* gene (Lillico et al. 2007). Though genetic manipulation of the chicken is challenging, the field is postively evolving with the incorporation of genome editing systems.

14.6 Production of Humanized Antibodies Through Transgenic Animal Bioreactors

The advancement of genome editing systems encourages designing genetically engineered livestock species harboring complex genetic modifications, such as animal models producing humanized antibodies for pharmaceutical purposes. Immunoglobulin (Ig), also known as antibody, is utilized in human medicine to boost the immune response in an immunocompromised state or to add in the clearance of pathogens and toxins (Butler et al. 2009). Although antibodies are an essential component of human medicine, difficulty lies in obtaining large amounts of human polyclonal Ig because availability is limited to healthy human blood donors (Novaretti and Dinardo 2011). In order to produce large amounts of functional human Ig for medical use, researchers seek to produce human Ig in animal models (Matsushita et al. 2015; Wu et al. 2019).

Immunoglobulin is a complex protein produced by B lymphocytes and is composed of a light chain and heavy chain, with both containing variable and constant regions (Fig. 14.3). The variable region serves as the antigen binding site and rearrangement of the genes coding for this region results in functional and diverse Ig, which aids in the immune response against pathogens. The distal constant region of the heavy chain can also be modified in a process termed class switch recombination (CSR), which promotes Ig isotype diversity and further aid in the immune response (Fig. 14.3) (Watson and Breden 2012; Chi et al. 2020). There are two variants of the Ig light chain, kappa and lambda, and variant selection for Ig formation is species specific. The mouse and human predominantly produce Ig using kappa light chains (Haughton et al. 1978; Katzmann et al. 2002), whereas cattle utilize the lambda light chain (Arun et al. 1996) (Fig. 14.3). The functionality of Ig is dependent on both heavy and light chains. Therefore, the expression of human heavy and light chain genes in animal bioreactor is necessary for the optimal production of humanized Ig in transgenic animals (Matsushita et al. 2014).

Designing transgenic animals which produce humanize antibodies requires complex genetic modifications. Specifically, endogenous genes coding for Ig heavy and/or light chain should be inactivated while genes coding for the human heavy and light chain genes need to be integrated into the genome of the transgenic animal. The Ig heavy and light chain loci are composed of multiple genes coding for the variable, diversity (heavy chain only), joining, and constant regions of the antibody. The complex segmented nature of the Ig chains results in large gene sequences; the human heavy chain is 1.5 Mb, the kappa light chain is 1 Mb; and the lambda light chain is 2 Mb (Ishida et al. 2002). When integrating substantially sized sequences of DNA into animals, artificial chromosome vectors are one of preferred choices. Artificial chromosomes contain the desired genetic sequence as well as a centromere, two telomeres, and origins of replication (Robl et al. 2003). Human artificial chromosomes (HACs) have been created to generate transchromosomic (Tc) animals that carry Ig genes (Kuroiwa et al. 2002; Matsushita et al. 2014, 2015; Robl et al. 2003).

Initial Tc cattle studies demonstrated that the integration and stable expression of a novel HAC containing the entire human heavy chain and the lambda light chain sequences (hIg-HAC) into the bovine genome was possible (Kuroiwa et al. 2002; Robl et al. 2003). While human Ig was produced in these cattle, the native cattle Ig expression was dominate and undesirable chimeric bovine-human antibodies were produced (Robl et al. 2003; Kuroiwa et al. 2002). The undesirable Ig production was likely a result of

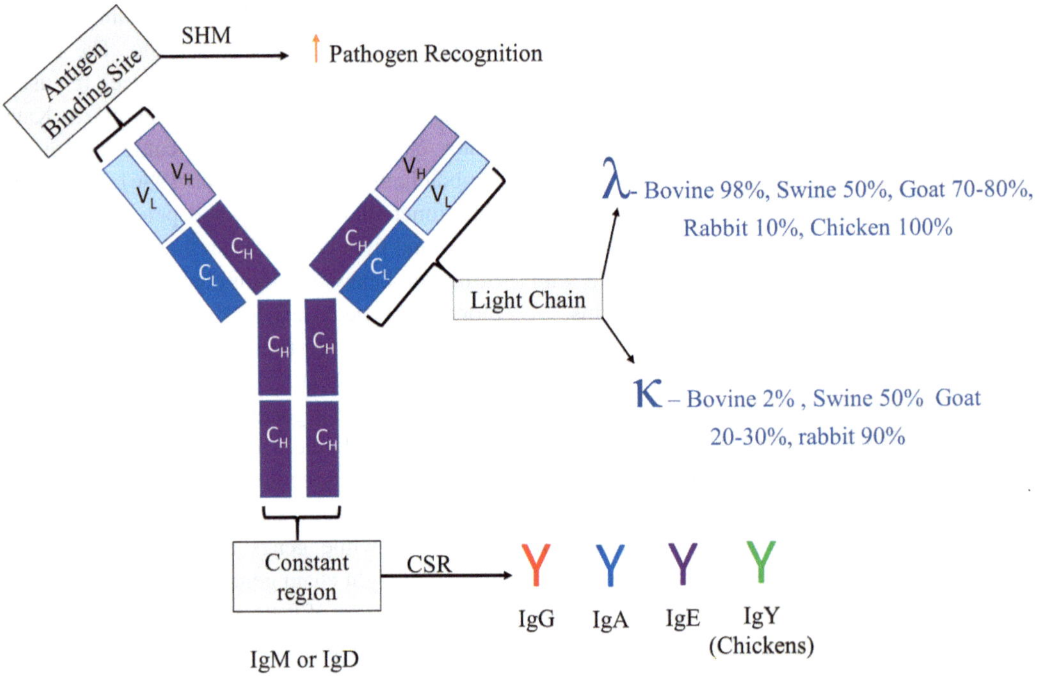

Fig. 14.3 General antibody structure. Both the light (L and depicted in blue) and heavy (H and depicted in purple) chains are composed of variable (V) and constant (C) regions. The variable regions serve as the antigen binding site and undergo the gene rearrangement process of somatic hyper mutation (SMH) to improve pathogen recognition and immune response against invading pathogen. The constant region of the antibody undergoes class switch recombination (CSR) to diversify antibody isotype (IgG, IgA, IgE and IgY) and improve the immune response. The light chain has two variants kappa (κ) and lambda (λ); there is only one variant per antibody, and variant usage is species specific. The heavy chain has two different loci in the cow but only one copy in the pig, goat, rabbit and chicken

the two functional Ig heavy chain loci present in the cattle genome and knockout of both Ig heavy chain genes improved human Ig production (Kuroiwa et al. 2009). To further improve human Ig production by Tc cattle, triple gene knockout cattle were generated by deleting both of the endogenous bovine Ig heavy chain gene sequences and the Ig lambda light chain gene sequence (Matsushita et al. 2014). Fibroblast cells collected from the Ig double or triple knockout cattle were introduced with the hIg-HAC and used for SCNT to generate Tc bovine capable of producing human Ig (Fig. 14.4). In addition to the deletion of endogenous genes, further refinement of the hIg-HAC structure was required to optimize CSR to obtain the desirable IgG isotype.

Improvements in the production of hIg after inactivating endogenous bovine genes suggest that the production platform can be dramatically improved by incorporating genome editing systems. A targeted genome editing system such as CRISPR/Cas9 can be utilized for targeted knockout of multiple genes (Khan et al. 2019; Cong et al. 2013; Wang et al. 2016; Lei et al. 2016). For instance, 62 copies of porcine endogenous retrovirus (*PERV*) genes could be inactivated using CRISPR/Cas9 in pigs (Yang et al. 2015). The multiplexing feature of genome editing systems can facilitate expedited production of Tc animals to produce human Ig. The use of genome editing systems for generating animals with multiple genes knocked out will eliminate extensive multistep breeding processes

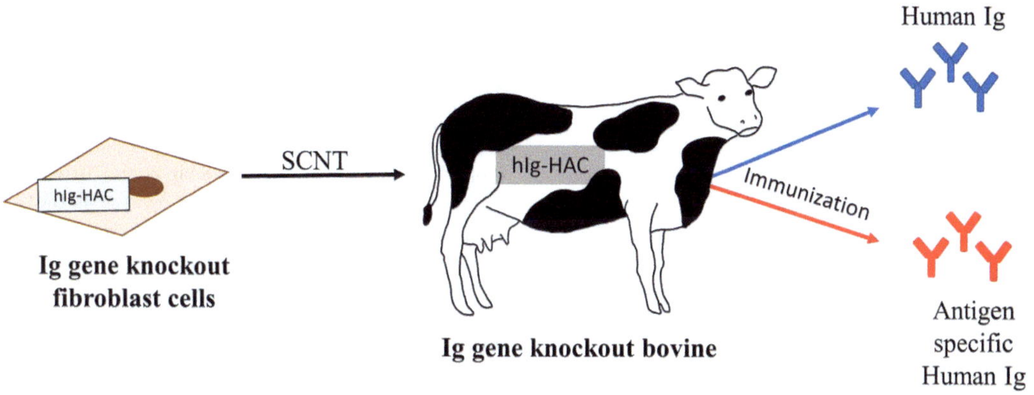

Fig. 14.4 A schematic to generate transchromosomic (Tc) cattle producing humanized antibodies. The human artificial chromosome (HAC) carrying the human immunoglobulin (Ig) light chain and heavy chain loci (hIg-HAC) is generated and fused to Ig gene knockout bovine fibroblast cells. Then, the modified fibroblast cells are used for somatic cell nuclear transfer (SCNT) to generate Tc cattle carrying the hIg-HAC with inactivated endogenous Ig genes. The resulting Tc cattle produce polyclonal human Ig. The Tc cattle can be immunized with a pathogen of interest to generate antigen specific human Ig for the treatment of diseases

that are often required when conventional genetic engineering methodologies are utilized for such animal production. Additionally, genetically engineered immunocompromised livestock animals often fail to thrive and do not reach puberty (Kang et al. 2016; Lee et al. 2014; Suzuki et al. 2012), and therefore, breeding Ig knockout livestock species is a near impossible task.

Animals producing humanized antibodies can be challenged with human pathogens to produce hyperimmunized pathogen-specific Ig, which can potentially aid in the treatment of infectious diseases (Fig. 14.4) (Wu et al. 2019; Luke et al. 2016). In a previous study, immunization of humanized antibody producing cows with anthrax protective antigen induced the production of antigen-specific hIgG that was capable of neutralizing the anthrax toxin in vitro and in vivo (Kuroiwa et al. 2009). Cattle engineered to produce humanized antibodies have been inoculated with human oral squamous carcinoma cells and were capable of mounting a robust Ig response to this cancer (Matsushita et al. 2015). Additionally, Tc cattle were engineered to produce neutralizing antibodies against the viral pathogen, MERS-CoV (Luke et al. 2016). These studies are important first steps towards producing antigen specific human polyclonal antibody for clinical use.

In addition to cattle, caprine hosts for hIg-HAC have also been explored by introducing HAC possessing the bovine switch regions into the goat genome (Wu et al. 2019). While this study did not extensively investigate the resultant hIg for chimerism, these goats did produce human Ig (Wu et al. 2019). When challenged with inactivated influenza virus, the Tc goats were able to mount a response of human antibodies capable of virus neutralization (Wu et al. 2019), indicating proper antibody functionality. This study highlights the goat as a promising host for hIg- HAC to produce large amounts of human polyclonal antibodies. Additional heavy or light chain gene knockout livestock species including the pig and chicken have also been created (Chen et al. 2015; Schusser et al. 2013; Yugo et al. 2018), suggesting their application towards producing humanized antibodies. For example, chickens have been genetically engineered to produce functional humanized antibodies expanding the use of transgenic animals for humanized antibody production (Ching et al. 2020, 2018).

While it is a challenging process to generate humanized antibody producing transgenic animals, these animals can be a valuable production source for hIg to treat human conditions. The production of the transgenic animals is complex

Table 14.1 Examples of transgenic animal produced recombinant human proteins in milk or eggs

Species	Protein	Genetic Engineering Technology	References
Rabbit	Human α-glucosidase Human plasminogen activator C1 esterase inhibitor	Pronuclear Injection Pronuclear Injection Pronuclear Injection	Van den Hout et al. (2001) Song et al. (2016) Van Veen et al. (2012)
Goat	Human Granulocyte-Colony Stimulating Factor Human Glycoprotein α-fetoprotein CuZn and EC- SOD Human Lactoferrin Antithrombin (ATryn)	Pronuclear injection SCNT SCNT TALENs/SCNT Pronuclear injection	Ko et al. (2000) Parker et al. (2004) Lu et al. (2018) Cui et al. (2015) Adiguzel et al. (2009)
Cow	Human Lactoferrin Human Serum Albumin	Pronuclear injection SCNT TALENs/SCNT	Van Berkel et al. (2002) Wang et al. (2017) Luo et al. (2016)
Pig	Human Blood Clotting Factor VIII Human Erythropoietin Human Furin Enzyme and Factor IX	Pronuclear injection Pronuclear injection SCNT	Paleyanda et al. (1997) Park et al. (2006) Zhao et al. (2015)
Chicken	Human lysosomal Acid Lipase (Kanuma) Human interferon α-2b Human Erythropoietin Human Interferon β	Viral vector Viral vector Viral vector PGCs/CRISPR/Cas9	Sheridan (2016) Rapp et al. (2003) Kwon et al. (2018) Oishi et al. (2018)

as both gene knockout steps and insertion of hIg coding genes should be performed. The application of genome editing systems facilitates inactivation of multiple Ig genes in host livestock species and improves integration of hIg genes into the genome of host livestock. The efficiency to produce transgenic animals supplying humanized antibodies should advance under the genome editing era, expanding contributions of the models in human medicine.

14.7 Conclusions

The use of genome editing systems has transformed efforts required to produce transgenic livestock species, extending the use of transgenic livestock species in different fields including the animal bioreactor industry. Recombinant proteins play an essential role in the biopharmaceutical industry for the treatment of numerous human diseases and can be stably produced in transgenic animal bioreactors. Unlike their single cell counterparts, bacteria and yeast, the transgenic animal can cost-effectively produce complex proteins. The advancements of genome editing systems have reduced the time and cost necessary to produce animal models bearing complicated genetic modifications such as ones synthesizing humanized antibodies. The application of genome editing systems will help overcome issues associated with genetically modifying livestock species, thus allowing for a broader impact of transgenic animal bioreactors in the biopharmaceutical industry.

Acknowledgements The authors' work reported herein was supported, in part, by NIH grant R21OD019934 and R21OD027062.

References

Adiguzel C, Iqbal O, Demir M, Fareed J (2009) European community and US-FDA approval of recombinant human antithrombin produced in genetically altered goats. Clin Appl Thromb Hemost 15:645–651

Arun SS, Breuer W, Hermanns W (1996) Immunohistochemical examination of light-chain expression (lambda/kappa ratio) in canine, feline, equine, bovine and porcine plasma cells. Zentralbl Veterinarmed A 43:573–576

Bertolini LR, Meade H, Lazzarotto CR, Martins LT, Tavares KC, Bertolini M, Murray JD (2016) The transgenic animal platform for biopharmaceutical production. Transgenic Res 25:329–343

Blomberg LA, Telugu BP (2012) Twenty years of embryonic stem cell research in farm animals. Reprod Domest Anim 47(Suppl 4):80–85

Bosze Z, Hiripi L, Carnwath JW, Niemann H (2003) The transgenic rabbit as model for human diseases and as a source of biologically active recombinant proteins. Transgenic Res 12:541–553

Butler JE, Zhao Y, Sinkora M, Wertz N, Kacskovics I (2009) Immunoglobulins, antibody repertoire and B cell development. Dev Comp Immunol 33:321–333

Carey K, Ryu J, Uh K, Lengi AJ, Clark-Deener S, Corl BA, Lee K (2019) Frequency of off-targeting in genome edited pigs produced via direct injection of the CRISPR/Cas9 system into developing embryos. BMC Biotechnol 19:25

Carvalho BP, Cunha ATM, Silva BDM, Sousa RV, Leme LO, Dode MAN, Melo EO (2019) Production of transgenic cattle by somatic cell nuclear transfer (SCNT) with the human granulocyte colony-stimulation factor (hG-CSF). J Anim Sci Technol 61:61–68

Cerbini T, Luo Y, Rao MS, Zou J (2015) Transfection, selection, and colony-picking of human induced pluripotent stem cells TALEN-targeted with a GFP gene into the AAVS1 safe harbor. J Vis Exp 96:52504

Chen F, Wang Y, Yuan Y, Zhang W, Ren Z, Jin Y, Liu X, Xiong Q, Chen Q, Zhang M, Li X, Zhao L, Li Z, Wu Z, Zhang Y, Hu F, Huang J, Li R, Dai Y (2015) Generation of b cell-deficient pigs by highly efficient Crispr/cas9-mediated gene targeting. J Genet Genomics 42:437–444

Chi X, Li Y, Qiu X (2020) V(D)J recombination, somatic hypermutation and class switch recombination of immunoglobulins: mechanism and regulation. Immunology 160:233–247

Ching KH, Berg K, Morales J, Pedersen D, Harriman WD, Abdiche YN, Leighton PA (2020) Expression of human lambda expands the repertoire of OmniChickens. PLoS One 15:e0228164

Ching KH, Collarini EJ, Abdiche YN, Bedinger D, Pedersen D, Izquierdo S, Harriman R, Zhu L, Etches RJ, van de Lavoir MC, Harriman WD, Leighton PA (2018) Chickens with humanized immunoglobulin genes generate antibodies with high affinity and broad epitope coverage to conserved targets. Mabs 10:71–80

Christian M, Cermak T, Doyle EL, Schmidt C, Zhang F, Hummel A, Bogdanove AJ, Voytas DF (2010) Targeting DNA double-strand breaks with TAL effector nucleases. Genetics 186:757–761

Clarissa Varajão Cardoso EVB, Maíra Halfen Teixeira Liberal, Evelize Folly das Chagas (2019) Transgenic technology: the strategy for the control and prevention of bovine staphylococcal mastitis? Biotechnol Res Innovation 3:291–297

Clark AJ, Ali S, Archibald AL, Bessos H, Brown P, Harris S et al (1989) The molecular manipulation of milk composition. Genome 31:950–955

Cong L, Ran FA, Cox D, Lin S, Barretto R, Habib N, Hsu PD, Wu X, Jiang W, Marraffini LA, Zhang F (2013) Multiplex genome engineering using CRISPR/Cas systems. Science 339:819–823

Cui C, Song Y, Liu J, Ge H, Li Q, Huang H, Hu L, Zhu H, Jin Y, Zhang Y (2015) Gene targeting by TALEN-induced homologous recombination in goats directs production of beta-lactoglobulin-free, high-human lactoferrin milk. Sci Rep 5:10482

Cutting GR (2005) Modifier genetics: cystic fibrosis. Annu Rev Genomics Hum Genet 6:237–260

Dai Y, Vaught TD, Boone J, Chen SH, Phelps CJ, Ball S, Monahan JA, Jobst PM, McCreath KJ, Lamborn AE, Cowell-Lucero JL, Wells KD, Colman A, Polejaeva IA, Ayares DL (2002) Targeted disruption of the alpha1,3-galactosyltransferase gene in cloned pigs. Nat Biotechnol 20:251–255

Doench JG, Fusi N, Sullender M, Hegde M, Vaimberg EW, Donovan KF, Smith I, Tothova Z, Wilen C, Orchard R, Virgin HW, Listgarten J, Root DE (2016) Optimized sgRNA design to maximize activity and minimize off-target effects of CRISPR-Cas9. Nat Biotechnol 34:184–191

Dyck MK, Lacroix D, Pothier F, Sirard MA (2003) Making recombinant proteins in animals–different systems, different applications. Trends Biotechnol 21:394–399

Ellis BL, Hirsch ML, Porter SN, Samulski RJ, Porteus MH (2013) Zinc-finger nuclease-mediated gene correction using single AAV vector transduction and enhancement by Food and Drug Administration-approved drugs. Gene Ther 20:35–42

Fan J, Watanabe T (2003) Transgenic rabbits as therapeutic protein bioreactors and human disease models. Pharmacol Ther 99:261–282

Gaj T, Gersbach CA, Barbas CF 3rd (2013) ZFN, TALEN, and CRISPR/Cas-based methods for genome engineering. Trends Biotechnol 31:397–405

Hammer RE, Pursel VG, Rexroad CE Jr, Wall RJ, Bolt DJ, Ebert KM, Palmiter RD, Brinster RL (1985) Production of transgenic rabbits, sheep and pigs by microinjection. Nature 315:680–683

Haughton G, Lanier LL, Babcock GF (1978) The murine kappa light chain shift. Nature 275:154–157

Hauschild J, Petersen B, Santiago Y, Queisser AL, Carnwath JW, Lucas-Hahn A, Zhang L, Meng X, Gregory PD, Schwinzer R, Cost GJ, Niemann H (2011) Efficient generation of a biallelic knockout in pigs using zinc-finger nucleases. Proc Natl Acad Sci USA 108:12013–12017

He Z, Lu R, Zhang T, Jiang L, Zhou M, Wu D, Cheng Y (2018) A novel recombinant human plasminogen activator: Efficient expression and hereditary stability in transgenic goats and in vitro thrombolytic bioactivity in the milk of transgenic goats. PLoS One 13:e0201788

Holkers M, Maggio I, Liu J, Janssen JM, Miselli F, Mussolino C, Recchia A, Cathomen T, Gonçalves MA (2013) Differential integrity of TALE nuclease genes following adenoviral and lentiviral vector gene transfer into human cells. Nucleic Acids Res 41:e63

Horvath P, Barrangou R (2010) CRISPR/Cas, the immune system of bacteria and archaea. Science 327:167–170

Ishida I, Tomizuka K, Yoshida H, Tahara T, Takahashi N, Ohguma A, Tanaka S, Umehashi M, Maeda H, Nozaki C, Halk E, Lonberg N (2002) Production of human monoclonal and polyclonal antibodies in TransChromo animals. Cloning Stem Cells 4:91–102

Jin XL, Guo H, Mao C, Atkins N, Wang H, Avasthi PP, Tu YT, Li Y (2000) Emx1-specific expression of foreign genes using "knock-in" approach. Biochem Biophys Res Commun 270:978–982

Jinek M, Chylinski K, Fonfara I, Hauer M, Doudna JA, Charpentier E (2012) A programmable dual-RNA-guided DNA endonuclease in adaptive bacterial immunity. Science 337:816–821

Kaiser GG, Mucci NC, González V, Sánchez L, Parrón JA, Pérez MD, Calvo M, Aller JF, Hozbor FA, Mutto AA (2017) Detection of recombinant human lactoferrin and lysozyme produced in a bitransgenic cow. J Dairy Sci 100:1605–1617

Kang JT, Cho B, Ryu J, Ray C, Lee EJ, Yun YJ, Ahn S, Lee J, Ji DY, Jue N, Clark-Deener S, Lee K, Park KW (2016) Biallelic modification of IL2RG leads to severe combined immunodeficiency in pigs. Reprod Biol Endocrinol 14:74

Katzmann JA, Clark RJ, Abraham RS, Bryant S, Lymp JF, Bradwell AR, Kyle RA (2002) Serum reference intervals and diagnostic ranges for free kappa and free lambda immunoglobulin light chains: relative sensitivity for detection of monoclonal light chains. Clin Chem 48:1437–1444

Khan FJ, Yuen G, Luo J (2019) Multiplexed CRISPR/Cas9 gene knockout with simple crRNA:-tracrRNA co-transfection. Cell Biosci 9:41

Kim JY, Kim YG, Lee GM (2012) CHO cells in biotechnology for production of recombinant proteins: current state and further potential. Appl Microbiol Biotechnol 93:917–930

Ko JH, Lee CS, Kim KH, Pang MG, Koo JS, Fang N, Koo DB, Oh KB, Youn WS, Zheng GD, Park JS, Kim SJ, Han YM, Choi IY, Lim J, Shin ST, Jin SW, Lee KK, Yoo OJ (2000) Production of biologically active human granulocyte colony stimulating factor in the milk of transgenic goat. Transgenic Res 9:215–222

Kodama D, Nishimiya D, Iwata K, Yamaguchi K, Yoshida K, Kawabe Y, Motono M, Watanabe H, Yamashita T, Nishijima K, Kamihira M, Iijima S (2008) Production of human erythropoietin by chimeric chickens. Biochem Biophys Res Commun 367:834–839

Komor AC, Badran AH, Liu DR (2018) Editing the Genome Without Double-Stranded DNA Breaks. ACS Chem Biol 13:383–388

Krimpenfort P, Rademakers A, Eyestone W, van der Schans A, van den Broek S, Kooiman P, Kootwijk E, Platenburg G, Pieper F, Strijker R et al (1991) Generation of transgenic dairy cattle using "in vitro" embryo production. Biotechnology (NY) 9:844–847

Kuroiwa Y, Kasinathan P, Choi YJ, Naeem R, Tomizuka K, Sullivan EJ, Knott JG, Duteau A, Goldsby RA, Osborne BA, Ishida I, Robl JM (2002) Cloned transchromosomic calves producing human immunoglobulin. Nat Biotechnol 20:889–894

Kuroiwa Y, Kasinathan P, Sathiyaseelan T, Jiao JA, Matsushita H, Sathiyaseelan J, Wu H, Mellquist J, Hammitt M, Koster J, Kamoda S, Tachibana K, Ishida I, Robl JM (2009) Antigen-specific human polyclonal antibodies from hyperimmunized cattle. Nat Biotechnol 27:173–181

Kwon DN, Lee K, Kang MJ, Choi YJ, Park C, Whyte JJ, Brown AN, Kim JH, Samuel M, Mao J, Park KW, Murphy CN, Prather RS, Kim JH (2013) Production of biallelic CMP-Neu5Ac hydroxylase knock-out pigs. Sci Rep 3:1981

Kwon MS, Koo BC, Choi BR, Park YY, Lee YM, Suh HS, Park YS, Lee HT, Kim JH, Roh JY, Kim NH, Kim T (2008) Generation of transgenic chickens that produce bioactive human granulocyte-colony stimulating factor. Mol Reprod Dev 75:1120–1126

Kwon MS, Koo BC, Kim D, Nam YH, Cui XS, Kim NH, Kim T (2018) Generation of transgenic chickens expressing the human erythropoietin (hEPO) gene in an oviduct-specific manner: Production of transgenic chicken eggs containing human erythropoietin in egg whites. PLoS One 13:e0194721

Lai L, Kolber-Simonds D, Park KW, Cheong HT, Greenstein JL, Im GS, Samuel M, Bonk A, Rieke A, Day BN, Murphy CN, Carter DB, Hawley RJ, Prather RS (2002) Production of alpha-1,3-galactosyltransferase knockout pigs by nuclear transfer cloning. Science 295:1089–1092

Lee J, Kim DH, Lee K (2020) Current approaches and applications in avian genome editing. Int J Mol Sci 21:3937

Lee K, Kwon DN, Ezashi T, Choi YJ, Park C, Ericsson AC, Brown AN, Samuel MS, Park KW, Walters EM, Kim DY, Kim JH, Franklin CL, Murphy CN, Roberts RM, Prather RS, Kim JH (2014) Engraftment of human iPS cells and allogeneic porcine cells into pigs with inactivated RAG2 and accompanying severe combined immunodeficiency. Proc Natl Acad Sci USA 111:7260–7265

Lei S, Ryu J, Wen K, Twitchell E, Bui T, Ramesh A, Weiss M, Li G, Samuel H, Clark-Deener S, Jiang X, Lee K, Yuan L (2016) Increased and prolonged human norovirus infection in RAG2/IL2RG deficient gnotobiotic pigs with severe combined immunodeficiency. Sci Rep 6:25222

Lillico SG, McGrew MJ, Sherman A, Sang HM (2005) Transgenic chickens as bioreactors for protein-based drugs. Drug Discov Today 10:191–196

Lillico SG, Sherman A, McGrew MJ, Robertson CD, Smith J, Haslam C, Barnard P, Radcliffe PA,

Mitrophanous KA, Elliot EA, Sang HM (2007) Oviduct-specific expression of two therapeutic proteins in transgenic hens. Proc Natl Acad Sci USA 104:1771–1776

Lu R, Zhang T, Wu D, He Z, Jiang L, Zhou M, Cheng Y (2018) Production of functional human CuZn-SOD and EC-SOD in bitransgenic cloned goat milk. Transgenic Res 27:343–354

Luke T, Wu H, Zhao J, Channappanavar R, Coleman CM, Jiao JA, Matsushita H, Liu Y, Postnikova EN, Ork BL, Glenn G, Flyer D, Defang G, Raviprakash K, Kochel T, Wang J, Nie W, Smith G, Hensley LE, Olinger GG, Kuhn JH, Holbrook MR, Johnson RF, Perlman S, Sullivan E, Frieman MB (2016) Human polyclonal immunoglobulin G from transchromosomic bovines inhibits MERS-CoV in vivo. Sci Transl Med 8:326ra21

Luo Y, Wang Y, Liu J, Cui C, Wu Y, Lan H, Chen Q, Liu X, Quan F, Guo Z, Zhang Y (2016) Generation of TALE nickase-mediated gene-targeted cows expressing human serum albumin in mammary glands. Sci Rep 6:20657

Maksimenko OG, Deykin AV, Khodarovich YM, Georgiev PG (2013) Use of transgenic animals in biotechnology: prospects and problems. Acta Naturae 5:33–346

Matsushita H, Sano A, Wu H, Jiao JA, Kasinathan P, Sullivan EJ, Wang Z, Kuroiwa Y (2014) Triple immunoglobulin gene knockout transchromosomic cattle: bovine lambda cluster deletion and its effect on fully human polyclonal antibody production. PLoS One 9:e90383

Matsushita H, Sano A, Wu H, Wang Z, Jiao JA, Kasinathan P, Sullivan EJ, Kuroiwa Y (2015) Species-specific chromosome engineering greatly improves fully human polyclonal antibody production profile in cattle. PLoS One 10:e0130699

McCreath KJ, Howcroft J, Campbell KH, Colman A, Schnieke AE, Kind AJ (2000) Production of gene-targeted sheep by nuclear transfer from cultured somatic cells. Nature 405:1066–1069

Mir B, Piedrahita JA (2004) Nuclear localization signal and cell synchrony enhance gene targeting efficiency in primary fetal fibroblasts. Nucleic Acids Res 32:e25

Monzani PS, Adona PR, Ohashi OM, Meirelles FV, Wheeler MB (2016) Transgenic bovine as bioreactors: challenges and perspectives. Bioengineered 7:123–131

Novaretti MC, Dinardo CL (2011) Immunoglobulin: production, mechanisms of action and formulations. Rev Bras Hematol Hemoter 33:377–382

Oishi I, Yoshii K, Miyahara D, Tagami T (2018) Efficient production of human interferon beta in the white of eggs from ovalbumin gene-targeted hens. Sci Rep 8:10203

Paleyanda RK, Velander WH, Lee TK, Scandella DH, Gwazdauskas FC, Knight JW, Hoyer LW, Drohan WN, Lubon H (1997) Transgenic pigs produce functional human factor VIII in milk. Nat Biotechnol 15:971–975

Park JK, Lee YK, Lee P, Chung HJ, Kim S, Lee HG, Seo MK, Han JH, Park CG, Kim HT, Kim YK, Min KS, Kim JH, Lee HT, Chang WK (2006) Recombinant human erythropoietin produced in milk of transgenic pigs. J Biotechnol 122:362–371

Parker MH, Birck-Wilson E, Allard G, Masiello N, Day M, Murphy KP, Paragas V, Silver S, Moody MD (2004) Purification and characterization of a recombinant version of human alpha-fetoprotein expressed in the milk of transgenic goats. Protein Expr Purif 38:1771–1783

Peng J, Wang Y, Jiang J, Zhou X, Song L, Wang L, Ding C, Qin J, Liu L, Wang W, Liu J, Huang X, Wei H, Zhang P (2015) Production of human albumin in pigs through CRISPR/Cas9-mediated knockin of human cDNA into swine albumin locus in the zygotes. Sci Rep 5:16705

Powell JS (2014) Lasting power of new clotting proteins. Hematology Am Soc Hematol Educ Program 2014:355–363

Rapp JC, Harvey AJ, Speksnijder GL, Hu W, Ivarie R (2003) Biologically active human interferon alpha-2b produced in the egg white of transgenic hens. Transgenic Res 12:569–575

Robl JM, Kasinathan P, Sullivan E, Kuroiwa Y, Tomizuka K, Ishida I (2003) Artificial chromosome vectors and expression of complex proteins in transgenic animals. Theriogenology 59:107–113

Ruan J, Li H, Xu K, Wu T, Wei J, Zhou R, Liu Z, Mu Y, Yang S, Ouyang H, Chen-Tsai RY, Li K (2015) Highly efficient CRISPR/Cas9-mediated transgene knockin at the H11 locus in pigs. Sci Rep 5:14253

Salter DW, Smith EJ, Hughes SH, Wright SE, Crittenden LB (1987) Transgenic chickens: insertion of retroviral genes into the chicken germ line. Virology 157:236–240

Sanchez-Garcia L, Martin L, Mangues R, Ferrer-Miralles N, Vazquez E, Villaverde A (2016) Recombinant pharmaceuticals from microbial cells: a 2015 update. Microb Cell Fact 15:33

Schilit SLP, Ohtsuka M, Quadros RM, Gurumurthy CB (2016) Pronuclear Injection-Based Targeted Transgenesis. Curr Protoc Hum Genet 91:15.10.1–15.10.28

Schnieke AE, Kind AJ, Ritchie WA, Mycock K, Scott AR, Ritchie M, Wilmut I, Colman A, Campbell KH (1997) Human factor IX transgenic sheep produced by transfer of nuclei from transfected fetal fibroblasts. Science 278:2130–2133

Schusser B, Collarini EJ, Yi H, Izquierdo SM, Fesler J, Pedersen D, Klasing KC, Kaspers B, Harriman WD, van de Lavoir MC, Etches RJ, Leighton PA (2013) Immunoglobulin knockout chickens via efficient homologous recombination in primordial germ cells. Proc Natl Acad Sci USA 110:20170–20175

Shepelev MV, Kalinichenko SV, Deykin AV, Korobko IV (2018) Production of recombinant proteins in the milk of transgenic animals: current state and prospects. Acta Naturae 10:40–47

Sheridan C (2016) FDA approves "farmaceutical" drug from transgenic chickens. Nat Biotechnol 34:117–119

Simons JP, McClenaghan M, Clark AJ (1987) Alteration of the quality of milk by expression of sheep beta-lactoglobulin in transgenic mice. Nature 328:530–532

Song S, Ge X, Cheng Y, Lu R, Zhang T, Yu B, Ji X, Qi Z, Rong Y, Yuan Y, Cheng Y (2016) High-level expression of a novel recombinant human plasminogen activator (rhPA) in the milk of transgenic rabbits and its thrombolytic bioactivity in vitro. Mol Biol Rep 43:775–783

Soto DA, Ross PJ (2016) Pluripotent stem cells and livestock genetic engineering. Transgenic Res 25:289–306

Suzuki S, Iwamoto M, Saito Y, Fuchimoto D, Sembon S, Suzuki M, Mikawa S, Hashimoto M, Aoki Y, Najima Y, Takagi S, Suzuki N, Suzuki E, Kubo M, Mimuro J, Kashiwakura Y, Madoiwa S, Sakata Y, Perry ACF, Ishikawa F, Onishi A (2012) Il2rg gene-targeted severe combined immunodeficiency pigs. Cell Stem Cell 10:753–758

Swift GH, Hammer RE, MacDonald RJ, Brinster RL (1984) Tissue-specific expression of the rat pancreatic elastase I gene in transgenic mice. Cell 38:639–646

Thomas KR, Capecchi MR (1987) Site-directed mutagenesis by gene targeting in mouse embryo-derived stem cells. Cell 51:503–512

Vajo Z, Fawcett J, Duckworth WC (2001) Recombinant DNA technology in the treatment of diabetes: insulin analogs. Endocr Rev 22:706–717

van Berkel PH, Welling MM, Geerts M, van Veen HA, Ravensbergen B, Salaheddine M, Pauwels EK, Pieper F, Nuijens JH, Nibbering PH (2002) Large scale production of recombinant human lactoferrin in the milk of transgenic cows. Nat Biotechnol 20:484–487

Van den Hout JM, Reuser AJ, de Klerk JB, Arts WF, Smeitink JA, Van der Ploeg AT (2001) Enzyme therapy for pompe disease with recombinant human alpha-glucosidase from rabbit milk. J Inherit Metab Dis 24:266–274

van Veen HA, Koiter J, Vogelezang CJ, van Wessel N, van Dam T, Velterop I, van Houdt K, Kupers L, Horbach D, Salaheddine M, Nuijens JH, Mannesse ML (2012) Characterization of recombinant human C1 inhibitor secreted in milk of transgenic rabbits. J Biotechnol 162:319–326

Wall RJ, Pursel VG, Shamay A, McKnight RA, Pittius CW, Hennighausen L (1991) High-level synthesis of a heterologous milk protein in the mammary glands of transgenic swine. Proc Natl Acad Sci USA 88:1696–1700

Walters EM, Agca Y, Ganjam V, Evans T (2011) Animal models got you puzzled?: think pig. Ann NY Acad Sci 1245:63–64

Wang H, Yang H, Shivalila CS, Dawlaty MM, Cheng AW, Zhang F, Jaenisch R (2013) One-step generation of mice carrying mutations in multiple genes by CRISPR/Cas-mediated genome engineering. Cell 153:910–918

Wang M, Sun Z, Yu T, Ding F, Li L, Wang X, Fu M, Wang H, Huang J, Li N, Dai Y (2017) Large-scale production of recombinant human lactoferrin from high-expression, marker-free transgenic cloned cows. Sci Rep 7:10733

Wang X, Cao C, Huang J, Yao J, Hai T, Zheng Q, Wang X, Zhang H, Qin G, Cheng J, Wang Y, Yuan Z, Zhou Q, Wang H, Zhao J (2016) One-step generation of triple gene-targeted pigs using CRISPR/Cas9 system. Sci Rep 6:20620

Wang Y, Fan N, Song J, Zhong J, Guo X, Tian W, Zhang Q, Cui F, Li L, Newsome PN, Frampton J, Esteban MA, Lai L (2014) Generation of knockout rabbits using transcription activator-like effector nucleases. Cell Regen (lond) 3:3

Wang Y, Zhao S, Bai L, Fan J, Liu E (2013) Expression systems and species used for transgenic animal bioreactors. Biomed Res Int 2013:580463

Watson CT, Breden F (2012) The immunoglobulin heavy chain locus: genetic variation, missing data, and implications for human disease. Genes Immun 13:363–373

West J, Gill WW (2016) Genome editing in large animals. J Equine Vet Sci 41:1–6

Wheeler MB, Walters EM (2001) Transgenic technology and applications in swine. Theriogenology 56:1345–1369

Wilmut I, Schnieke AE, McWhir J, Kind AJ, Campbell KH (1997) Viable offspring derived from fetal and adult mammalian cells. Nature 385:810–813

Woodfint RM, Hamlin E, Lee K (2018) Avian Bioreactor Systems: A Review. Mol Biotechnol 60:975–983

Wright G, Carver A, Cottom D, Reeves D, Scott A, Simons P, Wilmut I, Garner I, Colman A (1991) High level expression of active human alpha-1-antitrypsin in the milk of transgenic sheep. Biotechnology (NY) 9:830–834

Wu G (2022) Nutritionand metabolism: Foundations for animal growth, development, reproduction, and health.Adv Exp Med Biol 1354:1–24

Wu G, Bazer FW (2019) Application of new biotechnologies for improvements in swine nutrition and pork production. J Anim Sci Biotechnol 10:28

Wu H, Fan Z, Brandsrud M, Meng Q, Bobbitt M, Regouski M, Stott R, Sweat A, Crabtree J, Hogan RJ, Tripp RA, Wang Z, Polejaeva IA, Sullivan EJ (2019) Generation of H7N9-specific human polyclonal antibodies from a transchromosomic goat (caprine) system. Sci Rep 9:366

Yang D, Song J, Zhang J, Xu J, Zhu T, Wang Z, Lai L, Chen YE (2016) Identification and characterization of rabbit ROSA26 for gene knock-in and stable reporter gene expression. Sci Rep 6:25161

Yang D, Xu J, Chen YE (2019) Generation of Rabbit Models by Gene Editing Nucleases. Methods Mol Biol 1874:327–345

Yang D, Xu J, Zhu T, Fan J, Lai L, Zhang J, Chen YE (2014) Effective gene targeting in rabbits using RNA-guided Cas9 nucleases. J Mol Cell Biol 6:97–99

Yang L, Guell M, Niu D, George H, Lesha E, Grishin D, Aach J, Shrock E, Xu W, Poci J, Cortazio R, Wilkinson RA, Fishman JA, Church G (2015) Genome-wide inactivation of porcine endogenous retroviruses (PERVs). Science 350:1101–1104

Young LE, Sinclair KD, Wilmut I (1998) Large offspring syndrome in cattle and sheep. Rev Reprod 3:155–163

Yugo DM, Heffron CL, Ryu J, Uh K, Subramaniam S, Matzinger SR, Overend C, Cao D, Kenney SP, Sooryanarain H, Cecere T, LeRoith T, Yuan L, Jue N, Clark-Deener S, Lee K, Meng XJ (2018) Infection dynamics of hepatitis E virus in wild-type and immunoglobulin heavy chain knockout J(H) (-/-) gnotobiotic piglets. J Virol 92:e01208–18

Zakhartchenko V, Flisikowska T, Li S, Richter T, Wieland H, Durkovic M, Rottmann O, Kessler B, Gungor T, Brem G, Kind A, Wolf E, Schnieke A (2011) Cell-mediated transgenesis in rabbits: chimeric and nuclear transfer animals. Biol Reprod 84:229–237

Zetsche B, Heidenreich M, Mohanraju P, Fedorova I, Kneppers J, DeGennaro EM, Winblad N, Choudhury SR, Abudayyeh OO, Gootenberg JS, Wu WY, Scott DA, Severinov K, van der Oost J, Zhang F (2017) Multiplex gene editing by CRISPR-Cpf1 using a single crRNA array. Nat Biotechnol 35:31–34

Zhang XH, Tee LY, Wang XG, Huang QS, Yang SH (2015) Off-target Effects in CRISPR/Cas9-mediated Genome Engineering. Mol Ther Nucleic Acids 4:e264

Zhao J, Xu W, Ross JW, Walters EM, Butler SP, Whyte JJ, Kelso L, Fatemi M, Vanderslice NC, Giroux K, Spate LD, Samuel MS, Murphy CN, Wells KD, Masiello NC, Prather RS, Velander WH (2015) Engineering protein processing of the mammary gland to produce abundant hemophilia B therapy in milk. Sci Rep 5:14176

Zheng Q, Lin J, Huang J, Zhang H, Zhang R, Zhang X, Cao C, Hambly C, Qin G, Yao J, Song R, Jia Q, Wang X, Li Y, Zhang N, Piao Z, Ye R, Speakman JR, Wang H, Zhou Q, Wang Y, Jin W, Zhao J (2017) Reconstitution of UCP1 using CRISPR/Cas9 in the white adipose tissue of pigs decreases fat deposition and improves thermogenic capacity. Proc Natl Acad Sci USA 114:E9474–E9482

Zhu L, van de Lavoir MC, Albanese J, Beenhouwer DO, Cardarelli PM, Cuison S, Deng DF, Deshpande S, Diamond JH, Green L, Halk EL, Heyer BS, Kay RM, Kerchner A, Leighton PA, Mather CM, Morrison SL, Nikolov ZL, Passmore DB, Pradas-Monne A, Preston BT, Rangan VS, Shi M, Srinivasan M, White SG, Winters-Digiacinto P, Wong S, Zhou W, Etches RJ (2005) Production of human monoclonal antibody in eggs of chimeric chickens. Nat Biotechnol 23:1159–1169

15 Cows as Bioreactors for the Production of Nutritionally and Biomedically Significant Proteins

P. S. Monzani, P. R. Adona, S. A. Long, and M. B. Wheeler

Abstract

Dairy and beef cattle make a vital contribution to global nutrition, and since their domestication, they have been continuously exposed to natural and artificial selection to improve production characteristics. The technologies of transgenesis and gene editing used in cattle are responsible for generating news characteristics in bovine breeding, such as alteration of nutritional components of milk and meat enhancing human health benefits, disease resistance decreasing production costs and offering safe products for human food, as well as the recombinant protein production of biomedical significance. Different methodologies have been used to generate transgenic cattle as bioreactors. These methods include the microinjection of vectors in pronuclear, oocyte or zygote, sperm-mediate transgenesis, and somatic cell nuclear transfer. Gene editing has been applied to eliminate unwanted genes related to human and animal health, such as allergy, infection, or disease, and to insert transgenes into specific sites in the host genome. Methodologies for the generation of genetically modified cattle are laborious and not very efficient. However, in the last 30 years, transgenic animals were produced using many biotechnological tools. The result of these modifications includes (1) the change of nutritional components, including proteins, amino acids and lipids for human nutrition; (2) the removal allergic proteins milk; (3) the production of cows resistant to disease; or (4) the production of essential proteins used in biomedicine (biomedical proteins) in milk and blood plasma. The genetic modification of cattle is a powerful tool for biotechnology. It allows for the generation of new or modified products and functionality that are not currently available in this species.

Keywords

Transgenic cattle · Recombinant protein · Milk · Biopharmaceutics · Nutrition · Disease resistance · Transgenic methodologies

P. S. Monzani (✉)
Instituto Chico Mendes de Conservação da Biodiversidade/Centro Nacional de Pesquisa e Conservação da Biodiversidade Aquática Continental, Pirassununga, SP, Brasil

P. R. Adona
Saúde e Produção de Ruminantes, Universidade Norte do Paraná, Arapongas, PR, Brasil

S. A. Long · M. B. Wheeler
Departments of Animal Sciences and Bioengineering, University of Illinois at Urbana-Champaign, Urbana, IL, USA

M. B. Wheeler
Carl R. Woese Institute for Genomic Biology, University of Illinois at Urbana-Champaign, Urbana, IL, USA

G. Wu (ed.), *Recent Advances in Animal Nutrition and Metabolism*, Advances in Experimental Medicine and Biology 1354,
https://doi.org/10.1007/978-3-030-85686-1_15

15.1 Introduction

Cattle are an important nutritional source for humans. Cattle principally supply protein and functional nutrients from meat and milk (Wu 2020), but also provide lipids for butter, cheese, and oil (i.e., ghee in India) production (Rezaei et al. 2016). The breeding of cattle has economic impacts on agriculture, the food industry, the leather industry, meat processing, meat distributors, meat sales stores, and others (Bazer et al. 2020). The domestication of cattle began more than 10,000 years ago, and cattle have been continuously exposed to natural and artificial selection. Together with natural genetic drift, these processes have produced an array of breeds, phenotypes, and production characteristics (Magee et al. 2014). Cattle phenotypes selected by humans post-domestication include increased milk or meat production, high-quality protein production, docility, temperament, fertility, horns or absence of horns, and other relatively aesthetic traits (Zeder et al. 2006). The advance of technologies, principally in genetic engineering that provides tools for genetic modification of cattle, allows the delivery of new products for human needs. These tools have allowed scientists to specifically alter certain individual traits, which will impact the history of cattle domestication. Transgenic cattle have been generated for many different proposes including resistance to diseases, improved growth, alteration of meat and milk composition and quantity for nutritional, health and industrial needs, as well as bioreactors for the production of recombinant protein with biopharmaceutical proposes (Monzani et al. 2016; Wheeler 2007, 2013).

Several strategies have been applied to the generation of genetically modified cattle, including introducing new genes (knock-in), removing existing genes (knock-out), decreasing expression levels of existing genes (knock-down), editing of the sequence of existing genes (gene editing) as well as introducing artificial chromosomes into the genome of the cattle.

The knock-down of a gene was performed in cattle to silence that gene, thus removing its product that could pose risks to human health. Some examples of such modified cattle are the elimination of β-lactoglobulin, the primary allergen to humans in cow's milk (Sun et al. 2018) and the production of prion protein (C)-deficient cattle to circumvent spongiform encephalopathy in bovine and Creutzfeldt-Jakob disease in humans (Richt et al. 2007). New genes or modified genes have been introduced into cattle to alter milk composition for nutritional aspects and to humanize cow milk. Studies reported include the production of human α-lactalbumin (Wang et al. 2008), the alteration of casein composition in milk (Brophy et al. 2003), the production of lysine-rich proteins (Ma et al. 2019), and other studies presented in this review. Alteration in the composition of lipids in dairy and beef cattle has been reported (Wu et al. 2012). Cattle resistant to tuberculosis (Su et al. 2016) and mastitis (Wall et al. 2005) were generated, aiming to breed healthy cattle and increasing milk production.

The use of transgenic cattle as bioreactors to produce biopharmaceutical proteins for humans is an innovative and vital advance in applying biotechnology to human health. Because of the large amount of milk and blood that cows produce, these fluids are the primary vehicles to produce recombinant proteins in cattle. The average amount of milk produced by a dairy cow daily is about 7.5 gallons or 29 L, which translates to ~ 2,387 gallons or 8,890 L per year. Furthermore, the mammary gland can perform post-translational modifications of proteins necessary for the biological activity, including biopharmaceuticals proteins. Examples of transgenic cattle that have been generated to produce biopharmaceutical proteins in milk include cattle that produce (1) tissue-type plasminogen activator (An et al. 2004); (2) human recombinant growth hormone (Salamone et al. 2006); (3) recombinant human albumin (Echelard et al. 2009); and (4) recombinant anti-CD20 monoclonal antibody (Zhang et al. 2018). Transgenic cattle with an altered blood chemistry profile were produced by inserting a human artificial chromosome designed to express a bispecific single-chain antibody in their serum (Grosse-Hovest et al. 2004). In this review, we will discuss the main techniques to generate transgenic cattle and the biotechnological applications of transgenic cattle.

15.2 Driving Recombinant Protein Expression in Cattle

Cow's milk contains approximately 3.5% total protein, of which 80% are caseins and 15% whey proteins. Milk also contains vitamins, carbohydrates, lipids, and minerals; it is also one of the richest sources of dietary calcium. Casein and whey proteins constitute the main groups of proteins in milk. Casein includes four types, αs1-casein (CSN1S1), αs2-casein (CSN1S2), β-casein (CSN2), and κ-casein (CSN3) (Farrell et al. 2004). Whey proteins are composed of primarily α-lactalbumin (LALBA) and β-lactoglobulin (LGB) but also include other soluble protein fractions, such as serum albumin, immunoglobulins, lysozyme, lactoperoxidases, and lactoferrin among others (Farrell et al. 2004; Gupta et al. 2016). The mammary gland of the cow has been considered an important bioreactor for recombinant protein expression due to the high protein quantity in cow's milk, which along with the cow's high daily milk volume output, makes it an ideal "factory" for proteins.

Genetic modification in cattle involves two basic strategies, (1) loss of function (the removal, inhibition (down-regulation) or silencing of a specific gene/gene product) and (2) gain of function (the overexpression of endogenous genes (up-regulation), introduction of new exogenous DNA, or increasing multiple copies of a specific gene. These strategies may be used for specific sequence or random DNA insertion into the cattle genome. For tissue-specific expression of recombinant proteins, the exogenous DNA (transgene) must be inserted under the control of specific promoter regions that drive the expression into the desired specific tissue. Promoter sequences from several mammary-specific genes have been used in transgenic cattle to drive recombinant milk protein expression. In some cases, the whole gene region was used to direct protein production in milk (Bleck et al. 1998).

Due to the high concentrations of casein proteins in milk, the β-casein promoter has been used extensively to drive protein production in transgenic cattle. Using the β-casein promoter, transgenic cattle have been generated to produce milk that contains an altered casein composition. These animals harbor additional copies of β-casein and κ-casein (Brophy et al. 2003). The bovine β-casein promoter has been used to drive the production of recombinant human growth hormone (rhGH) (Salamone et al. 2006); human serum albumin (Luo et al. 2015); and human lysozyme using zinc finger nucleases (Liu et al. 2014) into transgenic cow milk. The goat β-casein promoter has also been employed to drive transgenic protein production into cow milk. These recombinant proteins include human serum albumin (Echelard et al. 2009), human lactoferrin, and lysozyme (Wang et al. 2017a; Yang et al. 2008) and recombinant anti-CD20 monoclonal antibody (Zhang et al. 2018). A commercially available mammary gland tissue-specific expression vector (pBC1) containing the goat β-casein promoter has also been used to introduce human lysozyme in the milk of transgenic cattle (Yang et al. 2011). Recently, the pBC1 vector carrying a lysine-rich gene was injected directly into the mammary glands of lactating cows, increasing the lysine level in the milk (Ma et al. 2019). In addition to the β-casein promoter, transgenic cattle have been generated using the bovine αS1-casein promoter to drive human lactoferrin into milk (van Berkel et al. 2002).

Promoters driving milk whey protein genes have also been used to produce transgenic cattle expressing recombinant proteins in milk. Transgenic cows secreting lysostaphin, a protein that enhances mastitis resistance, were generated using the ovine β-lactoglobulin promoter (Wall et al. 2005). Further, the entire genomic human *α-lactalbumin* gene, including the promoter, was inserted into the genome of cattle to produce recombinant human α-lactalbumin in milk (Wang et al. 2008). Transgenic cloned cows efficiently expressed human lysozyme, using the human lysozyme promoter (Yang et al. 2008), and human transferrin using the human lactoferrin promoter (Wang et al. 2017a) in milk. Wang et al. (2017b) generated transgenic cattle that produced active recombinant human bile salt-stimulated lipase in milk using the lactoferrin promoter. Whey protein promoters from cattle have been effectively used to generate transgenic animals of several other species that produce recombinant proteins in milk. These promoters include bovine α-lactalbumin used to create transgenic swine (Bleck et al. 1998) and the β-lactoglobulin promoter used to produce transgenic mice (Hyttinen et al. 1998). The mammary-specific gene promoters provide great utility for targeting recombinant proteins into milk.

Other tissue-specific promoters have also used to drive recombinant expression into different tissues in cattle. Transgenic cattle resistant to tuberculosis were produced expressing human β-defensin 3 in the lung alveolar epithelial cells and macrophages under control of bovine MUC1 promoter (Su et al. 2016). Further, it was confirmed that these cells secreted recombinant protein and possessed anti-mycobacterial capacity (Su et al. 2016). Some additional promoter methods used to produce transgenic cattle include (1) the site-specific knock-in of the transcription activator-like effector nickase (TALEN)-mediated SP110 nuclear body protein gene (*SP110*) via homologous recombination; (2) the bovine endogenous macrophage scavenger receptor 1 (MSR1) promoter used to direct mouse SP110 expression and express SP110 only in bovine macrophages (Wu et al. 2015); (3) the bovine natural resistance-associated macrophage protein-1 (NRAMP1) gene, including promoter region was used to generate transgenic cattle resistant to tuberculosis (Gao et al. 2017); (4) the production of human polyclonal antibodies in bovine plasma using a human artificial chromosome vector carrying the entire unrearranged human immunoglobulin heavy and kappa light chain loci to bovine fibroblasts (Kuroiwa et al. 2009); (5) the production of a bispecific antibody r28M to target human CD28 and the melanoma/glioblastoma-associated cell surface chondroitin sulfate proteoglycan 4 in bovine blood under control of murine kappa promoter (Spiesberger et al. 2015); (6) the use of a lentiviral vector expressing a short hairpin RNA (shRNA) to silence myostatin using the EF1-α promoter gene (Tessanne et al. 2012); and finally, (7) the production of transgenic cattle constitutively expressing a n-3 fatty acid desaturase using the CMV promoter (Cheng et al. 2015; Wu et al. 2012). Other regulatory elements, such as insulators, enhancers, suppressors, non-coding exons/introns, 5' and 3' UTRs, matrix attachment regions (MAR), have been added in the vector construction to ensure high-level and position-independent expression of the transgene (Liu 2013). These examples illustrate that there are many potential promoter and gene sequences that can be utilized to produce tissue specificity during the generation of transgenic cattle.

15.3 Methods for Generating Transgenic Cattle

To generate a transgenic cow, many different laboratory methodologies and skills are required. These methods and skills encompass a wide variety of disciplines including reproductive biology (in vitro embryo culture, embryo transfer, estrus synchronizing, obstetrics and post-natal care), genetic engineering (vector construction, transgene introduction, transgene analysis, gene editing, animal screening, and protein expression), cell culture (cell culture, cell synchronization, somatic cell nuclear transfer, and transgene analysis), and biochemistry (protein purification, protein structural/functional analyses, and histology). Because of the long gestation length of cattle, the number and care required for the surrogate dams (recipients), the low rates of live births, and the extended period needed to evaluate/analyze the resulting animals make the generation of transgenic cattle a laborious and very costly process. This section will focus broadly on the methods that have been used to produce transgenic cattle.

Three general methodologies have been used for gene transfer to produce stable genetic modifications in cattle. These methods are oocyte/zygote microinjection, sperm mediated DNA transfer, and somatic nuclear cell transfer (SCNT). Different vectors can be used for the construction of transgenes. The vectors successfully used to generate transgenic cattle were plasmid, artificial chromosomes, and lentiviral systems. Those vector systems were discussed in the previous section.

15.3.1 Pronuclear Microinjection

The first successful production of transgenic cattle was the result of pronuclear microinjection, or the injection of exogenous DNA into the pronucleus of a living cell, most typically a developing zygote. Early uses of microinjection had a very low success rate for the integration of the injected transgene into the host DNA of the zygote. Krimpenfort and his collaborators introduced a construct of human lactoferrin transgene via pronuclear injection, which was driven by the αS1-casein promotor, to produce animals secreting the recombinant protein in their milk (Krimpenfort et al. 1991). In this study, 2,500 oocytes were harvested, 19% of the microinjected embryos developed to the morula/blastocyst stage, 21 pregnancies were produced, and 19 calves were born. The analysis of the 19 calves born confirmed that only two of the calves contained the human lactoferrin gene (Krimpenfort et al. 1991). The efficiency of producing these two calves from 2,500 oocytes was 0.008%, illustrating that the low efficiency of pronuclear microinjection with random integration is very low (Krimpenfort et al. 1991). During zygote development, the metaphase II (MII) stage does not have a nuclear envelope and is in arrest for a more extended period compared to other stages of the cell cycle. By using the developmental anatomy of the phospholipid components of the cell membrane, higher integration of foreign DNA can be achieved (Chan et al. 1998). Techniques that Chan and collaborators developed produced higher transfection rates compared to standard microinjection, though the rates were still very low (Chan et al. 1998). Bovine oocytes injected with lentivirus vectors tagged with a green fluorescent protein (GFP) resulted in visual confirmation of integration and high efficiency of integration (83% in the blastocyst and 100% in live offspring). The use of lentiviral gene transfer when coupled with the Chan technique of injecting into the MII oocyte, lead to increased transfection and production of transgenic cattle (Hofmann et al. 2004). Alternatively, microinjection of lentiviral particles containing short interfering hairpin RNAs, targeting myostatin into the perivitelline space of bovine zygotes was used to produce transgenic calves with myostatin silenced (Tessanne et al. 2012).

Microinjection of bacterial artificial chromosomes (BAC) containing the human lactoferrin gene, into bovine fibroblast before SCNT, successfully produced a transgenic cow whose milk contained recombinant human lactoferrin. BACs are notoriously difficult to transfer into cells via electroporation or lipofection due to their large size; however, microinjection of the BAC into fibroblast cells has proved useful (Yang et al. 2008). The use of active and/or specific site gene transfer methodology combined with cytoplasmic or pronuclear microinjection has improved the microinjection methodology. An example of this is the ablation of an allergen. Milk allergy is a significant problem for many people and is most commonly caused by bovine β-lactoglobulin (BLG), an allergenic protein in cow's milk. The LGB gene of cattle produces bovine β-lactoglobulin. Cytoplasmic co-injection of TALENs with homology-directed repair (HDR) templates to disrupt the LGB gene, successfully knocked-out the *LGB* gene and prevented BLG production in the milk (Wei et al. 2018).

A non-viral method using active integration by a transposase is another important tool being used with cytoplasmic microinjection. The Sleeping Beauty and PiggyBac transposon plasmids with a green fluorescent protein (GFP) construct were microinjected and efficiently produced transgenic cattle (Yum et al.

2016). The transgenic animals were able to transmit the transgene into the germline illustrated by the production of GFP in oocytes and sperm (Yum et al. 2018). Furthermore, these animals secreted GFP in their milk.

As these examples illustrate, the use of microinjection, both pronuclear and cytoplasmic, continues to be a valuable tool and workhorse for the production of transgenic livestock. This technique has stood the test of time and will continue to be of significant utility.

15.3.2 Sperm-Mediated Transgenesis

Sperm-mediated transgenesis is another technique that has been used to generate transgenic cattle. Modified bovine sperm were produced by restriction enzyme-mediated insertion (REMI) of linearized plasmids and the corresponding restriction enzyme for integration into the sperm genomic DNA (Shemesh et al. 2000). The modified sperm was used in vitro fertilization and morulae expressing a GFP reporter gene produced calves expressing the GFP transgene in their lymphocytes (Shemesh et al. 2000). This method is less technologically demanding than microinjection or SCNT, but the disadvantage is that mosaicism occurred using this technique. One alternatively for using sperm as a vector for transgenesis is to combine intracytoplasmic sperm injection (ICSI) with recombinases or transposases, such as the PiggyBac transposon, to increase transfection efficiency (Shinohara et al. 2007). Although successful, this method has not been widely used to produce transgenic cattle.

15.3.3 Somatic Cell Nuclear Transfer (Cloning)

Somatic cell nuclear transfer (SCNT) is considered a more efficient technique for the generation of transgenic cattle. DNA microinjection into zygotes followed by embryo transfer into a recipient is a less technologically demanding procedure. However, in cattle, this approach results in severe mosaicism and has low efficiency. To overcome these issues, SCNT was introduced and successfully used to produce cloned transgenic cattle (Brophy et al. 2003). This methodology represents a significant advancement but has a low efficiency concerning the production of live offspring. The primary reason appears to be abnormal reprogramming of the embryonic/fetal genome (Ogura et al. 2013; Whitworth and Prather 2010). The gene targeting efficiency in donor cells for SNCT can be improved using new technologies of gene editing (Yum et al. 2018).

Different methods of cell modification, before SNCT, have been used successfully to insert the transgene into the host genome and produce transgenic cattle. These methods include (1) the use of linearized plasmids containing Cre-recombinase and human recombinant lysosome (Lu et al. 2016); (2) the use of a human artificial chromosome vector carrying the entire unrearranged, human immunoglobulin heavy and κ-light chain loci; (3) the use of purified microcells, from CHO cells, containing a human artificial chromosome with human polyclonal antibodies (Kuroiwa et al. 2009); (4) the use of lentiviral vectors carrying human factor IX cDNA and β-casein promoters (Monzani et al. 2013); and (5) the use of recombinant phiC31 integrase mRNA, without lentivirus, but with a human *β-defensin-3* donor plasmid (Yu et al. 2013) or a donor plasmid encoding human serum albumin and bovine β-casein promoter (Luo et al. 2015).

The development of the clustered regularly interspaced short palindromic repeats (CRISPR) and CRISPR associated protein 9 (Cas9), known as CRISPR/Cas9, and other gene-editing systems have facilitated the transfer of transgenes into the genome of livestock (Zhao et al. 2019). CRISPR/Cas9 has been used to knock-in of human fibroblast growth factor 2 (FGF2) into exon 3 of the bovine β-casein gene (Jeong et al. 2016). In another study, a single Cas9 nickase was used to introduce a single-strand break for the insertion of the natural resistance-associated macrophage protein-1 gene to confer increased resistance to tuberculosis (Gao et al. 2017). A combination of gene editing and SCNT can facilitate the generation of genetically engineered animals. One example of this is the use of TALEN-mediated microhomologous-mediated end-joining (MMEJ) to insert a lactase gene into a bovine β-casein locus. The vectors were introduced into bovine fibroblasts that were used for SNCT and resulted in one newborn that ultimately expressed lactase in milk to address lactose intolerance (Su et al. 2018).

Despite the low efficiency of production, transgenic cattle have been generated for several different biotechnological applications and can be considered a potential tool for the production of recombinant proteins in milk, plasma, and meat.

15.4 Biotechnological Applications of Transgenic Cattle

The use of biotechnology to produce transgenic cows has resulted in important advances in the nutritional composition of bovine milk and meat, resistance to disease, and biomedically significant protein production. In addition to livestock, production traits, transgenic cattle can be a tool for the industrial world as "factories" for recombinant proteins, textiles, biofuels, and biopharmaceuticals. Transgenic milk can be manipulated to be more nutritious for humans than natural bovine milk by changing the concentration of the major nutrients: protein (Krimpenfort et al. 1991; van Berkel et al. 2002; Wheeler 2007), fat (Cheng et al. 2015; Lai et al. 2006; Wu et al. 2012), and lactose (Su et al. 2018) while also removing allergenicity components such as BLG (Sun et al. 2018). Further, alterations of the lipid composition of milk and meat have been reported to have beneficial effects on human health (Lai et al. 2006; Świątkiewicz et al. 2015). The development of "medicinally" altered milk composition has important implications in human health and treatment(s). By utilizing the cow's natural ability to produce large amounts of proteins and her capacity to perform post-translational modifications of that protein has made transgenic cattle an attractive system to produce recombinant biopharmaceutical proteins at relatively low cost. The use of transgenic cattle to improve the nutritional composition of milk and meat, to enhance disease resistance, and to produce recombinant biopharmaceuticals is summarized in the following section.

15.4.1 Transgenic Cattle: Modification of the Nutritional Composition of Bovine Milk and Meat

The quality and composition of milk often bring premium pricing in the marketplace. This results in more money in the producer's pocket when the milk contains certain requirements, depending on the buyer. The cheese industry puts premiums on milk that has high protein content, specifically caseins, since that is the portion of milk used to make cheese. The larger percent of casein in the milk, the higher the premium. Transgenic cattle with additional copies of the β- and κ-casein gene produced milk with a 16% increase (8–20%) in β-casein and a two-fold increase in κ-casein (Brophy et al. 2003). The stability, size, and structure of the casein micelle and the ratio of the four types of casein greatly influence the production of cheese. Different ratios can decrease the micellar size, which leads to the increased thermal stability that is necessary for the production of cheese. Gene modification to achieve a desirable casein ratio can lead to increased milk value for the production of cheese (Kang et al. 1986).

Bovine milk is a world-wide nutrient source for humans, though the nutrient content between human milk and cow's milk is marginally different. α-Lactalbumin is a Ca^{2+}-binding whey protein that modifies β-1,4-galactosyltransferase to produce lactose in the mammary gland. The concentration of LALBA in bovine milk is typically 3.5%, whereas human concentrations reach upward of 25%. The LALBA in human milk contains a high proportion of nutritionally essential amino acids, specifically tryptophan, a precursor of serotonin, a chemical needed for proper development of young children (Layman et al. 2018; Nielsen et al. 2020). LALBA's combination of amino acids, along with other proteins and nutrients, stimulates mucus production in infants, preventing gastrointestinal infections (Ushida et al. 2007). Besides, LALBA has been shown to induce apoptosis in tumor cells (Fischer et al. 2004) and decreases lesions when used as a treatment for skin papillomas (Gustafsson et al. 2004). To humanize bovine milk, recombinant human α-lactalbumin (rhLALBA) was produced in transgenic cattle at concentrations up to 1.55 g/L (Wang et al. 2008). The rhLALBA showed similar physicochemical properties to the native protein but without the N-glycosylation, while the endogenous bovine α-lactalbumin was glycosylated (Wang et al. 2008). These results provide evidence that it is possible to humanize cow's milk.

The alteration of the amino acid composition in milk for enriched nutritional values is an important use for transgenic cattle. One example of this is lysine. Lysine is considered one of the most important essential amino acid for human nutrition because it is the most limiting in cereals grains (Baker 2007; Wu 2022). To address this issue, a vector, with goat β-casein promoter, carrying the gene for a lysine-rich protein was injected directly into the lactating mammary glands of cows. The resulting milk samples contained the lysine-rich protein indicating that the gene was successfully expressed (Ma et al. 2019). While in this case, only the mammary gland was transgenic and not the entire cow; it provides evidence that if the transgenic cow is produced, it should provide a system to increase the lysine content of milk (Ma et al. 2019).

Human and bovine milk produce different quantities of bioactive proteins (Lönnerdal 2013). Some bioactive proteins are highly expressed in human milk and weakly produced in cow milk (Zhang et al. 2017). Therefore, the alteration of cow milk composition to increase the concentration(s) of beneficial human bioactive proteins or nutraceuticals as a "better" substitute for human milk is a viable application of transgenic cattle. Two such bioactive proteins, lactoferrin and lysozyme, are antimicrobial and immunomodulatory proteins produced in high quantities in human milk and have beneficial effects when supplemented in human and animal diets. Cattle produce low levels of lactoferrin and lysozyme in milk, and genetically engineered cattle that produce recombinant human lactoferrin and human lysozyme in transgenic bovine milk have been reported (Lu et al. 2016; van Berkel et al. 2002; Yang et al. 2008).

Human lysozyme is an important natural immune protein with bactericide action that protects human infants from microbial infections. Due to the medicinal value and market demand for human lysozyme, the large-scale production of recombinant human lysozyme has been performed. Recombinant human lysozyme (rHLZ) was successfully cloned into a bovine β-casein promoter and produced transgenic calves. When mature, the transgenic cattle expressed milk with rHLZ levels of 25.96 mg/L (Yang et al. 2011).

Lactoferrin is an iron-binding glycoprotein involved in innate host defense against infection and excessive inflammation being beneficial multifunctional proprieties. Human recombinant lactoferrin has been produced at gram/liter concentrations in transgenic bovine milk. Natural and recombinant human lactoferrin showed identical iron-binding and iron-release properties and differences in N-linked glycosylation but were equally effective in three different in vivo infection models (van Berkel et al. 2002). Using the microinjection of bacterial artificial chromosomes (BACs) containing human lactoferrin, a transgenic cow was produced that provided high levels of human lactoferrin (2.5–3.4 g/L) in milk. This lactoferrin had a similar pattern of glycosylation and proteolytic susceptibility, as well as the iron-binding and releasing properties to the natural human counterpart (Yang et al. 2008). Another research group with the goal of large scale hLF production also utilized BACs carrying hLF but without the marker gene. They produced human lactoferrin in transgenic cattle that had similar enzymatic properties compared to wild type-LF and concentrations ranging from 4.5 to 13.6 g/L, offering the ability to purify hLF on a large scale with less than 100 L of milk a day (Wang et al. 2017a).

There have been several groups that have examined the chemical and physiological properties of the recombinant hLF from the line of transgenic hLF cattle produced by van Berkel and colleagues (van Berkel et al. 2002). One group analyzed the crystal structure of recombinant hLF expressed in milk and confirmed the structural integrity of the recombinant protein, validating this transgenic cow's mammary gland was a vehicle to produce recombinant human proteins (Thomassen et al. 2005). Another group studied the expression of recombinant hLF in the transgenic cow milk during the three months post-calving and showed the expression was fairly constant at about 2.9 mg/mL (Hyvönen et al. 2006a). Further, they showed that the native bovine lactoferrin decreased rapidly during the first days of lactation. The sustained high-level hLF also might suggest a role for hLF in improving disease resistance in the mammary gland of dairy cows by enhancing the content of recombinant lactoferrin the milk (Hyvönen et al. 2006a). This same research group looked at this aspect by experimentally infecting the mammary gland of hLF cows with *Escherichia coli* to evaluate the effects of the recombinant hLF in a mastitis model (Hyvönen et al. 2006b). They found that transgenic cows expressing recombinant hLF in their milk were not protected from *Escherichia coli*-induced mastitis. These results showed that recombinant hLF was not an efficient protein for increasing mastitis resistance of dairy cows (Hyvönen et al. 2006b).

The molecular and chemical structure of recombinant human lactoferrin and natural human lactoferrin need to be the same or incredibly similar for large scale production. A marker for the similarity in proteins is the glycosylation profile since glycan moieties play a role in the biological function of glycoproteins like lactoferrin. Two groups have studied the glycosylation patterns of the recombinant hLF from the van Berkel and Yang lines of hLF transgenic cattle (van Berkel et al. 2002; Yang et al. 2008). In the Yang cattle, the recombinant hLF had N-glycans of the high mannose-, hybrid-, and complex-type structures, with less N-acetylneuraminic acid fucose. This is in contrast to native human lactoferrin where the N-glycans consist of highly branched, sialylated, and fucosylated complex-type structures. These results (Yu et al. 2010) are consistent with the hypothesis that glycosylation patterns are species- and tissue/cell-specific (Raju et al. 2000). The other research group, studying the van Berkel line of hLF cattle, used Nano-LC-Chip-Q-TOF Mass Spectrometry and showed that recombinant human lactoferrin N-glycans share more similarities to bovine lactoferrin than human lactoferrin (Parc et al. 2017).

Alongside the humanization of bovine milk, human lactoferrin can be beneficial to other areas of animal production, specifically swine production. Bovine transgenic milk containing recombinant hLF has been shown to act as a modulator of intestinal flora in the neonatal pig enhancing the average daily weight gain. When cow's milk with rhLF was fed to neonatal pigs, the intestinal flora, microbial diversity, and daily weigh gain all increased (Hu et al. 2012). Piglets also had reduced incidences of diarrhea, enhanced antibody production (Li et al. 2014), and desirable gastrointestinal changes (Cooper et al. 2014, 2013; Li et al. 2014). One of these groups suggested that the combination of lactoferrin and lysozyme may have additive effects (Cooper et al. 2014). Subsequently, transgenic cattle co-expressing both hLF and rHLZ were produced (Kaiser et al. 2017). This resulted in the production of both recombinant proteins in colostrum and milk. However, the proteins were expressed at lower levels than in the transgenic cattle previously reported with only a single hLF or rhLZ transgene. This ability to introduce multiple transgenes into a single animal is an important step in the production of humanized milk (Kaiser et al. 2017).

Before humans can consume transgenic milk, the safety of such genetically modified food must be examined. The first step in such an assessment is a toxicity evaluation. Recently, a study to explore the toxicity of recombinant hLF was conducted. The first phase studied acute oral toxicity by feeding hLF to rats. The second phase examined the chronic toxicity of recombinant hLF by feeding it to mice for 90 days. Following the observation period, no toxic effects were reported and concluded that there was no food safety risk (Dai et al. 2018).

Transgenic cattle technology can also be applied to increase animal protein production in other tissues such as meat. The alteration of meat composition to increase nutritional components may allow for the generation of healthier food products. Myostatin is a negative regulator of muscle growth, and mutations in this gene result in an enhanced muscling phenotype, a desirable agricultural trait (Bennett et al. 2018). The production of transgenic offspring capable of stable myostatin suppression in vivo by shRNA targeting myostatin have been reported (Tessanne et al. 2012). Suppression of the myostatin gene via shRNA leads to an increase in muscle mass, and thus, meat production while the negative effects of myostatin silencing are decreased (Tessanne et al. 2012).

The alteration in the lipid composition of bovine milk and meat is another important application of transgenic cattle. Milk production includes a wide range of lipids that provide positive health benefits like oleic acid, conjugated linolenic acid, omega-3 fatty acids, and a variety of fatty acid chains. Modification of the lipid composition in milk along with meat can increase the benefits of animal product consumption. The increase in n-3 polyunsaturated fatty acids in the diet is associated with the reduction of the risk of major diseases, including cardiovascular diseases (Innes and Calder 2020), rheumatoid arthritis (Lee and Park 2013), diabetes (Tárnok et al. 2015), and cancer (Küllenberg de Gaudry and Massing 2014).

Mammals lack the n-3 fatty acid desaturase gene that converts omega-6 polyunsaturated fatty acids into n-3 polyunsaturated fatty acids; therefore, they cannot synthesize such long-chain n-3 polyunsaturated fatty acids as α-linolenic acid, eicosapentaenoic acid, and docosahexaenoic acid, which need to be obtained through their diet. The generation of transgenic cattle makes it possible to alter animal products that are rich in fatty acids. Transgenic cattle expressing n-3 fatty acid desaturase from *Caenorhabditis elegans* were observed to possess increased levels of n-3 polyunsaturated fatty acids in tissue and milk with decreased levels of n-6 polyunsaturated fatty acids (Wu et al. 2012). These investigators also showed that there was a significant reduction in the n-6/n-3 fatty acid ratios in both the tissue and milk (Wu et al. 2012). The expression of n-3 fatty acid desaturase, in transgenic cattle, altered the lipid metabolism, which resulted in cholesterol and triglyceride levels that were significantly decreased when compared to WT cattle (Liu et al. 2015).

Alteration of the lipid composition of animal products from transgenic cattle may be an exciting alternative to supply foods with additional health benefits. These findings may open up unlimited possibilities of modifying milk and meat composition to make it more supportive of human health and improve the functional properties of animal products.

15.4.2 Transgenic Cattle for Disease Resistance to Improve Animal Health and Increase Food Safety of Milk and Meat

The well-being and safety of the animals in cattle operations is a top priority for both the producer and the consumer. However, cattle are susceptible to disease, which leads to production loss and costly treatments. Studies are underway to help increase the well-being of cattle in livestock production systems through genetic engineering to offer additional options to traditional therapies. As an example, if cattle are more resistant to an infectious disease, farmers might be able to reduce the use of therapeutic antibiotics. This could benefit both animal and human health by potentially decreasing the opportunity for the development of antibiotic-resistant microbes.

One of the most costly diseases affecting dairy cattle is mastitis, with the yearly impact reaching billions of dollars (Aghamohammadi et al. 2018). Mastitis is caused by bacterial infection of the mammary gland, with the major causative organism being *Staphylococcus aureus* (*S. aureus*). Antimicrobial therapy can control *S.*

aureus, but a cure is difficult due to bacterial resistance and bacterial persistence in soil, water, and feces. Genetic engineering may provide a solution by producing transgenic dairy cattle with increased resistance to mastitis (Cardoso et al. 2019). Two examples of how this could work have been provided by transgenic cattle that produce either lysozyme (Liu et al. 2014) or lysostaphin in their milk (Liu et al. 2014; Wall et al. 2005). Human lysozyme is an antimicrobial protein, and lysostaphin is a peptidoglycan hydrolase naturally occurring in nature. Both have been recognized for their defense against bacterial infection, specifically *S. aureus* in mastitis infections. Transgenic cows expressing human lysozyme, at concentrations of 23–31 μg/mL, and lysostaphin, at levels of 0.9–14 μg/mL, in their milk have been produced. The milk from rhLZ cattle was shown to kill *S. aureus* in vitro (Liu et al. 2014). While the milk from the lysostaphin transgenic cattle also killed *S. aureus* in vitro, these investigators showed, besides, that lysostaphin transgenic cattle had reduced colonization of *S. aureus* when it was infused into the mammary gland (Wall et al. 2005). Further, there was no increase in milk somatic cells, no elevated body temperatures, and no induction of acute-phase proteins that were all observed in the control cattle. Both of these studies show that it may be possible to enhance resistance to disease and improve cattle's health and welfare by using genetic engineering.

Another economically important disease is foot and mouth disease virus (FMDV). Infection with FMDV results in weight loss, loss in mobility, decreased productivity, and reduced meat and milk production (Knight-Jones and Rushton 2013). Wang and colleagues developed a strategy whereby recombinant lentiviruses containing a short hairpin RNA (shRNA) were constructed against VP4 of FMDV (lenti-RNAi-VP4). The lenti-RNAi-VP4 in the transfected bovine fetal fibroblasts cells suppressed the production of the viral gene fusion protein. The lenti-RNAi-VP4 inhibited 98% of viral replication, demonstrating that this specific shRNA may have the potential for FMDV treatment and prevention (Wang et al. 2012).

Resistance to two other important diseases in cattle has also been investigated using the transgenic approach. These diseases are tuberculosis caused by *Mycobacterium Bovis (M. Bovis)* and bovine spongiform encephalopathy (BSE). Both diseases currently have no effective programs in place to eliminate or control the disease progression. Wu and colleagues employed a site-specific knock-in of a TALEN mediated SP110 nuclear body protein gene (Wu et al. 2015). This construct redirected *M. Bovis* to turn on apoptosis of the infected cells rather than necrosis once the infection occurred. The SP110 transgenic cattle were successfully able to control the growth and multiplication of *M. Bovis*, making them more resistant to tuberculosis than wild-type cattle. A different research group tried to address *M. Bovis* infection by introducing human β-defensin 3 controlled by the MUC1 promoter (Su et al. 2016). They showed the secretion of recombinant human β-defensin 3 protein, which exhibited anti-mycobacterial properties (Su et al. 2016). A third group used the CRISPR-Cas9 system to insert the natural resistance-associated macrophage protein-1 gene into cattle to introduce resistance to infection from *M. Bovis* (Gao et al. 2017).

Bovine spongiform encephalopathy (BSE) in cattle and Creutzfeldt-Jakob disease (CJD) in humans are diseases caused by aberrant folding of the normal prion proteins that exist in cells. Transgenic cattle were produced with the prion protein knocked out by sequential gene targeting events in cells before SCNT (Richt et al. 2007). This prion protein (C)-deficient cattle were shown to be "clinically, physiologically, histopathologically, immunologically and reproductively normal" (Richt et al. 2007). The generation of prion-free cattle opens the door to medical and industrial materials without contamination by prion proteins. Further, it has produced cattle that are resistant to BSE.

15.4.3 Transgenic Cattle as Bioreactors to Produce Biopharmaceuticals

The enormous need for biopharmaceuticals and the disparity in production speed have led the industry to explore alternative sources for these products. One possible solution is the use of transgenic animals, such as cattle, for the large-scale production of biopharmaceuticals (Monzani et al. 2016). However, the current state-of-the-art for producing transgenic cattle is fraught with low efficiencies and extended time frames. To capture the advantages offered by this methodology, the efficiency of producing transgenic cattle needs to be significantly improved. Transgenic cattle may be used to generate therapeutic proteins, including plasma proteins, monoclonal antibodies, and vaccines. Biopharmaceuticals from transgenic mammary glands have been produced, approved, released, and are commercially available for treating human diseases, they are antithrombin (ATryn®), used to prevent thromboembolic events in hereditary antithrombin deficient patients, and C1-esterase inhibitor (Ruconest®) to treat hereditary angioedema (HAE). Other biopharmaceuticals have been produced by diverse biopharmaceuticals companies using transgenic animals in the milk of transgenic animals (Niemann et al. 2012).

Several other recombinant human therapeutic proteins have been produced in transgenic cow milk. These proteins include (1) the production of tissue-type plasminogen activator (An et al. 2004); (2) the expression of recombinant hGH (Salamone et al. 2006); (3) the generation of recombinant human albumin (Echelard et al. 2009; Luo et al. 2015); (4) the production of bile salt-stimulated lipase for treatment of patients with fat malabsorption (Wang et al. 2017b); (5) the production of a monoclonal antibody against CD20 used to treat autoimmune diseases and lymphomas (Zhang et al. 2018); (6) the generation of α1-antitrypsin protein into milk using a GFP reporter gene (Jang et al. 2006); (7) the generation of human recombinant clotting factor IX (Monzani et al. 2013); (8) the production of human granulocyte colony stimulating factor (Carvalho et al. 2019; Yang et al. 2015); and (9) the generation of human fibroblast growth factor 2 using the CRISPR/Cas9 system (Jeong et al. 2016). These examples illustrate the power and utility of transgenic cow milk to produce biopharmaceuticals.

Along with milk, blood plasma has been used as a vehicle to generate recombinant antibodies in transgenic cattle. Examples of the use of plasma to produce antibodies include (1) the generation of transchromosomic cattle with a human artificial chromosome vector carrying the human immunoglobulin heavy and kappa light chain loci (Kuroiwa et al. 2009); (2) the production of fully human polyclonal IgG antibodies for the prevention and/or treatment of Middle East respiratory syndrome coronavirus (MERS-CoV) infection (Luke et al. 2016); and (3) the production of r28M, a bispecific scFv antibody directed against melanoma-associated proteoglycan and the human CD28 molecule on T cells for the treatment of melanoma (Grosse-Hovest et al. 2004; Spiesberger et al. 2015). The use of blood plasma is an important vehicle for generation biopharmaceutical proteins.

15.5 Conclusions

In the past 30 years, many advances have been made in the genetic engineering and generation of transgenic cattle. As highlighted in this review, a significant number of transgenic cattle have been produced with important biotechnological applications. In the area of nutrition, important advances were achieved in increasing the content of nutritionally essential amino acids in milk, increasing bioactive proteins, humanizing cow milk, increasing protein content, alteration of milk casein composition for cheese production, and production of n-3 fatty acids in dairy and beef cattle. Transgenic cattle resistant to diseases like mastitis, tuberculosis, and BSE is another important advance that could decrease the costs of animal production and increase food safety for consumers. Regarding the pharmaceutical industry, transgenic cattle, have great potential for producing biopharmaceuticals on large-scale at lower production costs. Industrial level production of albumin, growth hormone, antibodies, and bile salt-stimulated lipase are within reach. Transgenic cattle have proved to be an important biotechnological tool, but improving the efficiency of transgenic methodologies must increase to reap the tremendous potential transgenic cattle represent. The use of transgenic cattle for the production of recombinant proteins is the future for many industrial and consumer applications of this critical technology.

References

Aghamohammadi M, Haine D, Kelton DF, Barkema HW, Hogeveen H, Keefe GP, Dufour S (2018) Herd-level mastitis-associated costs on Canadian dairy farms. Front Vet Sci 5:100

An J, Li ZL, Huang WM, Huang WJ, Li J (2004) Expression of human tissue-type plasminogen activator in cow mammary gland. Di Yi Jun Yi Da Xue Xue Bao 24(546–548):552

Baker DH (2007) Lysine, arginine, and related amino acids: an introduction to the 6th amino acid assessment workshop. J Nutr 137:1599S-1601S

Bazer FW, Lamb GC, Wu G (2020). Animal Agriculture: Challenges, Innovations, and Sustainability. Elsevier, New York. pp 1–541

Bennett GL, Tait RG Jr, Shackelford SD, Wheeler TL, King DA, Casas E, Smith TPL (2018) Enhanced estimates of carcass and meat quality effects for polymorphisms in myostatin and μ-calpain genes1,2,3. J Anim Sci 97:569–577

Bleck GT, White BR, Miller DJ, Wheeler MB (1998) Production of bovine alpha-lactalbumin in the milk of transgenic pigs. J Anim Sci 76:3072–3078

Brophy B, Smolenski G, Wheeler T, Wells D, L'Huillier P, Laible G (2003) Cloned transgenic cattle produce milk with higher levels of β-casein and κ-casein. Nat Biotechnol 21:157–162

Cardoso CV, Barbosa EV, Liberal MHT, Chagas EFD (2019) Transgenic technology: the strategy for the control and prevention of bovine staphylococcal mastitis? Biotechnol Res Innov 3:291–297

Carvalho BP, Cunha ATM, Silva BDM, Sousa RV, Leme LO, Dode MAN, Melo EO (2019) Production of transgenic cattle by somatic cell nuclear transfer (SCNT) with the human granulocyte colony-stimulation factor (hG-CSF). J Anim Sci Technol 61:61–68

Chan AWS, Homan EJ, Ballou LU, Burns JC, Bremel RD (1998) Transgenic cattle produced by reverse-transcribed gene transfer in oocytes. Proc Natl Acad Sci USA 95:14028

Cheng G, Fu C, Wang H, Adoligbe C, Wei S, Li S et al (2015) Production of transgenic beef cattle rich in n-3 PUFAs by somatic cell nuclear transfer. Biotech Lett 37:1565–1571

Cooper CA, Maga EA, Murray JD (2014) Consumption of transgenic milk containing the antimicrobials lactoferrin and lysozyme separately and in conjunction by 6-week-old pigs improves intestinal and systemic health. J Dairy Res 81:30–37

Cooper CA, Nelson KM, Maga EA, Murray JD (2013) Consumption of transgenic cows' milk containing human lactoferrin results in beneficial changes in the gastrointestinal tract and systemic health of young pigs. Transgenic Res 22:571–578

Dai Y, Yang X, Liu S, Wang J, Zhou Q, Fang R, Yu T (2018) Acute oral toxicity and 90 days feeding test of recombinant human lactoferrin. Wei Sheng Yan Jiu 47:286–311

Echelard Y, Williams JL, Destrempes MM, Koster JA, Overton SA, Pollock DP et al (2009) Production of recombinant albumin by a herd of cloned transgenic cattle. Transgenic Res 18:361–376

Farrell HM, Jimenez-Flores R, Bleck GT, Brown EM, Butler JE, Creamer LK et al (2004) Nomenclature of the proteins of cows' milk—sixth revision. J Dairy Sci 87:1641–1674

Fischer W, Gustafsson L, Mossberg A-K, Gronli J, Mork S, Bjerkvig R, Svanborg C (2004) Human α-lactalbumin made lethal to tumor cells (HAMLET) kills human glioblastoma cells in brain xenografts by an apoptosis-like mechanism and prolongs survival. Cancer Res 64:2105–2112

Gao Y, Wu H, Wang Y, Liu X, Chen L, Li Q et al (2017) Single Cas9 nickase induced generation of NRAMP1 knockin cattle with reduced off-target effects. Genome Biol 18:13

Grosse-Hovest L, Müller S, Minoia R, Wolf E, Zakhartchenko V, Wenigerkind H et al (2004) Cloned transgenic farm animals produce a bispecific antibody for T cell-mediated tumor cell killing. Proc Natl Acad Sci USA 101:6858–6863

Gupta A, Jadhav JB, Gunaware KD, Shinde B (2016) Whey proteins and its impact on human health nutrition: review. J Anal Pharm Res 3:00083

Gustafsson L, Leijonhufvud I, Aronsson A, Mossberg A-K, Svanborg C (2004) Treatment of skin papillomas with topical α-lactalbumin–oleic acid. N Engl J Med 350:2663–2672

Hofmann A, Zakhartchenko V, Weppert M, Sebald H, Wenigerkind H, Brem G et al (2004) Generation of transgenic cattle by lentiviral gene transfer into oocytes1. Biol Reprod 71:405–409

Hu W, Zhao J, Wang J, Yu T, Wang J, Li N (2012) Transgenic milk containing recombinant

humanlactoferrin modulates the intestinal flora in piglets. Biochem Cell Biol 90:485–496
Hyttinen J-M, Korhonen V-P, Hiltunen MO, Myöhänen S, Jänne J (1998) High-level expression of bovine β-lactoglobulin gene in transgenic mice. J Biotechnol 61:191–198
Hyvönen P, Suojala L, Haaranen J, von Wright A, Pyörälä S (2006a) Human and bovine lactoferrins in the milk of recombinant human lactoferrin-transgenic dairy cows during lactation. Biotechnol J 1:410–412
Hyvönen P, Suojala L, Orro T, Haaranen J, Simola O, Røntved C, Pyörälä S (2006b) Transgenic cows producing rhLf in milk are not protected from experimental Escherichia coli intramammary infection. Infect Immun 74:6206–6212
Innes JK, Calder PC (2020) Marine omega-3 (N-3) fatty acids for cardiovascular health: an update for 2020. Int J Mol Sci 21:1361
Jang G, Bhuiyan MMU, Jeon HY, Ko KH, Park HJ, Kim MK et al (2006) An approach for producing transgenic cloned cows by nuclear transfer of cells transfected with human alpha 1-antitrypsin gene. Theriogenology 65:1800–1812
Jeong YH, Kim YJ, Kim EY, Kim SE, Kim J, Park MJ et al (2016) Knock-in fibroblasts and transgenic blastocysts for expression of human FGF2 in the bovine β-casein gene locus using CRISPR/Cas9 nuclease-mediated homologous recombination. Zygote 24:442–456
Kaiser GG, Mucci NC, González V, Sánchez L, Parrón JA, Pérez MD et al (2017) Detection of recombinant human lactoferrin and lysozyme produced in a bitransgenic cow. J Dairy Sci 100:1605–1617
Kang Y, Jimenez-Flores R, Richardson T (1986) Casein genes and genetic engineering of the caseins. Basic Life Sci 37:95–111
Knight-Jones TJD, Rushton J (2013) The economic impacts of foot and mouth disease-what are they, how big are they, and where do they occur? Prev Vet Med 112:161–173
Krimpenfort P, Rademakers A, Eyestone W, van der Schans A, van den Broek S, Kooiman P et al (1991) Generation of transgenic dairy cattle using 'in vitro' embryo production. Biotechnology 9:844–847
Küllenberg de Gaudry D, Massing U (2014) Importance of long-chain omega-3 fatty acids in prostate cancer. Urologe A 53:1620–1624
Kuroiwa Y, Kasinathan P, Sathiyaseelan T, Jiao JA, Matsushita H, Sathiyaseelan J et al (2009) Antigen-specific human polyclonal antibodies from hyperimmunized cattle. Nat Biotechnol 27:173–181
Lai L, Kang JX, Li R, Wang J, Witt WT, Yong HY et al (2006) Generation of cloned transgenic pigs rich in omega-3 fatty acids. Nat Biotechnol 24:435–436
Layman DK, Lönnerdal B, Fernstrom JD (2018) Applications for α-lactalbumin in human nutrition. Nutr Rev 76:444–460
Lee AL, Park Y (2013) The association between n-3 polyunsaturated fatty acid levels in erythrocytes and the risk of rheumatoid arthritis in Korean women. Ann Nutr Metab 63:88–95
Li Q, Hu W, Zhao J, Wang J, Dai Y, Zhao Y et al (2014) Supplementation transgenic cow's milk containing recombinant human lactoferrin enhances systematic and intestinal immune responses in piglets. Mol Biol Rep 41:2119–2128
Liu C (2013) Strategies for designing transgenic DNA constructs. Methods Mol Biol 1027:183–201
Liu X, Bai C, Ding X, Wei Z, Guo H, Li G (2015) Microarray analysis of the gene expression profile and lipid metabolism in fat-1 transgenic cattle. PLOS One 10:e0138874
Liu X, Wang Y, Tian Y, Yu Y, Gao M, Hu G et al (2014) Generation of mastitis resistance in cows by targeting human lysozyme gene to β-casein locus using zinc-finger nucleases. Proc Biol Sci 281:20133368
Lönnerdal B (2013) Bioactive proteins in breast milk. J Paediatr Child Health 49:1–7
Lu D, Liu S, Ding F, Wang H, Li J, Li L et al (2016) Large-scale production of functional human lysozyme from marker-free transgenic cloned cows. Sci Rep 6:22947
Luke T, Wu H, Zhao J, Channappanavar R, Coleman CM, Jiao J-A et al (2016) Human polyclonal immunoglobulin G from transchromosomic bovines inhibits MERS-CoV in vivo. Sci Trans Med 8:326ra321–326ra321
Luo Y, Wang Y, Liu J, Lan H, Shao M, Yu Y et al (2015) Production of transgenic cattle highly expressing human serum albumin in milk by phiC31 integrase-mediated gene delivery. Transgenic Res 24:875–883
Ma X, Zhang LSP, Zhang S, Tang B, Zhang X, Luan W, Li Z (2019) Expression and characterization of a lysine-rich protein in cow milk. Indian J Anim Res 53:874–879
Magee DA, MacHugh DE, Edwards CJ (2014) Interrogation of modern and ancient genomes reveals the complex domestic history of cattle. Anim Front 4:7–22
Monzani PS, Adona PR, Ohashi OM, Meirelles FV, Wheeler MB (2016) Transgenic bovine as bioreactors: Challenges and perspectives. Bioengineered 7:123–131
Monzani PS, Sangalli JR, De Bem TH, Bressan FF, Fantinato-Neto P, Pimentel JR et al (2013) Breeding of transgenic cattle for human coagulation factor IX by a combination of lentiviral system and cloning. Genet Mol Res 12:3675–3688
Nielsen CH, Hui Y, Nguyen DN, Ahnfeldt AM, Burrin DG, Hartmann B et al (2020) Alpha-Lactalbumin enriched whey protein concentrate to improve gut, immunity and brain development in preterm pigs. Nutrients 12:245
Niemann H, Kind A, Schienieke A (2012) Production of biopharmaceuticals in transgenic animals. In: Kayser O, Warzecha H (eds) Pharmaceutical biotechnology–drug discovery and clinical applications. Wiley-Blackwell, Weinheim, Germany, pp 71–111

Ogura A, Inoue K, Wakayama T (2013) Recent advancements in cloning by somatic cell nuclear transfer. Philos Trans R Soc Lond B Biol Sci 368:20110329

Parc AL, Karav S, Rouquié C, Maga EA, Bunyatratchata A, Barile D (2017) Characterization of recombinant human lactoferrin N-glycans expressed in the milk of transgenic cows. PLOS One 12: e0171477

Raju TS, Briggs JB, Borge SM, Jones AJS (2000) Species-specific variation in glycosylation of IgG: evidence for the species-specific sialylation and branch-specific galactosylation and importance for engineering recombinant glycoprotein therapeutics. Glycobiology 10:477–486

Rezaei R, Wu ZL, Hou YQ, Bazer FW, Wu G (2016) Amino acids and mammary gland development: nutritional implications for neonatal growth. J Anim Sci Biotechnol 7:20

Richt JA, Kasinathan P, Hamir AN, Castilla J, Sathiyaseelan T, Vargas F et al (2007) Production of cattle lacking prion protein. Nat Biotechnol 25:132–138

Salamone D, Barañao L, Santos C, Bussmann L, Artuso J, Werning C et al (2006) High-level expression of bioactive recombinant human growth hormone in the milk of a cloned transgenic cow. J Biotechnol 124:469–472

Shemesh M, Gurevich M, Harel-Markowitz E, Benvenisti L, Shore LS, Stram Y (2000) Gene integration into bovine sperm genome and its expression in transgenic offspring. Mol Reprod Dev 56:306–308

Shinohara ET, Kaminski JM, Segal DJ, Pelczar P, Kolhe R, Ryan T et al (2007) Active integration: new strategies for transgenesis. Transgenic Res 16:333–339

Spiesberger K, Paulfranz F, Egger A, Reiser J, Vogl C, Rudolf-Scholik J et al (2015) Large-scale purification of r28M: a bispecific scFv antibody targeting human melanoma produced in transgenic cattle. PLOS One 10:e0140471

Su F, Wang Y, Liu G, Ru K, Liu X, Yu Y et al (2016) Generation of transgenic cattle expressing human β-defensin 3 as an approach to reducing susceptibility to Mycobacterium Bovis infection. FEBS J 283:776–790

Su X, Wang S, Su G, Zheng Z, Zhang J, Ma Y et al (2018) Production of microhomologous-mediated site-specific integrated LacS gene cow using TALENs. Theriogenology 119:282–288

Sun Z, Wang M, Han S, Ma S, Zou Z, Ding F et al (2018) Production of hypoallergenic milk from DNA-free beta-lactoglobulin (BLG) gene knock-out cow using zinc-finger nucleases mRNA. Sci Rep 8:15430

Świątkiewicz S, Świątkiewicz M, Arczewska-Włosek A, Józefiak D (2015) The use of genetic engineering techniques to improve the lipid composition in meat, milk, and fish products: a review. Animal 9:696–706

Tárnok A, Marosvölgyi T, Szabó É, Györei E, Decsi T (2015) Low n-3 long-chain polyunsaturated fatty acids in newly diagnosed celiac disease in children with preexisting type 1 diabetes mellitus. J Pediatr Gastroenterol Nutr 60:255–258

Tessanne K, Golding MC, Long CR, Peoples MD, Hannon G, Westhusin ME (2012) Production of transgenic calves expressing an shRNA targeting myostatin. Mol Reprod Dev 79:176–185

Thomassen EAJ, Veen HAv, Berkel PHCV, Nuijens JH, Abrahams JP (2005) The protein structure of recombinant human lactoferrin produced in the milk of transgenic cows closely matches the structure of human milk-derived lactoferrin. Transgenic Res 14:397–405

Ushida Y, Shimokawa Y, Toida T, Matsui H, Takase M (2007) Bovine α-lactalbumin stimulates mucus metabolism in gastric mucosa. J Dairy Sci 90:541–546

van Berkel PHC, Welling MM, Geerts M, van Veen HA, Ravensbergen B, Salaheddine M et al (2002) Large scale production of recombinant human lactoferrin in the milk of transgenic cows. Nat Biotechnol 20:484–487

Wall RJ, Powell AM, Paape MJ, Kerr DE, Bannerman DD, Pursel VG et al (2005) Genetically enhanced cows resist intramammary Staphylococcus aureus infection. Nat Biotechnol 23:445–451

Wang H, Wu J, Liu X, He H, Ding F, Yang H et al (2012) Identification of short hairpin RNA targeting foot-and-mouth disease virus with transgenic bovine fetal epithelium cells. PLOS One 7:e42356

Wang J, Yang P, Tang B, Sun X, Zhang R, Guo C et al (2008) Expression and characterization of bioactive recombinant human α-lactalbumin in the milk of transgenic cloned cows. J Dairy Sci 91:4466–4476

Wang M, Sun Z, Yu T, Ding F, Li L, Wang X et al (2017a) Large-scale production of recombinant human lactoferrin from high-expression, marker-free transgenic cloned cows. Sci Rep 7:10733

Wang Y, Ding F, Wang T, Liu W, Lindquist S, Hernell O et al (2017b) Purification and characterization of recombinant human bile salt-stimulated lipase expressed in the milk of transgenic cloned cows. PLOS One 12:e0176864

Wei J, Wagner S, Maclean P, Brophy B, Cole S, Smolenski G et al (2018) Cattle with a precise, zygote-mediated deletion safely eliminate the major milk allergen beta-lactoglobulin. Sci Rep 8:7661

Wheeler MB (2007) Agricultural applications for transgenic livestock. Trends Biotechnol 25:204–210

Wheeler MB (2013) Transgenic animals in agriculture. Nat Educ Knowl 4:1

Whitworth KM, Prather RS (2010) Somatic cell nuclear transfer efficiency: how can it be improved through nuclear remodeling and reprogramming? Mol Reprod Dev 77:1001–1015

Wu G (2020) Important roles of dietary taurine, creatine, carnosine, anserine and hydroxyproline in human nutrition and health. Amino Acids 52:329–360

Wu G (2022) Nutrition and metabolism: foundations for animal growth, development, reproduction, and health. Adv Exp MedBiol 1354:1–24

Wu H, Wang Y, Zhang Y, Yang M, Lv J, Liu J, Zhang Y (2015) TALE nickase-mediated knockin endows cattle with increased resistance to tuberculosis. Proc Natl Acad Sci USA 112:E1530–E1539

Wu X, Ouyang H, Duan B, Pang D, Zhang L, Yuan T et al (2012) Production of cloned transgenic cow expressing omega-3 fatty acids. Transgenic Res 21:537–543

Yang B, Wang J, Tang B, Liu Y, Guo C, Yang P et al (2011) Characterization of bioactive recombinant human lysozyme expressed in milk of cloned transgenic cattle. PLOS One 6:e17593

Yang JS, Joe SY, Koo BC, Heo YT, Lee SM, Kang MJ et al (2015) Production of hGCSF and GFP Co-expressed transgenic cow embryo by somatic cell nuclear transfer technique. J Emb Trans 30:219–224

Yang P, Wang J, Gong G, Sun X, Zhang R, Du Z et al (2008) Cattle mammary bioreactor denerated by a novel procedure of transgenic cloning for large-scale production of functional human lactoferrin. PLOS One 3:e3453

Yu T, Guo C, Wang J, Hao P, Sui S, Chen X et al (2010) Comprehensive characterization of the site-specific N-glycosylation of wild-type and recombinant human lactoferrin expressed in the milk of transgenic cloned cattle. Glycobiology 21:206–224

Yu Y, Wang Y, Tong Q, Liu X, Su F, Quan F et al (2013) A site-specific recombinase-based method to produce antibiotic selectable marker free transgenic cattle. PLOS One 8:e62457

Yum S-Y, Lee S-J, Kim H-M, Choi W-J, Park J-H, Lee W-W et al (2016) Efficient generation of transgenic cattle using the DNA transposon and their analysis by next-generation sequencing. Sci Rep 6:27185

Yum S-Y, Lee S-J, Park S-G, Shin I-G, Hahn S-E, Choi W-J et al (2018) Long-term health and germline transmission in transgenic cattle following transposon-mediated gene transfer. BMC Genomics 19:387

Zeder MA, Emshwiller E, Smith BD, Bradley DG (2006) Documenting domestication: the intersection of genetics and archaeology. Trends Genet 22:139–155

Zhang L, van Dijk ADJ, Hettinga K (2017) An interactomics overview of the human and bovine milk proteome over lactation. Proteome Sci 15:1–1

Zhang R, Tang C, Guo H, Tang B, Hou S, Zhao L et al (2018) A novel glycosylated anti-CD20 monoclonal antibody from transgenic cattle. Sci Rep 8:13208

Zhao J, Lai L, Ji W, Zhou Q (2019) Genome editing in large animals: current status and future prospects. Natl Sci Rev 6:402–420

Use of Agriculturally Important Animals as Models in Biomedical Research

16

Brandon I. Smith and Kristen E. Govoni

Abstract

Livestock have contributed significantly to advances in biomedicine and offer unique advantages over rodent models. The human is the ideal biomedical model; however, ethical reasons limit the testing of hypotheses and treatments in humans. Rodent models are frequently used as alternatives to humans due to size, low cost, and ease of genetic manipulation, and have contributed tremendously to our understanding of human health and disease. However, the use of rodents in translational research pose challenges for researchers due to physiological differences to humans. The use of livestock species as biomedical models can address these challenges as livestock have several similarities to human anatomy, physiology, genetics, and metabolism and their larger size permits collection of more frequent and often larger samples. Additionally, recent advances in genetics in livestock species allow for studies in genomics, proteomics, and metabolomics, which have the added benefit of applications to both humans in biomedical research and livestock in improving production. In this review, we provide an overview of scientific findings using livestock and benefits of each model to the livestock industry and to biomedical research.

Keywords

Biomedical models · Livestock

Abbreviations

CFTR	Cystic fibrosis transmembrane conductance regulator
CRISPR	Clustered interspaced short palindromic repeat
DMD	Duchenne muscular dystrophy
FBP1	Fructose-1,6-bisphosphatase 1
HDL-C	High density lipoprotein cholesterol
IUGR	Intrauterine growth restriction
LDL-C	Low-density lipoprotein cholesterol
G6PC	Glucose-6-phosphatase
PCK1	Phosphoenolpyruvate carboxykinase 1
PDK4	Pyruvate dehydrogenase kinase 1
PGE_2	Progesterone E_2

B. I. Smith · K. E. Govoni (✉)
Department of Animal Science, University of Connecticut, Storrs, CT 06269-4040, USA
e-mail: kristen.govoni@uconn.edu

G. Wu (ed.), *Recent Advances in Animal Nutrition and Metabolism*, Advances in Experimental Medicine and Biology 1354,
https://doi.org/10.1007/978-3-030-85686-1_16

16.1 Introduction

Livestock are animals raised for an agricultural purpose, and for this review will include cattle (beef and dairy), swine, poultry, and sheep. Animal scientists have studied physiology, nutrition, anatomy, husbandry, genetics, and behavior of these species for over 100 years which provides a solid foundation for current and future studies that target disease, new technologies, and therapies in all species (Wu 2022). In biomedical research, the ideal model is the human; however, for obvious ethical reasons, it is not always possible to test hypotheses and/or novel treatments in humans. Non-human primates are also an excellent model for human biomedical research; however, they are costly to maintain and again, due to ethical concerns, sample collection is often limited. Rodents are well-accepted models to study disease, treatments, and genetics related to humans due to their size, reduced costs compared with humans and larger animal models, ability to develop genetically altered lines, and an abundance of well-defined reagents for these species. However, for many physiological processes including reproduction, nutrition, and metabolism, livestock often have anatomy, physiology, and genetics that are more similar to humans than rodent models, as will be discussed in this review. In addition, advances in genetics in livestock allows for mechanistic studies at genome, proteome, and metabolome level (Fig. 16.1), which have the added benefit of applications to humans in biomedical research and livestock in improving production. Transgenics and genetic selection and manipulation provide models to test roles of specific genes and pathways. The importance of using agriculturally relevant species in biomedical research are highlighted by numerous reviews in the literature

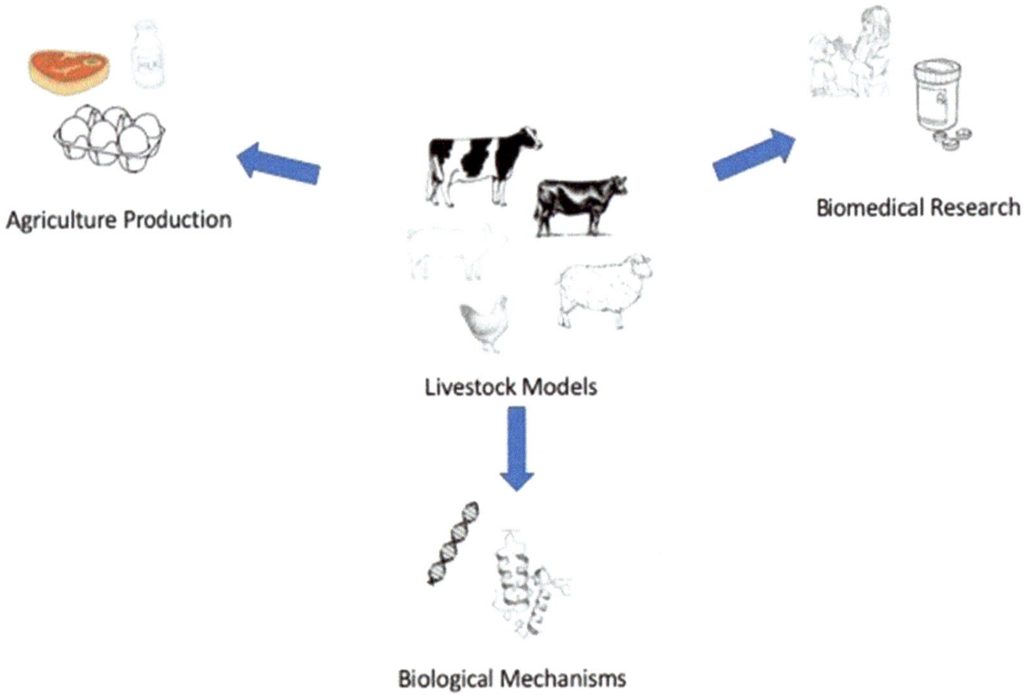

Fig. 16.1 **Livestock models for biomedical research**

that focus on specific uses and areas of research (e.g., Ireland et al. 2008; Reynolds et al. 2009; Polejaeva et al. 2016; Hamernik 2019), stakeholder meetings, and more recently a collaborative funding mechanism between United States Department of Agriculture (USDA) and the National Institutes of Health (NIH): Dual Purpose with Dual Benefit: Research in Biomedicine and Agriculture Using Agriculturally Important Domestic Species. In this review, we focus on an overview of key scientific findings using livestock (dairy and beef cattle, swine, sheep, and poultry) in the fields of fetal programming, nutrition, disease, reproduction, and muscle biology and the benefits of each model to the livestock industry and to biomedical research (Table 16.1).

16.2 Fetal Programming and Metabolism

Fetal programming is defined as changes to the maternal environment that induce alterations in the development of essential tissues to promote survival of the fetus, and can have lasting impacts on the developing fetus (Marciniak et al. 2017; Govoni et al. 2019). Fetal programming

Table 16.1 Summary of animal models and key findings of livestock species used as biomedical models

Livestock serve as excellent biomedical models due to their larger size, which allows for more frequent and larger sample collection, their similarities to humans in anatomy, physiology, and development, and extensive knowledge of their physiology, genetics, and behavior. The biological mechanisms are often similar across species and therefore, studies in livestock species not only benefit biomedical research, but also provide a dual benefit to improve agriculture production.

Animal	Study type	Findings	References
Pig	Maternal malnutrition	Overnutrition and undernutrition during gestation decreased HDL-C and increased LDL-C in obese offspring	Barbero et al. (2013)
		Low-protein diet during gestation increased G6PC enzyme activity in liver and increased FBP1 and G6PC mRNA expression	Jia et al. (2012)
	Placental insufficiency/IUGR	Increased hepatic insulin receptor abundance in IUGR offspring. Differential proteomic profiles of liver protein and carbohydrate metabolism	Ferenc et al. (2018)
		Postnatal high-fat diet and IUGR increased serum triglyceride, circulating leptin and hepatic fat concentration	Yan et al. (2017)
	Neonatal nutrition	Omega-3 fatty acid supplementation in infant formula improves neurogenesis and neural function	Innis (2008)
		Glycine supplementation increased daily weight gain and increased ant-oxidative capacity	Wang et al. (2014)
		Arginine supplementation increased plasma insulin concentrations and decreased plasma cortisol, ammonia, and urea	Yao et al. (2011)
	Disease	Genome editing technology, CRISPR/Cas9, generated DMD phenotypes	Klymiuk et al. (2013)
Cattle	Infertility	Acute nutrient restriction reduced induced failure of ovulation of dominant follicle in 60% of heifers	Mackey et al. (1999)
		Dexamethasone induced insulin resistance prevented ovulation and decreased plasma estradiol	Hackbart et al. (2013)
		High androstenedione in follicular fluid decreased androgen to estrogen conversion in ovaries	Summers et al. (2014)
	Disease	Development of protease-free assay to assess stability of prions	Vrentas et al. (2013)

(continued)

Table 16.1 (continued)

Animal	Study type	Findings	References
Sheep	Maternal malnutrition	Undernutrition and overnutrition during gestation increased lipid accumulation offspring at birth	Reed et al. (2014)
		Restricted nutrition reduced metabolites in glutamate, methionine and histidine pathways at birth and day 135 gestation	Martin et al. (2019)
		Overnutrition/obesity eliminated postnatal plasma leptin surge and increased plasma cortisol concentrations	Long et al. (2011)
	Placental insufficiency/IUGR	Decreased glucose utilization rates, decreased hepatic G6PC and PCK1 expression	Jones et al. (2019)
		Increased insulin sensitivity and visceral adiposity	De Blasio et al. (2007)
		Increased insulin stimulated glucose oxidation rates, increased PDK4 expression in muscle and liver	Brown et al. (2015)
	Disease	16MΔ*vjbR* vaccine reduced colonization of *Brucella melitensis* in placenta and was safer in pregnant animals	Hensel et al. (2020)
		Disruption of CFTR gene using CRISPR/Cas9 induced cystic fibrosis phenotypes	Fan et al. (2018)
Poultry	Disease	Aspirin treatment decreased the progression of ovarian cancer and decreased liver PGE_2 production	Urick et al. (2009)
		14 deletion mutations in the p53 gene in birds who experienced reduced light and caloric restriction increasing the risk of developing ovarian cancer	Hakim et al. (2009)

occurs when the uterine environment is disrupted by adverse conditions, such as poor maternal nutrition, placental insufficiency, stress, or disease (Marciniak et al. 2017). These phenomena were characterized in epidemiological studies, evaluating children born to mothers exposed to nutrient restriction during the Dutch winter famine, that linked an association between low birth weight (LBW) or intrauterine growth restriction (IUGR), and increased risk of developing insulin resistance, impaired glucose tolerance, and metabolic syndrome later in life (Hales and Barker 2001). This association may be the result of a programming during gestation to survive in an environment of nutrient deficiency. However, when the offspring are exposed to an environment of nutrient sufficiency, this programming effect led to negative metabolic adaptations later in life (Hales and Barker 2001; Barker 2005; Govoni et al. 2019). Animal models using poor maternal nutrition during gestation, including limited or excess intake of energy, protein, and/or vitamins and minerals, as well as, placental insufficiency have demonstrated fetal programming that influences fetal growth, offspring birth weights, and tissue development (Reynolds et al. 2022; Symonds et al. 2009; Hoffman et al. 2017; Govoni et al. 2019).

Due to ethical reasons and restrictions on quantity of sample collection in human research, the evaluation of the underlying mechanisms of fetal programming in humans is limited and, therefore, has necessitated the use of animal models (Reynolds et al. 2022; Satterfield et al. 2011; Wu et al. 2006). Rodent models of fetal programming and IUGR are used to evaluate the impacts of the maternal nutritional environment on metabolic adaptations in the offspring and are an advantageous model due to their small size, short gestation periods, and relatively low cost to house and manage (Swanson and David 2015; Reynolds and Vickers 2018). However, the translation of studies using small animal models to human medicine can pose challenges due to differences in physiology between humans and rodents. Livestock possess several advantages to traditional rodent models. Livestock species and humans have similar gestational periods, mother-to-fetus size ratio, fetal placenta vasculature development and morphology, and maternal–fetal nutrient exchange. Therefore, livestock are often used as models for humans for studies on developmental programming and reproductive biology (Barry and Anthony 2008). For example, sheep are often used in studies of placental insufficiency and fetal development as they have similar placental vasculature and relative maturity of the fetus at birth as humans, and also have similar metabolic, endocrine, and growth outcomes as reported in human IUGR (Bazer et al. 2021; Owens et al. 1994; Barry et al. 2008; Satterfield et al. 2013). In addition, swine are used as biomedical IUGR models as they possess similar anatomical and physiological characteristics to humans, and exhibit placental insufficiency, inducing naturally occurring IUGR (Wu et al. 2006; Gonzalez-Bulnes and Chavatte-Palmer 2017). Using agriculturally relevant animals as models of developmental programming, researchers have gained valuable knowledge and insights into metabolic adaptations in offspring prenatally and postnatally.

16.2.1 Glucose Metabolism and Insulin Sensitivity

Metabolic adaptations in response to poor maternal nutrition or placental insufficiency are well documented in human and animal models. Poor maternal nutrition and IUGR impact glucose homeostasis, insulin sensitivity, and insulin action, which may lead to developing hyperglycemia and hyperinsulinemia later in life (Symonds et al. 2009). Large animal models of poor maternal nutrition, especially sheep and swine, have contributed significantly to knowledge of outcomes of maternal nutrition and IUGR on glucose metabolism. Specifically, maternal undernutrition or overnutrition during pregnancy in sheep influences fetal plasma glucose concentrations, insulin sensitivity, glucose utilization rates, and glucose production rates that persist postnatally (Limesand et al. 2007; Camacho et al. 2017). Additionally, IUGR in sheep impacts fetal expression of pyruvate dehydrogenase kinase (PDK) 1 and PDK4, regulators of glucose homeostasis, in fetal offspring (Pendleton et al. 2019). Studies using a sheep model of IUGR induced by placental insufficiency as a result of reduced placentome number have shown increased insulin sensitivity associated with increased visceral adiposity in IUGR lambs (De Blasio et al. 2007). Additionally, IUGR induced by elevated humidity and temperature in sheep resulted in decreased glucose utilization and insulin sensitivity in offspring (Brown et al. 2015). Hypoxic conditions during gestation have also been used to induce IUGR in sheep to mimic developmental outcomes in populations at high altitudes and have been shown to impact glucose metabolism (Jones et al. 2019). Specifically, fetal sheep exposed to prolonged hypoxia exhibit decreased glucose utilization rates and decreased glucose-6 phosphatase catalytic subunit (G6PC) and PDK1 mRNA expression (Jones et al. 2019). Swine models of IUGR have also provided valuable insight into glucose metabolism and insulin action. For example, a study utilizing a swine model of IUGR showed increased expression of insulin receptors in IUGR neonates suggesting a compensatory mechanism of altered insulin response (Ferenc et al. 2018).

The duration and timing of fetal undernutrition can also impact fetal metabolism. Maternal nutrient restriction around the time of conception exhibited decreased hepatic protein abundance of insulin signaling molecules; cAMP response element binding protein (CREB) and cytoplasmic phosphoenolpyruvate carboxy kinase (PEPCK-C; Lie et al. 2014). In sheep, reduced maternal nutrient intake starting in early gestation showed decreased insulin and glucose concentrations in the fetal lamb (Luther et al. 2007). Studies of sheep maternal undernutrition have demonstrated that aged offspring of ewes nutrient restricted during early gestation (between day 28 and day 79) exhibited increased insulin concentrations when fed an ad libitum diet at 6 years of age (George et al. 2012). Maternal low-protein diets in pigs demonstrated alterations in glucose metabolism, specifically gluconeogenesis, with upregulated mRNA expression and enzyme activity of glucose-6 phosphatase, and altered epigenetic regulation of the GP6C promoter in the neonatal liver of male offspring (Jia et al. 2012). The use of animal models to evaluate the impacts of maternal nutrition on glucose metabolism and insulin sensitivity in the offspring are well-established in literature, have contributed significantly to the understanding of metabolic adaptations in utero and postnatally, and can be further used to develop methods to treat metabolic diseases in humans.

16.2.2 Leptin

Another metabolic adaptation induced by maternal malnutrition is alterations in plasma concentrations of leptin. Leptin is a protein hormone synthesized and secreted by white adipose tissue into circulation, and acts on the hypothalamus to regulate appetite and increase energy expenditure (Houseknecht et al. 1998; Ingvartsen and Boisclair 2001; Triantafyllou et al. 2016). Swine models of maternal over nutrition have demonstrated that a 30% increase in maternal dietary intake increases plasma leptin concentrations and leptin mRNA expression in subcutaneous adipose tissue (Eckert et al. 2000). Furthermore, in swine neonates with IUGR, plasma leptin concentrations increased when fed a high fat diet (Liu et al. 2012). In humans, prenatally, and in many animal species, postnatally, a plasma leptin surge occurs and is associated with fetal organ development (Long et al. 2011; Briffa et al. 2014). Maternal obesity during gestation in sheep eliminated this leptin surge which may alter development and life-time appetitive behavior (Long et al. 2011; Smith et al. 2018). Agriculturally relevant models can add to the understanding of leptin regulation and its involvement in fetal development for the benefit of human and animal health.

16.2.3 Lipid Metabolism

Proper lipid composition during gestation is vital for development of offspring (Herrera and Ortega-Senovilla 2014). Livestock have contributed significantly to knowledge of lipid metabolism with regards to poor maternal nutrition and IUGR. Specifically, poor maternal nutrition has been shown to induce alterations in lipid deposition and metabolism in offspring (Daniel et al. 2007; Yan et al. 2013, 2017). In a swine model of maternal malnutrition, obese offspring from mothers that were over- or undernourished had decreased plasma concentrations of high-density lipoprotein (HDL) cholesterol and increased concentrations of low-density lipoprotein (LDL) cholesterol, suggesting dyslipidemia in these offspring, increasing the risk of developing atherosclerosis (Barbero et al. 2013). Additionally, in sheep, maternal nutrient restriction increased adiposity, uncoupling protein (UCP) 2 mRNA expression and peroxisome proliferator-activated receptor alpha (PPARα) mRNA expression in fetal adipose tissue (Bispham et al. 2005). Our group has shown that in sheep, poor maternal nutrition, restricted-fed or overfed throughout gestation, increased lipid accumulation at birth in the offspring (Reed et al. 2014). However, at three months of age, the offspring of restricted-fed animals exhibited decreased lipid accumulation, suggesting alterations in lipid metabolism that are specific to the nutritional insult during gestation (Reed et al. 2014). Additionally, we have demonstrated that poor maternal nutrition in sheep induces changes in the abundance of lipid metabolites, specifically phosphatidyl-ethanolamines, lysophospholipids, phosphatidyl-choline, and sphingomyelin metabolites, in longissimus dorsi muscle in offspring at gestational days 90 and 135 and at birth (Martin et al. 2019). Others using sheep have shown that maternal nutrient restriction can also cause impaired hepatic lipid metabolism. For example, maternal nutrient restriction during early gestation decreased expression of peroxisome proliferator-activated receptor gamma coactivator 1-α (PGC-1α), PDK2, PDK4, and carnitine palmitoyltransferase-1 (CPT-1), regulators of fatty acid metabolism, in fetal hepatic tissue (Lie et al. 2016). Changes in lipid metabolism may increase the risk of developing metabolic diseases, such as cardiovascular disease, atherosclerosis, dyslipidemia, and metabolic syndrome. Transgenic pig models, such as the Yucatan miniature pig, have also been developed for studies of hypercholesterolaemia and atherosclerosis and could contribute significantly to understanding effects of maternal dyslipidemia on offspring development (Davis et al. 2014; Perleberg et al. 2018).

16.2.4 Maternal Obesity

Instances of obesity in women of reproductive age have increased significantly in recent years and can have adverse consequences on offspring development (McDonald et al. 2010; Catalano and Shankar 2017). Furthermore, offspring of obese mothers are more likely to develop childhood obesity and metabolic disease (Howell and Powell 2017). Studies on metabolic outcomes in the offspring, as a result of maternal obesity, are traditionally performed using small animal models; however, ovine models of maternal obesity and overnutrition have been developed (Ford et al. 2009; Satterfield et al. 2012). Ford and colleagues overfed ewes to induce obesity at conception and throughout gestation and demonstrated that maternal obesity increased mid-gestational glucose, insulin, cortisol concentrations, and increased pancreatic weight and β-cell numbers in fetuses of obese ewes (Ford et al. 2009). In a follow-up study using the same model, maternal obesity decreased fetal pancreatic weight and β-cell numbers at late gestation which may impact the production of insulin, potentially leading to diabetes later in life (Zhang et al. 2011). In another study using this model, offspring of obese ewes exhibited increased subcutaneous fat, perirenal adipose tissue, and plasma fatty acid concentrations, suggesting a predisposition to developing obesity later in life (Long et al. 2010, 2012). Furthermore, intravenous administration of arginine improved metabolic profiles and enhanced brown adipose tissue development in diet-induced obese sheep (Satterfield et al. 2012).

Certain breeds of pigs (Iberian, Minipigs and Mangalica) have also been used as a reliable biomedical model for human obesity and metabolic syndrome due to higher fat content and altered UCP expression in adipose tissue, compared with other domestic breeds, that may contribute to excess fat deposition (Dyson et al. 2006; Litten-Brown et al. 2010; Gonzalez-Bulnes et al. 2017, Kleinert et al. 2018). In an Iberian pig model of diet induced obesity, increased body weight and back fat thickness were observed in offspring at 120 days of age (Barbero et al. 2013). These offspring also exhibited an earlier age of puberty onset compared with control, similar with findings in humans and rodents (Barbero et al. 2013; Brix et al. 2019).

16.3 Nutrition

The piglet is a well-accepted model for pediatric nutrition and its use in identifying optimal nutritional needs has been discussed in several reviews (Wu et al. 2011; Odle et al. 2014, 2017). Odle et al. (2017) provided a detailed summary of use of livestock animals, in particular swine, to determine the role of nutrition in fetal and postnatal growth and health, and applications to humans. Due to similar digestive systems between swine and humans, both non-ruminants, findings in swine are easily translated to humans. In infants subjected to IUGR, small for gestational age (SGA), LBW, and pre-term delivery, there is a need for optimal nutrition immediately after birth and through the rapid growth period (Ball et al. 1986; Odle et al. 2014; Ji et al. 2017). The piglet model is ideal because the gastrointestinal development and nutritional requirements are similar to humans (Odle et al. 2017). The piglet has been used to identify the nutritional composition of formulas to optimize overall growth, gastrointestinal development, metabolism, and more recently brain development. Although the growth rate is slower in humans than larger livestock species, there is potential to extrapolate data from swine to humans for amino acid content and growth rate (Bergen 2022; Odle et al. 2017). Formula development is critical because it is the sole food source for many infants. It is well-established in livestock species that adequate colostrum is crucial to survival, growth, and overall health of offspring. Therefore, the piglet is an excellent model to develop and test colostrum and formula for infants unable to suckle and at risk due to LBW and/or IUGR. For example, the ability to measure passive transfer of immunoglobulins (Vallet et al. 2013) and intestinal growth and barrier function (Moeser 2015) in swine provides evidence needed for adequate colostrum consumption in human infants. As previously described in Odle et al. (2014), the piglet model has contributed to identification of optimum fatty acid composition, pre- and probiotics, total parenteral nutrition, and amino acid supplementation in infant formula. Although maternal milk lipid content is variable between swine and humans, the pig model can be used to study mechanisms and safety of consumption of human infant formulas. For example, lipid metabolism is highly regulated and studies in swine provide an opportunity to utilize molecular techniques to evaluate gene expression and post-translational regulation of key metabolic genes in lipid metabolism (Fernández-Hernanado et al. 2011; Gu et al. 2012; Odle et al. 2014). Long-chain polyunsaturated fatty acids, such as docosahexaenoic acid (DHA) and arachidonic acid (ARA), are important components of infant diet and critical for optimal growth and brain development (Innis and Presa 2001). Studies in piglets fed infant formula containing these long-chain fatty acids demonstrated the benefits and safety of utilizing these in infant formulas (Mathews et al. 2002), which led to supplementation with DHA and ARA in most infant formulas currently on the market (Odle et al. 2014).

Due to the decreased growth rate of pre-term and LBW infants, there is a need to optimize muscle growth. Numerous studies over the past 30 + years have identified several amino acids that are important for optimizing growth as well as muscle development. The availability of amino acids, their role, and supplementation have been extensively studied in the piglet model with applications to humans (Wu et al. 2011). Specifically, glycine supplementation in formula increases body weight (BW) gain and intestinal development in piglets demonstrating this could be an essential component of infant formula (Wang et al. 2014). Note that glycine is deficient in plant proteins (Hou et al. 2019; Li et al. 2021) but abundant in meat and collagen (Li and Wu 2020; Wu 2020). Arginine supplementation in piglet milk replacer also increased BW gain and is associated with increased circulating concentrations of insulin and growth hormone, two key factors in growth and metabolism (Kim et al. 2004), as well as enhanced intestine development (Yao et al. 2011). Furthermore, increased lysine in milk increased BW gain in piglets (Eisemann et al. 2014) and growing pigs (Krick et al. 1993). In addition to determining the effectiveness of various nutrient composition of formula, the piglet provides an excellent model to further understand the mechanisms of action of specific nutrients. The piglet has been used to determine the benefits and mechanisms of action of leucine supplementation in muscle growth when restricted energy is consumed (Manjarín et al. 2016; Manjarín et al. 2018). They demonstrated that leucine supplementation can increase muscle growth through activation of the mTOR pathway (Manjarín et al. 2016).

In addition to improved growth, nutrition studies in the piglet have been used to study neurodevelopment, in particular, in response to early nutrition. It is well-established that gastrointestinal development and nutrition are associated with brain development and the piglet provides an excellent model to determine the effects of diet on normal development. A review by Mudd and Dilger (2017), highlights recent key findings in perinatal brain development in response to various nutrients including fatty acids, cholesterol, milk bioactive compounds, and micro-nutrients. Studies as early as 1966 demonstrated that the pig is an excellent model for pediatric brain research (Glauser 1966). The piglet provides an excellent model of brain development because the anatomy, growth, and timing of development is more similar to humans than rodents (Lind et al. 2007). In addition, with recent advances in technology, magnetic resonance imaging (MRI) allows for in-depth study of the effects of nutrition on development of specific regions of the brain. With existing knowledge about swine behavior, this model provides the ability to further study the role of nutrition and preterm birth along with neurodevelopment on behavior. The ability to anesthetize the pig and use diagnostic techniques such as placing electrodes, catheters, and MRI provide the opportunity to study specific regions of the brain and response to different stimuli and diets. Ella et al. (2019) discussed these techniques and the use of other species, including sheep, for neuro development and understanding of neurological diseases. Although the pig is larger in size than rodents and requires more complex housing and care, the advantages in size of the brain and methods developed to evaluate brain development and behavior make it an excellent biomedical model (Lind et al. 2007).

Several studies using the piglet have demonstrated that fatty acids in infant formula improve neurogenesis and neural function (Innis 2008) by stimulating key signaling pathways. (Innis 2009). Cholesterol is a necessary component of diet for growth and development, in particular for proper brain development and behavior (Boleman et al. 1998; Pond et al. 2000). The role of key micronutrients (folate, iron, and choline) in fetal and postnatal brain development have been determined in the piglet model using nutrient-deficient formulas, demonstrating the need for these in infant formulation (Mudd and Dilger 2017). In pre-term infants, nutrition is critical for proper brain development as demonstrated using the piglet model with oral supplementation to mimic the environment and nutrients of full-term piglets. Piglets born before term had impaired growth unless they were provided a formula with nutrients similar to those in amniotic fluid during late gestation (Berding et al. 2016).

Food allergies are a health concern for humans as well as animals. Swine are an excellent model for food allergy research as they also suffer from food allergies and their anatomy, physiological, and digestive systems are similar to humans (Radcliffe et al. 2019). Soy protein is a primary source of many swine diets and therefore soy hypersensitivity can develop (Radcliffe et al. 2019). This model has been used to demonstrate changes in immunoglobulin (Ig) G and gastrointestinal growth and development in swine hypersensitive to soy (Li et al. 1990). Through genetic selection and/or manipulation, swine provide a model to better understand the molecular changes in response to food allergies and thereby develop therapies for individuals with food allergies (Calbrix et al. 2012; Hashimoto-Hill et al. 2019; Radcliffe et al. 2019).

16.4 Reproduction

Approximately 12% of women and 9% of men in the United States seek medical assistance due to infertility and/or impaired fecundity (CDC, 2019). Similar to humans, infertility is a problem in many production animals making them excellent models to understand the mechanisms contributing to infertility and develop technologies to improve fertility. Cattle, sheep, and swine are well-accepted models due to similarities in ovarian function, follicle development, and hormone regulation of the reproductive cycle with humans (Reynolds et al. 2022; Satterfield et al. 2011; Stenhouse et al. 2022).

Domestic ruminants provide an excellent model to understand folliculogenesis and have contributed to the development of assisted reproductive techniques that are used in both livestock and humans (Campbell et al. 2003). Cows and women are similar in ovarian size, hormone regulation of folliculogenesis, reproductive cycle length, and ovulation of a single egg. Both species also suffer from low egg reserve which leads to decreased response to fertility treatments (Ireland and Gleason 2017) and pathological conditions including polycystic ovarian syndrome (PCOS), follicular cysts, and luteinized anovulatory follicles (Adams and Pierson 1995). The cattle model is ideal for studies evaluating specific hormones such as follicle stimulating hormones, anti-Mullerian hormone, androstenedione, estradiol, and progesterone in follicle development and ovulation (Mossa and Ireland 2019). The ability to perform ultrasound of the ovaries and track follicular development in the same animal make the cow an excellent model for studies to determine hormone regulation, diagnostic tools, and assistive reproductive technologies (ART).

With the ability to collect whole ovaries from cattle, studies have focused on the development and role of specific cells such as theca and granulosa cells to elucidate the mechanisms by which they synthesize and secret hormones that regulate follicular development. Combined with recent advances in genomic sequencing there is evidence from cattle of cell-type specific gene expression and changes in gene expression with transformation of ovarian follicular cells into luteal cells (Romereim et al. 2017). Although there are some species differences in ovarian function, the mechanisms are highly conserved across species, therefore, findings in cattle are beneficial to understanding mechanisms that contribute to infertility in humans (Romereim et al. 2017).

Nutrition affects follicle development and cattle have provided an excellent model to determine the mechanisms by which nutritional deficits impact fertility (Rhodes et al. 1995; Diskin et al. 2003). Diskin et al. (2003) also reviewed studies using livestock models which demonstrated that reduced nutrient consumption reduces or delays ovulation and is associated with decreased concentrations of key reproductive hormones (luteinizing hormone and follicle stimulating hormone; Mackey et al. 1999), decreased insulin (Sinclair et al. 2002), IGF-I (Bossis et al. 2000), and leptin (Clarke and Henry 1999; Williams et al. 2002).

Reproductive technologies such as artificial insemination (AI), in vitro fertilization (IVF), and embryo transfer (ET) have been optimized in cattle and implemented in standard management practices in several production settings. Therefore, mechanisms and reproductive technologies developed in cattle can be translated to humans which is important since ethical reasons limit the ability to isolate tissue, collect multiple blood samples, and test new reproductive technologies in humans (Adams and Pierson 1995). Artificial insemination is used extensively in the dairy industry and provides an opportunity not available in humans to understand genetic contribution to male infertility. Bull DNA is publicly available and data regarding sperm quality and fertility rates are also available (Taylor et al. 2018). Research currently focuses on improving fertility in dairy cattle and understanding the relationship between phenotypic characteristics and genetics which can be applied to humans to evaluate new diagnostic tools and treatments for male infertility (Taylor et al. 2018). Although IVF originated in humans, it is used in livestock to combat infertility and improve production. Over the past 40 years, animal models have been used to optimize methods such as oocyte retrieval, culture conditions, and cryopreservation (Sirard 2018). Since environment can impact the early stage of the developing embryo and potential changes in the embryo occur with ART such as IVF, continuing to use animal models provides an opportunity to determine how in vitro conditions impact the developing embryo and epigenetic modifications that affect later growth and metabolism in adult life (Sirard 2018).

Polycystic ovarian syndrome is a fertility disorder with reproductive and metabolic complications (Padmanabhan and Veiga-Lopez 2013). Over the past 40 years diagnostic methods have improved; however, the etiology is not well-understood (Padmanabhan and Veiga-Lopez 2013). Several animal models have been used to study this disorder including non-human primates, rats, mice, and sheep. However, cattle, sheep, and swine provide advantages of more similar anatomy and endocrine regulation compared with humans than rodents (Abedal-Majed and Cupp 2019). Due to its larger size, ability to ultrasound the ovaries, non-liter bearing, and similarities in regulation of folliculogenesis, the sheep is often used as a model for PCOS (Padmanabhan and Veiga-Lopez 2013). Early studies demonstrated that fetal exposure to testosterone caused genitalial and behavioral masculinization in female offspring (Clarke et al. 1976; Wilson and Tarttelin 1978). Since then several studies have used the sheep to demonstrate the effects of testosterone exposure during gestation on female offspring reproductive development and PCOS as reviewed by Padmanabhan and Veiga-Lopez (2013).

Women with insulin resistance have a greater chance of developing PCOS. In cattle, insulin-resistance induced by dexamethasone (DEX) prevented ovulation and was associated with decreased progesterone and estrogen, but did not change follicle size (Hackbart et al. 2013). Further, a recent study demonstrated that expression of candidate genes determined by genome wide association study (GWAS) are similar in human and bovine fetal ovaries (Hartanti et al. 2020). Due to the limited availability of human fetal ovaries, utilizing bovine fetal ovaries provides an opportunity to determine the relationship between genetics and ovarian development. Findings from the bovine model can be used to elucidate mechanisms by which metabolic dysregulation and/or stress can impact reproduction, and how the fetal environment contributes to infertility, such as PCOS, later in life (Hartanti et al. 2020).

Anovulation due to hormone imbalance is another major cause of infertility in women (Smith et al. 2003; Abedal-Majed and Cupp 2019). Regulation of folliculogenesis is still not well understood and anovulation is often associated with a secondary diagnosis such as PCOS. Studies in sheep have provided a better understanding of steroidogenesis and follicle development (Comim et al. 2015; Smith et al. 2009). In cattle, a naturally occurring selection for a phenotype similar to PCOS due to excess androgen production has been identified (High A4; Summer et al. 2014). Using this model, researchers demonstrated increased expression of genes in the steroidogenesis pathway that lead to increased androstenedione, arrest in ovarian follicle development, and thus reduced fertility (Summers et al. 2014; Abedal-Majed and Cupp 2019). The advances in biotechnology and ART in both humans and livestock species has provided excellent models to address key biological issues in many species and further our understand of reproduction and infertility.

16.5 Muscle

Muscle is a highly metabolic tissue and critical for glucose metabolism. The primary focus of research in agricultural species is skeletal muscle development as a protein source for human consumption (Sandoval et al. 2020). However, recent studies in agricultural species focus on understanding the mechanisms regulating protein synthesis and metabolism of muscle as these are important in overall growth and health. Although nutrient utilization and carbohydrate and lipid metabolism differ between humans and ruminants, cellular mechanisms of protein synthesis and muscle metabolism are similar, allowing for translation of findings in livestock to humans (Zhao et al. 2019). Muscle is important in glucose and protein maintenance and the most abundant insulin-sensitive tissue (Martin et al. 2019). The ratio of skeletal muscle to intramuscular subcutaneous fat can affect overall body metabolism (Zhao et al. 2019). Oxidative and glycolytic metabolism are tightly regulated in muscle and there is evidence in sheep that maternal diet can alter muscle metabolism in offspring (Martin et al. 2019). Excessive lipid accumulation, as observed with obesity, an epidemic in the US, is associated with impaired glucose metabolism and metabolic dysregulation. Large animal models provide the opportunity to study disease states and collect samples to identify cellular and molecular mechanisms causing metabolic dysregulation (Govoni et al. 2019). In addition, due to extensive use of chickens to study somite development (Pourquié 2018), they provide an excellent model for studying muscle development. The chick embryo model is affordable due to low cost of commercially available fertile eggs, and provides an opportunity to monitor each stage of the developing embryo. This has allowed scientists to characterize the molecular pathways and genetic regulation of myogenesis as well as the unique stem cell population of skeletal muscle, satellite cells (Scaal and Marcelle 2018).

Understanding the role of the population of unique stem cells (satellite cells) in muscle growth, maintenance, metabolism, and aging can help improve quality of life in humans suffering from injury or age-related muscle loss. The role of satellite cells has been extensively studies in livestock due to the large sample size, ability to conduct multiple biopsies in the same animal, and ability to isolate cells and culture in the laboratory. These studies have contributed to understanding of muscle growth, maintenance, metabolism, and aging (Dayton and White 2008; Reed et al. 2015; Raja et al. 2016).

16.6 Disease

Studies on infectious disease in livestock have a strong emphasis on the prevention and treatment of zoonotic diseases, such as brucellosis, bovine tuberculosis, bovine spongiform encephalopathy, and others that pose considerable risk to human and animal health (Meurens et al. 2012; Greenlee and Greenlee 2015; Roth and Tuggle 2015; Michelotti et al. 2018). The transfer of many infectious zoonotic diseases to humans often occurs through the consumption of meat products and unpasteurized milk or dairy products of infected animals or handling of tissues of infected animals (Michelotti et al. 2018). In many developed countries, government regulation and utilization of pasteurization have reduced infection dramatically, but these zoonotic diseases are still prevalent in many developing countries, where such regulations have not been implemented (Michelotti et al. 2018). Therefore, research on zoonotic diseases in livestock has the dual benefit of improving our understanding of disease pathology, prevention, vaccine development, and treatment in both livestock species and humans (Roth 2011; Gerdts et al. 2015). Additionally, advances in gene-editing technologies can lead to novel biomedical models that have advantages over transgenic rodent models.

Mycobacterium bovis is the primary cause of tuberculosis in cattle and is a zoonotic pathogen that can cause severe disease in humans (Pollock and Neill 2002), but models of bovine tuberculosis are not typically regarded as a model for human tuberculosis. However, there are several advantages to using *M. bovis* infection in cattle as a biomedical model such as similarities in pathology, latency of infection, and immune responses (Van Rhijn et al. 2008; Waters et al. 2014). Furthermore, studies in livestock have increased our understanding of immunological responses to tuberculosis in cattle and humans, and have provided valuable insight that can aid in the development of tuberculosis vaccines for the control and prevention of this disease worldwide (Waters et al. 2012; Waters and Palmer 2015).

Brucellosis is a contagious disease caused by infection of the zoonotic pathogen, *Brucella*, that causes abortion and infertility in humans and animals, and is a public health concern worldwide (Pappas et al. 2006; Franc et al. 2018). In many developed countries, *Brucella* has been eradicated; however, in developing countries the disease is still prevalent and poses a risk to human health (Pappas et al. 2006). Vaccines have proven effective in the control and prevention of brucellosis in livestock species; however, they cause abortion in pregnant animals and are virulent for humans (Lalsiamthara and Lee 2017). Therefore, research in livestock to develop safer vaccines is relevant for controlling the disease. Recent studies in pigs and sheep have identified safe vaccines for pregnant animals that are effective against *brucella* infection (Zriba et al. 2019; Hensel et al. 2020). Furthermore, there is no vaccine for humans and this research could lead to advancements in the development of an effective vaccine.

Transmissible spongiform encephalopathies (TSEs) are prion-misfolding diseases that cause neurodegeneration in humans and animals and can be transferred from one host to another. Bovine spongiform encephalopathy (BSE) in cattle is generally accepted as a zoonotic agent that causes variant Creutzfeldt-Jakob disease in humans and is believed to transfer to humans through the ingestion of contaminated tissue (Bruce et al. 1997; Dormont 2002; Greenlee and Greenlee 2015). Recent studies in cattle have demonstrated advances in the prion identification and classification, disease pathogenesis, and zoonosis of TSE and BSE (Hamir et al. 2011; Vrentas et al. 2013; Mammadova et al. 2020). This knowledge will aid in the prevention and treatment of TSE in livestock and may also translate to human medicine to improve diagnostics and identify therapeutic targets for treatment of the disease.

Models of Duchenne muscular dystrophy (DMD), a debilitating disease induced by frameshift mutations of the X-linked DMD gene, have been developed in pigs by deletion of exon 52 in the DMD gene in pig cells and offspring were generated by somatic cell nuclear transfer (SCNT; Klymiuk et al. 2013). A recent study using pig models of DMD have demonstrated an intramuscular application of gene-editing technology, clustered interspaced short palindromic repeat (CRISPR)/Cas9 in vivo that restored the DMD reading frame and improve muscle function in a porcine model of muscular dystrophy (Moretti et al. 2020). Further research using this pig model could have the potential to advance our understanding of the underlying mechanisms of the disease and develop strategies for treatment.

Cystic fibrosis (CF) is a genetic disease caused by mutations in the gene encoding cystic fibrosis transmembrane conductance regulator (CFTR) gene (Stoltz et al. 2015). Transgenic pigs have been developed, utilizing gene-editing technologies, to disrupt the CFTR gene in pig cells and SCNT to generate CFTR knockout piglets (Rogers et al. 2008; Klymiuk et al. 2012). These models demonstrate phenotypes that mirror humans CF disease but unfortunately these pigs develop meconium ileus that leads to death in 100% of offspring, compared with 10% occurrence in humans. Sheep lung structure and function are similar to human lung, making them a potential model for human CF (Polejaeva et al. 2016). Due to the anatomical and physiological similarities an ovine model of CF has been generated using CRISPR/Cas9 technology (Fan et al. 2018). These models of CF could be beneficial for the determination of intrauterine development of the disease and could lead to further understandings of the mechanisms of disease progression prenatally which could lead novel therapeutic strategies (Fan et al. 2018).

Over the past few decades poultry models have been used in biomedical research and to develop numerous cell lines, vaccinations, and establish stem cell lines (Farzaneh et al. 2017). Poultry have been established as an ideal model for developing therapeutic proteins and antibodies for use in humans. Extensive scientific data and availability of resources for poultry models make them a well-accepted and valuable model for biomedical research. In particular, multiple eggs can be generated in a short period of time, maintained in control conditions in the laboratory, and specific pathogen-free (SPF) fertilized eggs are available (Kraus et al. 2011).

Avian models, such as chicken and quail, are well-established models for developmental biology research and therefore ideal for toxicology studies (Smith et al. 2012). They are advantageous due to lost cost of commercially available fertile eggs, ease of maintenance in the laboratory settings, and advances in technology and methodology for evaluating specific stages of embryonic development. In addition, molecular pathways of development in poultry are highly conserved with mammals (Stark and Ross 2019). Chickens have been used for decades to determine toxicity to environmental and synthesized compounds during embryonic development (Hill and Hoffman 1984). which is important for understanding exposure to developing fetus and development of safe and efficacious treatments such as vaccines.

Similar to humans, chickens develop ovarian cancer at a rate of 10–35% depending on age and genetic background (Hawkridge 2014). Therefore, studies in chickens translate to humans due to similarities in morphology and molecular regulation. For example, humans and chickens demonstrate similar mutations in expression of p53, a key regulator of cell cycle and cancer cell growth (Hakim et al. 2009). A recent review by Hawkridge (2014) provides a comprehensive review of the use chicken to characterize ovarian cancer and test potential therapies in humans.

16.7 Conclusion

Domestic livestock species, including cattle, sheep, pigs, and poultry, have contributed significantly and are critical for advancements in biomedicine. Studies using livestock as biomedical models have increased our understanding of nutrition, physiology, metabolism, reproduction, fetal development and disease, that is applicable to animal health, production and human medicine. Future biomedical studies require the use of animal models, of which the rodent model will continue to be vital. However, researchers may find livestock species more advantageous in translational research due similarities in anatomy, physiology, metabolism, and genetics. Furthermore, recent advantages in, and the use of, genome-editing technologies will continue to prove useful in studies utilizing livestock to benefit human health and may lead to the development of novel strategies and therapies for both humans and animals.

Acknowledgements This work was supported by USDA-AFRI grant 2016-08401 and the Storrs Agricultural Experiment Station.

References

Abedal-Majed MA, Cupp AS (2019) Livestock animals to study infertility in women. Anim Front 9:28–33

Adams GP, Pierson RA (1995) Bovine model for study of ovarian follicular dynamics in humans. Theriogenology 43:113–120

Ball RO, Atkinson JL, Bayley HS (1986) Proline as an essential amino acid for the young pig. Br J Nutr 55:659–668

Barbero A, Astiz S, Lopez-Bote CJ, Perez-Solana ML, Ayuso M, Garcia-Real I, Gonzalez-Bulnes A (2013) Maternal malnutrition and offspring sex determine juvenile obesity and metabolic disorders in a swine model of leptin resistance. PLoS One 8:e78424

Barker DJP (2005) The developmental origins of insulin resistance. Horm Res 64(Suppl 3):2–7

Barry JS, Anthony RV (2008) The pregnant sheep as a model for human pregnancy. Theriogenology 69:55–67

Barry JS, Rozance PJ, Anthony RV (2008) An animal model of placental insufficiency-induced intrauterine growth restriction. Semin Perinatol 32:225–230

Bazer FW, Seo H, Johnson GA, Wu G (2021) One-carbon metabolism and development of the conceptus during pregnancy: Lessons from studies with sheep and pigs. Adv Exp Med Biol 1285:1–15

Berding K, Wang M, Monaco M, Alexander L, Mudd A, Chichlowski M et al (2016) Prebiotics and bioactive milk fractions affect gut development, microbiota, and neurotransmitter expression in piglets. J Pediatr Gastroenterol Nutr 63:688–697

Bergen WG (2022) Pigs (Sus Scrofa) in biomedical research. Adv Exp Med Biol 1354:335–343

Bispham J, Gardner DS, Gnanalingham MG, Stephenson T, Symonds ME, Budge H (2005) Maternal nutritional programming of fetal adipose tissue development: differential effects on messenger ribonucleic acid abundance for uncoupling proteins and peroxisome proliferator-activated and prolactin receptors. Endocrinology 146:3943–3949

Boleman SL, Graf TL, Mersmann HJ, Su DR, Krook LP, Savell JW, Park YW, Pond WG (1998) Pigs fed cholesterol neonatally have increased cerebrum cholesterol as young adults. J Nutr 128:2498–2504

Bossis I, Wettemann RP, Welty SD, Vizcarra J, Spicer LJ (2000) Nutritionally induced anovulation in beef Heifers: ovarian and endocrine function during realimentation and resumption of ovulation. Biol Reprod 62:1436–1444

Briffa JF, McAinch AJ, Romano T, Wlodek ME, Hryciw DH (2014) Leptin in pregnancy and development: a contributor to adulthood disease? Am J Physiol 308:E335–E350

Brix N, Ernst A, Lauridsen LLB, Arah OA, Nohr EA, Olsen J et al (2019) Maternal pre-pregnancy obesity and timing of puberty in sons and daughters: a population-based cohort study. Int J Epidemiol 48:1684–1694

Brown LD, Rozance PJ, Bruce JL, Friedman JE, Hay WW, Wesolowski SR (2015) Limited capacity for glucose oxidation in fetal sheep with intrauterine growth restriction. Am J Physiol 309:R920–R928

Bruce ME, Will RG, Ironside JW, McConnell I, Drummond D, Suttie A et al (1997) Transmissions to mice indicate that "new variant" CJD is caused by the BSE agent. Nature 389:498–501

Calbrix R, Guernsey J, Schinckel A, Stewart T, Herman E, Helm R, Nielsen N (2012) Development of tools to study immune-mediated allergenic responses to food and feed. In: Wilson RF (ed) Designing soybeans for 21st century markets. AOCS Press, pp 239–252

Camacho LE, Chen X, Hay WW, Limesand SW (2017) Enhanced insulin secretion and insulin sensitivity in young lambs with placental insufficiency-induced intrauterine growth restriction. Am J Physiol 313: R101–R109

Campbell BK, Souza C, Gong J, Webb R, Kendall N, Marsters P et al (2003) Domestic ruminants as models for the elucidation of the mechanisms controlling ovarian follicle development in humans. Reproduction 61(Suppl):429–443

Catalano PM, Shankar K (2017) Obesity and pregnancy: mechanisms of short term and long term adverse consequences for mother and child. BMJ 356:j1

CDC (2019). https://www.cdc.gov/reproductivehealth/infertility/index.htm

Clarke IJ, Henry BA (1999) Leptin and reproduction. Rev Reprod 4:48–55

Clarke IJ, Scaramuzzi RJ, Short RV (1976) Effects of testosterone implants in pregnant ewes on their female offspring. Development 36:87–99

Comim FV, Hardy K, Robinson J, Franks S (2015) Disorders of follicle development and steroidogenesis

in ovaries of androgenised foetal sheep. J Endocrinol 225:39–46
Daniel ZCTR, Brameld JM, Craigon J, Scollan ND, Buttery PJ (2007) Effect of maternal dietary restriction during pregnancy on lamb carcass characteristics and muscle fiber composition. J Anim Sci 85:1565–1576. https://doi.org/10.2527/jas.2006-743
Davis BT, Wang X-J, Rohret JA, Struzynski JT, Merricks EP, Bellinger DA et al. (2014) Targeted disruption of LDLR causes hypercholesterolemia and atherosclerosis in Yucatan miniature pigs. PLoS One 9:e93457
Dayton WR, White ME (2008) Cellular and molecular regulation of muscle growth and development in meat animals. J Anim Sci 86:E217–E225
De Blasio MJ, Gatford KL, McMillen IC, Robinson JS, Owens JA (2007) Placental restriction of fetal growth increases insulin action, growth, and adiposity in the young lamb. Endocrinology 148:1350–1358
Diskin MG, Mackey DR, Roche JF, Sreenan JM (2003) Effects of nutrition and metabolic status on circulating hormones and ovarian follicle development in cattle. Anim Reprod Sci 78:345–370
Dormont D (2002) Prion diseases: pathogenesis and public health concerns. FEBS Lett 529:17–21
Dyson MC, Alloosh M, Vuchetich JP, Mokelke EA, Sturek M (2006) Components of metabolic syndrome and coronary artery disease in female Ossabaw swine fed excess atherogenic diet. Comp Med 56:35–45
Eckert JE, Gatford KL, Luxford BG, Campbell RG, Owens PC (2000) Leptin expression in offspring is programmed by nutrition in pregnancy. J Endocrinol 165:R1-6
Eisemann JH, Lewis HE, Broome AI, Sullivan K, Boyd RD, Odle J, Harrell RJ (2014) Lysine requirement of 1.5–5.5 kg pigs fed liquid diets. Anim Prod Sci 54:608–615
Ella A, Barrière DA, Adriaensen H, Palmer DN, Melzer TR, Mitchell NL, Keller M (2019) The development of brain magnetic resonance approaches in large animal models for preclinical research. Anim Front 9:44–51
Fan Z, Perisse IV, Cotton CU, Regouski M, Meng Q, Domb C et al (2018) A sheep model of cystic fibrosis generated by CRISPR/Cas9 disruption of the CFTR gene. JCI Insight 3:e123529
Farzaneh M, Attari F, Mozdziak PE, Khoshnam SE (2017) The evolution of chicken stem cell culture methods. Br Poult Sci 58(6):681–686. https://doi.org/10.1080/00071668.2017.1365354. Epub 2017 Sep 26. PMID: 28840744
Ferenc K, Pietrzak P, Wierzbicka M, Matyba P, Grzesiuk E, Gajewski Z, Zabielski R (2018) Alterations in the liver of intrauterine growth retarded piglets may predispose to development of insulin resistance and obesity in later life. J Physiol Pharmacol 69:211–218
Fernández-Hernando C, Suárez Y, Rayner KJ, Moore KJ (2011) MicroRNAs in lipid metabolism. Curr Opin Lipidol 22:86–92
Ford SP, Zhang L, Zhu M, Miller MM, Smith DT, Hess BW et al (2009) Maternal obesity accelerates fetal pancreatic β-cell but not α-cell development in sheep: prenatal consequences. Am J Physiol 297: R835–R843
Franc KA, Krecek RC, Häsler BN, Arenas-Gamboa AM (2018) Brucellosis remains a neglected disease in the developing world: a call for interdisciplinary action. BMC Public Health 18:125
George LA, Zhang L, Tuersunjiang N, Ma Y, Long NM, Uthlaut AB et al (2012) Early maternal undernutrition programs increased feed intake, altered glucose metabolism and insulin secretion, and liver function in aged female offspring. Am J Physiol 302:R795–R804
Gerdts V, Wilson HL, Meurens F, van Drunen Littel-van den Hurk SC, Wilson D, Walker S, et al (2015) Large animal models for vaccine development and testing. ILAR J 56:53–62
Glauser EM (1966) Advantages of piglets as experimental animals in pediatric research. Exp Med Surg 24:181–190
Gonzalez-Bulnes A, Chavatte-Palmer P (2017) Contribution of large animals to translational research on prenatal programming of obesity and associated diseases. Curr Pharm Biotechnol 18:541–551
Govoni KE, Reed SA, Zinn SA (2019) Poor maternal nutrition during gestation: effects on offspring whole-body and tissue-specific metabolism in livestock species. J Anim Sci 97:3142–3152
Greenlee JJ, Greenlee MHW (2015) The transmissible spongiform encephalopathies of livestock. ILAR J 56:7–25
Gu Y, Li M, Wang T, Liang Y, Zhong Z, Wang X, et al (2012) Lactation-related MicroRNA expression profiles of porcine breast milk exosomes. PLoS One 7: e43691
Hackbart KS, Cunha PM, Meyer RK, Wiltbank MC (2013) Effect of glucocorticoid-induced insulin resistance on follicle development and ovulation. Biol Reprod 88:153
Hakim AA, Barry CP, Barnes HJ, Anderson KE, Petitte J, Whitaker R et al (2009) Ovarian adenocarcinomas in the laying hen and women share similar alterations in p53, ras, and HER-2/neu. Cancer Prev Res 2:114–121
Hales CN, Barker DJP (2001) The thrifty phenotype hypothesis. Br Med Bull 60:5–20
Hamernik DL (2019) Farm animals are important biomedical models. Anim Front 9:3–5
Hamir AN, Kehrli ME, Kunkle RA, Greenlee JJ, Nicholson EM, Richt JA et al (2011) Experimental interspecies transmission studies of the transmissible spongiform encephalopathies to cattle: comparison to bovine spongiform encephalopathy in cattle. J VET Diagn Invest 23:407–420
Hartanti MD, Rosario R, Hummitzsch K, Bastian NA, Hatzirodos N, Bonner WM, et al (2020) Could perturbed fetal development of the ovary contribute to the development of polycystic ovary syndrome in later life? PLoS One 15:e0229351

Hashimoto-Hill S, Kim M, Friesen L, Ajuwon KM, Herman E, Schinckel A, Kim CH (2019) Differential food protein-induced inflammatory responses in swine lines selected for reactivity to soy antigens. Allergy 74:1566–1569

Hawkridge AM (2014) The Chicken model of spontaneous ovarian cancer. Proteomics Clin Appl 8:689–699

Hensel ME, Garcia-Gonzalez DG, Chaki SP, Hartwig A, Gordy PW, Bowen R, et al (2020) Vaccine candidate *Brucella melitensis* 16MΔvjbR is safe in a pregnant sheep model and confers protection. mSphere 5: e00120–20

Herrera E, Ortega-Senovilla H (2014) Lipid metabolism during pregnancy and its implications for fetal growth. Curr Pharm Biotechnol 15:24–31

Hill EF, Hoffman DJ (1984) Avian models for toxicity testing. J Am Coll Toxicol 3:357–376

Hoffman ML, Reed SA, Pillai SM, Jones K, McFadden KK, Zinn SA, Govoni KE (2017) The effects of poor maternal nutrition during gestation on offspring postnatal growth and metabolism. J Anim Sci 95:2222–2232

Hou YQ, He WL, Hu SD, Wu G (2019) Composition of polyamines and amino acids in plant-source foods for human consumption. Amino Acids 51:1153–1165

Houseknecht KL, Baile CA, Matteri RL, Spurlock ME (1998) The biology of leptin: a review. J Anim Sci 76:1405–1420

Howell KR, Powell TL (2017) Effects of maternal obesity on placental function and fetal development. Reproduction 153:R97–R108

Ingvartsen KL, Boisclair YR (2001) Leptin and the regulation of food intake, energy homeostasis and immunity with special focus on periparturient ruminants. Dom Anim Endocrinol 21:215–250

Innis SM (2008) Dietary omega 3 fatty acids and the developing brain. Brain Res 1237:35–43

Innis SM (2009) Omega-3 fatty acids and neural development to 2 years of age: do we know enough for dietary recommendations? J Pediatr Gastroenterol Nutr 48:S16

Innis SM, de la Presa OS (2001) Dietary fatty acid composition in pregnancy alters neurite membrane fatty acids and dopamine in newborn rat brain. J Nutr 131:118–122

Ireland JJ, Roberts RM, Palmer GH, Bauman DE, Bazer FW (2008) A commentary on domestic animals as dual-purpose models that benefit agricultural and biomedical research. J Anim Sci 86:2797–2805

Ireland J, Gleason S (2017) Cows may offer clues to improving fertility in women. MSUToday. http://msutoday.msu.edu/news/2017/cows-may-offer-clues-to-improving-fertility-in-women/

Ji Y, Wu Z, Dai Z, Wang X, Li J, Wang B, Wu G (2017) Fetal and neonatal programming of postnatal growth and feed efficiency in swine. J Anim Sci Biotechnol 8:42

Jia Y, Cong R, Li R, Yang X, Sun Q, Parvizi N, Zhao R (2012) Maternal low-protein diet induces gender-dependent changes in epigenetic regulation of the glucose-6-phosphatase gene in newborn piglet liver. J Nutr 142:1659–1665

Jones AK, Rozance PJ, Brown LD, Goldstrohm DA, Hay WW, Limesand SW, Wesolowski SR (2019) Sustained hypoxemia in late gestation potentiates hepatic gluconeogenic gene expression but does not activate glucose production in the ovine fetus. Am J Physiol 317:E1–E10

Kim SW, McPherson RL, Wu G (2004) Dietary arginine supplementation enhances the growth of milk-fed young pigs. J Nutr 134:625–630

Kleinert M, Clemmensen C, Hofmann SM, Moore MC, Renner S, Woods SC et al (2018) Animal models of obesity and diabetes mellitus. Nat Rev Endocrinol 14:140–162

Klymiuk N, Mundhenk L, Kraehe K, Wuensch A, Plog S, Emrich D et al (2012) Sequential targeting of CFTR by BAC vectors generates a novel pig model of cystic fibrosis. J Mol Med 90:597–608

Klymiuk N, Blutke A, Graf A, Krause S, Burkhardt K, Wuensch A et al (2013) Dystrophin-deficient pigs provide new insights into the hierarchy of physiological derangements of dystrophic muscle. Hum Mol Genet 22:4368–4382

Kraus B, Von Fircks S, Feigl S, Koch SM, Fleischanderl D, Terler K, Dersch-Pourmojib M, Konetschny C, Grillberger L, Reiter M (2011) Avian cell line-technology for large scale vaccine production. BMC Proc 5:1–3. https://doi.org/10.1186/1753-6561-5-S8-P52

Krick BJ, Boyd RD, Roneker KR, Beermann DH, Bauman DE, Ross DA, Meisinger DJ (1993) Porcine somatotropin affects the dietary lysine requirement and net lysine utilization for growing pigs. J Nutr 123:1913–1922

Lalsiamthara J, Lee JH (2017) Development and trial of vaccines against Brucella. J Vet Sci 18:281–290

Li P, Wu G (2020) Composition of amino acids and related nitrogenous nutrients in feedstuffs for animal diets. Amino Acids 52:523–542

Li DF, Nelssen JL, Reddy PG, Blecha F, Hancock JD, Allee GL, Goodband RD, Klemm RD (1990) Transient hypersensitivity to soybean meal in the early-weaned pig. J Anim Sci 68:1790–1799

Li P, He WL, Wu G (2021) Composition of amino acids in foodstuffs for humans and animals. Adv Exp Med Biol 1332:189–209

Lie S, Morrison JL, Williams-Wyss O, Suter CM, Humphreys DT, Ozanne SE et al (2014) Impact of embryo number and maternal undernutrition around the time of conception on insulin signaling and gluconeogenic factors and microRNAs in the liver of fetal sheep. Am J Physiol 306:E1013–E1024

Lie S, Morrison JL, Williams-Wyss O, Suter CM, Humphreys DT, Ozanne SE et al (2016) Impact of maternal undernutrition around the time of conception on factors regulating hepatic lipid metabolism and microRNAs in singleton and twin fetuses. Am J Physiol 310:E148–E159

Limesand SW, Rozance PJ, Smith D, Hay WW (2007) Increased insulin sensitivity and maintenance of glucose utilization rates in fetal sheep with placental insufficiency and intrauterine growth restriction. Am J Physiol 293:E1716–E1725

Lind NM, Moustgaard A, Jelsing J, Vajta G, Cumming P, Hansen AK (2007) The use of pigs in neuroscience: Modeling brain disorders. Neurosci Biobehav Rev 31:728–751

Litten-Brown JC, Corson AM, Clarke L (2010) Porcine models for the metabolic syndrome, digestive and bone disorders: a general overview. Animal 4:899–920

Liu J, Chen D, Yao Y, Yu B, Mao X, He J, et al (2012) Intrauterine growth retardation increases the susceptibility of pigs to high-fat diet-induced mitochondrial dysfunction in skeletal muscle. PLoS One 7:e34835

Long NM, George LA, Uthlaut AB, Smith DT, Nijland MJ, Nathanielsz PW, Ford SP (2010) Maternal obesity and increased nutrient intake before and during gestation in the ewe results in altered growth, adiposity, and glucose tolerance in adult offspring. J Anim Sci 88:3546–3553

Long NM, Ford SP, Nathanielsz PW (2011) Maternal obesity eliminates the neonatal lamb plasma leptin peak. J Physiol 589:1455–1462

Long NM, Rule DC, Zhu MJ, Nathanielsz PW, Ford SP (2012) Maternal obesity upregulates fatty acid and glucose transporters and increases expression of enzymes mediating fatty acid biosynthesis in fetal adipose tissue depots. J Anim Sci 90:2201–2210

Luther J, Aitken R, Milne J, Matsuzaki M, Reynolds L, Redmer D, Wallace J (2007) Maternal and fetal growth, body composition, endocrinology, and metabolic status in undernourished adolescent sheep. Biol Reprod 77:343–350

Mackey DR, Sreenan JM, Roche JF, Diskin MG (1999) Effect of acute nutritional restriction on incidence of anovulation and periovulatory estradiol and gonadotropin concentrations in beef heifers. Biol Reprod 61:1601–1607

Mammadova N, West Greenlee MH, Moore SJ, Sakaguchi DS, Greenlee JJ (2020) Experimental study using multiple strains of prion disease in cattle reveals an inverse relationship between incubation time and misfolded prion accumulation, neuroinflammation, and autophagy. Am J Pathol 190:1461–1473

Manjarín R, Columbus DA, Suryawan A, Nguyen HV, Hernandez-García AD, Hoang NM et al (2016) Leucine supplementation of a chronically restricted protein and energy diet enhances mTOR pathway activation but not muscle protein synthesis in neonatal pigs. Amino Acids 48:257–267

Manjarín R, Columbus DA, Solis J, Hernandez-García AD, Suryawan A, Nguyen HV et al (2018) Short and long-term effects of leucine and branched-chain amino acid supplementation of a protein and energy reduced diet on muscle protein metabolism in neonatal pigs. Amino Acids 50:943–959

Marciniak A, Patro-Małysza J, Kimber-Trojnar Z, Marciniak B, Oleszczuk J, Leszczyńska-Gorzelak B (2017) Fetal programming of the metabolic syndrome. Taiwanese J Obstetr Gynecol 56:133–138

Martin DE, Jones AK, Pillai SM, Hoffman ML, McFadden KK, Zinn SA et al (2019) Maternal restricted- and over-feeding during gestation result in distinct lipid and amino acid metabolite profiles in the longissimus muscle of the offspring. Front Physiol 10:515

Mathews SA, Oliver WT, Phillips OT, Odle J, Diersen-Schade DA, Harrell RJ (2002) Comparison of triglycerides and phospholipids as supplemental sources of dietary long-chain polyunsaturated fatty acids in piglets. J Nutr 132:3081–3089

McDonald SD, Han Z, Mulla S, Beyene J (2010) Overweight and obesity in mothers and risk of preterm birth and low birth weight infants: systematic review and meta-analyses. BMJ 341:c3428

Meurens F, Summerfield A, Nauwynck H, Saif L, Gerdts V (2012) The pig: a model for human infectious diseases. Trends Microbiol 20:50–57

Michelotti JM, Yeh KB, Beckham TR, Colby MM, Dasgupta D, Zuelke KA, Olinger GG (2018) The convergence of high-consequence livestock and human pathogen research and development: a paradox of zoonotic disease. Trop Med Infect Dis 3:55

Moeser A (2015) Gut barrier function in pigs: influence of stress, management, and nutrition. Minnesota Nutrition Conference. University of Minnesota, St. Paul, MN

Moretti A, Fonteyne L, Giesert F, Hoppmann P, Meier AB, Bozoglu T et al (2020) Somatic gene editing ameliorates skeletal and cardiac muscle failure in pig and human models of Duchenne muscular dystrophy. Nat Med 26:207–214

Mossa F, Ireland JJ (2019) Anti-Müllerian hormone: a biomarker for the ovarian reserve, ovarian function, and fertility in dairy cows. J Anim Sci 97:1446–1455

Mudd AT, Dilger RN (2017) Early-life nutrition and neurodevelopment: Use of the piglet as a translational model. Adv Nutr 8:92–104

Odle J, Lin X, Jacobi SK, Kim SW, Stahl CH (2014) The suckling piglet as an agrimedical model for the study of pediatric nutrition and metabolism. Annu Rev Anim Biosci 2:419–444

Odle J, Jacobi SK, Boyd RD, Bauman DE, Anthony RV, Bazer FW et al (2017) The potential impact of animal science research on global maternal and child nutrition and health: a landscape review. Adv Nutr 8:362–381

Owens JA, Kind KL, Carbone F, Robinson JS, Owens PC (1994) Circulating insulin-like growth factors-I and -II and substrates in fetal sheep following restriction of placental growth. J Endocrinol 140:5–13

Padmanabhan V, Veiga-Lopez A (2013) Sheep models of polycystic ovary syndrome phenotype. Mol Cell Endocrinol 373:8–20

Pappas G, Papadimitriou P, Akritidis N, Christou L, Tsianos EV (2006) The new global map of human brucellosis. Lancet Infect Dis 6:91–99

Pendleton AL, Humphreys LR, Davis MA, Camacho LE, Anderson MJ, Limesand SW (2019) Increased pyruvate dehydrogenase activity in skeletal muscle of growth-restricted ovine fetuses. Am J Physiol 317: R513–R520

Perleberg C, Kind A, Schnieke A (2018) Genetically engineered pigs as models for human disease. Dis Model Mech 11:dmm030783

Polejaeva IA, Rutigliano HM, Wells KD (2016) Livestock in biomedical research: history, current status and future prospective. Reprod Fertil Dev 28:112–124

Pollock JM, Neill SD (2002) Mycobacterium boviss infection and tuberculosis in cattle. Vet J 163:115–127

Pond WG, Boleman SL, Fiorotto ML, Ho H, Knabe DA, Mersmann HJ et al (2000) Perinatal ontogeny of brain growth in the domestic pig. Proc Soc Exp Biol Med 223:102–108

Pourquié O (2018) Somite formation in the chicken embryo. Int J Dev Biol 62:57–62

Radcliffe JS, Brito LF, Reddivari L, Schmidt M, Herman EM, Schinckel AP (2019) A swine model of soy protein-induced food allergenicity: implications in human and swine nutrition. Anim Front 9:52–59

Raja JS, Hoffman ML, Govoni KE, Zinn SA, Reed SA (2016) Restricted maternal nutrition alters myogenic regulatory factor expression in satellite cells of ovine offspring. Animal 10:1200–1203

Reed SA, Raja JS, Hoffman ML, Zinn SA, Govoni KE (2014) Poor maternal nutrition inhibits muscle development in ovine offspring. J Anim Sci Biotechnol 5:43

Reed SA, LaVigne EK, Jones AK, Patterson DF, Schauer AL (2015) The aging horse: effects of inflammation on muscle satellite cells. J Anim Sci 93:862–870

Reynolds CM, Vickers MH (2018) Utility of small animal models of developmental programming. In: Guest PC (ed) Investigations of early nutrition effects on long-term health: methods and applications. Springer, New York NY, pp 145–163

Reynolds LP, McLean KJ, Kacie L. McCarthy KL, Diniz WJS, Menezes ACB, Forcherio JC, Scott RR, Ward AK, Dahlen CR, Caton JS (2022) Nutritional regulation of embryonic survival, growth and development. Adv Exp Med Biol 1354:63–76

Reynolds LP, Ireland JJ, Caton JS, Bauman DE, Davis TA (2009) Commentary on domestic animals in agricultural and biomedical research: an endangered enterprise. J Nutr 139:427–428

Rhodes FM, Fitzpatrick LA, Entwistle KW, De'ath G, (1995) Sequential changes in ovarian follicular dynamics in *Bos indicus* heifers before and after nutritional anoestrus. J Reprod Fertil 104:41–49

Rogers CS, Hao Y, Rokhlina T, Samuel M, Stoltz DA, Li Y et al (2008) Production of *CFTR*-null and *CFTR-ΔF508* heterozygous pigs by adeno-associated virus–mediated gene targeting and somatic cell nuclear transfer. J Clin Invest 118:1571–1577

Romereim SM, Summers AF, Pohlmeier WE, Zhang P, Hou X, Talbott HA et al (2017) Gene expression profiling of bovine ovarian follicular and luteal cells provides insight into cellular identities and functions. Mol Cell Endocrinol 439:379–394

Roth JA (2011) Veterinary vaccines and their importance to animal health and public health. Proc Vaccinol 5:127–136

Roth JA, Tuggle CK (2015) Livestock models in translational medicine. ILAR J 56:1–6

Sandoval C, Wu G, Smith SB, Dunlap KA, Satterfield MC (2020) Maternal nutrient restriction and skeletal muscle development: Consequences for postnatal health. Adv Exp Med Biol 1265:153–165

Satterfield MC, Dunlap KA, Keisler DH, Bazer FW, Wu G (2012) Arginine nutrition and fetal brown adipose tissue development in diet-induced obese sheep. Amino Acids 43:1593–1603

Satterfield MC, Dunlap KA, Keisler DH, Bazer FW, Wu G (2013) Arginine nutrition and fetal brown adipose tissue development in nutrient-restricted sheep. Amino Acids 45(3):489–499

Satterfield MC, Edwards AK, Bazer FW, Dunlap KA, Steinhauser CB, Wu G (2021) Placental adaptation to maternal malnutrition. Reproduction 162:R73–R83

Scaal M, Marcelle C (2018) Chick muscle development. Int J Dev Biol 62:127–136

Sinclair KD, Molle G, Revilla R, Roche JF, Quintans G, Marongiu L et al (2002) Ovulation of the first dominant follicle arising after day 21 post partum in suckling beef cows. Anim Sci 75:115–126

Sirard M-A (2018) 40 years of bovine IVF in the new genomic selection context. Reproduction 156:R1–R7

Smith S, Pfeifer SM, Collins JA (2003) Diagnosis and management of female infertility. JAMA 290:1767–1770

Smith P, Steckler TL, Veiga-Lopez A, Padmanabhan V (2009) Developmental programming: differential effects of prenatal testosterone and dihydrotestosterone on follicular recruitment, depletion of follicular reserve, and ovarian morphology in sheep. Biol Reprod 80:726–736

Smith SM, Flentke GR, Garic A (2012) Avian models in teratology and developmental toxicology. In: Harris C, Hansen JM (eds) Developmental toxicology: methods and protocols. Humana Press, Totowa, NJ, pp 85–103

Smith AM, Pankey CL, Odhiambo JF, Ghnenis AB, Nathanielsz PW, Ford SP (2018) Rapid communication: reduced maternal nutrition during early- to mid-gestation elevates newborn lamb plasma cortisol concentrations and eliminates the neonatal leptin surge. J Anim Sci 96:2640–2645

Stark MR, Ross MM (2019) The chicken embryo as a model in developmental toxicology. In: Hansen JM, Winn LM (eds) Developmental toxicology: methods and protocols. Springer, New York, NY, pp 155–171

Stenhouse C, Seo H, Wu G, Johnson GA, Bazer FW (2022) Insights into the regulation of implantation and placentation in humans, rodents, sheep, and pigs. Adv Exp Med Biol 1354:25–48

Stoltz DA, Meyerholz DK, Welsh MJ (2015) Origins of cystic fibrosis lung disease. N Engl J Med 372:351–362

Summers AF, Pohlmeier WE, Sargent KM, Cole BD, Vinton RJ, Kurz SG et al (2014) Altered theca and cumulus oocyte complex gene expression, follicular arrest and reduced fertility in cows with dominant follicle follicular fluid androgen excess. PLoS One 9: e110683

Swanson AM, David AL (2015) Animal models of fetal growth restriction: considerations for translational medicine. Placenta 36:623–630

Symonds ME, Sebert SP, Hyatt MA, Budge H (2009) Nutritional programming of the metabolic syndrome. Nat Rev Endocrinol 5:604–610

Taylor JF, Schnabel RD, Sutovsky P (2018) Identification of genomic variants causing sperm abnormalities and reduced male fertility. Anim Reprod Sci 194:57–62

Triantafyllou GA, Paschou SA, Mantzoros CS (2016) Leptin and hormones: energy homeostasis. Endocrinol Metab Clin North Am 45:633–645

Urick ME, Giles JR, Johnson PA (2009) Dietary aspirin decreases the stage of ovarian cancer in the hen. Gynecol Oncol 112:166–170

Vallet JL, Miles JR, Rempel LA (2013) A simple novel measure of passive transfer of maternal immunoglobulin is predictive of preweaning mortality in piglets. Vet J 195:91–97

Van Rhijn I, Godfroid J, Michel A, Rutten V (2008) Bovine tuberculosis as a model for human tuberculosis: advantages over small animal models. Microbes Infect 10:711–715

Vrentas CE, Greenlee JJ, Baron T, Caramelli M, Czub S, Nicholson EM (2013) Stability properties of PrPScfrom cattle with experimental transmissible spongiform encephalopathies: use of a rapid whole homogenate, protease-free assay. BMC Vet Res 9:167

Wang W, Dai Z, Wu Z, Lin G, Jia S, Hu S et al (2014) Glycine is a nutritionally essential amino acid for maximal growth of milk-fed young pigs. Amino Acids 46:2037–2045

Waters WR, Palmer MV (2015) Mycobacterium bovis infection of cattle and white-tailed deer: translational research of relevance to human tuberculosis. ILAR J 56:26–43

Waters WR, Palmer MV, Buddle BM, Vordermeier HM (2012) Bovine tuberculosis vaccine research: historical perspectives and recent advances. Vaccine 30:2611–2622

Waters WR, Maggioli MF, McGill JL, Lyashchenko KP, Palmer MV (2014) Relevance of bovine tuberculosis research to the understanding of human disease: historical perspectives, approaches, and immunologic mechanisms. Vet Immunol Immunopathol 159:113–132

Williams GL, Amstalden M, Garcia MR, Stanko RL, Nizielski SE, Morrison CD, Keisler DH (2002) Leptin and its role in the central regulation of reproduction in cattle. Dom Anim Endocrinol 23:339–349

Wilson PR, Tarttelin MF (1978) Studies on sexual differentiation of sheep. I. Foetal and maternal modifications and post-natal plasma LH and testosterone content following androgenisation early in gestation. Acta Endocrinol 89:182–189

Wu G (2020) Important roles of dietary taurine, creatine, carnosine, anserine and hydroxyproline in human nutrition and health. Amino Acids 52:329–360

Wu G (2022) Nutrition and metabolism: foundations for animal growth, development, reproduction, and health. Adv Exp Med Biol 1354:1–24

Wu G, Bazer FW, Wallace JM, Spencer TE (2006) Board-invited review: intrauterine growth retardation: implications for the animal sciences. J Anim Sci 84:2316–2337

Wu G, Bazer FW, Burghardt RC, Johnson GA, Kim SW, Knabe DA et al (2011) Proline and hydroxyproline metabolism: implications for animal and human nutrition. Amino Acids 40:1053–1063

Yan X, Huang Y, Zhao JX, Rogers CJ, Zhu MJ, Ford SP et al (2013) Maternal obesity downregulates microRNA let-7g expression, a possible mechanism for enhanced adipogenesis during ovine fetal skeletal muscle development. Int J Obesity 37:568–575

Yan H, Zheng P, Yu B, Yu J, Mao X, He J et al (2017) Postnatal high-fat diet enhances ectopic fat deposition in pigs with intrauterine growth retardation. Eur J Nutr 56:483–490

Yao K, Guan S, Li T, Huang R, Wu G, Ruan Z, Yin Y (2011) Dietary l-arginine supplementation enhances intestinal development and expression of vascular endothelial growth factor in weanling piglets. Br J Nutr 105:703–709

Zhang L, Long NM, Hein SM, Ma Y, Nathanielsz PW, Ford SP (2011) Maternal obesity in ewes results in reduced fetal pancreatic β-cell numbers in late gestation and decreased circulating insulin concentration at term. Dom Anim Endocrinol 40:30–39

Zhao L, Huang Y, Du M (2019) Farm animals for studying muscle development and metabolism: dual purposes for animal production and human health. Anim Front 9:21–27

Zriba S, Garcia-Gonzalez DG, Khalaf OH, Wheeler L, Chaki SP, Rice-Ficht A et al (2019) Vaccine safety studies of *Brucella abortus* S19 and S19ΔvjbR in pregnant swine. Vaccine: X 3:100041

Pigs (*Sus Scrofa*) in Biomedical Research

17

Werner G. Bergen

Abstract

Much of biomedical oriented research is conducted with animal models. Over the years, rodents (primarily rats and mice) have emerged as the preferred species for basic biochemistry, cell biology, physiology and nutrition studies. In the past, dogs have been used for the evaluation of dietary protein quality and other aspects of animal nitrogen metabolism and physiology, cardiovascular and endocrine research. At an increasing rate, pigs have also been used as a model species in biomedical research. Pigs are readily available in various mature sizes and genotypic/phenotypic traits, and there are many anatomic, nutritional and physiologic similarities between human beings and pigs. Many notable reviews summarizing the role of pigs in biomedical studies have already been published and these are cited below. The present review focuses on characteristics that make pigs an excellent biomedical animal model in particular in obesity, diabetes and cardiovascular research. To procure an animal model for obesity, irrespective of species used, these animals must be fed a dense caloric diet (high fat) to achieve an experimental working model within a reasonable period. This review also focuses on a putative role of gastrointestinal microbiota in obesity as obese animals exhibit a shift in the distribution of gastrointestinal microbial phyla from lean animals. But to date such results have not pinpointed a treatable cause for obesity. Sometimes, the choice of sampling sites for microbial assessment in many reports can be questioned as the microbial content and phyla distribution in easily collected fecal samples may differ from those obtained directly from the small intestine and upper colon. While pigs are still utilized in many countries for medical surgery practice, this has been discontinued in US medical schools.

Keywords

Pigs · Biomedical research · Experimental animal model species · Gut microbiota · Obesity

17.1 Introduction

Pond (1991) wrote an introductory chapter named "Of pigs and people" in the classical treatise "Swine Nutrition" edited by Miller et al. (1991). In his chapter, Pond (1991) briefly summarizes pig domestication, demographic relationships such as ratios of pigs to people, pigs as a key animal food source and of the clearly

W. G. Bergen (✉)
Department of Animal Sciences, Auburn University,
AL Auburn 210 Upchurch Hall, 36854, USA
e-mail: bergewg@auburn.edu

G. Wu (ed.), *Recent Advances in Animal Nutrition and Metabolism*,
Advances in Experimental Medicine and Biology 1354,
https://doi.org/10.1007/978-3-030-85686-1_17

vast differences in genetic backgrounds, body size and shape and body fat and lean ratios in pigs across the world. He then reiterates the anatomic, nutritional and physiologic similarities between people and pigs. Pond (1991) highlighted the following similarities, which became relevant to the impetus of using pigs for biomedical research: skin physiology, dental characteristics, kidney morphology and function, eye structure and visual acuity, cardiovascular anatomy, physiology, nutritional requirements and body composition. Notable articles and reviews on the role of pigs in biomedical research include Pond (1982), Stanton and Mersmann (1986), Tumbleson (1986), Swindle et al. (1992), Baker (2008) and Gutierrez et al. (2015). Spurlock and Gabler (2008) and Walters et al. (2017) have authored more recent excellent exposés on pigs as a model species for biomedical research. Nafikov and Beitz (2007) in a review focused on comparative carbohydrate and lipid metabolism in farm animals and on pigs as a model for biomedical research in humans. They also noted that hundreds of publications using pigs as human models have provided scientific data on cardiovascular physiology, skeletal muscle growth, intestinal microbial metabolism, obesity, immunology, diabetes, renal physiology, lipoprotein and cholesterol metabolism and many other biomedical topics (Dai et al. 2022; Nafikov and Beitz 2007; Smith and Govoni 2022). Finally, researchers have used the commercial pigs to explore fundamental aspects of mammalian protein metabolism. Notable examples here are the work by Teresa Davis and her group on the influence of leucine administration in low birth-weight pigs on enhancing skeletal muscle protein synthesis and the role of the mTOR signal pathway in these metabolic processes (Rudar et al. 2019). Further, Wu and co-workers described the role of arginine in urea metabolism in pigs and the roles of other amino acids (e.g., glutamine, glycine, 4-hydroxyproline, and tryptophan) in nutrient signaling in the intestine of pigs (Dillon and Wu 2021; Hu et al. 2021; Rhoads and Wu 2009; Wu 2021; Wu and Morris 1998; Wu et al. 2009). Sulfur amino acids metabolism, particularly methionine trans-methylation and its trans-sulfuration to cysteine was extensively explored in pigs by Baker (2006) and Benevenga (1984). The aim of this chapter is to reiterate early examples of using pigs in biomedical research and then explore pig use from studies that are more contemporary.

17.2 Characteristics of Pigs that Make Them a Model for Biomedical Research

Since pigs are a major human dietary meat source, agricultural/food production research continues to use swine focusing on nutrition, reproductive biology and fat and lean gains patterns (Bazer et al. 2021; Bergen and Brandebourg 2016; Smith and Govoni 2022; Stenhouse et al. 2022). Commercial pigs (USA) by the early twenty-first century had also changed from typically fat animals in the past, to now highly muscled, low fat content animals. This notable body composition shift was a direct consequence of dedicated genetic selection for lean pigs over the last 50 years. However, many farm-type/commercial swine breeds around the world still exhibit traditional body composition traits. Furthermore, so called miniature pig breeds (such as Yucatan, Ossabaw, Sinclair (Hormel Institute), Goettingen and others) are firmly established as excellent model species for obesity, diabetics and cardiovascular research.

Much of experimental research with pigs has focused on various aspects of obesity and in particular as related to cardiovascular maladies. As pigs mature, they may continue to consume excess feed and progressively get fatter, and in time to exhibit cardiovascular system related symptomologies. However, the time needed for obtaining obese pigs for chronic cardiovascular studies using commercial pigs will necessitate longer-term animal trials and handling of large animals. The feeding of young pigs with high fat/cholesterol diets (40% fat calories and 2% cholesterol) can significantly reduce the length of such experiments. This approach has become relatively standard for many obesity studies with pigs (Zhang and Lerman 2016). Bergen and

Merkel (1991a) noted that human diets often contain 40% fat calories, and are not typically high carbohydrate diets *(Zea maize* starch plus protein supplements) as usually consumed by pigs. With regard to the cardiovascular system, numerous physiological and anatomical characteristics shared between humans and pigs are not shared with typical laboratory rodent models (Swindle and Smith 2013).

When studying obesity and lipid metabolism in pigs as a model for humans, it must be recognized that there are also some significant differences between humans and pigs (Mersmann 1986; Bergen and Mersmann 2005; Nafikov and Beitz 2007) in de novo fatty acid synthesis and some aspects of lipoprotein trafficking. Among these differences are the primary site of de novo fatty acid synthesis and in triacylglycerol/lipoprotein metabolism. In humans, de novo fatty acid synthesis occurs primarily in the liver while adipose tissue is the principal site for fatty acid synthesis in pigs (Bergen and Mersmann 2005). This in turn also changes the pattern of lipid transport among tissues between pigs and humans. In humans, fatty acids are synthesized in the liver and incorporated into triacylglycerol and then exported via lipoproteins to other organs such as adipose tissues or skeletal and cardiac muscles. In pigs, de novo synthesized fatty acids when released from adipose tissues as non-esterified fatty acids and when arriving at the liver may either be oxidized or be re-incorporated into triacylglycerol and then exported from the liver as lipoproteins. This dichotomy of fatty acid synthesis between humans and pigs does not, however, appear to change other lipid transport dynamics nor has any apparent influence on cardiovascular biology.

Instead of listing all available porcine models, especially miniature pigs, in this chapter, the author point the readers to the following citations (Pond 1982; Tumbleson 1986; Swindle et al. 1992; Vodicka et al. 2012; Spurlock and Gabler 2008; Gutierrez et al. 2015; Wu 2022). In addition, at the University of Missouri, Prather and co-worker have been producing transgenic and knockout pigs for many specific research needs (Walters et al. 2017). As noted above, in most obesity pig or rodent models, high fat and high cholesterol diets are fed to reach the state of obesity in a reasonable timeframe for experimental work. Brandebourg and his group (Bergen and Brandebourg 2016) have been working with an exotic pig line “Mangalica” from the Balkans (Europe). Mangalica pigs may be a porcine model for hyperphagic juvenile obesity, early onset cardiovascular diseases and other metabolic diseases related to obesity. These pigs develop the obesity related symptoms without the necessity of feeding high fat/high cholesterol diets.

17.3 Why Animal/Pig Model Research?

As omnivores, both swine and humans consume plant- and animal-sourced foods (Hou et al. 2019; Li and Wu 2020; Li et al. 2021; Sarkar et al. 2021). A primary objective of studying various aspects of lipid metabolism in pig and other animal models is to get an understanding of what causes obesity followed by delineation of physiologic/metabolic changes and health consequences of obesity. This then, in a perfect world, was to be followed up by research on how to ameliorate these clinical conditions and significantly lower associated mortality and morbidity first in animal models and then in human patients. Below I describe a short summary of what we have learned to date from animal/pig models and how this work can be applied in clinical approaches.

In reality, the root-cause of obesity is well known. The cause of overweight is the long-term overconsumption of dietary energy in relation to the energy requirements of humans. This fact has often been clearly stated, but the concept of an ideal “weight”, and how it may reflect the health of people, is not a simple clear-cut benchmark. Further, the ability to control food intake also varies widely among people. While many obese patients suffer from compromised cardiovascular function, diabetes and insulin resistance, not all overweight people do. All animal model studies indicate that excess calories must be consumed to initiate fat cell development and adipose tissue expansion in a reasonable period. Therefore, a solution would seem to be "Do not overeat"; but when do we start such a regime in humans? In experimental animal studies, energy intake can be controlled, but such restrictions are not imposable in a general sense on humans. Often extreme self-discipline is necessary to eat just "right" and in addition to exercise to maintain body weight to achieve reasonable, healthy standards. From animal model work, researchers have identified and manipulated physiological/biochemical processes using genomic tools, knockout models and managed feeding regimens involved in weight/fat gain and loss. These findings are often not quickly translated into treatments and patient management schemes in the human clinical conditions of obesity, diabetes and metabolic syndrome.

What we have learned from obesity models is that once significant amounts of fats are deposited in the body, and liver and endocrine functions are compromised, reversing overweight and reversing arising morbidity and mortality are not an easy task (Maruvada et al. 2017). This reality is a significant drawback to developing therapies for established obesity. Once obesity is established, the countervailing mechanisms controlling energy intake and energy retention are modified (Maruvada et al. 2017). While research has also pointed to the potential roles of growth hormone, epinephrine and other molecules in regulating energy intake and fattening in pig models (Bergen and Merkel 1991b; Etherton 2000), there is a general hesitancy to apply such pharmacological approaches to clinical practice of humans. A "lean pill" breakthrough has also not materialized (Dodson et al. 2011, 2012, 2013). Therapeutic treatments would presumably have an effect on food intake and rebalancing body composition and endocrine functions.

17.4 The Gut Microbiome in Obesity Research

A putative role of the digestive tract microbiome/microbiota in development of human obesity is also emerging from animal work. There is an association between the digestive tract microbiome and feed efficiency in pigs (Quan et al. 2019) and most likely a negative role of pathogens on inflammations in the gut. More efficient dietary energy use would lower dietary energy needed to maintain a "healthy" weight in humans. While such emerging knowledge may certainly be of use in animal-agricultural applications (Wu 2022), conversely humans blessed with this type of microbiota are facing a possible necessity of even eating less.

Ever since the role of the gastrointestinal microbiota had been recognized as a major influencer of health and disease (Ley et al. 2005; Turnbaugh et al. 2006), a plethora of papers have been published indicating that the microbiota may be managed to alleviate almost all serious human chronic metabolic diseases. Concurrently, many papers were published which were less positive and more cautious in their data analysis about the role of the gut microbiota in obesity and metabolic diseases. A very common trend in many of these papers is that there is a high frequency of claimed associations and correlations between changes in number of certain organism/phyla of the microbiota with diets and obesity. While such data are useful for furthering thinking and designing new research approaches, statistical associations and correlations are not "causative" factors. As will be described below, diet and likely obesity in animal models and humans will affect the relative presence of the anaerobic macro-phyla of *Firmicutes* and *Bacteroidetes* in the gastrointestinal tract. There is also documented evidence that feeding fat above 5% (of dry-matter) suppresses fermentative capacity in pre-gastric anaerobic microbial ecosystems often characteristic in herbivores (Boggs et al. 1987) and by analogy the small intestinal and cecum/colon microbial ecosystem in pigs. Others have reported that *Firmicutes* and *Proteobacteria* are more abundant in the proximal GI tract, while *Bacteroidetes*, *Verrucomicrobia* and *Akkermansia* are more abundant in the distal GI tract (Gu et al. 2013; Donaldson et al. 2016; Scheithauer et al. 2016). *Firmicutes* are usually gram positive and acid-producing organisms while *Bacteroidetes* are gram-negative fiber digesters. Finally, many research efforts have only assessed anaerobic biota in fecal samples rather than in more difficult to obtain small intestinal, colon and large intestinal contents. This results in some interpretative problems with such data.

Nevertheless, research over the last quarter century has clearly identified an association between the gut microbiome and metabolic maladies (Dabke et al. 2019). A very typical finding in gut microbiota and obesity studies is that in obese experimental animals the gut *Bacteroidetes* abundance declines whereas the *Firmicutes* abundance increases (Chen and Devaraj 2018; Ley et al. 2005; Baeckhed et al. 2005) when compared to lean animals. Others demonstrated that this pattern of variable gut organisms' abundance is also observed in the human intestine (Baeckhed et al. 2004; Ridaura et al. 2013). Fecal microbiota transplants (FMT) are an experimental tool that has highlighted the role of gut microbiota in obesity and diabetes (Wargo 2020). In numerous studies, FMT from lean and obese animals have been cross-administered to lean and obese animals. FMT from obese animals to lean animals would result generally in an obese animal and the reverse (lean FMT to obese resulting in lean phenotype) was true as well (Ridaura et al. 2013; Chen and Devaraj 2018). FMT will, however, need to be repeated in a long-term strategy of obesity control, as gut microbiomes will eventually revert to the original pattern of *Firmicutes* and *Bacteroidetes* (Koottee et al. 2017; Dabke et al. 2019). Another caveat is that most studies to date have sampled gut contents or feces at specific time points, thereby often ignoring any time-trend variations in the microbiota. At the time of writing this review, there are no experimentally described causative associations between *Firmicutes* and *Bacteroidetes*, proteolytic bacteria nor any other gut microbes with any specific aspect of obesity (Cani and Van Hul 2020; Wargo 2020).

Workers have attempted to define which animal models would be most appropriate for basic and clinical studies into the role of the gut microbiome in lipid metabolism and obesity in animals. Much work has been completed and published with rodents and humans. A critical point is the source of the microbiota, that is, from gastrointestinal (GIT) or fecal samples. Numerous studies were conducted with either commercial (farm) or miniature pigs to explore the relationship between the gut microbiome and obesity, metabolic syndrome, and other related co-morbidities. In particular, specific interactions between dietary components, the microbiome and host animals are still not well delineated, as reviewed above.

The most common models have been mice and rats and miniature or full-size pigs. The gastrointestinal tract of pigs is anatomically and physiologically alike to the human GIT (Bergen and Wu 2009; Mu et al. 2022). Other workers have also introduced dogs as a further model species in microbiome, host weight gain and obesity and diet compositions research (Oberbauer and Larsen 2021). I believe that the US public would have an overwhelming negative reaction to using dogs as experimental biomedical animal models when slaughter, tissue harvesting and sampling gut contents are involved; however, routine feeding studies, fecal collections for microbiome assessment and body composition (lean/fat) determinations using weight or imaging technology would be considered non-invasive. Coelho et al. (2018) conducted comparative nutritional studies with fecal microbial phyla assessments in dogs, and then compared their results to published studies with humans (Li et al. 2014), pigs (Xiao et al. 2016) and mice (Xiao et al. 2015, 2016). They found that the GIT microbiome taxonomic annotation by phylum of the four host species contained the same major phyla (Firmicutes, Bacteroidetes, and Proteobacteria) and overall were similar; the dog microbiome was more comparable to humans, while pigs and mice ranked next and last, respectively. In their nutritional studies, comparing a high protein: low carbohydrate (HPLC) diet with a low protein: high carbohydrate (LPHC) diets in dogs, the HPLC diet had a Clostridiales-enriched (Firmicutes), while the LPHC had a Bacteroidiales enriched microbiota. These findings reflect the multiple similar results from studies with other species. Since dogs have been humankind's best friends for ages, it is my contention that pigs would be better suited for intensive microbiome studies requiring sampling of proximal and distal GIT sites.

Breeds of pigs exhibit different propensities to fatten. For example, Reiter et al. (2007) showed that in lean-type pigs (Pietrain) and in fat-type pigs (Duroc), the expression of transcription factors related to de novo fatty acid synthesis (DNL) and fatty acid oxidation differed. Fat-type pigs had a higher expression of DNL related genes and lower expression of fatty acid oxidation genes, while the complete opposite was observed in lean-type pigs. Yang et al. (2018) utilized fecal transplantation from adult Jinhua (Chinese obese pig) and Landrace (lean) pigs to gnotobiotic mice to evaluate effects of obese and lean gut microbiota on metabolism and body composition of the mice. After a 4 weeks feeding period, the microbiota of mice (mice-obese) transplanted with feces from the obese pig line differed from the gut microbiota of mice (mice-lean) transplanted with fecal preparations from the lean pig line. The mice-obese had an elevated ratio of *Firmicutes* to *Bacteroidetes* compared to the mice-lean in their fecal samples. This shift in major gut phyla has also been noted repeatedly in many previous studies. In addition, mice-ob showed increased expression of key lipogenic genes and their livers had elevated lipid stores and lipoprotein lipase activities. This study shows that microbiota transplants from obese or lean host pigs had dramatic effects on lipid metabolism in another species, i.e., mice and provides further evidence of the role of the gut microbiota on subsequent metabolic phenotypes (Yang et al. 2018).

For gut microbiota studies in non-ruminants, researchers frequently use fecal samples. This is an efficient and non-invasive approach; however, it ignores any differences in microbial species diversity in the various GIT compartments (ileum, cecum and colon). Quan et al. (2018) compared the microbiome compositions in three GIT locations in two groups of Duroc X (Landrace x Yorkshire) pigs. One group exhibited a high feed efficiency (feed conversion ratio, FCR) and the other group a significantly lower feed efficiency. Luminal GIT samples were taken from the ileum, cecum and colon of each group after a 140-day feeding trial and then submitted to high-through-put sequencing. The average number of OTU differed significantly among GIT locations with incremental increases from ileum > cecum > colon. Based on principal component analysis, microbial compositions between the three GIT sites were significantly different. The taxonomic distributions of numerically abundant bacteria within each GIT site was determined. In the ileum, Firmicutes and Proteobacteria were most abundant with a low abundance of Bacteroidetes. In the cecum and colon, *Proteobacteria* were of much lower abundance while Bacteroidetes abundance was much higher in cecum and colon samples. At the genus level, *Escherichia-Shigella, Terrispobacter, Romboutsia and Chlostidium* were most prevalent in the ileum. In the cecum, *Alloprovotella, Lactobicillus and Prevotellaceae NKB31 group* were most prevalent. *Streptococcus, Lactobacillus and Clostridium* were the most prevalent genera in the colon. Quan et al. (2018) concluded that upon analysis of microbial abundance, variability and metabolic pathways in pigs with different feed efficiencies, they discovered that high feed efficiency pigs generally exhibited higher abundance of microbes that are beneficial for feed digestion and fermentation than low feed efficiency pigs. Finally, these findings should caution researchers that fecal microbiome analysis might not present a very definitive picture of microbial phyla compositions in different gut segments.

17.5 Pigs in Toxicology and Organ Transplantation Applications

Wide varieties of pigs are widely used as models for biomedical research. More recently, use of non-rodent models for toxicological studies increased with the reasoning that broad comparative animal studies might be very helpful in describing toxicological processes and pathologies (Helke and Swindle 2013). Yet today, the overall understanding of xenobiotic metabolism in pigs is not as expansive as in rodent models. The Cytochrome P 450 (CYP) family of enzymes is large, functions to detoxify xenobiotics, eicosanoids and vitamins, and participates in the disposal of many drugs from the body (Guengerich 2019). This family of enzymes (many isozymes) are monooxygenases and usually found in endoplasmic reticulum membranes. The liver exhibits the greatest amount of CYP activity with additional activity in the GIT, skin and kidneys. Pigs are not uniform, depending on traits and breeds, in their CYP isozymes distribution and their kinetic/catalytic characteristics of xenobiotics metabolism. In this regard, commercial breeds of pigs (now very lean) appear less uniform in CYP activities than for example miniature pigs (Helke et al. 2016). On balance, the use of pigs for assessment of drug safety and toxicological research may introduce additional variants of P 450 enzymes. Presently, based on genomic analysis, there are 57 human P450 genes; 88 and 103 P450 genes in rats and mice, respectively (Guengerich 2019).

Another emerging area for involvement of pigs in clinical practice and research is transplanting of organs from pigs to humans (Sykes and Sachs 2019). It is easily argued, that based on organ sizes and physiological similarities to human beings, pigs are an appropriate source of xenografts. Pigs also are a readily available source of tissues and organs. Early pig to non-human primates/human transplantation resulted in rejections (Platt et al. 1991) and was not possible until issues of high levels of natural antibodies to porcine antigens in humans were resolved (Mc Morrow et al. 1997). By the present time, advances in genomic editing (example, Lai et al. 2002) and with increasing in-depth understanding of human immune reactions to porcine organs, in the not too distant future pigs to human organ transplantation may become a common occurrence (Sykes and Sachs 2019).

17.6 Pigs in Medical Instruction and Translational Research

In US Medical Schools, use of live animals in surgical clerkships was once a well-accepted practice. Initially, surgical practice as well as translational clinical research used dogs (example Banting and Best (1922), from research on diabetes). By the mid-twentieth century, public sentiments rose against this extensive use of dogs and surgical clerkship live animal training procedures also started to utilize live pigs. Simkin et al. (2017) relate the history how John Hopkins University School of Medicine started to utilize pigs in surgical practice from 1895 until the practice was discontinued in 2016. By that time, the use of live animals in surgical training in standard medical curricula was also abandoned in all US medical schools. While many surgeons and physicians did not necessarily agree with these decisions, medical schools adopted other approaches and simulations procedures for early surgical training. Simkin et al. (2017) noted that today when first time surgical trainees will operate on a live subject, these would be human beings. The likelihood is high that such patients may be individuals less able to pay for costly medical procedures. Basic and translational biomedical research with pigs is still conducted in the US; however, while some other countries are apparently still using/developing surgical training procedures with pigs (Kobayashi et al. 2012).

References

Baeckhed F, Sonnenburg JL, Peterson DA, Gordon JL (2005) Host-bacterial mutualism in human intestine. Science 307:1915–1920

Baeckhed F, Ding H, Wang T, Hooper LV, Koh GY, Nagy A et al (2004) The gut microbiota as an environmental factor that regulates fat storage. Proc Natl Acad Sci USA 101:15718–15723

Baker DH (2008) Animal models in nutrition research. J Nutr 138:391–396

Baker D (2006) Comparative species utilization and toxicity of sulfur Amino Acids. J Nutr 136:1670S-1675S

Banting FG, Best CH (1922) The internal secretions of the pancreas. J Lab Clin Med 7:251–266

Bazer FW, Seo H, Johnson GA, Wu G (2021) One-carbon metabolism and development of the conceptus during pregnancy: lessons from studies with sheep and pigs. Adv Exp Med Biol 1285:1–15

Benevenga NJ (1984) Evidence for alternative pathways of methionine catabolism. Adv Nutr Res 110:1–18

Bergen WG, Merkel RA (1991a) Body composition of animals treated with partitioning agents: implications for human health.FASEB J 5:2591–2597
Bergen WG, Merkel RA (1991b) Protein accretion in: growth regulation in farm animals. Adv Meat Res 7, Pearson AM, Dutson TR (eds) Elsevier Applied Science, London pp 169–202
Bergen WG, Brandebourg TD (2016) Regulation of lipid deposition in farm animals: parallels between agriculture and human physiology. Exp Biol Med 241:1272–1280
Bergen WG, Wu G (2009) Intestinal nitrogen recycling and utilization in health and disease. J Nutr 139:821–825
Bergen WG, Mersmann HJ (2005) Comparative aspects of lipid metabolism: impact on contemporary research and use of animal models. J Nutr 135:2499–2502
Boggs DL, Bergen WG, Hawkins DR (1987) Effect of tallow supplementation and protein withdrawal on ruminal fermentation, microbial synthesis and site of digestion. J Anim Sci 64:907–914
Cani PD, Van Hul M (2020) Gut microbiota and obesity: causally linked? Exp Rev of Gastroent Hepatol 14:401–403
Chen X, Devaraj S (2018) Gut microbiome in obesity, metabolic syndrome and diabetes. Curr Diabetes Rep 18:129–135
Coelho LP, Kultima JR, Costea PI, Fournier C, Pan Y, Czarnecki-Maulden G et al (2018) Similarity of the dog and human gut microbiomes in gene content and response to diet. Microbiome 6:72–82
Dabke K, Hendrick G, Devkota S (2019) The gut microbiome and metabolic syndrome. J Clin Invest 129:4050–4057
Dai ZL, Wu ZL, Zhu WY, Wu G (2022) Amino acids in microbial metabolism and function. Adv Exp Med Biol 1354:127–143
Dillon EL, Wu G (2021) Cortisol enhances ctrulline synthesis from proline in enterocytes of suckling piglets. Amino Acids. https://doi.org/10.1007/s00726-021-03039-y
Dodson MV, Mir PS, Hausman, Guan LL, Du M, Jiang Z, Fernyhough, Bergen WG (2011) Obesity, metabolic syndromes, and adipocytes. J Lipids 2011:721686
Dodson MV, Boudina S, Albrecht E, Bucci L, Fernyhough M, Wei S, Bergen WG et al (2013) A long journey to effective obesity treatments: Is there light at the end of the tunnel? Exp Biol Med 238:491–501
Dodson MV, Wei S, Duarte M, Do M, Jiang Z, Hausman GJ, Bergen WG (2012) Cell Supermarket: adipose tissue as a source of stem cells. J Genomics 1:39–44
Donaldson GP, Lee SM, Mazmanian SK (2016) Gut biogeography of the bacterial microbiota. Nat Rev Microbiol 14:20–32
Etherton TD (2000) The biology of somatotropin in adipose tissue growth and nutrient partitioning. J Nutr 130:2623–2625
Guengerich FP (2019) Cyochrome P450 Res the J Biol Chem. J Biol Chem 294:1671–1680
Guitierrez K, Dicks N, Glanzner WG, Agellon LB, Bordignon V (2015) Efficacy of porcine species in biomedical research. Front Genet 6:293
Gu S, Chen D, Chang J-N, Wang K, Duan L-P et al (2013) Bacterial community mapping of the mouse gastrointestinal tract. PLosOne 8(10):e74957. https://doi.org/10.1371/journal.pone.0074957
Helke KL, Swindle MM (2013) Animal models of toxicology testing: the role of pigs. Expert Opin Drug Metab Toxicol 9:127–139
Helke KL, Nelson KN, Sargeant AM, Jacob B, McKeag S, Haruna J, Vemireddi V et al (2016) Background pathological changes in minipigs: a comparison of the incidence and nature among different breeds and populations of minipigs. Toxicol Pathol 44:325–337
Hou YQ, He WL, Hu SD, Wu G (2019) Composition of polyamines and amino acids in plant-source foods for human consumption. Amino Acids 51:1153–1165
Hu SD, He WL, Wu G (2021) Hydroxyproline in animal metabolism, nutrition, and cell signaling. Amino Acids. https://doi.org/10.1007/s00726-021-03056-x
Kootte RS, Levin E, Salojarvi J, Smits LP, Hartstra AV, Udayappan SD et al (2017) Improvement of insulin sensitivity after lean donor feces in metabolic syndrome is driven by baseline intestinal microbiota composition. Cell Metab 26:611–619
Kobayashi E, Hishikawa S, Teratani T, Lefor AT (2012) The pig as a model for translational research: overview of porcine animal models at Jichi Medical University. Transplant Res 1:8–16
Lai L, Kolber D, Park KW, Cheong HT, Greenstein JL, Im GS et al (2002) Production of alpha-1,3-galactosyltransferase knockout pigs by nuclear transfer cloning. Science 295:1089–1092
Ley RE, Baeckhed F, Turnbough PJ, Lozupone CA, RD Knight RD, Gordon JL (2005) Obesity alters gut microbial ecology. Proc Natl Acad Sci USA 102:11070–11075
Li J, Jia H, Cai Z, Zhong H, Feng Q, Sunagawa S et al (2014) An integrated catalog of reference genes in the human gut microbiome. Nat Biotechnol 32:834–841
Li P, He WL, Wu G (2021) Composition of amino acids in foodstuffs for humans and animals. Adv Exp Med Biol 1332:189–209
Li P, Wu G (2020) Composition of amino acids and related nitrogenous nutrients in feedstuffs for animal diets. Amino Acids 52:523–542
Maruvada P, Leone V, Lee M, Kaplan M, Chang EB (2017) The human microbiome and obesity: Moving beyond associations. Cell Host Microbe 22:589–599
McMorrow IM, Comrack CA, Sachs DH, Der Simonian H (1997) Heterogeneity of human anti-pig natural antibodies cross reactive with the Gal1 (alpha1,3) Galactose epitope. Transplant 64:501–510
Mersmann HJ (1986) Lipid metabolism in swine. In: Swine in cardiovascular research (Stanton HC and HJ Mersmann, eds). CRC Press, Boca Raton, FL
Miller ER, Ullrey DE, Lewis AJ (eds) (1991) Swine nutrition. Butterworth-Heinemann, Stoneham, MA

Mu C, Pi Y, Zhang C, Zhu WY (2022) Microbiomes in the intestine of developing pigs: Implications for nutrition and health. Adv Exp Med Biol 1354:161–176
Nafikov RA, Beitz DC (2007) Carbohydrate and lipid metabolism in farm animals. J Nutr 137:702–705
Oberbauer AM, Larsen JA (2021) Amino acids in dog nutrition and health. Adv Exp Med Biol 1285:199–216
Platt JL, Fischel RJ, Matas AJ, Reif SA, Bolman RM, RM, Bach FH (1991) Immunopathology of hyper acute xenograft rejection in a swine swine-to-primate model. Transplant 52:214–220
Pond WG (1982) In: Pigs model for biomedical research (Roberts HR and Dodds WJ, eds) Maoli, Taiwan, ROC: Pigs Research Institute of Taiwan, p 4
Pond WG (1991) Of pigs and people. In: Miller ER, Ullrey DE, Lewis AJ (eds) Swine Nutrition. Butterworth-Heinemann, Stoneham, MA, pp 3–23
Quan J, Cai G, Yang M, Zeng Z, Ding R, Wang X, Zhuang Z et al (2019) Exploring the fecal microbial composition and metagenomics functional capacities associate with feed efficiency in commercial DLY pigs. Front Microbiol 10:52
Quan J, Cai G, Ye Y, Yang M, Ding R, Wang X et al (2018) A global comparison of the microbiome ncompositions of three gut locations in commercial pigs with extreme feed conversion ratios. Nat Sci Rep 2018
Reiter SS, Halsey CHC, Stronach BM, Bartosh JL, Owsley WF, Bergen WG (2007) Lipid metabolism related gene-expression profiling in liver, skeletal muscle and adipose tissue in crossbred Duroc and Pietrain pigs. Comp Biochem Physiol D 2:200–207
Rhoads JM, Wu G (2009) Glutamine, arginine, and leucine signaling in the intestine. Amino Acids 37:111–122
Rudar M, Fiorotto ML, Davis TA (2019) Regulation of muscle growth in early postnatal life in a swine model. Annu Rev Anim Biosci 7:309–335
Ridaura VK, Faith JJ, Ray FE, Cheng J, Duncan AE, Kau AL et al (2013) Gut microbiota from twins discordant for obesity modulate metabolism in mice. Science 341:1241214
Sarkar TR, McNeal CJ, Meininger CJ, Niu YB, Mallick BK, Carroll RJ, Wu G (2021) Dietary intakes of amino acids and other nutrients by adult humans. Adv Exp Med Biol 1332:211–227
Smith BI, Govoni KE (2022) Use of agriculturally important animals as models in biomedical research. Adv Exp Med Biol 1354:315–333
Stanton HC, Mersmann HJ (1986) Swine in Cardiovascular research. CRC Press, Boca Raton, FL
Spurlock ME, Gabler NK (2008) The development of porcine models of obesity and the metabolic syndrome. J Nutr 138:397–402
Stenhouse C, Seo H, Wu G, Johnson GA, Bazer FW (2022) Insights into the regulation of implantation and placentation in humans, rodents, sheep, and pigs. Adv Exp Med Biol 1354:25–48
Swindle MM, Moody DC, Phillips LD (1992) Swine as model in biomedical research. Iowa State University Press, Ames, IA
Swindle MM, Smith AC (2013) Best practices for performing experimental surgery in swine. J Invest Surg 26:63–71
Scheithauer TP, Dallinga-Thie GM, de Vos VM, Nieuwdorp N, van Raalte DH (2016) Causality of small and large intestional microbiota in weight regulation and insulin resistance. Mol Metab 5:759–777
Sykes M, Sachs DH (2019) Transplanting organs from pigs to humans. Sci Immunol 4:eaau6298
Simkin DJ, Greene JA, Jung JJ, Sacks BC, Fessler HF et al (2017) The death of animals in medical school. N Engl J Med 376:713–715
Tumbleson ME (1986) Swine in biomedical research, Vol 1, 2 and 3. Plenum Publishing Co. New York, NY
Turnbaugh PJ, Ley RE, Mahowald MA, Magrini V, Mardis ER, Gordon JL (2006) An obesity-associated gut microbiome with increased capacity for energy harvest. Nature 444:1027–1031
Vodicka P, Smetana Jr K, Dvorankova B, Emerick T, Xu YZ, J Ourednik J, Ourednik V, J Motlik J (2006) The miniature pig as an animal model in biomedical research. Ann NY Acad Sci 1049:161–171
Walters EM, Wells KD, Bryda EC, Schommer S, Prather RS (2017) Swine models, genomic tools and services to enhance our understanding of human health and diseases. Lab Anim (NY) 46:167–172
Wargo JA (2020) Modulating gut microbes. Science 369:1302–1303
Wu G (2021) Amino Acids: biochemistry and nutrition. 2nd ed. CRC Press, Boca Raton, Florida
Wu G (2022) Nutrition and metabolism: foundations for animal growth, development, reproduction, and health. Adv Exp Med Biol 1354:1–24
Wu G, Morris SM Jr (1998) Arginine metabolism: nitric oxide and beyond. Biochem J 336:1–17
Wu G, Bazer FW, Davis TA, Kim SW, Li P, Rhoads JM, Satterfield CM et al (2009) Arginine metabolism and nutrition in growth, health and disease. Amino Acids 37:153–168
Xiao L, Feng S, Liang S, Sonne S, Xia Z, Qiu X et al (2015) A catalog of the mouse metagenome. Nat Biotechnol 33:1103–1108
Xiao L, Esttele J, Kiilerich P, Ramayo-Caldes Y, Xia Z et al (2016) A reference gene catalogue of the pig gut microbiome. Nat Microbiol 1:1616
Yang H, Xiang Y, Robinson K, Wang J, Zhang G, Zhao J (2018) Gut microbiota is a major contributor to adiposity in pigs. Front Microbiol 9:3045
Zhang X, Lerman LO (2016) Investigating the metabolic syndrome: Contribution of swine models. Toxicol Pathol 44:358–366

Index

G. Wu (ed.), *Recent Advances in Animal Nutrition and Metabolism*,
Advances in Experimental Medicine and Biology 1354,
https://doi.org/10.1007/978-3-030-85686-1

MIX
Papier aus verantwortungsvollen Quellen
Paper from responsible sources
FSC® C105338

If you have any concerns about our products,
you can contact us on
ProductSafety@springernature.com

In case Publisher is established outside the EU,
the EU authorized representative is:
Springer Nature Customer Service Center GmbH
Europaplatz 3, 69115 Heidelberg, Germany

Printed by Libri Plureos GmbH
in Hamburg, Germany